"This book offers clear and concise articulations of the critical environmental policy issues in various developing and developed countries. Paying particular attention to the interface between human behavior and environmental degradation, the book also exposes its readers to the full scope of issues predicated on a risk-based framework for environmental policy analysis that encourages them to judge for themselves the significance of the future trends in ecological discourse."

—**Felix M. Edoho,** *Lincoln University*

"The chapters in this book have been carefully written to address policy concerns from a social science perspective in order to provide a thorough understanding of sustainable development. Governmental and nongovernmental organizations will find this book informative and instructive as they wrestle with country-based programs that impinge upon environmental variables and have ramifications that could be irreversible with regard to the negative consequences. This book is highly recommended to university libraries, conservation and preservation scholars, and mid-management policy makers researching or working in the emerging nations of the world."

—**Valentine James,** *Clarion University*

Comparative Perspectives on Environmental Policies and Issues

Comparative Perspectives on Environmental Policies and Issues presents tools and concepts about environmental policies in several developed and developing countries. It explores a broad survey of ecological modernization theory, ecological feminism theory, environmental justice theory, the concept of sustainability, and research on environmental policies. Data were collected through surveys, interviews, and focus groups, and are used to analyze social, economic, and environmental impact on people.

The book specifically discusses how the earth's basic life-supporting capital (soils, forests, species, fresh water, and oceans) is degraded or depleted to provide for human needs, and how air pollution and acid precipitation are causing widespread injury to humans, forests, and crops. Realistically, overtaxing of natural resources and ecological systems throughout the world has promoted economic growth and created increasing opportunities for people while also advancing social injustice. The use of the environment to accomplish social and economic transformation raises fundamental issues for the study of environmental policy and the natural ecological system. As human beings exploit the natural environment to meet present needs, they often will destroy resources needed for the future generations. Thus, environmental policies are enacted to ensure that social and economic impacts of the environment are compatible with the limits of natural systems.

Offering an intuitive and crystal-clear explanation of the key concepts and principles of environmental policies and sustainable development, this volume is suitable not only for environmental science students but also for instructors, practitioners, researchers, and academics.

Robert Dibie, PhD, is Dean and Professor of Public Policy and Public Management at Indiana University Kokomo's School of Public and Environmental Affairs (SPEA). Previously, he served as the director of graduate programs in public administration at Western Kentucky University. Professor Dibie is the author of several books, book chapters, and more than 70 research articles in peer-refereed journals. He has presented more than 100 academic papers in national and international conferences, focusing on issues of sustainable development, public management, public policy, women empowerment, environmental policies, development administration, and non-governmental organizations (NGOs). As a nationally recognized leader in higher education, Professor Dibie has presented seminars, workshops, and lectures in the areas of higher education leadership, public policy, environmental policy, and sustainable development in a number of universities around the world. He is currently the editor of the *Journal of International Politics and Development*. Professor Dibie has also consulted for several NGOs and universities in the United States, Europe, Africa, and the Caribbean Islands. Professor Dibie earned a bachelor's and master's degree in Business Administration. In addition he earned a master's degree in Social Sciences (public administration and international relations) from the University of Detroit Mercy. His PhD in Public Policy and Public Management is from Western Michigan University. He also completed several professional certification diplomas from the Charted Institute of Marketing and the Association of Business Executives, both in England.

Routledge Research in Environmental Policy and Politics

1 **Green vs. Green**
The Political, Legal, and Administrative Pitfalls Facing Green Energy Production
Ryan M. Yonk, Randy T. Simmons, and Brian C. Steed

2 **The Lilliputians of Environmental Regulation**
The Perspective of State Regulators
Michelle C. Pautz and Sara R. Rinfret

3 **Globalization, Political Institutions, and the Environment in Developing Countries**
Gabriele Spilker

4 **Comparative Perspectives on Environmental Policies and Issues**
Edited by Robert Dibie

Comparative Perspectives on Environmental Policies and Issues

Edited by Robert Dibie

Routledge
Taylor & Francis Group
LONDON AND NEW YORK

First published 2014
by Routledge

2 Park Square, Milton Park, Abingdon, Oxfordshire OX14 4RN
711 Third Avenue, New York, NY 10017

Routledge is an imprint of the Taylor & Francis Group, an informa business

First issued in paperback 2018

Library of Congress Cataloging-in-Publication Data

Comparative perspectives on environmental policies and issues / edited by Robert Dibie.
pages cm. — (Routledge research in environmental policy and politics ; 4)
1. Environmental policy—Developing countries. 2. Environmental policy—Africa. 3. Environmental health—Developing countries. 4. Environmental health—Africa. 5. Environmental degradation—Developing countries. 6. Environmental degradation—Africa. 7. Sustainable development—Developing countries. 8. Sustainable development—Africa. I. Dibie, Robert A., 1956– author, editor. II. Series: Routledge research in environmental policy and politics ; 4.
GE149.C664 2013
363.70561091724—dc23
2013031025

ISBN: 978-0-415-81391-4 (hbk)
ISBN: 978-1-138-38323-4 (pbk)

Typeset in Sabon
by Apex CoVantage, LLC

This book is dedicated to my children, Ajiri Oghenemaro Dibie and Kome Ajiroghene Dibie

Contents

List of Figures xi
List of Tables xiii
Preface xv
Acknowledgments xvii
List of Contributors xix
List of Acronyms xxiii

1 A Preview of Environmental Policies and Issues 1
ROBERT DIBIE, PHD

2 Environmental Concepts and Applications 20
ROBERT DIBIE, PHD

PART I
Developed Countries

3 Environmental Issues and Policies in the United States 59
ROBERT DIBIE, PHD AND MARY BOURKE, PHD

4 Analysis of Environmental Health Issues and Policies in Western Europe 86
EMEKA NWAGWU, PHD

5 Environmental Health Policies and Issues in Canada 117
EMMANUEL A. IYIEGBUNIWE, PHD AND ROBERT DIBIE, PHD

PART II
Developing Countries—Africa

6 Environmental Governance in Sub-Saharan Africa 159
ROBERT DIBIE, PHD

7 Environmental Policies and Issues in Ghana 191
ALBERT ASSIBEY-MENSAH, PHD AND ROBERT DIBIE, PHD

8 Environmental Policy and Impacts in South Africa 222
ROBERT DIBIE, PHD AND KEALEBOGA MAPHUNYE, PHD

9 Environmental Policy and Issues in Ethiopia 239
ROBERT DIBIE, PHD

10 Environmental Policy and Issues in Nigeria 268
OFFIONG J. OFFIONG, PHD AND ROBERT DIBIE, PHD

PART III
Developing Countries—Other Continents

11 Environmental Health Policy in Brazil 303
ROBERT DIBIE, PHD, WILLIAM MELLO, PHD, AND GALIA BENITEZ, PHD

12 Environmental Policies and Issues in Jamaica 329
ROBERT DIBIE, PHD

13 Environmental Policies and Issues in Trinidad and Tobago 351
ROBERT DIBIE, PHD

14 Environmental Policies and Issues in Belize 374
EMMANUEL IYIEGBUNIWE, PHD

15 Environmental Policies and Issues in Some Arab Countries 400
ROBERT DIBIE, PHD AND MAUWYA HUSSEIN, MA, MBA

16 Environmental Sustainability and Solutions 424
ROBERT DIBIE, PHD

Index 459

Figures

2.1	Compartment Model	31
3.1	Relationships between Environment and Human	65
3.2	Environmental Policy Development Actors	68
4.1	Levels of Influence in the Social Ecological Model	90
4.2	Life Expectancy in Selected European Countries and the United States	110
4.3	Healthcare GDP Spending—Western European Countries versus USA	110
5.1	Model of the Relationship between OSH Policy, Health, and the Environment	122
5.2	Health Behavior Lifestyle Model	123
6.1	The Relationship between Sustainability and Society	164
6.2	The Relationship between Human Development and Political Freedom	176
7.1	Reflective Self-Regulation towards the Environment	195
7.2	The Relations between Economic Growth and Natural Resources	196
8.1	Model on the Relationship among Biological, Economic, and Social Systems	226
8.2	Nature and Animals in Cape Town, South Africa	228
9.1	The Framework for Achieving Sustainable Development	243
9.2	Dump Site by the Side of a Street in Ethiopia	248
10.1	Example of Human Relations with the Ecosystem in Nigeria	276
10.2	Human and Industrial Waste Dumped on the Street in Ibadan, Nigeria	296
11.1	Summary of Survey Findings in Brazil	320
12.1	Environmental Impact Model	333
12.2	Human Waste Dumped in the Gullies in Jamaica	336

13.1 Natural Capital Degradation Model 357
13.2 Photo of Garbage Collection in a Street inTrinidad, 2010 360
14.1 Typical Pit Latrine in Mayan Village of Red Bank, Belize 384
15.1 Arab Countries' Internal Freshwater Resources 407
15.2 Desertification Trend in Selected Arab Countries 410

Tables

2.1	World Populations and Growth Rate	38
2.2	Top Twenty-five Fossil Fuel Carbon Dioxide Emitters in 2003	39
2.3	Per Capita Wealth with Rate of Fertility and Population Growth	45
3.1	Economic and Social Impact of Environmental Issues	71
3.2	Environment and Costs to Human Welfare	72
3.3	Environmental Variables and New Policy Goals	81
4.1	Life Expectancy Table for EU-27	109
5.1	Percentage Diagnosed with Asthma in Canada, 2001–2010	137
5.2	Federal Legislation Relevant for Protecting Children's Environmental Health	138
6.1	Human Development Index in Africa	168
6.2	Carbon Dioxide Emissions Level in Selected African Nations	172
6.3	Environmental Problems due to Human Activities	173
7.1	List of Environmental Policies, Agencies, and Ministries	204
7.2	Impact of Liquid Biofuel Plantations and Their Mitigation Measures	205
7.3	Negative Impacts and Their Mitigation Measures	207
9.1	Human Development Index in Ethiopia	246
9.2	Projected Budgets of NGOs Working in Southern Nations Nationalities and People Region (SNNPR) in 2006–2008	254
9.3	Actual and Planned Project Expenditures, 2004–2008	254
9.4	Major Ethiopian Environmental Policies and Institutions	257
9.5	Other Environmental Policies Enforced Outside the EPA	258
10.1	Major Oil Companies Involved in the Exploration for and Marketing of Oil in Nigeria	283

10.2 Environmental Policies Enacted between the 1930s and 1990s 285
10.3 Ranking of Major Environmental Problems, Social Issues, and Priorities 287
10.4 Listed Laws Relevant to Environmental Issues in Nigeria 288
11.1 Environmental Impact on Health 318
11.2 Awareness of Environmental Policy Impact 319
12.1 Social and Economic Impact of Environmental Issues 337
12.2 Economic and Social Impact of Environmental Issues 339
12.3 Jamaican Production of Mineral Products 340
12.4 Jamaican Structure of Mineral Industry 341
13.1 Economic and Social Impact of Environmental Issues 362
14.1 Belize: Factsheet at-a-Glance 380
15.1 Selected Environmental Policies in Arab Countries 405
15.2 Level of Water Stress in Four Arab Countries 406
15.3 Arab Region's National Per Capita CO_2 Emissions from the Consumption and Flaring of Fossil Fuels 412
15.4 Average Annual Damage Costs of Environmental Degradation, in MENA Countries (Percentage of GDP, 2002) 416
15.5 Environmental Performance Index for Arab Countries, 2006 and 2008 417
16.1 Impact of Green Economy 435

Preface

This book is about environmental policies and issues, with a focus on comparative environmental policies. It presents a broad survey of theory and research on environmental policies and issues. Topics of special interest are economic impact of environmental policies and issues in several countries around the world. Since the book is not intended to be a history of environmental policy, it seemed appropriate to reduce the amount of details about early research programs and old theories that are no longer popular. Instead the book focuses more closely on what I now know about the economic impact of environmental issues. This book discusses the various activities of individuals and society that create environmental health hazards. Survey data is used to analyze social, economic, and environmental issues' impact on people in several countries. This book also discusses how earth's basic life-supporting capital of soils, forests, species, and fresh water and oceans are been degraded or depleted to provide for human needs and aspirations. In addition, it examines how air pollution near ground level and acid precipitation are already causing widespread injury to humans, forests, and crops. Overtaxing of natural resources and ecological systems throughout the world has promoted economic growth and created increasing opportunities for billions of people while also advancing social injustice. The use of the environment to accomplish social and economic transformations raises fundamental issues for the study of environmental policy and the natural ecological system. As human beings exploit the natural environment to meet present needs, they may be destroying resources needed for future generations. Thus, environmental policies are enacted to ensure that social and economic impacts of the environment are compatible with the limits of natural systems.

Readers who desire to improve their knowledge about environmental policy impact will find the book very useful. I believe it is important for legislators, policy analysts, and students to understand the complexity of environmental policies, the source of our knowledge about the ecosystem, and the limitation of this knowledge. Further, I believe it is important for academics to think more about how their theories and research can be used to improve the implementation of environmental policies all over the world. I often observe that too much of our environmental research is designed to examine narrow esoteric questions that interest only a few other scholars who publish in the same journals.

For students and practitioners who desire to become environmental specialists, I attempt to convey a better appreciation of the complexity of environmental issues in several countries. The importance of having theoretical knowledge about environmental issues and the need to be flexible and pragmatic in applying this knowledge is explored in the book. This book also provides guidelines and recommendations for improving environmental degradation and pollution as well as how nations can better implement their environmental policies. Most chapters have one or two cases designed to help the reader gain a better understanding of the concepts, guidelines, and theories presented in the book.

This book is appropriate for use as the primary text in undergraduate and some graduate courses in environmental science, environmental policy, or environmental health. Such courses are found in many different universities or colleges or community colleges. This book was written because of my concern about the vital importance of environmental policies and issues in today's world. This concern makes it imperative to engage, educate, and inspire a broad audience of today's students, who will become the citizens and leaders of tomorrow. This book was written for both environmental policy practitioners and students who are planning to work in environmental regulatory agencies. It will also be very useful for policy analysts, senior public administrators, and practitioners of economic development at national, regional, city, or rural government agencies. I hope that this book will be a reference for elected officials and economic development or planning specialists as they carry out their responsibilities on environmental issues. In addition, this book is intended to provide insight to industrial executives who are engaged in the struggle to find viable means to develop appropriate environment-friendly communities. Policy makers at either the national and regional government or the community level will find the book useful in exploring the role of environmental regulatory agencies and civil society institutions in the sustainable development process.

Robert Dibie, PhD

Acknowledgments

I would like to express my appreciation to many individuals who directly or indirectly played a role in the development of this book. First, I wish to thank the many people at Routledge, in particular my editor Natalja, who has strongly supported this project and provided exemplary leadership throughout the revision process. Collectively, they are an extraordinary team that demonstrated very high standards of excellence in their work.

I am also grateful to the reviewers who took time to review early drafts of my work and provided helpful suggestions for improving the book. Their comments undoubtedly made the book better. My heartfelt thanks to all the scholars who contributed chapters or coauthored chapters with me: Professor Emeka Nwagwu, Jackson State University; Dr. Emmanuel Iyegbuniwe, Western Kentucky University; Dr. Mary Bourke, Indiana University Kokomo; Professor Offiong Offiong, University of Benin, Nigeria; Dr. Galia Benitez, Michigan State University; Professor Albert Assibey-Mensah, Kentucky State University; Dr. William Mello, Indiana University–Purdue University, Indianapolis; Dr. Muawya Hussein, Dhofar University, Oman; and Professor Kealeboga Maphunye, University of South Africa.

I wish to thank my colleagues and graduate students at Indiana University Kokomo, Babcock University, Nigeria, and the University of the West Indies for their continued encouragement and interest in environmental policy, environmental health, and occupational health and safety issues.

A special acknowledgment goes to Professor Valentine James, Clarion University of Pennsylvania, and Professor Ishenkumba Kahwa, the University of the West Indies, Mona Campus, Kingston, Jamaica, for their exceptional support throughout this project. Their insights and contributions have added considerably to the quality of this book.

Finally, I am especially thankful for the efforts of my wife and best friend, Dr. Josephine Dibie, for her time, energy, commitment, and contribution to the quality and success of this book.

Contributors

Albert O. Assibey-Mensah, PhD, is a Professor of Economics and Director of Graduate Programs at Kentucky State University, Frankford, USA. Dr. Assibey-Mensah has published widely in the areas of economic development, and he is currently specializing in sustainability and management planning as it relates to emerging economics. His research interests are in areas of macro- and microeconomics, political economy in Africa, multinational corporations, sustainable development, and entrepreneurship.

Galia Benitez, PhD, is an Assistant Professor in Political Science at Michigan State University, East Lansing, USA. She received her PhD degree from Indiana University in Bloomington with a specialty in comparative policy. Her research interests are in international trade agreements in South America, sustainable development, public policy, and comparative environmental policies.

Mary Bourke, PhD, is an Associate Professor of Nursing and Assistant Dean of Graduate Programs at Indiana University Kokomo School of Nursing, USA. Her research interests are in the areas of nursing, community health, diversity management in nursing, health policies, and environmental health. She has presented her research papers in national and international conferences in China, Malaysia, Australia, and the United States.

Robert Dibie, PhD, is Dean and Professor of Public Policy and Public Management at Indiana University Kokomo's School of Public and Environmental Affairs (SPEA). Previously, he served as the director of graduate programs in public administration at Western Kentucky University. Professor Dibie is the author of several books, book chapters, and more than seventy research articles in peer-refereed journals. He has presented more than 100 academic papers in national and international conferences, focusing on issues of sustainable development, public management, public policy, women empowerment, environmental policies, development administration, and non-governmental organizations (NGOs). As a nationally recognized leader in higher education, Professor Dibie has presented seminars, workshops, and lectures in the areas of higher education leadership, public policy, environmental policy, and sustainable development in a number of universities around the world. He is currently the Editor of the *Journal of International Politics and Development*. Professor Dibie has also consulted for several NGOs and universities in the United States, Europe, Africa, and the Caribbean Islands. Professor Dibie earned a bachelor's and master's degree in Business Administration. In addition he earned a master's degree in Social Sciences (public administration and international relations) from the University of Detroit Mercy. His PhD in Public Policy and Public Management is from Western Michigan University. He also completed several professional certification diplomas from the Charted Institute of Marketing and the Association of Business Executives, both in England.

Muawya Hussein, MA, MBA, is a Senior Lecture and Acting Chairperson of the Department of Accounting and Finance in the College of Commerce and Business Administration at Dhofar University, Salalah, Sultanate of Oman in the Middle East. He has presented his research papers at several national and international conferences in Ethiopia, Sudan, United Emirates, and South Africa. His research interests are in accounting, finance, international environmental policy, sustainable development, public policy, and comparative trade policies.

Emmanuel Iyiegbuniwe, PhD, is an Associate Professor of Environmental and Occupational Health at Western Kentucky University, Bowling Green, USA. His research interests are in areas of public health, environmental health, and epidemiology. His research has been published in leading public health and sustainable development journals including *International Journal of Politics and Development*.

Kealeboga Maphunye, PhD, is a Professor and Brigalia Bam Chair in Electoral Democracy at the University of South Africa. He is the coauthor of several books on public administration in South Africa. He has also published several peer-reviewed journal articles on issues related to civil service in Africa, sustainable development, public policy, and environmental policy in South Africa.

William Mello, PhD, is an Associate Professor in Labor Studies at Indiana University–Purdue University Indianapolis, USA. He has served as the project coordinator of a labor union–based management team in Brazil. His research interests are in the areas of labor unions, collective bargaining in Brazil, and sustainable development in South America. He is the author of a couple of books on labor relations.

Emeka Nwagwu, PhD, is a Professor of Public Health, Interim Associate Dean, and Director of Assessment, School of Health Sciences, Jackson

State University, Jackson, Mississippi, USA. He is also Chair, Department of Health Policy and Management. He has also done some work on resource economics and health institutional change. He regularly advises national, state, regional, and municipal leaders on health policy reforms. His research interests are in the area of public health, environmental health, and health policies.

Offiong Offiong, PhD, is a Professor of Public Administration at the University of Benin, Nigeria. He was the formal chairperson of the Department of Political Science and Public Administration at the same institution. His research interests are in the areas of organizational behavior, public administration, public policy, environmental policy, gender empowerment, and sustainable development.

Acronyms

ADB	Asian Development Bank
ADF	African Development Fund
AKST	Agricultural knowledge, science, and technology
BAU	Business as usual
BBC	British Broadcasting Corporation
BCI	Better Cotton Initiative
BONGO	Business organization nongovernmental organization
BSI	Better Sugar Initiative
CAADP	Comprehensive Africa Agriculture Development Program
CADEC	Catholic Development Commission
CBO	Community-based organizations
CEDAW	Communique for the Eradication of All Discrimination against Women
CGIAR	Consultative Group on International Agricultural Research
CIDA	Canadian International Development Agency
CIG	Common Initiative Group
COWAN	Country Women's Association of Nigeria
CSIRO	The Commonwealth Scientific and Industrial Research Organization
DEFRA	Department for Environment, Food and Rural Affairs (United Kingdom)
DRC	Democratic Republic of Congo
EEA	European Environment Agency
EEC	European Economic Community
EU	European Union
FAO	Food and Agriculture Organization of the United Nations
FAOSTAT	Food and Agriculture Organization Statistical Databases
FiBL	German Research Institute of Organic Agriculture
FONGO	Funder organization nongovernmental organization
FWCW	Fourth World Conference on Women
G8	Group of Eight
GAP	Good agricultural practices
GDP	Gross domestic product

GEF	Global Environmental Facility
GHG	Greenhouse gas
GMO	Genetically modified organism
GONGO	Government-organized nongovernmental organization
GRID	Global Resource Information Database
GRO	Grassroots organizations
GSO	Grassroots support organizations
HICs	High income countries
HRWWR	Human Right Watch World Report
IAASTD	International Assessment of Agricultural Knowledge, Science, and Technology for Development
ICARDA	International Centre for Agricultural Research in the Dry Areas
IDH	Dutch Sustainable Trade Initiative
IEA	International Energy Agency
IFAD	International Fund for Agricultural Development
IFOAM	International Federation of Organic Agriculture Movements
IFPRI	International Food Policy Research Institute
ILO	International Labor Organization
IMF	International Monetary Fund
IP	Intellectual property
IPCC	Intergovernmental Panel on Climate Change
IPM	Integrated Pest Management
ITC	International Trade Centre
IITA	International Institute of Tropical Agriculture
INGO	International nongovernmental organization
KANGO	Kenya HIV/AIDS NGOs Consortium
LICs	Low income countries
LMICs	Lower middle income countries
MDG	Millennium development goal
MSCI	Morgan Stanley Capital International
MSO	Membership support organizations
NANGO	National Association of Non-Governmental Organization
NCAR	National Centre for Atmospheric Research
NFGO	Non for Gain Organization
NGO	Nongovernmental Organization
NORAD	Norwegian for Development
OAU	Organization of African Unity
ODA	Oversees Development Assistance
ODA	Official Development Assistance
OECD	Organization for Economic Co-operation and Development
PAHM	Plant and animal health management
PES	Payment for Ecosystem Services
PICS	Purdue Improved Cowpea Storage
PONGO	Political Non-governmental Organization

QUANGO	Quasi-autonomous nongovernmental organization
R&D	Research and development
ROI	Return on investment
RSPO	Roundtable on Sustainable Palm Oil
RTRS	Round Table on Responsible Soy
SAM	Sustainable Asset Management AG
SAP	Structural Adjustment Program
SIDA	Swedish International Development Agency
SOM	Soil organic matter
SRI	System Rice Intensive
SWFs	Sovereign wealth funds
TANGO	Tanzania Association of Non-Governmental Organizations
UMICs	Upper middle income countries
UNCTAD	United Nations Conference on Trade and Development
UNDESA	United Nations Department of Economic and Social Affairs
UN	United Nations
UNDP	United Nations Development Program
UNEP	United Nations Environment Program
UNESCECA	United Nations Economic and Social Council, Economic Commission for Africa
WDR	World Development Report
WIPO	World Intellectual Property Organization
WTO	World Trade Organization
WWAP	World Water Assessment Program

1 A Preview of Environmental Policies and Issues

Robert Dibie, PhD

INTRODUCTION

This book will offer strategies for creating healthier, more vibrant communities, and learning how to thrive in a sustainable changing future. The goal is to offer a text to enrich instructors and students in undergraduate and some graduate courses on environment science, environmental policy, or environmental health. According to the World Health Organization, environmental health is "concerned with all aspects of the natural and built environment that affect human health and diseases" (Maxwell, 2014). This text would give life to the ways human interact with the natural world (environment). It will provide a refreshing vision of climate change and how to minimize the risks and benefit from the opportunities a changing environment can create with the use of environmental policies.

Environmental policy explores the relationship between human beings and the ecosystem around them. Whether we recognize it or not everything human beings do—for example, the cars we drive, the products we use, the waste that we throw away, virtually every choice we make and action that we take—has a certain environmental impact and a certain consequence for the future generation (Wright and Boorse, 2014). It is often argued that the moral test of a nation or government depends on how it treats those who are in the dawn of life, its children; those who are in the twilight of life, its aged; and those who are in the shadow of life (Vice President Hubert H. Humphrey). This includes those who suffer from exposure to environmental pollution and environmental health diseases; those who are needy, sick, and handicapped. A government or nation that cannot prevent its citizens from exposure to environmental pollution, educate its children, care and sustain its elderly, or provide hope and meet the needs of its infirmed sick, poor, and disabled is a government without compassion and a country without a soul. Through an examination of the contemporary involvement of government and politics in the organization, financing, and implementation of environmental policies in both developing and developed countries, this book attempt to underscore the important role political factors play in the development of appropriate environmental policies in all nations.

According to Michael Kraft and Scot Furlong (2012) and Guy Peters (2013), the United States as well as many other industrialized nations continues to face two environmental and development problems. On one hand, these countries are virtually insatiable in their demand for energy. On the other hand, there is the need to manage the affluent of an industrialized society and to preserve as much of the natural environment as possible. These two environmental policy issues are closely related to the issue of climate change because of the widespread burning of fossil fuel that in turn results in the continued warming of Earth's atmosphere. Heavy reliance worldwide on use of fossil fuels that contribute to the risk of climate change is a major environmental problem (Kraft, 2011).

Niko Roorda (2012) contends that climate change and global warming constitute a serious challenge to the existing ecological pattern in the world. As a result any negative impact on the ecological pattern often affects low agricultural productivity and thus shortage of food. It is interesting to note, however, that most industrial countries cherish the desire to find the means to achieve energy independence, as well as reduce the level of energy consumption, especially fossil fuel consumption. The irony of these desires is that the high level of imports of oil from several developing countries also creates another set of ecological problems that often result in oil spillage and the destruction of farm lands and fishing rivers, and the depletion of agricultural forest. In short, the consumption of energy in both the developing and developed countries has a great cost on the earth's basic life-supporting capital of forests, species, and soils, which are being depleted, and on its fresh waters and oceans, which are being degraded at an accelerated rate.

In addition to the environmental problems just discussed, Michael Kraft (2011) and Thomas Dye (2011) identified the following environmental problems. The problems include public exposure to toxic chemicals and hazardous waste; the destruction of ecologically critical land and forests, which in turn hastens the loss of biological diversity; the production of prodigious quantities of solid waste that wind up in landfills; air and water pollution; and the deterioration of the quality of life in creating congested cities around the world. One other very important problem associated with the use of the environment is the growth of population. It is reported that the world population has multiplied more than three times since 1900 (Holmberg, 1992; Kraft, 2011; Roorda, 2012). Therefore to house, feed, and otherwise provide for people's needs and aspirations will tax natural resources and ecological systems throughout the world more than we could have imagined. When it comes to the carbon dioxide discharge into the atmosphere, we transfer the damage our energy consumption causes onto the natural environment and onto future generations.

The same holds for rubbish dumps all over the world (Roorda, 2012). The emission of greenhouse gases cannot continue unabated for centuries, even though oil and natural gases can probably still be extracted for much longer at acceptable cost. It should be noted here that the developing countries producing oil and natural gas tend to pay most of the environmental depletion cost due to weak environmental policies. As a result, the conflict between states of ecological equilibrium and economic development in the long term are still not answered.

The relationship between environmental policy and politics is essential to the development of effective public policy. Further, democratic environmental politics are paramount to the formulation and implementation of effective environmental policy in countries all over the world. The health, social, and economic well-being of citizens of all nations in the world is fundamentally linked to the quality of their environment (Taylor, 2012). The bone of contention, however, is that most national governments often employ formidable power to promote their strategic plans for achieving sustainable environmental policy goals. Historically, the relationships between the environment and human health have always been of keen interest in environmental health policy (Dibie and Offiong, 2012; Gupta, 2011).

Environmental health policy encompasses a subset of policies that address the relationship between public health and the environment (Skolnik, 2008; Maxwell, 2014). The two core goals of environmental policy are environmental health (measurement, evaluation, and control of environmental stressors to human health) and ecosystem vitality (measurement and assessment of ecosystem health and natural resource management). Although the governments of several developed and developing countries have enacted environmental policies, they have not been able to effectively achieve their goals for preventing their citizens and ecosystems from experiencing constant exposure to hazardous environment conditions.

There is no doubt that the existing affluent society and complex lifestyle of human beings have greatly affected the incidence of environmental degradation. Several international environmental studies reported in this book illustrate that several workers including farmers have suffered dermatologic conditions, hearing loss, low back disorders, lung diseases, fertility and pregnancy abnormalities, infectious diseases, musculoskeletal disorder of the upper extremities, traumatic injuries, and adverse mental health in their various places of work due to exposure to dangerous environmental conditions. Nancy Maxwell (2014) contends that work-related stress can also cause metal health disease. Occupational injuries in several countries are caused by acute exposure in the workplace to physical agents such as mechanical energy, electricity, chemicals, and ionizing radiation, or from the sudden lack of essential agents such as oxygen or heat (Castillo, Pizatella, and Stout, 2011). Workers' injuries are also caused by motor vehicle crashes, assaults, falls, being caught in parts of machinery, being struck by tools or objects, and electrocutions. In most cases injuries associated with these agents include lacerations, fractures, abrasions, burns, amputations, poisoning, and damage to internal organs. For example, the Caribbean Environmental Health Institute Report (2006) indicated that more than

2,000 workers died from occupational injuries in Trinidad and Tobago in 2005. As a result, occupational and nonoccupational injuries present a serious public health problem in the country.

The citizens and societies around the world live in extraordinary times due to the challenges posed by climate change, air and water pollution, deforestation, environmental degradation, and environmental health. The human impact on the environment issues in the past fifty years has never been so intensive or far-reaching. Although, the citizens and people of the nations of the world must collectively accept the responsibility for the increases in causes of sickness that are the direct result of deteriorating environmental conditions, individuals and corporations must accept blame for the deleterious effect that their personal or corporate lifestyles and practices have had on their health and environment. For example, smoking, overreacting, impaired driving, high living, and so on all add to self-provoking illness, and hence people and corporations must accept a greater degree of self-responsibility if these trends are to be changed. In many developing countries covered in this book, it seems environmental health no longer correlates with per capita income, and in fact higher income seems to do more harm than good with respect to waste disposal and management.

The idea that people can spend their way to better health and environmental conditions without government regulations and public policy is vastly overrated. There is little that even modern medical science can do for a lung that has been destroyed by smoking or exposure to hazardous environmental conditions, or a liver that has ceased to function because of too much exposure to asbestos, or a heart attack or ulcer that is the result of too much stress and air pollution. These are some of the negative impacts of environmental issues that are, in a majority of cases, being presented to the respective governments. Further, in many cases, illnesses can be classified as resulting from ineffective implementation of environmental policies or lack of policy.

There are many environmental challenges facing the people of the world in the present time. In some countries, the governments have not been prepared to implement penalties for those people and corporations who voluntarily pollute the environment. On the other hand, the governments of many countries have not been able to reward those who try to prevent environmental pollution. Perhaps it is the right time to introduce incentives to reward environmental pollution and maintenance, as well as penalties against those who persist in a lifestyle that brings on continuous environmental degradation and pollution. Although it is very difficult to evaluate and analyze the effectiveness of lifestyle change and eco-friendly behavior of citizens to date, governments have not been as successful as hoped in many developing countries when compared with developed nations. Nevertheless, the growth of recycling waste suggests that the public is perhaps beginning to recognize the problem and support the drive toward a better lifestyle and sustainable development practices.

The new environmental and environmental health policies that are being enacted all over the world constitute a sustainable approach to the recognition, diagnosis, treatment, and prevention of environmental degradation, pollution, illnesses, injuries, and other adverse health conditions resulting from hazardous environmental exposures in the workplace, the home, and the community (Dibie, 2010). Many hazardous exposures occur in both the workplace and the general environment, such as contamination of the ambient air and water near a chemical factory, where its workers are also exposed to hazardous substances; application by agricultural workers of pesticides that may contaminate surface and ground water; inadvertent transportation of lead, asbestos, and other hazardous substances to the home on workers' clothes, shoes, skin, and hair; exposure of workers and community residents to hazardous wastes from an industrial facility (Friis, 2012).

The World Health Organization report on medical care and public health in many countries (World Health Organization [WHO], 2012) reveals that the workplace, school, home, and community settings present many challenges to environmental health (Dibie, 2010). The WHO report also reveals that the coal and bauxite mining industry has been creating major environmental health problems for its workers in many countries (WHO, 2012). In addition to the bauxite industry the oil and gas companies have been responsible for over 80 percent of the industrial pollution and occupational health and safety issues in countries where petroleum is drilled. Prolonged exposure to noise from the environment may result in psychological reactions that adversely impact the immune system and physical well-being. Sound is produced by oscillating waves of various frequencies. In many countries, construction and industrial workers who use bulldozers, heavy trucks, and loading machines tend to suffer from partial hearing loss. Workers in the health care industry who use high-speed bone-cutting drills and suctioning devices also tend to suffer from hearing loss after a long exposure to such conditions (Friis, 2012).

The World Health Organization (2012) reports that food-processing plants in different industries have been emitting their waste in a dangerous manner, and their workers are constantly been exposed to the adjacent water courses. Some industrial refineries have also been emitting phosphate-based industrial waste into the lower reaches of the rivers. In many developing countries hazardous wastes are still being disposed of the same way as domestic solid waste (Skolnik, 2008; WHO, 2012).

Most of the industries operating in both developing and developed nations use fossil fuel as their energy source. The main pollutants from the combustion of fossil fuel include carbon monoxide (CO), nitrogen oxide, sulfur dioxide, ground level ozone, volatile organic compounds (VOCs), and lead. There are several health issues or diseases associated with these pollutants. Some of the health problems include irritation of the lungs, respiratory infections like influenza, skin allergies, and reproductive dilemmas (Kraft, 2011). In addition, lead is reported to hamper fetus development and can cause retardation in children after chronic exposure (Dye, 2011; Kraft and Furlong, 2013).

According to Robert Friis (2012) and Nancy Maxwell (2014), toxic heavy metals (arsenic, lead, mercury, cadmium, chromium, and nickel) are potential hazards to human health. Processing and milling of heavy metals put workers at risk of breathing fumes and dusts that contain toxic levels of these metals. In some of the factories in many countries all over the world, an odorless, hazardous, toxic gas has caused several deaths by depriving the body of oxygen (WHO, 1999). Workers suffering from health conditions such as lung disease, heart disease, and anemia are at increased risk of the effects of CO. Toxic heavy metals (arsenic, lead, mercury, cadmium, chromium, and nickel) are potential hazards to human health (Kraft, 2011).

Processing and milling of heavy metals put workers at risk of breathing fumes and dusts that contain toxic levels of these metals. Workers in the bauxite industry and in oil and gas companies who do not following appropriate safety measures are likely to suffer from the impact of these occupational and environmental health problems (Kraft 2011; Maxwell, 2014). Guy Peters (2013) and Nancy Maxwell also identify ambient air pollution as a risk factor for respiratory and cardiovascular morbidity and mortality in numerous epidemiologic assessments in many industrial environments.

As a result of increasing awareness of the negative impact of environment problems all over the world, very few educated people would question the desirability of a clean environment. However, some people would like to see that value balanced more carefully with other important values, such as economic growth, jobs, and controlling inflation (Peters, 2013; Kraft and Furlong, 2013). While it would be difficult to find a group that definitely favors environmental degradation, the politics of pollution is generally linked to acceptable trade-offs between environmental values and other economic values. Guy Peters (2013) argues that there is sufficient public concern in many industrial countries about the environmental that it would be almost impossible to make wholesale retreat from existing environmental problems.

In most cases the attention in environmental policy is placed in the jurisdiction of the national government. The regional governments also play an active role in the formulation and implementation of environmental policies in several democratic political systems. As a resulting making environmental policy is an ideological exercise on the part of many participants especially political parties, environmental groups and nongovernmental organizations (NGOs). The process of making environmental policy also involves a number of complex technical questions about the nature of environmental problems and the feasibility of solutions offered for them (Kraft, 2011; Peters, 2013). In practice, however, the technical dimension is often used to mask the economic interests and ideologies involved in the policy-making debacle. Therefore making environmental policy often results in finding trade-offs among technical feasibility, environmental values, economic growth, and sustainable development that can satisfy multiple constituencies within a country. It is very important to note, however, that environmental policy decision making involves assessment of what governments should do as much as it involves the feasibility question of what government can do (Dye, 2011).

As a result, the range of technical possibilities before policy formulation is frequently broader than the range of ethical justification for acting in the public interest. Despite the earlier assumption, the values that should affect policy formulation in government institutions often conflict with one another and with the utilitarianism of cost-benefit analysis (Peters, 2013). The utilitarian concept postulates that government should do only what would create the greatest good for the greatest number of people in a society or nation. This means that government should be responsible for creating the greatest economic values for the society rather than worrying too much about the minor values. Governments all over the world may have to understand the relationship between human beings and the earth in order to effectively address all the major environmental issues that confront our societies, as well as help to define the subject matter of environmental science.

According to Richard Wright and Dorothy Boorse (2014), in order to achieve sustainable development the government, the private sector, and civil society groups in a country will have to work together to promote the common good. The values associated with the common good include compassion for the less well-off and concern for justice, honesty, sufficiency, humility, and neighborliness. Many people (including traditional environmental organizations and other sectors) will have to collaborate with the government to bring relief to people in need and make the kind of decisions that are needed for effective stewardship (chapter 16 will expand on this discussion). The desire for common goods could galvanize citizens' participation to work toward a sustainable future for their nation (Roorda, 2012; Wright & Boorse, 2014). Inclusiveness and the commitment to sustainable development could help promote changes in individual lifestyle choices, political involvement, career choices, membership and participation in nongovernmental environmental organizations, and volunteer work to improve local environmental conditions.

In addition to striving to provide the common good, government must also do its part to establish and maintain a society infrastructure without which no complex society can survive and no economy can function. Norman Myers and Scott Spoolman (2014) and Richard Skolnik (2008) contend that stagnating nations or societies are frequently those in which the government has been unable to make the necessary improvement in the implementation of environmental policies and the administration of other key public functions. In most cases these stagnating governments may have been involved instead in things that that they do not have to do. Such unnecessary functions include running marketing boards, hotels, fertilizer plants, and domestic transportation. As a result of the negligence of the major sustainable development instruments in the country, the power and energy sector, roads and other forms of transportation, and communications infrastructure become inadequate to support increased investment and production industries. Further, this negligence also makes the educational infrastructure inadequate to provide the skilled labor for industrial expansion. If all graduates have the knowledge, understanding, and skills required for converting sustainable attitude into deeds, an unstoppable, society-wide process of sustainable development will arise. Therefore, what a nation really requires is that every company, and also every public institution, school, health care institute, and NGO, operates in a socially responsible and future-oriented manner.

KEY TERMS

The concept of environmental policy, sustainable development, socioeconomics, eco-friendly behavior, occupational safety, environmental health, and environment policy are pivotal to this research; therefore its context needs to be defined. In order to better understand the framework of this book it is very important that words used, such as *accident*, *injury*, *hazard*, *safety*, *health*, *risk*, and *public policy*, be defined. The definitions will help to give common and acceptable meanings to readers of the book.

An Act, for example (Act, 2014) is a document formally stating what has been done or what has been made into law, by a legislature, executive, court, or constitutional body etc.

Environment is here conceptualized as the summation of all surroundings of a living organism, including natural forces and other living things, which provide conditions for development and growth as well as of danger and damage. Omisore and Akande (2003) define environment under two broad classifications: the natural and the cultural. The former covers the atmosphere (gasses such as nitrogen, oxygen, and carbon dioxide) and hydrosphere (water bodies such as lakes, rivers, oceans, and seas as well as underground water). The latter is the cultural environment, which is man's partial unit. Olorunfemi (2000) saw it in two forms: lithosphere (rocks and minerals) and biosphere (man and animals). Jeje and Adesina (1996) submitted that within this environmental sphere are a number of interactions that propel the different types of human-related activities. As a result, interactions easily translate into several environmentally related challenges, which impart some negative influence on man.

Environmental health is the health consequences of interactions between the human population and the whole range of natural and built and social environment factors (WHO, 1999). Environmental factors include a very broad range of biophysical and built/human-made influences as determinants. Paul Samson (2002) identifies five key determinants of health: living and working conditions, physical environment, personal health practices and coping skills, health services, and biology and genetic endowment. Thus, environmental factors could be seen as only one of the number of determinants of health within a much broader context.

Public policy is what a government chooses to do or not to do (Dye, 2011). Public policies are instruments through which governments can wield their power and provide directives for action (Anderson 2006; Davies and Mazumder, 2003). According to David Easton (1953) public policy is the authoritative allocation through the political process of values to groups or individuals in the society. Public policy is basically a program of actions enacted by government in order to exact control over a particular issue or range of issues within its domain (Peters, 2013). Public policy is an establishment of specific procedures and rules that in turn are used by national and regional governments. Policies should be based on the best available information and result in action related to these principles (Kraft and Furlong, 2012). Hence, policy in this context is the action taken or planned to be taken or implemented by the government in order to control environment, natural or cultural, social, or economic issues as well as workplace safety and health risks. Thus, a policy is a principle plan, or course of action, as pursued by a government, organization, individual, and so on.

Accident is an unplanned event that may or may not result in damage, loss or injury.

Common good or public goods are public resources or services that are shared by everyone in a society. No one can own or use public goods by themselves. In most cases common goods are provide with tax payers' money or a major donation to the government. Common goods are associated with public policies that are enacted to improve human welfare and to protect the natural environment. The values associated with the common good include compassion for the less well-off, concern for justice, honesty, sufficiency, humility, neighborliness (Wright and Boorse, 2014). Public goods are goods or services that are once produced and consumed by a relatively large number of individuals and whose consumption is difficult or impossible to control. As a matter to fact, it is difficult for any individual or private company to produce public goods (Peters, 2013). Common goods could lead to an inclusive value system that could galvanize the bases on stewardship and help people in the society in decisions involving both public and personal lifestyle.

Injury is damage to the body resulting from a delivery of energy to the body above the capacity of the body to cope with that energy or an interference with the normal function and systems within the body.

Hazard is a source of unwanted or excess energy with the capacity to cause damage, loss, or injury. Hazard could also be defined as anything or any condition that has the potential to cause injury or harm to health, or a source of potential damaging energy.

Health is the degree of physiological and psychological well-being of the individual. According to Thomas Dye (2011) no modern government should overlook the responsibility for the health of its people, nor would it wish to do so. To a large extent, each nation's health care system is a reflection of its own particular legacy of traditions, organizations, and institutions, and the United States, as well as several other European countries, has been no exception.

Safety is an individual's perception of risks. Two alternative definitions are (1) safety is a state of mind whereby workers are made aware of the possibility of injury at all times, and (2) safety is a state in which the risk of harm to a person or damage is limited to an acceptable level. Some scholars would argue for tolerable, not acceptable, because no risk is acceptable (Taylor, Easter, and Hegney, 2004).

Socioeconomic is here conceptualized as the combination of social and economic conditions and activities of government as well as citizens, which interact with the environment and create developmental challenges.

Risk is the combination of the likelihood that a hazard will actually result in an accident and the consequences of that accident, often expressed as the product of the two. According to Taylor et al. (2004), risk is the potential for realization of unwanted negative consequences from an event. Technically competent people such as doctors, engineers, and chemists are trained to measure risks objectively. However, no accurate prediction can be made for the future.

Unitarianism concept was propounded by Hosmer (2006). He argued that compassion and a sense of community would be ideal if everyone was compassionate, kind, and worked for the good of the community rather than for themselves. Utilitarianism proposes that an act is then "right" if it leads to greater net social benefits than social harm. This is the rule that is often summarized as "greatest good for the greatest number." The bone of contention is that government should do what would lead to the greatest protection of the environment for the greatest number of people in the country or society.

WHAT IS SPECIAL ABOUT EACH CHAPTER?

Within this book, I have included all the important and traditional topics found in environmental policy and other environmental science courses. I have also added many case studies on environmental policies and sustainability issues. Each chapter includes unique materials that support sustainable development now and prepares one to work effectively toward preventing the depletion of the earth without consideration for the next generations. The chapters in the book show that environmental science stands at the interface between human beings and the earth. The book also explores the interactions and relationships between human being and the earth. It argues that this relationship will need to be considered in virtually all future sustainable development and environmental policies.

Chapter 1 provides a preview of the current state of affairs of changes in the environment around the world. It discusses how changes are taking place in the atmosphere and climate, the human population, and its well-being. It argues that the earth's natural resources are used more effectively in several countries compared to few decades ago. There are also evidences of climate change that have resulted in the increased fossil fuel burning. The chapter examines the crisis in food and agricultural production due to increases in the heat waves and extended droughts in several parts of the world. Drought is also leading to greater number of hungry and impoverished people, especially in developing countries. Despite these negative impacts of the environmental changes, the chapter examines some positive effects as well. It discusses how renewable energy is ramping up swiftly in its share of the world's energy supply. Further good news involves an increasing major reduction in greenhouse gas emissions, as well as decreases in death by tobacco use, which has been addressed in a global campaign. Public health agencies all over the world have been playing a crucial role in the past two decades to expand treatment to reduce diseases such as HIV/AIDS, small pox, malaria, and tuberculosis. The good news about declining growth in population in developing countries will result in less stress on natural resources. The chapter provides evidence that human well-being seems to be improving while the environment continues to decline. It also provides evidence of environmental policies' impact on people, lifestyle, and behavior toward the environment and its ecosystems.

Chapter 2 provides an introduction to environmental concepts and their application. It explores the relationship between living organisms and their environment. It also provides a vivid conceptualization of environmental problems and issues such as air pollution, water pollution, waste management, deforestation, loss of soil productivity, poverty, and degradation of the natural environment. The chapter examines ecology, the subfield of biology that explains the interrelationships among people, other living things, and their environment. The chapter defines an ecosystem as any group of plants, animals, or nonliving things interacting within their external environment. It points out that all organisms have nutrient needs that the ecosystem and its environment must provide for the organisms to thrive. Therefore, the need to thrive requires all human beings and species to have space, clean air, water, food, and other essential nutrients to survive and maintain a quality existence. The chapter also examines the importance of environmental policy. It argues that environmental policies are very important because human beings must protect the very environment that gives us life and at the same time protect the interests of those who exploit it. Thus, understanding ecological interrelations is very crucial in protecting the environment. The concept of sustainability and the common pool predicament provide the background necessary for appreciating the complexities in environmental policy formulation and implementation.

In chapter 3 Professor Robert Dibie and Dr. Mary Bourke examine the economic impact of environmental policies and issues in the United States. The chapter offers perhaps the most complex challenges to the environment and how public policies have been an instrument of sustainability. The chapter argues that natural environmental issues and policies continue to affect the social and economic affairs of the citizens. It discusses the evolution of environmental issues in the country. It suggests that an understanding of the conflict and struggles to rebuild and redirect environmental policy in the country is necessary to determine appropriate solutions in the twenty-first century. With the aid of questionnaires administered, the chapter provides a view that although the citizens of the United States have been affected socially, economically, and health wise, pollution is no longer a national question. The United States has been very effective in providing and regulating clean water and air. The chapter also points out that the United States has taken on a pronounced international dimension, including widespread concern about transborder pollution, renewable energy, and decreasing biological diversity. The concluding section recommends that national economic development should go hand in hand with environmental development.

Chapter 4, by Professor Emeka Nwagwu, focuses on how environmental factors affect the health of individuals, local communities, and the public at large in several Western European countries (England, Germany, Italy, France, Sweden, Finland, Netherlands, and Denmark). These countries were selected because they seem to have enacted similar environmental policies. The chapter explores the health impacts of pollution, climate change and global warming, education, poverty, and the physical environment, including physical activities (exercises). The chapter examines diet, citizens' health, and the ability of individuals to maintain and promote health and healthy lifestyles. It explores linkages between health expenditures, environmental factors, and life expectancy. It discusses examples from Western Europe (England, France, Germany, Ireland, Sweden, Finland, Italy, Netherlands, and Denmark). These are countries where improved socio-environmental factors, rather than expenditures, have been determinants of health outcomes. The chapter also explores the impact of the nexus of unstable political environments and natural disasters (earthquakes, tsunamis, and flooding), man-made disasters (wars, nuclear explosions, toxic gas leakages) that create environmental hazards, and their impact on the general public health. The chapter concludes with some policy implications and recommendations.

Chapter 5 examines the impact of selected environmental health policies in Canada. It provides an extensive review of the Canadian Environmental

Protection Act (CEPA), which emphasizes the use of precautionary principles. CEPA is the main environmental policy for public health protection, pollution prevention, and management of environmental risks posed by harmful substances. The chapter discusses three case studies of environmental public health importance (children's environmental health, Bisphenol A, and the tar sands of Alberta) and the research that reveals ongoing challenges to human health protection and environmental risk management with emphasis on their policy implications. The chapter advocates the use of participatory research to resolve ongoing environmental health problems in local communities with deep distrust of government and industry experts. To be involved in the improvement and protection of its local environment, every community should be empowered by government through the provision of an enabling environment and the tools necessary to identify, analyze, and solve prevailing environmental problems. Furthermore, the chapter recommends the development and implementation of simple exposure and hazard assessment tools to set priorities for systematically identifying substances of highest priority due to their potential to cause adverse public health and environmental effects.

In chapter 6 Professor Robert Dibie examines the significance of transboundary environmental problems in sub-Saharan Africa that are global in nature. The economy of any sub-Saharan African country is a production and distribution system that affects not only producers and consumers but also communities and governments all over the subcontinent. The chapter argues that every country in sub-Saharan Africa (i.e., Nigeria, Ghana, South Africa, Kenya, Sudan, Senegal, Botswana, and Tanzania) needs to develop pragmatic environmental policies to effectively balance economic development and environmental protection. It challenges the public, private, and NGO institutions to collaborate and bear the responsibilities of protecting the environment. The chapter suggests that an evaluation of alternative environmental policy instruments all over the African continent will help to improve the economic and human well-being.

Chapter 7 examines the nature and impact of environmental problems in Ghana and the policies enacted to address the problems. The chapter explains that the pervading condition of underdevelopment in Ghana may require the national government to immediately address poverty, low living standards, poor education, high unemployment, and poor health care systems. The condition and management of ecosystem services is a dominant factor influencing prospects for reducing poverty. Ghana's dominant social paradigm has negatively influenced the nature of environmental policies, public attitudes toward the environment, and environmental regulations in the country. Despite globalization, the need to greatly enhance the environment is crucial in the nation's strategic policies. In fostering good governance and environmental programs, in Ghana the budding institutions of democracy have not effectively promoted sustainability and social equity. Structural capacity of the government could be enhanced to become more instrumental to promoting strategic growth. The chapter stresses the need to further promote broader participation of the citizens in environmental issues. In light of the revitalized milieu, the structure of the political economy needs to be transformed by creating regionally autonomous strategic development units. Thus, policies in Ghana could be considered to be paradoxical because of the limited knowledge of public administrators who enact and implement environmental policies. The chapter recommends that public administrators and political officials should be more responsive to protecting the needs of the people as well as their environment.

In chapter 8 Professor Robert Dibie and Professor Kealeboga Maphunye examine the social impact of environmental policies in South Africa. They argue that the connection between sustainable development and environmental issues works in two ways. On the one hand, unmitigated growth in emissions has the potential to undermine sustainable development in South Africa. The projected impact of climate change may affect agriculture, water, food security, the coastal system, health, and the ecosystem and thus the social and economic lives of the citizens. On the other hand, making development paths more sustainable can contribute to climate mitigation. The chapter concludes by recommending some environmental policies for South Africa.

Chapter 9 provides an analysis of the economic and social impact of environmental policies in Ethiopia. It examines the current environmental policies and how they have been implemented over the years. The chapter discusses the current dynamics of urbanization and how they impact daily life and the complexity of the environmental management processes, and the implementation of conventional and nonconventional in Ethiopia. Increased diversity and partnership with environmental stakeholders could contribute to positive economic and social impact of environmental policies. An evaluation of the role of NGO programs in preventing environmental degradation in the rural areas is presented. The analysis of environmental policies and population growth control programs in Ethiopia reveals that they have not been effective. The chapter contends that the choice of a community-based approach and related options could help to positively improve the social impact of environmental policy in Ethiopia. The chapter concludes by recommending some policies that could help alleviate poverty and improve environmental protection awareness as well as engage citizens in appropriate contributions to solving the nation's sustainable development problems.

Chapter 10 examines the nature and impact of environmental policies in Nigeria. The chapter stresses that the probability of the Nigerian government and corporations succeeding in resolving the environmental problems of this decade will ultimately depend on public understanding, support, and action. The chapter uses survey data to explain the economic and social impacts of environmental policy in the nation. It points out that to be effective over the next several decades, environmentalists, non-government organizations and the governments in Nigeria will need to improve their political skills and forge broader alliance with more citizen groups around common interests in sustainability. The chapter calls on scholars and NGOs to analyze Nigeria's ecological problems by reviewing evidence of the risk they present to human health and well-being as well as to the ecosystem. An extensive analysis of government reports and policy assessment is used in discussing the nature of environmental problems and impacts in Nigeria. In addition, the chapter explores the following questions: To what extent do environmental problems affect the social activities and lives of citizens in Nigeria? What are the barriers to effective implementation of environmental policies in Nigeria? How can these barriers be overcome? What factors could help foster a sustainable environmental policy in Nigeria? Some policy recommendations are provided at the end of the chapter. The chapter concludes with a review of the progress that the Nigerian government has made in dealing with environmental problems and the implications of keeping or revising present policies.

Chapter 11 examines the social impact of environmental policies and health issues in Brazil. Although several studies have explored environmental issues in Brazil, few have critically examined the social impact of environmental health policy in the nation. The chapter draws on data from several communities in Brazil to conduct a cross-national comparison of the social impact of environmental health policy. In addition, the study explores the following questions: To what extent does environmental health affect the social activities and lives of citizens in major cities in Brazil? What are the barriers to effectively implementing environmental health policies in Brazil? How can these barriers be overcome? What factors could help foster a sustainable environmental health policy in the nation? Although environmental scientists have learned about the behavior of contaminants in the environment, their chemicals, their physical transformation, and their movement with or between environmental media, such contaminants in air and water continue to affect the social welfare and health of citizens. Therefore, without practicing sustainable development, Brazil faces a deteriorating environment and may even invite ecological disaster in the near future.

In chapter 12 Professor Robert Dibie examines the social and economic impact of environmental issues in Jamaica. The chapter uses primary and secondary data to discuss the variety of environmental challenges facing Jamaica. It argues that there is a close relationship between population, consumption, and technology pollution in Jamaica. Dibie contends that although environmental problems created by industrialization, heavy automobiles, large quantities of waste, and informal-sector pollution activities are now being addressed, the Jamaica government faces challenges with respect to adequate policy implementation. Despite the efforts of the government, various environmental problems continue to affect the social and economic affairs of the citizens. Dibie observed that there is still no clear relationship between urban size, prosperity and environmental problems. The chapter identifies the most pressing environmental problems in Jamaica and proposes how citizens, communities, and government could respond to them. The concluding section provides some recommendations on how national economic development policies could be implemented with environmental development simultaneously.

In chapter 13 Professor Robert Dibie examines sustainable environmental issues in Trinidad and Tobago. The chapter has two objectives. The first objective is to investigate the economic and social impact of environmental issues and policies in Trinidad and Tobago. The second objective is to evaluate the nature of sustainable environmental policies in Trinidad and Tobago. It points out that Trinidad and Tobago has been experiencing depletion of resources and mounting problems of poverty. It discusses how environmental degradation, social inequality, and population explosion are still common problems in the country. The chapter contends that the lack of rigid sustainable environmental policies in the country has exacerbated partial eco-friendly human activities toward the ecosystem. It explores how environmental changes have been one of the most complex challenges in Trinidad and Tobago. Interviews and survey data are used to present an argument that Trinidad and Tobago will bear the brunt of the effect of environmental change, even as it strives to overcome poverty and advance economic growth. Inadequate sustainable environment policies are inextricably related to modern civilization in the nation. Sustainable transformation in the future would require the industrial sector, governments, NGOs, and citizens to act differently, and act together. Sustainable changes cannot be attainable without these major stakeholders cooperating to improve energy efficiencies, developing and deploying clean technologies, and expanding the desire to grow green by absorbing gases. The chapter concludes by discussing the survey results, limitations, and directions for future environmental research in the country.

Chapter 14 discusses the impact of environmental health policies in Belize, a country in the Caribbean Sea. Dr. Emmanuel Iyiegbuniwe uses survey data to analyze the social impact of environmental issues and policies in Belize. The challenges of biodiversity loss in Belize and pollution must be addressed by policy makers with equal priority and in close coordination, if the most severe impact of each is to be avoided in the near future. The chapter reviews social and economic evidence that speaks to the various environmental problems and the policy implications. It describes the natural resource policies and the many conflicts between natural resources preservation and economic development in Belize. New directions for policy change in the country are presented.

In chapter 15 Professor Robert Dibie and Muawya Hussein identify different types of organizations and people that pollute the environment in selected Arab countries. The countries examined include Egypt, Oman, Bahrain, Jordan, Kuwait, and United Arab Emirates. These countries where selected because they have enacted environmental policies to ensure that development programs embrace ecological concerns as part of the overall planning strategies. Environmental concerns are reflected in their respective industrial, agricultural, urban, transportation, energy, and tourism development plans. The chapter focuses attention on the scope and complexity of addressing environmental pollutants and the difficulty of effectively implementing policies in these Arab countries. The main environmental problems and issues are identified as degradation of forest, destruction of habitats, and loss of species. Other causes of the environmental problems are identified as rapid growth of population, agricultural productivity, urban expansion into preserved forest lands, poverty, unsustainable use of biota, and industrial pollution. The chapter explores the emergence of urban air pollution as a serious threat. Cities and citizens are experiencing air pollution from gases such as particulates and lead at levels often exceeding global standards. The main sources of air pollution in the region are energy plants, industrial production factories, and automobile emissions, which account for almost 90 percent of air pollution in urban centers. Research reveals that pollution from vehicles was mainly due to poor maintenance, aged cars, low-quality fuel, and poor traffic management. Stationary sources, such as outdated power generation stations, are also a major source of the environment problems in the Arab countries examined. Apart from air pollution problems, the region is experiencing ecological changes as a result of the burning of fossil fuel. Economic growth in Arab countries tends to depend much on input from the environment, thereby resulting in severe negative impact on citizens' lives in the region.

Chapter 16 provides a discussion of the nature and challenges of environmental sustainability and solutions. Environmental issues are becoming a dominant political force in nations all over the world. In the nations covered in this book, environmental issues have manifested themselves as major policy dilemmas. The chapter also presents an argument that governments all over the world are still not properly equipped to resolve many of the short- and long-term environmental problems that they face. This is why institutional reform and new leadership methods are critical to successfully achieving sustainable development in the near future. The chapter examines obstacles and leadership challenges that nations could face as they create new approaches to environmental management. Promoting sustainability at local and national levels will require paying attention to many factors such as economic development, transportation, social justice, land use, environmental planning, and housing. It concludes by describing sustainable development as human development that is ecologically sustainable. Economic and environmental development should be directed toward human freedom, opportunity, and higher quality of life for both present and future generations. The chapter suggests that through appropriate planning, national governments can achieve both traditional development and environmental protection or the restoration of the ecosystem. Additional recommendations are made about what environmental agencies and political leaders will need to do, be, and become to help our ecosystems and world.

REFERENCES

Anderson, J. E. (2006). *Public Policy Making: An Introduction* (2nd ed.). New York, NY: Houghton Mifflin.

Caribbean Environmental Health Institute Report. (2006). "A National Inventory for the Generation of Hazardous Waste in Trinidad and Tobago." Environmental Management Authority Case Study Series.

Castillo, D., Pizatella, T., and Stout N. (2011). "Injuries and Occupational Safety." In *Occupational and Environmental Health* (6th ed., 121–122). New York, NY: Oxford University Press.

Davies J. M., and Mazumdar, A. (2003). "Health and Environment Policy Issues in Canada: The Role of Watershed Management in Sustaining Clean Drinking Water Quality at Surface Source." *Journal of Environmental Management, 68*, 273–286.

Dibie, R. (2010). "Social Impact of Environmental Policies in Ethiopia." *Journal of International Politics and Development, 8*(1), 1–23.

Dibie, R., and Offiong J. (2012). "The Nature and Impact of Environmental Policy in Nigeria." *Journal of International Politics and Development, 10*(1–2), 113–144.

Dye, T. (2011). *Understanding Public Policy* (13th ed.). New York, NY: Longman.

Easton, D. (1953). *The Political System*. New York, NY: Knopf.

Friis, R. (2012). *Essentials of Environmental Health* (2nd ed.). Burlington, MA: Jones & Bartlett Learning.

Gupta, D. K. (2011). *Analyzing Public Policy: Concepts Tools and Techniques*. Washington DC: Congressional Quarterly Press.

Holmberg, J. (1992). *Making Development Sustainable: Redefining Institutions, Policy, and Economics*. Washington DC: Island Press.

Hosmer, L. (2006). *The Ethics of Management* (5th ed.). Boston, MA: McGraw-Hill.

Kraft, M. (2011). *Environmental Policy and Politics*. New York, NY: Longman Press.

Kraft, M., and Furlong, S. (2012). *Public Policy: Politics, Analysis, and Alternatives*. Washington DC: Congressional Press.

Maxwell N. (2014). *Understanding Environmental Health* (2nd ed,). Burlington, MA: Jones & Bartlett Learning.

Myers N., and S. Spoolman. (2014). *Environmental Issues and Solutions*. Belmont, CA: Brooks/Cole Press.

Olorunfemi, J. F., and Jimoh, H. I. (2000). "Anthropogenic Activities." In E. O. Omisore and C. G. Akande, *The Role of All Trier of Government and Non-Governmental Organizations (NGO) in Environment Conservation*.

Omisore, E. O., and Akande, C. G. (2003). "The Role of all Ties of Government and Non-Governmental Organizations, in Environmental Conversation." A paper presented at the National Conference on Environmental Sustainability in a Democratic Government. Report Nordiska Afrikainstitutet No. 11.9

Peters, G. (2013). *American Public Policy*. Washington DC: Congressional Quarterly Press.

Samson, P. R. (2002). "Developing Environmental Public Health Indicators: A View from Canada." *Statistical Journal of the United Nations ECE, 19*, 105–116.

Skolnik, R. (2008). *Essential of Global Health*. Burlington, MA: Jones & Bartlett Learning.

Taylor, G., Easter K., and Hegney R. (2004). *Enhancing Occupational Safety and Health*. Burlington, MA: Elservier Butterworth-Heinemann.

Taylor, R. (2012). *Taking Sides: Clashing Views in Sustainability.* New York, NY: McGraw-Hill.

World Health Organization. (1999). *Environmental Health Indicators: Framework and Methodologies*. Geneva, Switzerland: WHO Technical Report Series, WHO/SDE/OEH/ 99.10.

World Health Organization. (2012). *Quantifying Selected Risks to Health: Reducing Risks Promoting Healthy Life*. WHO Technical Report Series, Geneva, Switzerland.

Wright, R., and Boorse, D. (2014). *Environmental Science: Towards A Sustainable Future*. Upper Saddle River, NJ: Prentice Hall.

2 Environmental Concepts and Applications

Robert Dibie, PhD

INTRODUCTION

Environmental challenges and issues are considered to be worldwide problems. The media report nearly every day some of the complex environmental issues and how human beings live and relate with their ecosystem. The media news tends to cover how forest, water, and soil resources are under stress. Forests are also reported to be disappearing. Human activities all over the world tend to inflict harsh and often irredeemable damage on the environment and on critical natural resources (Kendall, 2012). Human beings are creators of pollution and garbage waste. The earth's climate is reported to be getting significantly warmer (Kraft and Furlong, 2013). On one hand the environment includes the earth, such as air, atmosphere, soil, water, rock, and all living things. On the other hand, an ecosystem means the things that live and interact in parts of the geophysical environment (Kraft, 2011; Harper, 2008).

The most basic way for understanding the world's complex environmental issues is the notion of a system that is interdependent and interconnected. Human beings traditionally have established relationships with nature (Taylor, 2012). Men and women all over the world have recognized the importance of air, land, and water management because without these natural resources they cannot survive. The human traditional communities valued the ethics of not taking more than is needed from the environment (Withgott and Brennan, 2009). According to Myers and Spoolman (2014), human environmental impacts include large-scale cutting of tropical forests, drought (or long term, excessively dry weather), topsoil erosion, species extinction, and air and water pollution. They contend that some forms of degradation such as drought are natural, but can be worsened by human activities. Further, human activities such as the clearing of forests and the expansion of cities can eliminate habitats for endangered species such as eagles and Sumatran tigers.

The complex environmental world makes it imperative for all nations to integrate environmental protection with other social and economic activities. For instance, the air that people breathe in the United States may have been polluted by multinational corporations in Canada, some countries in the African continent, or in Mexico. At the same time the air that is contaminated in Australia, New Zealand, or Japan may spread to the Caribbean Island countries and all over Western Europe. Further, the chlorofluorocarbons (CFCs) that are used in China, Nigeria, Russia, or Brazil may break down the ozone layer, which might in turn affect ecosystems all over the world. Since all human beings in the world depend on these ecosystems, they may be affected equally. The 2011 tsunami and the subsequent nuclear disaster in Japan caused radiation fallout that spread all the way to California in the United States, which is thousands of miles away. The aftermath of oil spillage that occurred in the Gulf of Mexico in 2010 had an impact on the biodiversity of various countries in the Caribbean Islands and as far as Argentina in South America. According to Sachs (2005), the poverty and hunger in one country due to climate change may cause social and economic pressures in another country. The famine in Darfur in Sudan caused the world to donate funds to help save the children and women affected. In addition climate change in one country may cause immigration to other nations. Excessive immigration may also spill over into urbanization problems, health issues, waste management problems, unemployment, and other social problems in an event of trying to survive.

Most of the petroleum products drilled in some African and the Middle Eastern countries are used in Western Europe and North America to generate energy and manufacturing industries that in turn pollute the air that the people in South America may be exposed to. Such exposure may cause environmental health challenges that could have severe impact on the social and economic livelihood of the people in the region. Thus, human beings all over the world and the ecosystems are closely related to the natural laws. At the same time human culture and spiritual heritage could reinforce our social and economic interests and survival. Although natural disasters have contributed to some of the world environmental problems, human beings have also played a major role in creating the current environmental crisis. In general human beings have been grappling with natural causes such as earthquakes, volcanic eruptions, storms, droughts, diseases, and the like (Kraft and Furlong, 2012). It has been argued that human beings all over the world have exacerbated the environmental crisis (Peters, 2007; Taylor, 2012). For instance, human beings have played a crucial role in air and water pollution, depletion of the ozone layer, and extinction of fish and animal species and forest. The world environmental crisis emanates from these factors: (1) greed by developed countries; (2) ignorance and poverty by people in developing countries; (3) inappropriate use of science and technology; and (4) conflict over the world political economy (Rosenbaum, 2014; Taylor, 2012). The bone of contention is that if we are interested in understanding the social and economic impact of environmental policies and issues in the world, we must consider both the developed and developing countries and their contributions to the ecosystem hazards that we now face.

Edoho and Dibie (2000), Charles Okidi (1984) and Michael Kraft (2011) contend that while multinational corporations from the developed countries continue to pollute the environment in developing nations at an alarming rate, it is interesting to observe that when developing nations' elites are fortunate to establish similar industries, they also have been culprits of the same pollution crime. As you will read from most of the chapters in this book, political leaders in both developed and developing nations have failed to either enact or effectively implement appropriate environmental policies to address the crisis at hand. According to Guy Peters (2007) a gradual erosion of biodiversity has been observed in the United States and in the rest of the world. The United States both exports acid rain to Canada and imports some water and air pollution from Mexico, Canada, and China. These examples show that it is no longer sufficient to address environmental problems within the context of a single nation (Gore, 1992). Environmental problems around the world now require concerted action and policies by all national governments.

The interdependent relationship between living organisms and their environment entails that all organisms have nutrient needs that the ecosystem must provide for the organisms to thrive. Therefore, if a population gets too large, the ecosystem may become overloaded and unable to provide the basic needs of everybody in the complex world environment. The need to thrive means all human beings and species need space, clean air, water, food, and other essential nutrients to survive (Harper, 2008). In a situation when the human population gets too large relative to its environment, there may be the consequence of the ecosystem not being able to carry the capacity. This effect may threaten human welfare (Buchholz, 1993; Dye, 2011; Steel, Clinton, and Lovrich, 2003). Therefore, fundamental changes are urgent if human beings are to avoid the collision of their present course of harsh infliction on the environment.

ENVIRONMENTAL POLICY

Most people around the world do not understand what environmental policies are all about. Some people refer to them as the relationship between human beings and nature. Others see environmental policies as government control of parks, open spaces, natural resources wilderness, wildlife, and forest. According to Michael Kraft and Scott Furlong (2012) a more useful way to understand the environment is to see it as a set of natural systems that interact in complex ways to supply human beings and other species with the necessities of life. The necessities of life include clean water, food, breathable air, fiber, energy, and the recycling of waste. Thomas Dye (2011) contends that animals and human beings are very dependent on the environmental system to meet their needs in order to survive. It is very important to note that human beings and animals cannot survive without these environmental systems. Despite the natural resources that human beings derive from the environment, we have not been able to fully recognize our responsibility to protect the environment or to place a reasonable value on the natural services that we all take for granted in protecting our source of survival. When human beings use or pollute a renewable resource faster than natural processes can replace it or clean it up, we say it has become degraded or less useful. Further, when human beings degrade or deplete natural resources, we also tend to reduce or disrupt the ecosystem services they provide, such as natural water purification and cycling of nutrients (Myers and Spoolman, 2014).

Environmental policy is whatever a government or a nation chooses to do or not do about its environment. As a result a government may regulate industries to decrease pollution, give society members incentives to properly manage their waste, or change behavior toward the environment. A government may also distribute a variety of symbolic rewards and material services to citizens of the country. At the same time, a great deal of environmental policy has been enacted by setting rules or limits and threatening punishment for violating these rules or limits in what is often called a command-and-control approach (Dye, 2011; Peters, 2013; Withgott and Brennan, 2009). Most industrialized countries often adopt creative use of economic incentives to encourage desired outcomes as well as discourage undesired goals and results. Some developed countries also set market dynamics in motion to achieve goals in an economically efficient manner. Taxation is often used by some countries to discourage undesirable activities. The governments of some developed countries also use a marketable permit approach to provide companies an economic incentive to find ways to reduce emissions. Thus, environmental policy is a problem-solving means that government uses to enforce science, ethics, and economic goals. Environmental policy also requires an astute understanding of the political process.

Guy Peters (2013) contends that governments attempt to manage or change behavior toward nature by enacting environmental policy. He described these governments' actions as different kinds of fixes for environmental problems that involve using public policy to change behavior, technology ideas, and law. From this perspective environmental policy could be enacted to stimulate the adoption of new technology in a variety of ways, such as public investment, subsidies, tax policies, or regulatory mandates. In addition, environmental policy may be adopted to provide incentives to get citizens to change their behavior. Behavior fixes in respect to protecting the environment may include getting people to stop smoking, reduce waste by eating lower on the food chain, turn down thermostats in the winter, install ceiling fans and use less air conditioners during the hot season, and carpool or use public transportation as an alternative to driving (Kendall, 2012; Harper, 2008). Another public policy method that a government may adopt is creating awareness. This method often relies on public education and media campaigns. Further, the government may adopt a legal fix that requires a mandated change through laws and regulations rather than incentives, subsidies, or persuasion. Catton and Dunlap (2002) suggested that the best way to achieve an effective environmental protection goal is to simultaneously provide better technical means and incentives to changing peoples' behavior. By providing material incentives and using regulatory restrictions that target the right constituencies, more awareness could be achieved. It is important to note, however, that even if all citizens are in favor of a clean environment it is difficult to agree on how such a trade-off among values could be in made individual cases.

Environmental policy as used in this book refers to all government actions that affect or attempt to affect environmental quality and the use of natural resources (Kraft and Furlong, 2012). Environmental policies are means used to achieve collective choices of the society as well as ways citizens adapt to govern the environment. They are the sum of government activities whether pursued directly or through agents to achieve environmental goals. These government activities may positively influence the lives of citizens (Peters, 2007). Environmental policies are not stagnant; nations of the world may respectively maintain their current policies and practices or they may envision a better future and design the institutions and policies necessary to help bring the needed change (Gore, 1992; Kraft, 2011). Steel et al. (2003) pointed out that environmental policy is what the government says and does about perceived environmental problems. Policy making is how the government decides what will be done about perceived environmental problems. Policy making is a process of interaction among governmental and nongovernmental actors. Environmental policy is the outcome of that interaction (Ripley and Franklin, 1987). Broadly speaking, one can group the actors in the environmental policy-making process into two broad categories: institutional and noninstitutional. Most environmental policies are regulated by an appointed public agency. It is very important to note that regulatory policies in some countries attempt to reduce or expand the choices available to citizens and industries to achieve a social goal. Michael Kraft (2011) contends that the most common approach to regulation is the setting and enforcement of standards such as the amount of pollutants that a factory or corporation may emit into the water or air. Environmental protection policies such as the Clean Water Act, Clean Air Act, and Hazardous Waste Disposal Act in several countries are regulated. Research in the countries covered in this book reveals that politics are often associated with the regulation of public policies. In general, it was observed that regulatory policies tend to pit environmental stakeholders and public health groups against industries.

According to Steel et al. (2003) and Thomas Dye (2011), institutional actors involved in the environmental policy process are governments and nongovernmental agencies that deal with public affairs. Such institutions include executive departments, legislatures, and, in some democratic countries such as the United Kingdom, Canada, the United States, Brazil, or Nigeria, the judicial branch. The United Nations Development Programme (UNDP, 2000) defines nongovernmental organizations as noninstitutional actors. According to UNDP (2000) nongovernmental organizations (NGOs) are any nonprofit organizations, groups, or institutions that operate independently of a government and have humanitarian or cooperative, rather than commercial, objectives. Steel et al. (2003) contend that the UNDP generally collaborates with NGOs representing poor and vulnerable groups. They often work in the areas of thematic priorities such as sustainable livelihood, gender equality, poverty eradication, natural resource management, and governance. NGOs could be large-scale membership groups, community-based groups, and international organizations. These organizations are commonly referred to as international nongovernmental organizations (INGOs) if they have operations in more than one country.

The United Nations Environment Program Report (1997) stipulates that as nations make progress in their development efforts, different sets of environmental and natural resource policy concerns tend to galvanize priority. However, the role a country plays in international environmental politics depends on both domestic and international factors.

SUSTAINABLE DEVELOPMENT

The word *sustainable* is adopted from the basic term to sustain or maintain over a long period of time. In this book the word *sustainable* refers to the ability of human beings to maintain their natural ecosystem while improving human social and economic welfare over a long period of time. Robert Taylor (2012) observed that sustainable development is currently a "catchword" and as such is often misused. Because there is no universally accepted definition of the concept, I will adopt my own definition. In the narrowest sense, however, sustainability means indefinite survival of the human species across all the regions of the world (Kendall, 2012; Smith-Sebasto, 2012).

The meaning of sustainable development covers virtually all human beings, from childhood to adulthood. It also entails that human beings should have quality lives beyond mere biological survival. In the broadest sense, sustainability includes all components of the biosphere, even those with no apparent benefit to humanity (B. J. Brown, Hanson, Liverman, and Meredith, 1987). Sustainable development is defined as a pattern of social and structured economic transformation or modernization that optimizes the economic and societal benefit available in the present, without jeopardizing the likely potential for similar benefits in the future (Rosenbaum, 2007; Kraft, 2011; Harper, 2008). The major goal of sustainable development is to achieve a reasonable and equitably distributed economic well-being that can be perpetuated continually for many human generations.

The United Nations promoted the idea of sustainable development for several years until the concept was eventually acknowledged at the 1972 Stockholm Convention. The 1982 World Commission on Environment and Development report, which was published in 1987, reinvigorated the term *sustainability*. The prime minister of Norway at that time, Gro Harlem Brundtland, chaired the group that was put together by the General Assembly of the United Nations (Taylor, 2012). The Brundtland Commission laid out the most famous definition of *sustainable development* as

> development that meets the needs of the present without compromising the ability of future generations to meet their own needs (World Commission, 1987, 43).

This Brundtland Commission definition contains two key concepts: (1) the concept of needs, in particular the essential needs of the world's poor, to which overriding poverty should be given; and (2) the idea of limitations imposed by the state of technology and social organization on the environment's ability to meet present and future needs (United Nations The World Commission on Environment and Development (WCED, 1987). The 1992 Rio Summit Declaration sought to reflect on efficient and sustainable use of factors of production in the formation of commodity prices, including the reflection of environmental, social, and resources costs. The post-Rio deliberations at the United Nations Commission on Sustainable Development did not make any significant progress. However, the United Nations Environment Programme (UNEP) and the United Nations Conference on Trade and Development (UNCTAD) declared in 1997 that they worked together to provide an integration of trade and environmental issues. The most important promoter of these issues is the institution of the General Agreement on Tariffs and Trade (GATT). The World Trade Organization (WTO) superseded this. The key entity under the WTO is the Committee on Trade and Environment (CTE), which was constituted in 1995.

Stephen Wheeler (2004) contends that a number of other specific themes could be derived from the term *sustainability*. These specific themes include:

1. Sustaining human livelihood—Sustainability is the ability of a system to sustain the livelihood of the people who depend on that system for an indefinite period (Otto Soemarwoto cited in Taylor, 2012).
2. Not making things worse—Sustainability is the ability of a system to sustain the livelihood of the people who depend on the system for an indefinite period (Rees, 1988).
3. Maintaining natural capital—Sustainability requires a least a constant stock of natural capital, construed as a set of all environmental assets (Pearce et al., 1988).
4. Grab-bag approach—Sustainable development seeks to respond to the following five requirements: (1) provision of social self-determination

and cultural diversity; (2) integration of conservation and development; (3) maintenance of ecological integrity; (4) satisfaction of basic human needs; (5) achievement of equity and social justice (International Union for the Conservation of Nature, 1986, cited in Taylor, 2012).

5. Oppose exponential growth—Sustainability is the fundamental root metaphor that can oppose the notion of exponential material (Callenbach, 1972, cited in Taylor, 2012).
6. Protecting and restoring the environment—Sustainability equals conservation plus stewardship plus restoration (Van der Ryn, 1994, cited in Myers and Spoolman, 2014)).
7. Maintenance and improvement of systems—Sustainability implies that the overall level of diversity and the overall productivity of components and relations in systems are maintained or enhanced (Norgaard, 1988, cited in Taylor, 2012).
8. Carrying capacity of ecosystems—Sustainable development means improving the equity of human life while living within the carrying capacity of supporting ecosystem (World Conservation Union, 1991, cited in Myers and Spoolman, 2014).

These eight definitions of sustainability show that environmentalist must work closely with economists because their problems and solutions are interrelated. A strong collaboration between government, NGOs and the private sectors could help to promote and achieve a healthy environment in the future. Some scholars contend that the perceived dichotomy between human and nature is the root of all human environmental problems (Withgott and Brennan, 2009; Dye, 2011; Harper, 2008). However, environmental protection need not lead to economic stagnation; instead it is likely to enhance economic opportunities. Environmental protection on the contrary may increase a nation's attractiveness by drawing more immigrants or residents. If a nation embarks on environmental protection, such policy may increase property values and the tax revenues that help fund social services. Some of the constituents of sustainable development universally recognized are health and environment, economic development, equity, environment protection, sustainable livelihoods, population control and education, climate change, high pollution levels, deforestation, and depletion of natural resources (Taylor, 2012; Smith-Sebasto, 2012).

Sustainability is a necessary but not sufficient condition for sustainable development (Kraft and Furlong, 2012). Sustainable development warrants a delicate balance of policies and programs that are aimed at social development. All the programs may ensure the resistance of the ecosystem to remain a fundamental requirement. One of the greatest flaws in the Brundtland Report (United Nations World Commission on Environment and Development: Brundtland Commission Report, 1987) is its failure to address the urgent need for a huge increase in funding universal family planning and birth control as well as the removal of any incentives rewarding population growth (Cassil, 2012). The report, however, advocated more equality, education, and opportunities for women. The report also contributed significantly to the long-term awareness and the implications of birth control.

Sustainable development must include human values, democratic institutions, economic growth with social justice, uninterrupted processes of institutions, nature conservancy, and modest consumption habits. In addition, it must include reformed international financial and development institutions for unforced effectiveness, enhanced role of people's institutions, environmental literacy and respect for all forms of life. The sustainable development term fails to make clear that a vital variable in the development equation is population growth. Projected population growth may exacerbate regional disparity, fuel resentment, and possibly give rise to more (Cassil, 2012). This is the problem that has alluded to population limits, which has led most social scientists to dismiss the term *sustainable development* as a contradictory term. According to Withgott and Brennan (2009) sustainable solutions to environmental problems are numerous. The challenges that most countries face lie in the effort to think outside the box and being shrewd and dogged enough to overcome political or economic obstacles that they confront in the implementation process. Here are some of the major strategies that nations all over the world may adopt:

1. Reduce unnecessary consumption
2. Refine their idea about economic growth and quality of life.
3. Limit population growth
4. Encourage green technologies
5. Promote research and education
6. Mimic natural systems by promoting closed-loop industrial processes
7. Enhance local self-sufficiency, and embrace some aspects of globalization
8. Vote with their political wallets
9. Be politically active
10. Think in the long term (Withgott and Brennan, 2009)
11. Reduce and eventually eliminate poverty
12. Manage resources crucial to human welfare more effectively.
13. Ensure sexual equality and guarantee women control over their own reproductive decisions (Smith-Sabesto, 2012)

The availability of resources and their distribution makes a difference to the sustainability of environment (Taylor, 2012). Therefore, poverty and gender issues become important in developing and developed countries. These are the main criteria along with some other sociocultural ones that influence the distribution of resources. According to Steel et al. (2003) planning for sustainable development while directly addressing environmental challenges has to alleviate socioeconomic and gender differential issues. In addition, it is important to note that there are various obstacles to the process of development and the importance of planning is, therefore, crucial.

There are a number of issues and factors that have to be kept in mind while planning the development process and employing inadequate measures. This chapter discuses some of the important factors with respect to sustainable development. It is important to note, however, that not all these social and environmental factors are consistent with sustained growth in an interdependent world over the long term. Given the social and environment processes caused by past development strategies, the goal of raising human well-being worldwide must be pursued through a development process that "does better" in eliminating the poverty path. It is also very important to integrate social and environmental concerns in pursuit of the goal of sustained improvements in human well-being.

According to Zachary Smith (2004) development does not imply growth that tends to continue indefinitely in unchecked capitalism. As a result of this fact, sustainable development requires economic improvements and efficiencies guided by democracy, where constant growth only occurs in human and environmental health education and welfare. It is very important to note that without practicing sustainable development, humanity may face a deteriorating environment that could escalate to ecological disaster.

HUMAN AND ENVIRONMENTAL INTERCONNECTION

The world social and environmental patterns depend on consistent, sustained growth in an interdependent world over the long period. Given the social and environment processes caused by development strategies, the goal of raising human welfare standards worldwide must be pursued through a development process that does better. This should entail a poverty-eliminating quality path that integrates social and environmental concerns in pursuit of the goal of sustained improvements in human livelihood. Ecological sustainability is a system approach to capitalism and democracy. Economic growth and ecological sustainability must be such that people's livelihoods, basic needs, political participation, and rights do not work against them. Environmental sustainability must be conceptualized so as to support and sustain livelihoods and basic needs. It should not be perceived in a way that automatically counterpoises nature against the survival needs of the most vulnerable. Where trade-offs among these different goals exist or are inevitable, the costs and burdens must not fall on the poorest and most vulnerable. Further, all people must have a voice in negotiating resolution through open and genuinely participatory political processes. In addition, environmental strategies that enhance livelihoods and fulfill needs can probably help lay the basis for reduced rates of mortality and fertility.

Smith (2004) contends that since businesses only respond to money, they must be shown how environmental goods can help them earn more. On one hand, less pollution can come from more efficient production that uses fewer raw materials. On the other hand, ignoring environmental problems will not save money, but will be much more costly in the future. Therefore, spending a few thousand pounds or dollars today on some clean-up efforts is better than a billion euro or dollar lawsuit tomorrow. As a result a cleaner environment also means happier, healthier, and more productive employees (Taylor, 2012). Environmental development, growth, and protection could be regarded as a saving cost or an investment, because it is not a luxury. In most cases it is a necessary cost for continued economic progress and a progressive quality of life. As a result of a better quality of life for the citizens, the role of government is not to let economic agents act as they please. The government must introduce policies to ensure the collective will of everyone in the country or world. This means the internationalization of environmental protection issues.

THE COMPARTMENT RELATIONSHIP MODEL

Odum (1971) used a compartment model to describe the relationship between different kinds of environments and ecosystems. He identified four broad types of natural settings that are partitioned according to their biotic function and life cycle criteria. Odum called the functional settings (1) compromise or multiple use environments and ecosystems that combine both types and functions; (2) environments with young, relatively immature and rapidly growing ecosystems; (3) urban-industrial protective equilibrium; and (4) environments with more mature, diverse, or climax ecosystems that tend toward protective equilibrium.

The compartment model presents an argument that the growth of human settlements and communities obviously decreases the proportion of other types of environments and ecosystems at the expense of the more mature, protective ones (Harper, 2008). From this perspective it could be argued that humans' activity and their great expansion may negatively prevent the growth of the ecosystem. Charles Harper (2008) contends that human beings' expansion in the form of cutting of forest and acquiring land for agriculture and other uses may equally have some negative impact on other organisms in the environment. As a result human constant expansion may lead to loss of biodiversity in monocultures. It is interesting to note that after almost four decades of Odum's proposed compartment model, human beings in the nations of the world still continue to expand for intensive agriculture and urban sprawl at the expense of the protective landscape. According to Odum (1971) and Harper (2008), a sociocultural system is a network of interdependent actors that comprise of subsystems, organizations, and individuals that are in relatively stable patterns of interaction

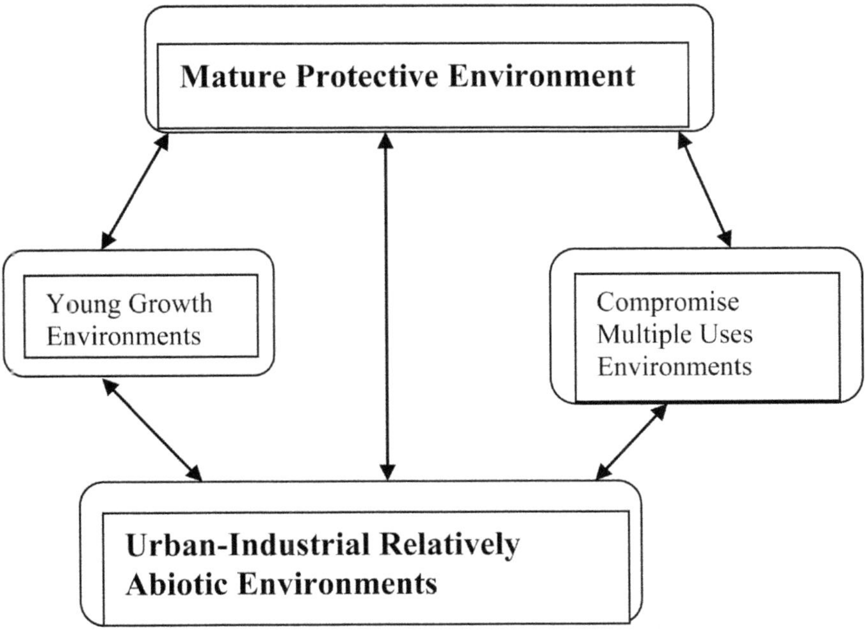

Figure 2.1 Compartment Model

Source: Adopted from Odum, E. P. (1971). *Fundamentals of Ecology* (3rd ed.). Philadelphia, PA: W.B. Saunders.

and intercommunication. Figure 2.1 shows the compartment model of environments and ecosystems according to function and life cycle criteria.

ECOLOGICAL MODERNIZATION THEORY

The theory of ecological modernization by (Mol, 1995) explains the actual transformation in social practice and institutional development. The theory points out that while modernizing, companies could become more efficient by minimizing the use of the ecosystem. Mol (1995) argues that through feedback systems and recycling by using waste from one process to supply or fuel another economic process, the environment could be preserved. Recycling rather than continuously extracting raw materials may reserve the environment better, rather than extracting raw materials, consuming them, and throwing them away. Mol and Sonnenfeild (2000) contend that the prevention of constant extraction of raw materials would generate profit, and the cost could be incorporated in cost accounting and pricing practices. Charles Harper (2008) observed that ecological modernization originated from nonclassical economists. The main bone of contention is that ecological modernization has the potential to promote positive environmental change.

In the past two decades such potential has in some ways been evident in segments of the economy. Ecological modernization theory has led to corporate policy shift among some of the largest energy multinational companies, such as British Petroleum and Royal Dutch Shell, that have broken ranks with other fossil fuel firms to take climate change more seriously.

As a result of the ecological modernization theory, some logging companies in Canada and the western part of the United States have started practicing selective cutting instead of the traditional clear-cutting methods. Selective cutting involves leaving trees to check runoff and soil erosion and to provide wildlife habitats to regenerate the forest (Harper, 2008). Further, Kraft and Furlong (2012) made an observation that about 57 percent of the steel produced now in the United States is recycled. The steel factories built in the past three decades in the United States are no longer located in Pennsylvania, where coal and iron ore are close; rather they are scattered all over the nation in California, Nebraska, and North Carolina. These new steel companies are now feeding on local supplies of iron scrap. It is also observed that the new steel factories produce steel with less energy and far less pollution than the older mills producing from newly extracted iron ore from the environment (Harper, 2008). Similar patterns of ecological modernization have taken place in the paper manufacturing companies that were previously established near forested areas. In the past three decades paper factories have been built near cities feeding on local supplies of scrap or recycled paper. The ecological modernization theory has thus created the basis for a growing confidence that environmental reform can no longer be viewed in an imaginative and exclusive way.

ECOLOGICAL THEORY

The ecological theory by Odum (1971) and Rogers (1994) stipulates that ecosystems evolve as different species continue to compete for available space and energy in the physical environment in order to survive. The result over a long period if the ecosystem is not interrupted could be a large and more complex as well as inclusive structure of species. These species could also be interconnected in food chain niches. According to Charles Harper (2008), this chain often causes a symbiotic relationship that ranges from mutualistic to parasitic interdependent. This relationship also leads to social exchanges of reciprocity in the biological world, which in turn produces egalitarian, mutual benefits in nonhierarchical contexts. Further the established relationship may galvanize a social exchange of redistribution whereby goods and services are shifted upward to persons or centers that reallocate those (Rogers, 1994). For example, human labor has incrementally become a commodity for exchange at a fixed monetary rate. In addition money has become a symbolized value of more concrete boil-physical resources such as land and minerals.

ECOLOGICAL FEMINISM THEORY

The ecological feminism theory stipulates that women and the environment are treated similarly. According to d'Eaubonne (1994) and Mies and Vandana (1993), women and the environment have been oppressed and are in need of liberation. Ecofeminism theory presents an argument that the patriarchal male dominated structure of society that traditionally grants more power and prestige to men than women is the cause of both social and environmental problems. The theory postulates that a male tendency to try to dominate and conquer what men hate, fear, or do not understand has historically been exercised against both women and the natural environment (Withgott and Brennan, 2009). The ecofeminism theory proposes that women can free themselves and nature at the same time by promoting equity in economic markets (Clifford and Edward, 2012). It is interesting to note that male power is identified as centered cantered on economic domination of women and nature (Merchant, 1996).

Cultural, religious, social, economic, political and even colonial legacy have constituted factors that play dominant roles in shaping the participation of women in every aspect of human endeavor in Africa, Asia and South American countries and society (Usua and Osuagwu, 2010). As a result there is a rigid division between male and female roles with the males being encouraged to participate in the more aggressive, competitive and independent ventures. The overall effect is that the males in these countries are properly positioned to benefit maximally from social opportunities for government facilities, environmental issues and infrastructures (Adeleke, 2004).

In some African and Arab countries for example, gender inequality involves three common concepts: inequality in prestige, social inequality, inequality in power(political inequality and inequality in access to or control over resources) and economic inequality (Okojie, 1995). As Okojie (1995) depicts the picture above, men and women are unequal along many social, economic, environmental and political dimensions.

In their effort to help women seek social equilibrium with their male counterparts, non-governmental organizations (NGOs) are embarking on a number of remediation measures to help in bringing about social, economic and political empowerment of woman. Although several NGOs has been networking internally and internationally to ensure women's emancipation and empowerment in various sectors in several developing countries, the socialization process in some nations is still strongly based on patriarchal and matriarchal values that tend to promote polygamy (Ako-Nai, 2013). Islam is a major religion in Arab and African countries and it supports men having more than one wife. Just like their Muslim counterparts, some Christian sects also embrace polygamy (Ako-Nai, 2013). Although the theory emphasizes the domination of men over women and the natural environment, it also argues that since men and women are tied together, they must both be activists for the liberation of the environment.

Social ecology and ecological feminism tend to look to the society to find the underlying causes of the environmental crisis. According to Murray Bookchin (1990) the root of our ecological crisis lies in certain social factors. Specifically, social ecology and ecofeminism theories suggest that the domination and degradation of that natural environment arise from social patterns of domination and hierarchy patterns of social life in which some human beings exercise control or domination over others. On one hand, social ecologists link environmental destruction to what they see as general and widespread forms of domination and hierarchy. They tend to include such social practices as economic status and social class structures, sexism, racism, private ownership, and capitalism (Warren, 1987 & 1990). According to Murray Bookchin (1982) these factors tend to establish social hierarchy in which some human beings exercise power and domination over others.

On the other hand, ecological feminism tends to identify the oppression of women as a comprehensive method of social domination. Ecological feminists also identify a close relationship between the oppression of women and of the environment (Des Jardins, 1997). As a result of this belief they contend that the goals of the feminist movement closely parallel those of the ecological movement. Both the social ecologist and ecological feminism approaches postulate that a failure to see the connection between women and nature oppression will result in the continued exploitation of both women and nonhuman nature in the development of policy, theory, and practices that are grossly inadequate from a feminist perspective. In addition, one of the goals of feminism is the eradication of all oppressive sex-gender and related race, class, age, and affective preference categories and the creation of a world in which difference does not breed domination.

ENVIRONMENTAL JUSTICE THEORY

The environmental justice theory postulates the equitable distribution of the environment among peoples and the use of natural resources (White, 2008). The environmental justice approach promotes the social distribution of environmental benefits and burden. It explains that a society that distributes these benefits and burdens unequally is likely to be unjust (Des Jardins, 1997). The main goal of this proposition is how human health and well-being are influenced by particular types of production and consumption. As a result environmental justice is the major component in pursuing a society that meets its triangular needs of economic, environmental, and the promotion of sustainability. According to White (2008) environmental justice could be analyzed from three perspectives: (1) anthropocentric; (2) biocentric; and (3) ecocentric. On one hand, the anthropocentrism approach suggests that human beings dominate the world, including animals. The anthropocentrist denies the notion that nonhuman entities can have rights. In the anthropocentric perspective, anything not providing benefit to human beings is considered to be of negligible value (Withgott and Brennan, 2009). A biocentric approach argues that human beings morally and ethically are equal to other entities. In this perspective, nonhuman life has ethical standing. As a result human beings should not destroy other creatures in the ecosystem even if it would create jobs, generate economic growth, and pose no threat to human health (Clifford and Edwards, 2012; White, 2008).

On the other hand, ecocentrism postulates actions in terms of their benefit or harm to the integrity of the whole ecological system that consists of living and nonliving elements and their relationships with each other. The ecocentric approach suggests an enlightened human self-interest, informed by the notion of the interrelationship between humans and the natural environment (White, 2008; Clifford and Edwards, 2012). The ecocentrism approach is a more holistic perspective than the anthropocentrism because it encompasses a wider variety of variables and stresses the connection that ties the variables together into a functional ecosystem.

In the past three decades one common source of environmental injustice among nations is the dumping of hazardous waste. For instance, a few years ago an Italian ship dumped several tons of toxic waste at Koko in the Delta State of Nigeria (Okidi, 1984). It was reported that the Italian government was contacted by the Nigerian federal government about the incident, and the toxic waste was cleaned (Ogungbemi, 1998). The aftermath of the incident was that the local citizens who used the toxic drums for water storage died. Several fish and different species of water animals in the territory were also reported to have died due to the toxic waste that was dumped in the Koko area. It is interesting to note that while the developed countries of the world have polluted the international waters of developing nations, there are several industries in the poor countries that have been dumping pollutants in the local rivers and lands. These examples show that a significant amount of environmental policy does distribute benefits and burdens unjustly. It is a common practice for most society to place the environmental pollution burden upon people in the least advantaged position, such as the poor people in developing countries or poor minority people in developed nations.

In another instance, countries with relaxed health and environmental regulations often do not inform their workers and residents about the dangers from hazardous waste that are falsely labeled as harmless or beneficial material such as fertilizers, fill dirt, or plastics for recycling (Withgott and Brennan, 2009). Citizens who have been exposed to such hazardous waste may fall ill or die prematurely. Despite several international treaties on environmental protection—including the Basel Convention, which exits to prevent the international export and dumping of hazardous waste and trade—the illegal dumping of harmful toxins still continues. Environmental justice proposes that action should be delivered from a balanced anthropocentric and ecocentric perspective. The bone of contention is that ecocentric actions should reflect the opportunity to be concerned with issues of general ecological equality (White, 2008). As a result of this kind of approach, individuals should be concerned about the environment because they are better off. Therefore, the need to build policy regiments that produce greater compliance, less impact on economic performance, and smaller administrative costs is very essential for the future. Governments all over the world should reassess the profound impact of man's activities on the interrelationships of all components of the natural environment. The critical importance of restoring and maintaining environmental quality for human welfare in both developing and developed countries should be the universal goal.

ENVIRONMENTAL, ECONOMY, AND EQUITY IMPACT

The concept of sustainability is closely related to three factors: (1) economy; (2) environment; and (3) equity. These factors are very crucial to the capacity of a country to balance its development needs. According to Kraft and Furlong (2012) business and its various industries tend to place great stress on the natural environment and deal directly with issues of social justice and economic development. The bone of contention, however, is when the goals of business are to make profit and return value to their shareholders or owners: does this not conflict with the goals of sustainability? These issues are highly sensitive and need to be addressed if countries of the world can implement sustainability goals and practices. In addition cities and their urban settlements are the place where people concentrate and where most global resources and energy are consumed (Taylor, 2012). The issues raised suggest that there are problems at several levels in the path toward sustainable development. If we focus on Nigeria, Brazil, India, Mexico, or other developing countries, the problems in these nations emerge from the fact that they are poor and are part of the global system. The rapid development in developed and rich nations leads to imbalances in the economies of poorer developing nations (Steel et al., 2003). An instance of this is the recent trend toward the introduction of non-fuel-efficient technology by the rich nations in the developing countries with a motive of profit. Further, since the consumption of fuel by these developed nations, such as the United States, Canada, England, Germany, and Japan, is far more than in the developing countries, the latter may not need the new technology. Often the environmental agenda is given short attention by the developing nations because they feel that the developed countries have created the problems from which they are suffering.

The preceding observation is only one part of the problem. The serious challenges to sustainable development come from within the developing country itself (Dibie, 2009). Some of these are rapid population growth, growing pollution, and impoverishment of large sections because of unplanned economic development and gender inequalities, among others. For example, since India, Nigeria, and Brazil are countries of several sociocultural and regional diversities, the other important issue is to keep the process of development even and balanced, which is often difficult to achieve.

It is important to discuss some significant trends or challenges that create obstacles in sustaining the process of development. It is noteworthy that most of the trends are characteristic not only of developing nations but of some developed countries as well. Depending on the ecological specificities of each of the countries, their respective governments may have developed strategies to address the problems.

POPULATION GROWTH

The trends as of 2009 in the world population growth was about 1.2 percent annually (Kraft and Furlong, 2012). These trends are expected to add about 80 million people each year to its base of 6.8 billion. This trend also constitutes a growth rate of about 220,000 people every day. A major concern is that the population of the poorest countries in the world is projected to double in the next four decades or forty years. Despite the fact that fertility rates continue to decline slowly, the United Nations projects that the world population may reach 8 billion people in 2025. The United Nations also projects the world population may increase to about 9.4 billion people in 2050 (United Nations, 2010). The implication, however, is that an increase in population may intensify the impact on the environment as more individuals take up space, use resources, and generate more waste (Withgott and Brennan, 2009).

The density of population is very high in developing countries. The burden on the ecosystem capacity is unquestionable, especially in the regions with high population density. The growth of population also has its impact on the increase in the number of the poor and, thus, indirectly affects the development process. According to Charles Harper (2008) there are several sociocultural and religious factors in developing countries that make it difficult to reduce fertility rates to a great extent. One of these is the traditional value placed on children. They are considered as gifts of God, and so the people of the older generations still do not comprehend the need for fewer children for better lifestyles. In this context, it becomes even more relevant to focus our efforts on better distribution of available resources. Further, a new school of thought argues that the problems of development in these countries are not directly an outcome of population growth but rather inequitable distribution of resources. Some scholars argued that technology that enhances human ability to exploit minerals, fossil fuels, old-growth forest, or fisheries may generally increase impact. However, technology to reduce smoke stacks, harness renewable energy, or improve manufacturing efficiency can decrease impact (Withgott and Brennan, 2009; Rosenbaum, 2014).

Table 2.1 World Populations and Growth Rate

Region or Country	Mid-2009	2025	2050	Rate of Increase (2010) (%)
World Total	6,810	8,887	9, 9421	1.2
More Developed Nations	1,232	1,282	1,318	0.2
Less Developed Nations	5,578	6,805	8,103	1.4
United States	309	357.5	439	1.0

Source: United Nations. (2010). *Human Development Report*. New York, NY: Oxford University Press; Population Reference Bureau. (2009). *World Population Data Sheet*. Washington DC: Author.

The data in table 2.1 show that annual population growth in less developed countries will increase in 2050 in comparison to mid-2009. This is, of course, a healthy trend. However, the migration from rural to urban areas has been on the increase from 2009 to 2025 and is expected to further increase by the year 2050. Table 2.1 shows some of the United Nations world population projections.

According to Henry Kendall (2012), if human beings are compelled to stop the destruction of the environment, it is necessary that they accept limits to the growth of population. A World Bank report (2011) estimated that the world population will not stabilize at less than 12.4 billion. Another United Nations report (2010) stated that the world population may eventually reach 14 billion, nearly tripling of today's 5.4 billion people. The sad part of it all is that even at this moment, one person in five lives in absolute poverty without enough to eat, and one out of ten people in the world suffer serious malnutrition (Kendall, 2012).

ENVIRONMENTAL POLLUTION

Pollution is the addition to the environment of any substance or energy form (e.g., heat) at a rate faster than the environment can accommodate it by dispersion, breakdown, recycling, or storage in some harmless form. Pollution of the natural environment is a largely unintended and unwanted consequence of human activities in manufacturing, transportation, agriculture, and waste disposal (Vig and Kraft, 2005). High levels of pollution are largely a consequence of unplanned industrialization and urbanization, and the rapid increase of human populations in modern times (Steel et al., 2003; Rosenbaum, 2014; Dye, 2011). Further, air pollution near ground levels and acid precipitation are already causing widespread injury to human beings, forests, and crops (Kendall, 2012). Table 2.2 shows the fossil fuel emission level of the top twenty-five countries in the world.

Table 2.2 shows that the United States accounted for almost one quarter of the entire stock of carbon dioxide emission in the world in 2003. Pollutants are classified according to the part of the environment primarily

Table 2.2 Top Twenty-five Fossil Fuel Carbon Dioxide Emitters in 2003

Countries	Carbon Dioxide Emitted (Millions of Tons of Carbon)	Percentage of Annual Global Total
United States	1,580	23
China	900	15
Russia	410	8.0
Japan	300	6.0
India	210	5.0
Germany	200	4.0
Canada	182	3.00
United Kingdom	176	3.00
South Korea	173	2.97
Italy	170	2.90
Australia	165	2.75
France	150	2.60
Ukraine	146	2.30
South Africa	143	2.10
Mexico	140	2.00
Iran	137	1.95
Brazil	135	1.90
Spain	134	1.70
Saudi Arabia	133	1.35
Indonesia	130	1.30
Poland	128	1.10
Netherlands	126	1.00
Taiwan	124	0.80
Turkey	100.2	0.70
Thailand	100	0.50

Source: Gruber, J. (2009). *Public Finance and Public Policy*. Richmond, United Kingdom: Worth Publishers.

affected by them: air, water, or land. Subgroupings depend on characteristics of the pollutants themselves: chemical, physical, thermal, and others (Kraft, 2011). Many pollutants affect more than one resource.

AIR POLLUTION

Air Pollution is a very broad term, which actually covers all different types of pollution. According to Kraft and Furlong (2012) the use of motor vehicles powered by internal combustion engines is the major source of urban air pollution in most developed countries. Other sources of air pollution include (1) windblown dust and volcanic eruption; (2) factories and power plants; (3) automobiles; (4) dry cleaners and degreasing operations. The emissions from these sources often causes reduced visibility and mild irritation. Further, when sulfur dioxide and nitrogen dioxide that are emitted into the air react with water, oxygen, and oxidants, they create a compound called acid rain. Apart from acid rain, other compounds in the air include the greenhouse effect, particulate, ozone layer depletion, smoke and smog. Each of which these compounds has serious consequences for the environment. Acid rains can cause a lot of damage to lakes and rivers while particulate matter, smoke, and smog can create havoc in the health of people. One of the two problematic long-term consequences of air pollution is the global warming process often referred to as the greenhouse effect, which could further degrade plant and animal life. The other is the depletion of the ozone layer that serves as a protective layer in the atmosphere, exposing us to harmful radiation. Air pollution has been reported to cause a heavy toll on public health. In some countries the annual death caused by air pollution is to the tune of several thousands (Maxwell, 2009).

Scientists continue to provide evidence that the depletion of the ozone layer or the decline in the stratospheric ozone could result in human health problems such as depression of the immune system (Friis, 2012). Guy Peters (2007) and Michael Kraft (2011) observed that an increase in ultraviolet radiation could affect both plants and animals, and adversely reduce agricultural production in several countries around the world.

WATER POLLUTION

Overcoming the shortage of fresh water constitutes a major challenge for globalization because human being and animals existence relies on liquid. Water is one of the most important natural resources for survival of all species (Dolayar and Gray, 2000). Water covers 71 percent of the earth's surface. Scientists also observe that water makes up about 65 percent of the human body (Smith, 2004; Withgott and Brennan, 2009). In most parts of the world, fresh water is becoming a scarce commodity. The United

Nations report (2010) indicates that twenty countries in the developing world suffer from water stress and projects that this will rise to twenty-five countries by 2050. As a matter of fact, many people see water as purely a commodity. Some organizations, such as the World Bank, pragmatically believe that water must be properly priced to reflect true costs and to promote efficiency, conservation, and extensions of services, life, and agriculture productivity. Pollution of lakes, rivers, and ground water further limits the supply of fresh water. Excessive exploitation of ground water supplies often endangers food production and other essential human systems. A United Nations report (2010) indicates that due to rapid population growth, about 1.2 billion people lack potable water. This number constitutes 20 percent more than in 1990. The UN (2010) report further indicates that about 3 billion people in the world lack adequate sanitation facilities compared with 2 billion in 1990.

ENVIRONMENTAL NOISE AS POLLUTION

Environmental noise is defined as any form of unwelcome sound. Exposure to noise may occur indoors or out. Environmental noise is a form of pollution (Smith-Sebasto, 2012). When considering the effects of noise on human health and quality of life, it is very important to take into account the intensity of the sound in question, its duration, and the time and place at which it is heard. Sound intensity is usually measured on the logarithmic decibel (dB) scale (Friis, 2012). The threshold level for the human ear is between 3 and 45 dB (Maxwell, 2009). The decibel is a measure of sound intensity; that is, the magnitude of the fluctuations in air pressure caused by sound waves. In most developing countries the decibel levels are very high.

According to Nancy Maxwell (2009) most environmental noises studied are related to traffic, airport, and industrial production noise. The research indicates that health impacts of noise are not limited to hearing loss. Maxwell (2009) observed that the sounds of airport and traffic noise clearly cause annoyance in both adults and children. It has also been reported that exposure to airport noise may result in major cardiovascular illness (Stansfeld, Matheson, and Mark, 2003). Hearing loss has been linked to sound exposure in musicians who play the steel pan percussion instrument used in Caribbean music (Karmody and Simon, 2004). In addition, teenagers who attend night clubs regularly and use personal music players are reported to have some hearing impairments in both reading comprehension and long-term memory (Stansfeld et al., 2003). There are signs of concern that a whole generation of teenagers who listen regularly to music at high volume through headsets may suffer hearing loss. In some developing countries, where generator sets are used regularly for electricity, the sound from multiple generators could cause some level of impairment after a long time. Workers who are employed in factories with machines that create heavy sound have been found to suffer from limited loss (Kieszak et al., 2001).

ECOLOGICAL IMBALANCES

According to Catton and Dunlap (2002) the function of the environment is that of providing for the well-being of the human society as well as other species. On one hand, human impact on the environment may become so large that they threaten human ability to see clearly. The discussion on environmental pollution talks of the impact it has on the survival of different forms of life—plant, animal, and human beings (Kendall, 2012). In fact, it is in this context that pollution is the foremost challenge to sustainable development (Rosenbaum, 2007). This in turn has a long-term effect of causing ecological imbalances and a disruption of the natural ecological cycle. On the other hand, the ecosystem provides living space for all activities, and overuse of this function produces crowding, congestion, and the destruction of habitats for other species (Harper, 2008). Withgott and Brennan (2009) contend that ecological imbalances are caused by (1) depletion of natural resources, forest cover, wildlife, water, agricultural land, and livestock and (2) ecosystem conversion.

Depletion and degradation of natural resources is occurring as a result of the various types of influences. Industrialization without proper planning, the burden of an increasing population, and increases in automobile traffic are some of the factors. The aquatic environment and its inhabitants are on the decline (Steel et al., 2003). Seventy-one percent of the world's commercial fisheries are fully exploited or overexploited and experiencing declining yields (Kendall, 2012). If we allow depletion of the environment to intensify, the exploitation of natural resources could increase in order to fulfill basic needs of all human beings, thereby creating a vicious cycle.

Michael Kraft and Scott Furlong (2012) point out that the pace of degradation of cultivable land and grazing land is alarming. Deforestation, overgrazing, faulty agricultural practices, and firewood needs have resulted in landslides, flash floods, desertification, and water logging. This in turn leads to deforestation, which causes soil erosion. Deforestation is proceeding at a significant rate. One fifth of all tropical forests have been cleared since the 1960s (Withgott and Brennan, 2009). According to the Food and Agriculture Organization (FAO) of the United Nations (2010), deforestation has been concentrated in the developing world, which lost nearly 200 million hectares between 1980 and 1995. As a result, there is flooding during the monsoons and severe water shortage in summer. The frequency of floods has increased not only in Africa but also in Asia.

The incidence of natural disasters is on the rise in Asia, Africa, Australia, and South America (United Nations, 2010). Some of these are earthquakes, cyclones, floods, and droughts (Harper, 2008; Smith-Sebasto, 2012). Natural disasters are clearly linked to the intensified exploitation and unplanned use of natural resources. With the occurrence of each of these disasters, there is a further challenge to the even distribution of resources to all human beings, resulting in increased poverty (Taylor, 2012). The process of rehabilitation takes up so many resources that development comes to a standstill.

The urban areas of some countries represent complex environmental problems. The living conditions of millions of urban poor are such that they pose a threat to their health and have potentially catastrophic social consequences (Kraft, 2011). If these problems are not addressed in an adequate and timely manner, serious environmental and associated health consequences may result. Burgeoning urban population beyond the carrying capacity of the different components of the ecosystems coupled with indifferent urban governance are the root causes of environmental problems in some countries (Dibie, 2009). Air pollution in urban areas can be clearly evidenced in the increasing presence of smog (Steel et al., 2003; Smith, 2004). Air pollution can cause chronic and acute respiratory diseases, ventilator malfunction, heart disease, cancer of the lungs, and even death (Maxwell, 2009). The blood lead levels of persons in some developing countries have been reported to be higher than the corresponding levels of persons in lead-free gasoline areas (Kraft, 2011). The rural population of some developing countries uses substantial quantities of noncommercial fuel, such as crop residues, animal dung, or wood. Although their share in total fuel consumption is decreasing, these local sources of fuel provide 80 percent of rural energy for cooking (Mol & Sonnenfield, 2000). Several adverse health effects are suspected to occur as a result of indoor pollution, especially where there are conventional houses that have not been replaced with smokeless block cement houses (Friis, 2012). Respiratory infections in children, chronic lung diseases, lung cancer in adults, and adverse pregnancy outcomes, such as low birth weight and stillbirth of the child, for women exposed during pregnancy are some of the diseases associated with indoor pollution (Smith, 2004; Segerfeldt, 2012).

POVERTY

Poverty continues to remain an important public issue all over the world, yet the governments of most countries have given different amounts of attention to it. The United Nations International Conference on Population and Development, held in September 2002 in Cairo, focused on the alleviation of poverty. One of the reasons for this attention on poverty is that poorer nations tend to show a higher population growth rate than developed countries. According to Richard Tobin (2010) almost half of the world's population lives on less than two dollars a day. Tobin (2010) contends that there are over 1 billion people living in severe poverty all over the world, with especially high rates of poverty in sub-Saharan Africa, Asia, and some regions of South America. Zachary Smith (2004) points out that on one hand, severe poverty could galvanize poor people to engage in destructive environmental behavior. On the other hand, wealth could produce even more far-reaching environmental impact. The environmental impact of human activities, however, depends not only on the number of people involved but also on the way the people live. This is why the impact of habitats, biodiversity, air, water quality, energy and water use, and other aspects of environmental quality are likely to be enormous. One point that is clear, however, is that some countries in Asia, Africa, the Caribbean Islands, and South America will be affected far more than others by food and water shortages, poor health, environmental degradation, economic dislocation, and large scale unemployment (Kraft, 2011; Rosenbaum, 2014). According to Hinrichsen and Robey (2012), in the 1990s the number of people in poverty increased to about 1 billion. During that same period about 2 billion people were subsisting on the equivalent of US$2 per day or less. The United Nations report (2010) indicates that by 2000 that the number of people in poverty or living with less than US$2 per day had risen to about 3 billion. This number constitutes half of the current world population.

The future poverty problems and challenges all over the world could be projected as follows. First, a large percentage of the world population lives in abject poverty. Over one quarter of the world's population lives in industrialized countries and consumes 83 percent of the world's goods (Tobin, 2010). Second, in the next fifty years the world population is likely to reach 11 billion people (Kendall, 2012; Smith-Sebasto, 2012; Peters, 2007). This will constitute a very difficult problem because with 11 billion people, economic activity to meet the basic needs of every human being on this earth will need to increase significantly. In addition, the only source for all this economic activity is the natural environmental. At the present time, the world environment is already under stress from current economic activity and the future looks bleak (Des Jardins, 1998; Taylor, 2012).

Each nation needs its own program to eradicate such root causes of poverty as hunger, illiteracy, inadequate medical and child care, lack of employment, and population pressures (Kendall, 2012; Withgott and Brennan, 2009). According to Michael Kraft (2011) we can hardly place all the blame on the poor countries, where nearly all of the future growth in human numbers will occur. This is because the industrial countries consume a far greater proportion of the world's resources and have a much higher per capita effect on the environment. Therefore, poor nations' governments must receive support, including financial assistance, because the struggle against poverty is the shared responsibility of all countries. The increasing pattern of the human consumption rate that has accompanied the rapid industrialization of India, China, and other populous countries makes it all more urgent and crucial for all the nations of the world to

Table 2.3 Per Capita Wealth with Rate of Fertility and Population Growth

Nations	Per capita GNP (US$)	Population. Increase Rate % per Year	Children Born per Woman
Tanzania	740	2.6	5.4
Niger	830	3.4	7.1
Ethiopia	1,190	2.5	5.4
Haiti	1,490	1.8	4.0
Pakistan	2,500	2.3	4.1
India	3,800	1.6	2.9
Syria	3,920	2.5	3.5
China	7,730	0.5	1.6
Brazil	8,800	1.4	2.3
Romania	9,820	−0.2	1.3
Mexico	11,330	1.7	2.4
Spain	28,420	0.3	1.4
Japan	33,730	0.0	1.3
Canada	34,610	0.3	1.5
United Kingdom	35,690	0.3	1.8
United States	44,260	0.6	2.1

Source: Population Reference Bureau. (2009). *World Population Data Sheet*. Washington DC: Author.

find a path to global sustainability (Withgott and Brennan, 2009). While establishing the relationship with sustainable development, poverty alleviation is essential for environmental sustainability. Table 2.3 shows how poorer nations tend to have higher fertility and growth rates.

Some countries have started to focus their attention on different measures for poverty alleviation. They have introduced various developmental initiatives, which at times appear to be ineffective because of lack of developmental apparatus and other related factors. For example, Trinidad and Tobago and Brazil have initiated social development programs that could help to improve education, provide a higher status for women and improved gender equity, and provide better reproductive health care and nutrition. Further adoption of old age security programs and economic reform are all as important as provision of family planning services. In addition, the movement toward creating sustainable communities in many areas of some nations indicates a strong potential for such integrated assessment of local and regional environmental data as well as for citizens to be more involved in decisions that affect their lives (Kraft, 2011; Rosenbaum, 2014).

ENVIRONMENT AND GENDER INEQUALITIES

The environmental justice and ecological feminism theories discussed previously stipulate that any society or country that distributes environmental benefits and burden unequally among men and women is unjust. An important question in development is the evenness of the spread of its impact. The much existing diversity in our society and economy contribute to the benefits of development reaching the privileged. The privileged include the more prosperous sections, some regions, and men. Women in communities—urban, rural, or tribal—across several countries in the world benefit less from the fruits of development (Dibie and Dibie, 2008).

Since the United Nations declaration of the women's decade in the 1970s and especially in the 1990s when nongovernmental organizations began to proliferate on the African continent, women and their issues have been at the center-stage of national and international policy discussions. Most of the literature has painted a very bleak picture of the social, economic, political and cultural position and status of African and Asian women (Dibie and Dibie, 2008). The literature often describes sub-Saharan African women as subjugated and subordinated to their menfolk (Ezeilo, 2006; Kalu, 1996; Pam Sha, 2007; Warren 1990) . Makinwa and Jensen (1995) state that the majority of women in sub-Saharan Africa are in a subordinate position relative to that of men. They also argue that gender stratification generally results in gender inequality, inequality in prestige, inequality in decision-making power, and inequality in access to resources. More recently Robert Dibie and Josephine Dibie (2008) have reiterated the point that women in Africa and Asia are consciously, deliberately, and perpetually considered as subordinate to their male counterparts in many traditions and cultures.

Specific examples of this gender inequality include the harsh Sharia laws that govern women's lives in North Africa and some Arab and Asian countries. The "Trokosi" system in which young girls are given to fetish priests to atone for their parents transgressions in the Volta Region of Ghana is another example of unequal treatment of gender. The painful widowhood rites and female circumcision in many African and Asian cultures, and inheritance traditions that dispossess women of their property at the death of their husbands are also injustice to women (Dibie and Dibie, 2008). Other examples include women's lack of property rights in some societies (especially rights to land) and their subjection to domestic violence without any recourse to legal redress. Women in some African countries also suffer from unequal access to education, health care services, land ownership and in their preponderance in the unstable and marginalized informal economic sector of their countries (Dibie and Dibie, 2008). Put differently, the factors underlying the subordinate position of many sub-Saharan African and Asian women are (1) religious; (2) cultural; (3) political; and (4) economic.

Other scholars argue, however, that the issue of African and Asian women's subordination is more complex than is usually acknowledged. They contend that those who portray African and Asian women as a marginalized group across the board are simply transposing Western ideas of social organization into an African and Asian context where "complementarity," instead of "equality," is the organizing principle of gender relations (Ezeilo, 2006; Kalu, 1996; Oyewumi, 2002). According to Oyewumi (2002) theories that use gender to explain African women's subordination are based on Western family structures that are ill-suited to the African context. The burden of work and environmental degradation continues to fall on women and children, particularly the girl child in several developing countries. In the long run this injustice may have severe negative impact on their status as well as reflecting poorly in terms of human development index. Further, such injustice against female children could raise doubts about the value of development.

This observation calls for a more nuanced analysis of gender relations in African, Asian, and South American societies, women's status and empowerment, and their role in development. Gender analysis in Africa and Asia should be based on precolonial and colonial African history and postcolonial African realities, rather than concepts based on foreign gender theories. Oyewumi (2002) aptly points out that when African and Asian realities are interpreted based on Western claims, it causes distortions and obfuscations in language, and often results in a total lack of comprehension by women due to the incommensurability of social categories and institutions. Getting a handle on the real factors undermining Asian and African women's social and economic empowerment is important because the effectiveness of environmental policies and programs to promote women's empowerment depends on an accurate identification of the problem.

HEALTH CONSEQUENCES

Environmental health consequences are firmly rooted in the understanding of the world and recognizing that the human species creates hazards to the health of the environment (Dye, 2011). Throughout the discussion in the preceding paragraphs it has been pointed out that each of these hazards has significant consequences on health of humans, animals, and plants. In the long run, these consequences lead to the degradation of the environment, posing serious challenges to the sustainability of development and living standards that have been achieved presently. It is important to note that deteriorating health, as seen in several acute and chronic diseases in several developing countries, indicates low quality of life. Poor health is also often associated with negative impacts on the environment (Friis, 2012). The infant mortality rate (IMR) reveals the health status of women in the reproductive age group and the care available for infants to a great extent.

According to Nancy Maxwell (2009) the interaction between genetic traits and environmental hazards, such as the effect of a woman's genetic makeup on her breast cancer risk, can be linked to exposure to man-made chemicals. It is also very important to note that several environmental hazards created by human beings cause chromosomal damage with a wide range of health effects. Therefore, an interaction between the physiological effects of a social stressor and the asthma risk of air pollution is well within the scope of environmental consequences. Robert Friis (2012) contends that some hazards that are fundamentally social have effects that could cause environmental health problems. Notably, international conflicts that have resulted in war have exposed soldiers and civilians to chemical weapons. On the other hand, farmers that often used pesticides have infected both the crops harvested and citizens who may purchase such agricultural produce (Maxwell, 2009). Exposure to pesticides has been linked most clearly to neurologic effects, cancer, and reproductive and developmental outcomes (Kraft, 2011). The neurologic effects of acute pesticide exposure are well known and include headaches, dizziness, nausea and vomiting, muscle weakness, and even convulsions and coma at high exposures (Alvanja, Hoppin, and Kamel, 2004; Maxwell, 2009). Radiation from nuclear weapons and plants has caused serious health problems for people who come around such locations. A host of other infectious diseases have been associated with environmental pollutants. Each time there is a global disaster, we are often reminded of the importance of deterring the potential harm.

There is another important aspect that emerges from the preceding discussion. Not only the availability of resources but also their distribution makes a difference to the sustainability of the environment. Therefore, poverty and gender issues become important. These are the main criteria along with some other sociocultural ones that impact the distribution of resources. Planning for sustainable development while directly addressing environmental challenges has to alleviate socioeconomic and gender differentials.

PRESENT ENDEAVORS

There have been several planned endeavors toward sustainable development in many developed and developing nations around the world to tackle the depletion of resources and mounting problems of poverty, environmental degradation, social inequalities and population explosion. Efforts have been taken up by several developing and developed countries at the following levels: (1) national; (2) regional; (3) local governments; (4) NGOs and civil society organizations; and (5) community-based organizations. Some illustrations of measures toward environmental protection and sustainability include the following:

1. Provision of loans, direct payments, tax credits, and price support or low fees for mineral and water rights, agricultural and irrigation subsidies, and financial incentives to protect wetlands under the national wetland reserve policy of some countries (Kraft, 2011).

2. Assessment for damages by using tort laws to award damages to plaintiffs for harm suffered or government imposition of damage assessment. This also includes private lawsuits against corporations for damages and natural resources damage assessment (Dye, 2011)
3. Investing in river projects that aim at cleaning and also checking erosion in the banks (Dibie, 2009).
4. Discouraging anti-environmental behavior by taxing an activity at a level that encourages or discourages environmental pollution; for example, government giving citizens tax credits to encourage purchase of fuel efficient vehicles, giving green taxes to discourage corporate pollution or spending money on preferred programs such as energy research and development (Smith-Sebasto, 2012).
5. Establishing wildlife sanctuaries in many parts by the governments of these different countries to give endangered wildlife a secure place for survival (Dibie, 2009).
6. Practicing afforestation, which is the positive measure of planting and nurturing trees to stop the ill-effects of deforestation. Several nations' governments have adopted strategic plans to plant new trees as they cut down mature ones (Dibie, 2009).
7. Using natural gas (the least environmentally damaging of the fossil fuels) and water as sources of energy as well as developing renewable energy sources (Dibie, 2009).
8. Rationing by limiting access to scarce resources; for example, requesting permits for campsites, fishing, and hunting, and enforcing restrictions on energy or water use during periods of scarcity (Kraft and Furlong, 2012).
9. Creating awareness about disposal of wastes and other environmental issues among people in general (Rosenbaum 2014).
10. Establishing enforcement of standards with sanctions for noncompliance; for example, environmental regulations under the Clean Air Act, Clean Water Act, and energy efficiency requirements (Kraft, 2011).
11. Providing information to the public through formal programs or other actions; for example, disclosure of toxic chemical releases, labels indicating auto fuel efficiency, formal environmental education programs (Kraft, 2011).
12. Self-regulation by voluntary adoption of performance standards of industry or others; for example, state programs for environmental self-regulation (Dye, 2011).

The increasing awareness of dangers of environmental degradation and pollution has led to successful public campaigns against many wrong practices, for instance, the recent public media against dumping waste on the streets in several countries. This was done because a strong feeling about its polluting effects was generated in those campaigns. There have been similar campaigns against the excessive use of plastic and careless disposal of the same (Dibie and Atibil, 2012). To encourage responsibility of the stakeholders, permits and punitive measures such as fines have also been introduced by some countries to enforce policy effectiveness (Kraft, 2011; Rosenbaum, 2014).

A good illustration of such an endeavor is industries having pollutants as discharge/effluents from their factories. Industries located in areas where they are not connected to a sewer can discharge directly into a waterway, but they need a permit and may be asked to have their own treatment plants. Even industries that are connected to sewers may have to pretreat their wastewaters before discharging them into the sewers, because they may contain materials that will harm the sewers or the treatment plants—or may be a danger to the people who work in maintaining the sewer system. According to Michael Kraft (2011) governments could also adopt alternatives to regulation such as market-based incentives, provision of information, public education, negotiation, voluntarism, stakeholder collaboration, and public-private partnership strategies in implementing environmental policies. As a result of frequent concern over environmental equity forces, policy makers should find a more balanced ethical approach in choosing environmental policy strategies. Steps taken toward a more just and ecologically resilient world should be larger, faster, and more broadly rooted for them to have greater positive impact on people in both the developing and developed countries.

The transformation of the world has contributed to substantial net gain in human well-being and economic development. However, it could be argued that not all countries of the world and people have equally benefited from the process. In fact, many people in both the developing and developed countries have been harmed (Smith-Sebasto, 2012). There is no simple solution to these environmental problems since they arise from the interaction of many recognized development challenges, including biodiversity loss and degradation and climate change. Each of these three environmental variables is too complex to address by any country on its own. One point that is clear, however, is that most changes to the environment have been made to meet a dramatic growth in the demand for food, fiber, fuel, timber, and water.

SUMMARY

This chapter has examined the concepts and applications of environmental issues and policies. It also discusses environmental impacts within the framework of sustainable development and social justice. The chapter provides an argument that no responsibility of government and society will be more difficult over the next generation than accommodating the level of economic development needed to meet human needs and aspirations while simultaneously averting devastating effects on the environmental systems that sustain life and the ecosystem. There is a strong feeling that environmental problems need to be addressed at different levels at the same time in order to not cause ad hoc solutions. Therefore, while stricter systems of licenses and permits, market-based incentives, provision of information, public education, negotiation, voluntarism, stakeholder collaboration, and public-private partnership strategies are being developed, there should be greater emphasis on public awareness. The involvement of the public and community is seen as the cornerstone to success in the effort toward achieving sustainable development.

Some problems associated with sustainability are already urgent and require immediate action. This is because one of the most important national impacts of sustainability is a change in behavior toward the ecosystem. Many nations all over the world have embraced the sustainability concept, by operationalizing new ways of behavior as well as practicing recycling of their consumable products, downsizing their homes, and localizing their consumption (Andre, 2011). This chapter begs to ask the question, What are corporations doing to embrace the concept of sustainability? It is argued that although individuals have embraced the concept of sustainability and subsequent changes, their corporate behaviors have not been as effective in implementing the concept as expected. Therefore, the sustainability concept has not been adopted by companies, especially multinational corporations, in such a way as to drive them to do more than they are required to do. The main implication of promoting sustainability around the world is that of understanding organizational types. This type of understanding is paramount for knowing how a sustainable agenda could be adopted by both the public and private sectors in all nations. The urgency of some of these problems has been overlooked because the people most affected are physically remote from the center of power or because their voices are not heard, or both. On one hand, some issues call for immediate action because there are good prospects for reversing the damage to the environment at relatively low cost. For example, some countries have implemented policies against air and water pollution. Even then, undoing some of the damage of the affected population may not be fully possible. On the other hand, knowing the health impact does create a moral imperative to protect those affected from further exposure, to compensate them to the extent possible, and to prevent others from becoming victims.

This chapter has suggested the need for governments of all nations to adopt a sustainable development plan, so that development in some sectors does not lead to other resources being destroyed. The natural environment and its degradation are some of the chief concerns in this respect. Problems associated with the degradation of the physical environment often result in inequalities in society. These inequality problems within societies and nations also need to be addressed. In order to truly adopt conservation, nations all over the world may need to design new organizational types that are less driven by short-term thinking and competition for increasing scarce environmental resources. It is essential to get ahead of the curve and prevent worsening the crisis before it is too costly. There is a need to correct behavior, but there is also additional scope for mitigation, though not for complacency. It may be that only when a future generation addresses the design of both the public and private sectors that their sustainability concerns will be adequately met. Finally, promoting sustainability at the local level requires developed and developing countries to pay attention to environmental planning, transportation, land use, housing, energy use, economic development, and social justice.

REFERENCES

Adeleke, V. (2002). "Nigerian Women and National Development". *Babcock Journal of Management and Social Sciences*, *1*(1), 73–89.

Ako-Nai, Ronke (2013). *Gender and Power Relations in Nigeria*. Lanham, MD: Lexington Books.

Alvanja, M., Hoppin, J., and Kamel, F. (2004). "Health Effects of Chronic Pesticide Exposure: Cancer and Neurotoxicity." *Annual Review of Public Health*, *25*, 155–197.

Andre, R. (2011). "Organizational Responsiveness to Sustainability Paradigm: A Comparison of Government Departments, Government Sponsor Enterprises and Private Firms." *International Journal of Business Insights and Transformation*, *3*(3), 8–13.

Bookchin, M. (1982). *The Ecology of Freedom*. Palo Alto, CA: Chesire Books.

Bookchin, M. (1990). *The Philosophy of Social Ecology*. Montreal, Canada: Black Rose Books.

Brown, B. J., Hanson, M. E., Liverman, D. M., and Meredith, R. W., Jr. (1987). "Global Sustainability: Towards Definition," *Environmental Management*, *11*(6), 87.

Buchholz, R. A. (1993). *Principles of Environmental Management: The Greening of Business*. Englewood Cliffs, NJ: Prentice Hall.

Cassil, A. (2012). "Overpopulation, Sustainable Development, and Security: Developing and Integrated Strategy." In *Taking Sides: Clashing Views in Sustainability*, edited by R. Taylor. New York, NY: McGraw-Hill, 177–181.

Catton, W. R. and Dunlap, R. E. (2002). "A New Ecological for Post-Exuberant Sociology," *American Behavioural Scientist*, *24*, 15–47.

Clifford, M., and Edward, T. (2012). *Environmental Crime*. Burlington, MA: Jones & Bartlett Learning.

d'Eaubonne, F. (1994). "The Time for Ecofeminism," trans. R. Hottell. In *Ecology: Key Concepts in Critical Theory*, eds. C. Merchant and R. Gottlieb. Atlantic Highlands, NJ.: Humanities Press International.

Des Jardins, J. (1997). *Environmental Ethics*. Albany, NY: Wadsworth.

Des Jardins, J. (1998). *Environmental Ethics: An Introduction*. New York, NY: McGraw-Hill.

Dibie, J., and Dibie, R. A. (2008). "Current and Future Roles of NGOs in the Empowerment of Women in Africa." In *Non-Governmental Organizations (NGOs) and Sustainable Development in Sub-Saharan Africa*, edited by R. A. Dibie. Lanham, MD: Lexington, 229–251.

Dibie, R. (2009). "Environmental Governance and the Political Economy of Sustainable Development in Sub-Saharan Africa." *Journal of International Politics and Development*, *7*(1), 1–24.

Dibie, R. and C. Atibil. (2012). "NGOs and the Social and Economic Empowerment of Women In Ghana and Nigeria." *Journal of International Politics and Development, 10*(1–2), 27–64.

Dye, T. (2011). *Understanding Public Policy* (13th ed.). New York, NY: Longman.

Edoho, F. and Dibie, R. (2000). *Executing Environmental Policy and Waste Management in Ghana and Nigeria. African Journal of Sustainable Development.* www.jsd-africa.com. Accessed May 19, 2011.

Ezeilo, J. 2006. "Feminism and Islamic Fundamentalism: Some Perspectives from Nigeria and Beyond." *SIGNS: Journal of Women in Culture and Society, 32*(1), 31–52.

Friis, R. (2012). *Essentials of Environmental Health.* Burlington, MA: Jones and Bartlett.

Gore, A. (1992). *Earth in the Balance.* Boston, MA: Houghton Mifflin.

Harper, C. (2008). *Environmental and Society: Human Perspectives on Environmental Issues.* Upper Saddle River, NJ: Prentice Hall.

Hinrichsen, D., and Robey, B. (2012). "Population and the Environment," in *Sustainability*, edited by N. Smith-Sebasto. New York, N.Y: McGraw Hill.

Kalu, A. (1996). "Women and the Social Construction of Gender in African Development." *Africa Today, 43*(3), 269–288.

Karmody, J., and Simon, D. (2004). "Hearing Loss in Steel Band Musicians." *Otolaryngol Head Neck Surgery, 232*, 461–465.

Kendall, H. (2012). "World Scientists' Warning to Humanity." In *Sustainability*, edited by N. Smith-Sebasto. New York, NY: McGraw-Hill, 73–98.

Kiskar A. S., Kieszak S. M., Holmes A. E., Esteban, E., Rubin, C., and Brody, D. J. (2001). "Estimated Prevalence of Noise induced Hearing Threshold Shift among Children 6 to 19 Years of Age: The Third National Health and Nutrition Examination Survey, 1988–1994, United States." *Paediatric, 108*(1), 40–43.

Kraft, M. (2011). *Environmental Policy and Politics.* New York, NY: Longman.

Kraft, M., and Furlong, S. (2012). *Public Policy: Politics, Analysis, and Alternatives.* Washington DC: Congressional Quarterly Press.

Makinwa, P., and Jensen, A-M. (Eds.). (1995). "Introduction." In *Women's Position and Demographic Change in Sub-Saharan Africa.* Liège, Belgium: International Union for the Scientific Study of Population, 1–14.

Maxwell, N. (2009). *Understanding Environmental Health.* Burlington, MA: Jones and Bartlett Learning.

Mies, M, and Vandana, S. (1993). *Ecofeminism.* London and Atlantic Highlands, NJ: Zed Books.

Merchant, C. (1996). *Earthcare: Women and the Environment.* London and New York: Routledge.

Mol, A. (1995). *The Renewment of Production: Ecological Modernization Theory and the Chemical Industry.* Utrecht: Van Arkel.

Mol, A., and Sonnenfield, D. (Eds.). (2000). *Ecological Modernization around the World: Perspectives and Critical Debates.* Ilford, UK: Frank Cass.

Myers. N., and Spoolman, S. (2014). *Environmental Issues & Solutions.* Belmont, CA: Brooks/Cole.

Odum, E. P. (1971). *Fundamentals of Ecology* (3rd ed.). Philadelphia: W. B. Saunders.

Ogungbemi, S. (1998). "An African Perspective on the Environmental Crisis." In *Environmental Ethics: Readings in Theory and Application*, edited by L. Pojman. Albany, NY: Wadsworth, 263–270.

Okidi, C. (1984). "Management of Natural Resources and the Environment for Self-Reliance." *Journal of Eastern African Research and Development, 14*, 98–121.

Okojie, C. E. E. (1995). "The Relationship between Women's Status, Proximate Determinants and Fertility in Nigeria." In *Women's Position and Demographic Change in Sub-Saharan Africa*, M. Kofo and An-Magritt (eds). Belgium: International Union for Scientific Study, Conference Proceedings, 28–44.

Oyewumi, O. (2002). "Conceptualizing Gender: The Eurocentric Foundation of Feminist Concepts and the Challenge of African Epistemologies." http://codesria.org/IMG/pdf/OYEWUMI.pdf [accessed October 2013].

Pam Sha, D. (2007). *Globalization and Challenges of Organizing among Women in the Informal Sector in Nigeria*. Markurdi, Nigeria: Benue State University Monograph Series No. 2.

Pearce D., Markandya, A., and Barbier, E. (1988). *Blueprint for a Green Economy*. London: Earthscan.

Peters, G. (2007). *American Public Policy* (7th ed.). Washington DC: Congressional Quarterly Press.

Peters, G. (2013). *American Public Policy*. Los Angeles, CA: Sage.

Population Reference Bureau. (2009). *World Population Data Sheet*. Washington DC: Population Reference Bureau.

Rees, W. (1988). "A Role for Impact Assessment in Achieving Sustainable Development." *Environmental Impact Assessment Review*, *8*, 273–291.

Rogers, R. (1994). *Nature and the Crisis of Modernity*. Montreal, Canada: Black Rose Books.

Rosenbaum, W. (2014). *Environmental Politics and Policy*. Los Angeles, CA: Sage.

Sachs, J. *The End of Poverty: Economic Possibilities for Our Time*. New York, NY: Penguin.

Segerfeldt, F. (2012). "Water for Sale: How Business and the Market Can Resolve the World's Water Crisis." In *Taking Sides: Clashing Views in Sustainability*, edited by R. Taylor. New York, NY: McGraw-Hill, 299–315.

Smith, Z. (2004). *The Environmental Policy Paradox* (4th ed.). Upper Saddle River, NJ: Prentice Hall.

Smith-Sebasto, N. (2012). *Sustainability*. New York, NY: McGraw-Hill.

Stansfeld, S. A., Matheson, M. P., and Mark, P. (2003). "Noise Pollution: Non-auditory Effects on Health." *British Medical Bulletin*, *68*, 243–257.

Steel, B., Clinton, R., and Lovrich, N., Jr. (2003). *Environmental Politics and Policy: A Comparative Approach*. New York, NY: McGraw-Hill.

Taylor, R. (2012). (Ed.) *Taking Sides: Clashing Views in Sustainability*. New York, NY: McGraw-Hill.

Tobin, R. (2013). "Environment, Population, and the Developing World." In *Environmental Policy* (8th ed.), edited by N. J. Vig and M. E. Kraft. Washington D.C.: Congressional Quarterly Press, 299–318.

United Nations. (1987). *The World Commission on Environment and Development: Brundtland Commission Report*. New York, NY: Oxford University Press.

United Nations. (1997). *Environment Program Report*. New York, NY: Oxford University Press.

United Nations. (2010). *Human Development Report*. New York, NY: Oxford University Press.

United Nations. (2011). *Human Development Report*. New York, NY: Oxford University Press.

Usua, A., and Osuagwu, C. (2010). "Women and Politics in Male-Dominated Sphere: Getting the Wolves and the Sheep to Lie Together through Media Advocacy." In *Political Communication and Nigerian Democracy: A Book of Reading*, edited by G. Okon and A. Udoudo. Port Harcourt, Nigeria: Amethyst and Colleagues Publishers, 123–132.

Vig, N. J., and Kraft, M. (2005). *Environmental Policy* (6th ed.). Washington DC: Congressional Quarterly Press.

Warren, K. (1990). "The Power and Promise of Ecological Feminism." *Environmental Ethics*, *12*, 37–62.

Warren, K. (1987). "Feminism and Ecology: Making Connections." *Environmental Ethics*, *9*, 1–23.

Wheeler, S. (2004). *Planning for Sustainability: Towards More Livable and Ecological Communities*. New York, NY: Routledge.
White, R. (2008). *Crime against Nature: Environmental Criminology and Ecological Justice*. Devon, UK: Willan.
Withgott, J., and Brennan, S. (2009). *Environment: The Science behind the Story* (3rd ed.). Boston, MA: Pearson.
World Bank. (2011). *Toward Environmental Sustainable Development in Sub-Saharan Africa*. Washington DC: The World Bank.
World Bank. (2011). *World Development Report*. New York: Oxford University Press.
World Commission on Environment and Development. (1987). *Our Common Future*. New York, NY: Oxford University Press.
World Development Report. (2003). *Achievement and Challenges: Sustainable Development in a Dynamic World—Transforming*. New York, NY: Oxford University Press.

Part I

Developed Countries

3 Environmental Issues and Policies in the United States

Robert Dibie, PhD and Mary Bourke, PhD

INTRODUCTION

An emerging national agenda in the United States of interrelated economic and social development issues such as growth, poverty, environment, and political pluralism has become the central focus for national cooperation for the past four decades. In the last thirty years the United States has operated an all-inclusive environmental legislation in addition to several environmental policies. Despite this exclusive environmental legislature, more than 60 percent of the population of the United States (approximately 186 million Americans) live in 186 counties where healthful levels of either ozone or particulate pollution still prevails (Rosenbaum, 2014). In addition, more than half the total area of the nation's biologically essential estuaries and almost half the nation's rivers miles are considered unacceptably polluted (Baumer & Van Horn 2014; Rosenbaum, 2014).

The level of poverty in the United States has been growing fast in the past three decades (McConnell and Abel, 2008). Despite marked progress over the last three decades in improved human well-being, the environment has been polluted more than ever before in the history of the country (Dye, 2011). Environmental stress at both the local and national levels continue to threaten irreversible deterioration of the nation's already fragile environment. Continued high levels of population and economic growth exacerbate both poverty and environmental degradation. Although the United States has witnessed technology development and political openness, these development achievements have also come with both dangers and opportunities. The federal government estimates suggest that information on public exposure is available for less than 6 percent of more than 1,400 naturally occurring and manufactured chemicals considered to pose a human threat in the country (Rosenbaum, 2014; American Lung Association, 2009).

U.S. citizens have experienced various environmental problems. The list of environmental problems includes water and air pollution, public exposure to toxic chemicals and hazardous wastes, the production of large quantities of solid waste that winds up in landfills, heavy reliance on fossil fuels that contribute to the risk of climate change, and the destruction of ecologically critical lands and forest. These environmental problems often spill over to hasten the loss of biological diversity (Kraft, 2011; Kraft and Furlong, 2013). In addition to these problems is the alarming growth of the population. Jay Withgott and Scot Brennan (2008) present an argument that there is a strong correlation between increased human population and high levels of consumption of food, energy, and natural resources. These variables also have great impact on the social and economic well-being of the citizens of the United States. An aftermath of the excessive increase in population is the deterioration of the quality of life. Examples of the deterioration of life can be found in most congested cities in the United States (Guber and Bosso, 2010; Smith, 2004; Vig and Kraft, 2013).

United States is a country where all human activity produces waste. There seem to be a phenomenon of activities that continue to cyclically create pollution problems. Most companies and local governments discharge their waste into rivers, and this activity in turn shifts the costs to individual firms or local governments downstream, who must forgo using the river for recreation and water supply. In such cases the local government downstream would have to undertake the costs of cleaning up the rivers themselves. Thomas Dye (2011) contends that the United States tends to suffer more from environmental externalities. Thus, a coal-burning electricity-generating plant that discharges waste into the air shifts its costs to others, who must endure irritating smog. By shifting these costs to others, polluting companies lower their production costs, which allows them to lower prices to customers or citizens in the nation as well as increase their profit (Drummond, 2010; Hudson et al., 2007). On the other hand, polluting governments have lower costs of disposing their community's waste, which allows them to lower taxes for their own citizens (Boswell, 2010). As long as these costs of production can be shifted to others, polluting individuals, companies, and governments have no incentive to minimize waste or develop alternative techniques of production. In the process individual citizens are exposed to health problems (Dye, 2011). Further, businesses and consumers pay over three quarters of the environmental bill in the United States.

According to Walter Rosenbaum (2014) important and reliable environmental data should be fundamental in a durable national consensus that environmental protection must now be a first-order public concern, but this is not the case in the United States. It is increasingly apparent that the scope and scale of the United States' ecological degradation were often gravely underestimated and that the social and economic cost of pollution regulation was frequently miscalculated badly when the nation's major environmental policies were enacted (Rosenbaum, 2014; U.S. General Accounting Office, 2001). The appropriate correction of these miscalculated mistakes is essential to public policy making in a democratic society such as the United States. Therefore, environmental protection should be a work in progress in the nation.

Michael Kraft and Scott Furlong (2013) contend that climate changes in the United States have far reaching consequences for the nation's environment. These effects include an increase in stifling heat waves; threats to alpine meadows, coral reefs, and other vulnerable ecosystems; and public health threats. Other effects of environmental pollution include heat stress, air pollution, extreme weather, and diseases that spread easily in warm climate (Lauren et al., 2010; Myers & Spoolman, 2014). It is interesting to note that the cause of these environmental predicaments is blamed on human activities, especially the burning of fossil fuels such as coal and natural gas. Guy Peters (2013) and Jay Withgott and Scott Brennan (2008) underscore the strong relationship between energy and environmental policy in the United States. They argue that the amount of energy the United States citizens use and its sources, especially fossil fuel and nuclear energy, can have profound environmental impacts, including increased urban air pollution, the production of toxic chemicals and hazardous waste, and damage to ecosystems.

The United States' taste for high consumption of energy, especially fossil fuel, produces huge quantities of pollution, and the periodic shortages and high prices of petroleum and natural gas place pressure on industries to burn cheaper coal, which in turn creates even higher levels of pollution (Vig and Kraft, 2013). The Exxon Valdez spill in Alaska reveals that the transportation of fossil fuel, especially oil, presents a visible threat to the environment (Peters, 2013). Failure to adequately address the energy problem by exploring nontraditional sources of energy or by engaging in extensive conservation efforts will almost certainly exacerbate the environment problem (Kraft, 2010). The 2010 oil spillage off the coast of Louisiana in the Gulf of Mexico created a huge social and economic impact for the citizens who depended on the seawater for the livelihood.

As the United States' population has grown from a land of indigenous natives and foreign colonies to a large multicultural populous, capitalism has evolved and allowed the nation to prosper and grow. However, there has been a dark side to this growth, one that has allowed degradation of the environment (Salin and Pierskalla, 2005). The creation of automobiles, industrial factories, pesticides, fertilizers, and deforestation, to name a few, has resulted in egregious environmental abuses in the United States. Companies were concerned with profits, ignoring the effects of their products or processes on the environment. As the proliferation of environmental abuses occurred with long- term deleterious effects on the health of all living organisms, it was a struggle to overcome special interest groups' influence on the political arena, thus silencing concerned scientists. As citizens and scientists espoused current and future harm to all living things, the demand for public policy grew.

Exposure of the environmental abuses resulted in a powerful movement that fueled the demand for environmental policy development (Sonnenfeld, 2000). Once dormant environmental values were recognized and expressed by mothers finding that they were living in an area that was a threat to their health and that of their children, a grassroots movement was born (Hays, 2000). Fiorino (1995) refers to policy making in the United States as reactive, realistic, and incremental in response to an event or disclosure instead of being anticipatory, future oriented, and comprehensive. As environmental abuses were uncovered, citizen outrage resulted in the formation of the Environmental Protection Agency (EPA) in 1970. Programs of the EPA were enhanced by Congress with the passing of the National Environmental Policy Act in 1969, the Clean Air Act in 1970, and the Federal Water Pollution Control Act in 1972. Since the establishment of the EPA in the United States, some regulations issued by the regulatory agency to reduce pollution, such as emission control on automobiles, have required using more energy than would have otherwise been used (Kraft, 2011; Peters, 2007). Several scholars have argued that change in human behavior and consumption pattern in the United States, including things a simple as energy conservation and recycling use paper, plastic, and metal, may be more important than technological change in producing improvements in the environment (Vig and Kraft, 2013; Withgott and Brennan, 2008).

Environmental policy as used in this chapter refers to the extent to which government action affects human health and safety, energy use, transportation and urban design, agriculture and food production, population growth, and the protection of vital global ecological, chemical, and geophysical systems (Vig and Kraft, 2013). Thus environmental policies in the United States are going to have a pervasive and growing impact on the social and economic affairs of the citizens in the country.

The chapter will describe and evaluate the United States' environmental and energy policies and how they affect the social and economic affairs of the citizens. It discusses the evolution of environmental issues in the country. It argues that it is necessary to understand the conflict and struggles to rebuild and redirect environmental policy in the country in order to come up with appropriate solutions in the twenty-first century. It contends that although the citizens of the United States have been affected socially, economically, and health wise, pollution is no longer a national question about clean water and clean air. Pollution has taken on pronounced international dimensions, including widespread concern about global warming, ozone depletion, trans-border pollution, and decreasing biological diversity.

ENVIRONMENT AND PHILOSOPHICAL APPROACHES

The United States is the most industrialized and developed country in the world, yet it suffers from severe environmental problems—both in terms of the dimensions of the problem, which include global warming, national destruction of the ozone layer, national deforestation, and national overpopulation, and in terms of the sheer number of issues, for example, air, water, ground, and noise pollution; radioactivity; toxic waste; pesticides; and endangered species and the possible solutions to these issues. According to Jessica Adolino and Charles Blake (2011), food safety, environmental protection, biodiversity, protection of children against misuse of laborer, human rights democracy, and opposition to capitalism tends to be one reason for modernization.

The compassionate environmental theory is concerned not only with caring for the natural environment, but also with establishing a just relationship among human beings. According to the compassionate environmental theory, it is essential for people to care for each other and for the natural world (Wright and Boorse, 2014). The compassionate environmental theory depicts elements that contribute to inequitable environmental health and safety outcomes in developing countries as well as low-income communities in developed countries (Adler and Newman, 2002).

The theory argues that the environment encompasses the social determinant of human health and the ecosystem, which includes root factors such as poverty, misuse of natural resources, discrimination, other forms of oppression, and the economic, social, and physical environment of community. These factors shape exposures and behaviors that are the manifestation of the environment in the population and individual. Exposures and behaviors in an unhealthy environment include breathing polluted or contaminated air and exposure to their toxins, experiencing stressors associated with roots factors such as living in impoverished community environments and practicing risky and unhealthy behaviors (Giles and Liburd, 2007).

In order to reduce uncompassionate human behavior toward the environment as well as significantly reduce disparities, intervention should not only occur to improve medical care but also occur as early as possible to ensure that people are not becoming sick or injured in the first place. People should also be taught how to do things that are compatible with the care of the environment. Wright and Boorse (2014) contend that such human behavior change should include turning off lights in an empty room, recycling of used products, purchasing cars that pollute less and use less energy, refusing to engage in the conspicuous consumption constantly being urged on them by commercial advertising, supporting organizations that promote sustainable development practices, staying informed on environmentally sensitive issues, and expressing their citizenship by voting for candidates who are sympathetic to environmental concerns and the need for sustainable development.

Institutional restructuring in the United States in the past four decades has stimulated direct democracy in matters relating to environmental issues. The proliferation of environmental policy professionals representing federal and state agencies, industries, advocacy groups, and nongovernmental organizations has created a sizable base of talent and ideas for policy innovation in the United States and several other developed countries (Vig and Kraft,

2013). Further, these patterns are especially evident in environmental policy, where numerous areas of specialization place a premium on expert ideas and allow for considerable innovation within agencies (Karch, 2007).

The determinants of compassionate human relations to the environment are multifaceted in nature. The forces that shape health and environment are lifestyle, riches, poverty, and awareness of consequences of exposure to environmental toxins, risky or unhealthy behaviors, and experiencing chronic stressors associated with poverty or witnessing unsafe health outcomes. Education and justice are especially crucial for the development where unjust relationships often leave people without land, with inadequate food, and with poor health. It will also take significant government reform intervention in developed and developing countries to turn this situation around so that the poor can achieve justice in securing property rights and gain greater access to ecosystem goods and services. It is also critical to address the reason why people from low-income communities and developing nations become disproportionately exposed to dangerous environmental conditions. Developing nations should also be mindful that their practice of discriminating against developing countries by taxing and restricting imports from third world countries and by flooding the world market with agricultural products that are subsidized has caused critical consequences (Wright and Boorse, 2014). Such practices often deprive people in developing countries of jobs and money that could go far to improve their living conditions, prosperity, and global modernity. Figure 3.1 shows the relationship between environment and human beings in the sustainable development process.

Robert Dibie (2010) argues that a compassionate environmental strategy in the United States as well as other developed countries has not yet been taking seriously because of the net withdrawal of natural resources from and the net addition of pollution to developing nations by the more developed ones. Both developed and developing nations of the world must also address the spread of persistent organic pollutions, the radioactive fallout from nuclear accidents, the exploitation of oceanic fisheries, the destruction of the ozone layer, and the global climate change (Bulkley and Moser, 2007). Thus, a new ethic and partnership for the caring of human beings and ecological creation is long overdue.

The risk society theory was propounded by Ulrich Beck (1986). The theory postulates the truly global character of the twenty-first-century environmental risks, and the inherent apocalypse-blindness of contemporary economic and political institutions. The risk theory states that global environmental risks are democratic in that they do not distinguish between classes, and traditional class differences are no longer adequate for understanding the distribution of these risks among the population. The bone of contention about the risks theory is the unequal distribution of environmental risks (Mol, 2001). The reaction to the theory, however, led to somewhat new distributional patterns in late modern society. As stated

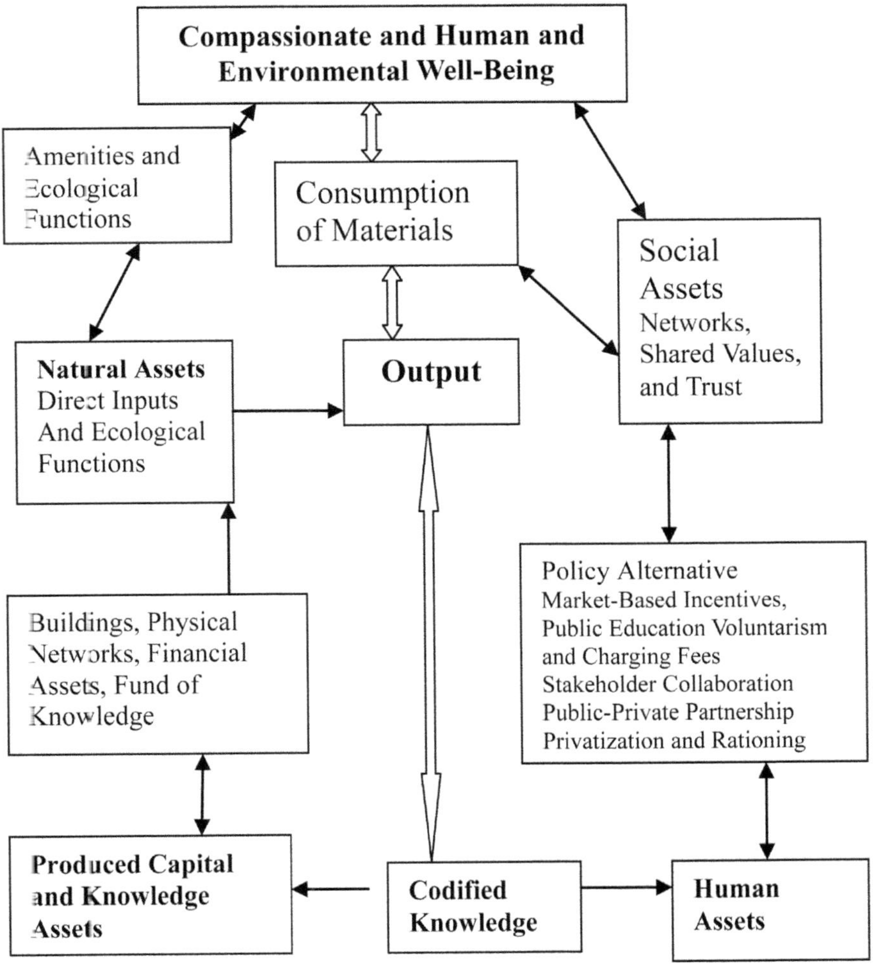

Figure 3.1 Relationships between Environment and Human

Source: The World Bank. (2002). *World Development Report 2003*. New York, NY: Oxford University Press.

earlier, environmental pollution and risk do affect everyone in that all face uncertainties on how to cope with them. This is especially true in those cases where a growing complexity of interdependencies on an international or global level occurs, as evidenced in cases of high-consequence risk (Beck, 1986; McConnell and Abel, 2008). Given the vital connection between the environment and the economy, there is a clear need for a different kind of economy. Robert Dibie (2010) contends that environmental business as usual does not work anymore. Thus, the path to a sustainable future will require the United States to further enhance its sustainable development economy and processes.

The ecological economist argues that the classical view sees the environment simply as one set of resources within the larger sphere of the human economy (Kraft, 2010; Withgott and Brennan, 2008). The ecological economist contends that the environment's vital role in supplying the goods and services on which human activities depend is largely ignored. Ecological economists present an argument that the classical approach looks at things backward and that the natural environment actually encompasses the economy, which is constrained by the limits of resources in the environment. According to Bassett and Shandas (2010) and Myers and Spoolman (2014), the production and consumption that articulate the classical economic view are just part of the system. On one hand, without the vital raw materials provided by the environment and without the capacity of the environment to absorb waste pollution, there is no economy. Therefore, the ecosystems and natural resources found in the United States provide the context for the country's economy.

In response to the theories discussed in this chapter and the environmental consequences, scientist and policy makers in the United States are being to work together to protect habitats, slow extinction, and safeguard endangered species. Recycling is helping relieve the nation's waste disposal problems, and alternative renewable energy sources are being developed to take the place of fossil fuels. These are a few of many solutions being explored in the United States (Withgott and Brennan, 2008). Thus, this chapter attempts to clarify the causes and consequences of human impact on the environment, and is it up the citizens, companies, and all levels of government in the United States to help develop more solutions.

ENVIRONMENTAL EVOLUTION AND IMPACT

The first environmental movement was started in the 1930s by John Muir, Theodore Roosevelt, and Gifford Pinchot (O. L. Graham, 2000). The movement focused on the loss of forests, soil erosion, and water and air pollution. In the late 1930s the movement gained momentum and power through the efforts of Franklin D. Roosevelt. In the 1960s a third environmental movement exploded with the revelations of Rachel Carson (1962/1990) in *The Silent Spring*. Her book fueled debate concerning the use of chemical pesticides and the responsibility of science. Her revelations, from a purely scientific perspective, set in motion a ban on the production of DDT and more importantly a grassroots movement demanding protection of the environment through state and federal regulation. Carson (2002) chronicles our historical environmental roots that have forever changed the chemical makeup of all living things. During World War II, scientists created industrial agents for chemical warfare. During this process they found that the chemical properties also controlled insects. After further development post–World War II, commercial uses were identified as insecticides. With commercialization, synthetic forms of the insecticides were created that were potent and lethal. Human contact with the potent insecticides resulted in an alteration of protective human enzymes and an alteration in cell structure leading to cancers. The production of synthetic pesticides in the United States soared from 124,259,000 pounds in 1947 to 637,666,000 in 1960 (Dye, 2011; Smith, 2004).

The United States environmental policies have change significantly over the past five decades. The 1960s and 1970s policies focused on air and water pollution control. The period also witnessed the preservation of natural resources such as parks, wilderness, and wildlife (Kraft, 2011). The environmental agenda between the 1970s and 1980s was somewhat pragmatic, but very difficult to accomplish. Vig and Kraft (2013) indicated that the frustration during this period galvanized economists, business groups, and conservatives to rethink a how best they could rejuvenate appropriate environmental policies to achieve better results. The U.S. president in the 1980s, Ronald Reagan, proposed a more efficient regulation and the promotion of alternatives to regulation, such as the use of market incentives, voluntary pollution control initiatives, public-private partnership, and collaboration among various stakeholders (Mazmanian and Kraft, 2009; Fiorino, 1995 Adolino and Blake, 2011). Several scholars argue that the environmental policies that were enacted in the 1980s are still currently in place. Although politics has been blamed for the lack of change to the environmental policies, there has been no consensus on how to transform the core environmental statutes. That notwithstanding, there has been modest modification in the safe Drinking Water Act in 1996, the Clean Air Act in 1990, and the pesticide control policy in 1996 through the Food Quality Protection Act (Dye, 2011; Eisner, 2007; Kraft, 2011).

Figure 3.2 shows the complexity of policy development in the United States. The political institutions such as Congress, the courts, the president and executive branch, research and development agencies, environmental groups and citizen participation, industry and trade associations, and the media all influence the process.

These political institutions all have competing agendas; consequently, they contribute to the complexity of environmental change and reform in the country. The history of the environmental movement and subsequent policy development is substantial and complex. Current literature barely contains the content necessary to fully discuss the complexity (Peters, 2013). The following highlights provide some insight into the birth of our awareness in the United States.

With the addition of political power, change occurred and the EPA gained power to ensure a safe environment for everyone. However, the very political power that gave substance to the EPA also threatened the progress of the EPA. Political ideology changes after elections and thus, the changing of political power. For example, President Reagan in 1981 started a process to decrease environmental regulation by reducing the

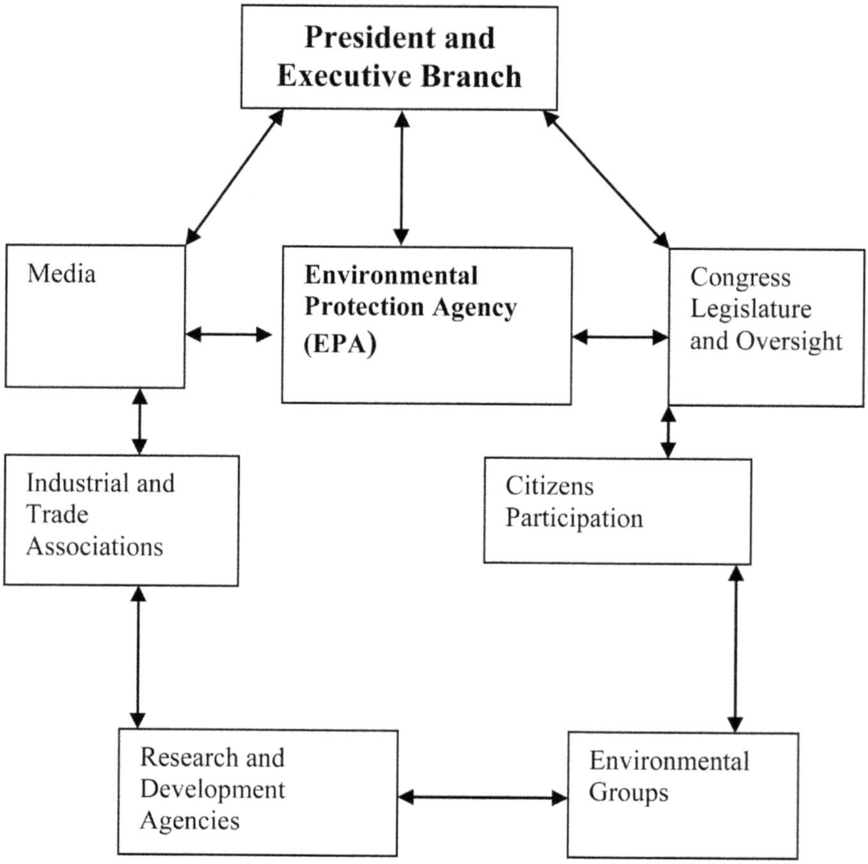

Figure 3.2 Environmental Policy Development Actors

Source: Adopted from Dibie, R. (2010). "Social Impact of Environmental Policies in Ethiopia." *Journal of International Politics and Development, 8*(1), 1–23.

EPA's budget and programs. Luckily, public support for the environment was stronger than the political influence of the Reagan administration. Congress was not about to give up the gains from the past. The future of millions of Americans was at stake. The EPA continued to focus on "risk" in setting priorities (Fiorino, 1995). By looking back at our environmental history, a window into the complexity of protecting the environment emerges and gives us an understanding of the social impact.

The National Commission on the Environment (1993) stated that the natural processes that support life on Earth are increasingly at risk. As a solution the National Commission on the Environment endorsed sustainable development policies. The National Commission also pointed out that if the United States continues in its current path, primarily reacting to environmental injuries and trying to repair them, the quality of our environment will continue to deteriorate, and the economy may decline if care is not taken. The statement of the Commission calls for change in behavior of the citizens of the United States. In 2008 when the price of oil increased tremendously, the consequent social and economic impact was astonishing; citizens that own trucks and SUVs scrambled to find fuel-efficient vehicles (Kraft, 2011). The lack of purchase of SUVs and other big vehicles spilled over into automobile manufacturers' financial solvency. This led to mass layoffs of automobile plant workers. Unemployment rates increased across the nation. The housing industry also almost collapsed.

Guy Peters (2013) and Michael Kraft (2011) contend that the United States has and continues to have an insatiable appetite for energy to power its cars, heat and cool its homes, and enjoy the many convenience of the modern life. It is interesting to note that when high price of oil went off the roof, major manufacturers tried to speed up the production of new fuel-efficient cars. On the other hand, those who could not afford the new price of oil changed their social and economic way of life. For example, they started using mass transit as an alternative to driving their own vehicles. Thus, in most parts of the United States life-supporting capital of forest, species, and soil is being depleted and its fresh waters and oceans are being depleted at an accelerating rate (Kraft, 2011). Citizens who depend on these forest, species, soil, and fresh waters had to change their social and economic lifestyle in order to continue to survive.

RESEARCH METHODOLOGY

The data for this study was derived from both primary and secondary sources. It was difficult to obtain all the data required across the United States. Data were collected from local and state governments. An extensive literature review and search for existing data and records about the social and economic impact of environmental issues was conducted. The secondary sources come from local government offices, relevant state and federal governments departments and agencies, nonprofit organizations, community-based organizations, books, journal articles, and corporate organizations. Other EPA reports and outputs of several environmental activities were consulted. Some content analysis of the social and economic impact of climate change was also selected as a stratified random sample.

Following an exhaustive search for existing secondary data on the indicators of social and economic impact of environmental issues, primary data were collected through surveys and focus group discussion. Fifteen hundred questionnaires were administered in twenty-five of the fifty states in the United States. Nine hundred and sixty completed questionnaires were returned. Specific information on the impact of environmental issues was obtained from the twenty-five states.

The results of the study are organized into an analysis matrix to allow us to summarize and develop typologies and descriptive statistics to link our findings to our research goals. We evaluated the impact of environmental issues or the array of climate change and relevant environmental policies identified for adoption and justified and operationalized each of the impacts. This study report sour findings related to the various social and economic impacts of environmental issues in the United States.

ANALYSIS AND DISCUSSION OF FINDINGS

Our results are discussed in the following section. It is assumed that the rapid rate of change in climate and its subsequent impact on the respective environments or locations of the United States make our evaluation highly time sensitive and appear to play a major role in the government leading climate change action plans. Our results are based on a review of secondary data and results from questionnaires administered in 2009–2011.

Table 3.1 and 3.2 shows the economic and social impact of environmental issues in most parts of the United States. Human and natural variables have the potential for doing severe damage to the environmental, and such damages have a direct impact on the present and future human welfare. In some states respondents indicated that the costs of human welfare are felt in the areas of economic, productivity, health, and the ability of the environment to support the citizens social and economic life style. Climate change for example will also widen the gap between rich and poor people all over the United States. This is because poor people have less wealth and technology with which to adapt to climate change, and poorer people rely more on natural resources derived from local plants and water.

Social Impact of Air Pollution

Eighty-five percent of the respondents in twenty-five states indicated that air pollution has affected their eyes and caused some form of skin irritation in the past five years. In addition they have suffered from one form medical problem that may be connected to exposure to air pollution. Michael Kraft (2011) and Kallman (2008) summarized the following health effects from air pollution: (1) Sulfur dioxide, a by-product from industry, causes respiratory and cardiovascular illness. From an environmental perspective it is a precursor to acid rain that damages lakes, rivers, and trees. (2) Nitrogen oxides are produced by vehicles and industry and cause respiratory and cardiovascular illness. (3) Particulate matter, which is fine particles in air caused by vehicles and industry, enters the bloodstream via the lungs causing damage. (4) Carbon monoxide is a by-product of vehicle exhaust and causes headaches and fatigue. It is especially harmful

Table 3.1 Economic and Social Impact of Environmental Issues

Environmental Problem	Nature of Social and Economic Impact	States Affected
Air Pollution	Several chronic health impacts exist across the nation. The high level of particulate matter and smoky indoor air are responsible for well over 1.9 million premature deaths every year.	Michigan, Illinois, Indiana, New York, Wisconsin, New Jersey, Minnesota, Missouri, Georgia, Pennsylvania, Ohio, Texas, and California
Atmospheric Change	Atmospheric change results in shift in vector-borne diseases, risks from climate natural disasters, and skin cancers attributable to depletion of ozone shield.	California, Washington, Louisiana, Mississippi, Florida, Georgia, North and South Carolina
Water Pollution	More than 2.5 million deaths and over 5 million illnesses a year are attributed to water pollution and poor household hygiene caused by water scarcity and pollution	Michigan, Illinois, Indiana, New York, Wisconsin, New Jersey, Minnesota, Missouri, Georgia, Pennsylvania, Ohio, Texas, and California
Deforestation	In two states localized flooding leads to death and diseases. Economic impact includes reduced logging, increased erosion and watershed instability, and diminished carbon storage capability for forest. In most cases deforestation result in the loss of nontimber products.	Oregon, Washington
Solid Degradation	In several states land use for crops, grazing, and forestry has been degraded by erosion. There have been reported field productivity losses. This in turn has affected gross national product.	Arkansas, Louisiana, Texas, Missouri, Ohio, Indiana, Tennessee, Michigan, Pennsylvania, New York, New Jersey

(*Continued*)

Table 3.1 (Continued)

Environmental Problem	Nature of Social and Economic Impact	States Affected
Solid and Hazardous Waste	Respondents in some states indicated that ground water is often polluted. When pollution occurs diseases are spread by rotting garbage drains are blocked. There is also evidence of risks from hazardous wastes due to polluted groundwater.	Indiana, Illinois, Missouri, Tennessee, Kentucky, Michigan, Florida, California, Texas
Loss of Biodiversity	In some of the very hot states ecosystem adaptability is reduced and generic resources are lost. The economic and social impact includes potential loss of new drugs.	New Mexico, Arizona, Nevada, California, Texas, Louisiana
Tornadoes	Destroy homes, farms, industries sites, and kill over 2,000 to 3,000 people each year	Alabama, Indiana, Missouri, Tennessee, Oklahoma, Illinois Arkansas, Kentucky, Florida, Mississippi, North and South Carolina, Iowa, Kansas, Texas

Source: Derived from surveys 2009–2013.

Table 3.2 Environment and Costs to Human Welfare

Environmental Problems	Economic and Social Impact	States Where People Are Most Affected
Hurricanes	Destroy homes, farms, industries sites, and kill over 2,000 to 3,000 people each year	Florida, Texas, Louisiana, Mississippi, North and South Carolina, Georgia, Virginia, Alabama
Mood Slide due to Rain	Destroy homes, roads, streets, farms, and industrial area	California, Colorado, Nevada, Pennsylvania
Flood due to Excessive Rain	Destroy farms, homes, industrial areas, roads	Indiana, Ohio, Louisiana, Mississippi, Missouri, Illinois, North Dakota, Texas, Florida, Tennessee, Kentucky

Duration of Heating degree days in Winter	Prolonged energy needs use heating home during winter. Excessive heating bills may cause economic hardship	New York, Maine, Rhode Island, Wisconsin, Minnesota, Michigan, Indiana, Illinois, North and South Dakota, Montana, Colorado, New Jersey, Ohio
Duration of Cooling degree days	Prolonged energy use for cooling homes and businesses may cause economic hardship.	Nevada, California, New Mexico, Texas, Louisiana, Florida, Arizona, Alabama, Georgia, Mississippi, North and South Carolina
Urbanization and Creating livable cities	Congestion in cities and diseases related to smoking, alcohol, drugs, and obesity are the greatest killers in urban societies; urbanization may also facilitate epidemics.	Los Angeles, California; Chicago, Illinois; New York, New York; Philadelphia, Pennsylvania; Atlanta, Georgia; Huston, Texas
Environmental health and Toxicology	Countless natural chemical substances that may pose health risks, for example, oil oozing naturally from the ground, radon gas seeping up from bedrock; and toxic chemicals stored or manufactured in the tissues of living organisms, that is, toxins that plants use to ward of herbivores and toxins that insects use to defend themselves from predators.	All over the United States
Fossil fuel	Pollution imposes significant costs on human and ecosystem health.	All over the United States
Biotechnology		
Public Policy effect	The effects of pollution and misuse of resources are seen most clearly in the states where environmental public policy is often not well established and implemented.	Alabama, Mississippi, Montana, North and South Dakota, Delaware, Rhode Island, Maine

Source: Derived from Survey in 2009–2013.

to people with cardiovascular disease. (5) Lead used in the manufacturing process is an air contaminate that can cause an accumulation over time of lead in the blood stream, which can lead to damage of the nervous system. Lastly, (6) VOCs, or volatile organic compounds, include hydrocarbons, alcohols, aldehyde, and ethers. VOCs cause eye and skin irritation, nausea, headaches, and cancer (Kraft, 2011). For a person suffering from asthma or lung disease, any one of the described air pollutants can lead to death.

A World Health report indicated that about 1,000 deaths each year are attributable to air pollution in the United States (World Health Organization [WHO], 2002b). Radon, a naturally occurring gas found in soil, is also considered a significant health threat. It is one of the leading causes of lung cancer. Another known cause of lung disease is asbestos. The American Lung Association describes the danger of asbestos as follows: "Asbestosis is a disease that involves scarring of lung tissue as a result of breathing in asbestos fibers. The disease causes the lung tissues and the lining of the chest wall to thicken and harden. Asbestos is one of the leading causes of lung cancer" (Dye, 2011; Kraft, 2011).

Policy development related to air pollution has less of a reactionary stimulus for discourse. In 1952 London, England, suffered from years of smog secondary to the Industrial Revolution. In an unusual weather situation, London experienced a serious episode of fog that caused approximately 4,000 more deaths than normal. This created the need to pass the Air Pollution Control Act of 1955 in the United States. In 1963 the Clean Air Act was enacted to give federal authority to monitor and control air pollution by setting emission standards. Thomas Dye (2011) and O. L. Graham (2000) state that the EPA reported that in 2010 the nation's air was the cleanest since the earliest record keeping in 1997.

According to this 2010 EPA report, six major pollutants were down 31 percent, smog levels improved, carbon monoxide dropped 38 percent, sulfur dioxide dropped 39 percent, and particulates dropped from 25 million tons to 10 tons in twenty years. The EPA announced in March 2010 the following: "Since 1990, nationwide air quality has improved significantly for the six common air pollutants: ground-level ozone, particle pollution, lead, nitrogen dioxide, carbon monoxide, and sulfur dioxide. Emissions of toxic air pollutants, such as benzene, have declined about 40 percent nationwide between 1990 and 2005. Despite this progress, about 127 million Americans live in counties violating at least one of the national air quality standards. The EPA has taken recent actions to tighten air quality standards to help ensure improvements in air quality for everyone" in the United States (U.S. Environmental Protection Agency [EPA], 2010, 49).

The World Health Organization (2002a) estimates that 2.4 million people die each year from air pollution. Sources of air pollution include smoke stacks from manufacturing, burning of fossil fuels, and exhaust from automobiles. The most commonly used fossil fuels in the United States are oil, coal, and natural gas (Vig and Kraft, 2013; Withgott and Brennan, 2008).

Carson (1962/1990) captivated her audience with revelations that "contamination of man's total environment with substances that accumulate in the tissues of plants, animals, and humans and have the potential to alter the genetic structure of organisms" (cited in Carson, 2002, 78). She further explained that the human body was permeable and therefore vulnerable to toxic substances in the environment, and subsequently exposures resulted in human cancers. Although she influenced the demise of DDT production in the United States, its exportation continued resulting in the pollution of the earth's atmosphere, oceans, streams, and wildlife. DDT is found in the livers of birds and fish and in the breast milk of every mother. According to Adami and colleagues (1995), in a study of 159 women with breast cancer it was identified that they were five times as likely to have DDT pesticide residues in their blood. These residues of organochlorines (DDT) contain estrogen, the female hormone that has been shown to contribute to higher risk of breast cancer (Dye, 2011; Wright and Bcorse, 2014).

In 1995, 14 million fish were found dead floating in the Neuse River on Pamlico Sound, North Carolina. In recent times, a similar mass death of fish occurred from the shores of Maryland to North Carolina. The cause was determined to be nutrient overload from human waste, fertilizer runoff, and a huge expansion of poultry and hog factory farming in Maryland, Virginia, and North and South Carolina, which caused waste leaching in underground water supplies (Withgott and Brenna, 2008). In the 1990s fish were dying in mass. Also, in 1995 the UN Food and Agriculture Organization reported that "demand for fish protein due to the worlds increasing population had decimated natural stocks and jeopardized the future of the seas as a source of food" (Graham, 2000, 5).

Social Impact of Water Pollution

Seventy-three percent of the respondent states indicated that they have experienced some form of water pollution in the community. They stated that the water pollution was a result of excessive rain, which led to flooding. Others indicated that melting snow was the cause of the flooding and polluted water that they experienced. Polluted rivers and oceans became the next movement, causing alarm among environmentalists as they were now called. Our rivers and oceans were polluted from point and nonpoint sources of pollutants. Point sources of pollution include air emissions and water effluents from a pipe or ditch. Nonpoint sources are harder to identify and include pesticides and fertilizer run off from farmland to streams (Kraft and Furlong, 2013; Peters, 2013). For example, the pollution of the Potomac River killing thousands of fish in the 1950s created a repair movement by President Lyndon B. Johnson. After three decades the Potomac was cleaner. In 1969, oil and garbage clogged the Cuyahoga River in downtown Cleveland, Ohio, and caught fire. It brought attention to the seriousness of water pollution experienced by the waterways in the United States. It took until the 1990s for the Cuyahoga River to become odorless. The 1990s brought control of the point-source pollution of factory pipes and municipal sewage plants (Graham, 2000). The pollution did not end there, though; a nonpoint source of pollution was created by farms (animal feces contamination, insecticides, and fertilizers), city streets, and suburbs that leached pollutants into America's creeks, rivers, and lakes. As a consequence, rivers flow into seas causing a pollution of estuaries and bays.

Cordy (2001) found that "Giardia and cryptosporidium are pathogens that have been found in public-water supplies and have caused illness in a large number of people. Pathogens can enter our water from leaking septic tanks, wastewater-treatment discharge, and animal wastes." Giardiasis, a gastrointestinal illness, is caused by the protozoan parasite Giardia intestinalis. (Centers for Disease Control, 2007; Dye, 2011)

Schueler (2000) discussed what he identified through research as the following pathogens (bacteria that cause harm to humans) in storm water runoff: Shigella (causes dysentery, an inflammation of the colon that results in abdominal cramps and pain, loose stools containing mucous and blood), Salmonella (causes gastrointestinal illness such as diarrhea), and Pseudomonas auerognosa (can lead to pneumonia, septicemia, urinary tract infection, gastrointestinal infection, and skin and soft tissue infections) (Centers for Disease Control, 2007). Some pathogens can cause cholera (acute diarrhea), typhoid fever (fever, rash, diarrhea), and staph infections. (Centers for Disease Control, 2009).

Colborn, Dumanoski, and Meyers (1996) published "Our Stolen Future" exposing how even small amounts of pesticides and other manufactured chemicals in the water supply act as endocrine disrupters that lead to decreased sperm counts, developmental problems, and some cancers in wildlife and humans. This revelation facilitated the tightened pesticide regulation in 1996.

Social Impact of Fossil Fuels

Eighty-one percent of the respondent in twenty-five states indicated that exposure to fossil fuel pollution has affected their health. They contend that air pollution, thermal pollution, land devastation, groundwater pollution, acidification of streams and rivers, erosion, subsidence of land caused by underground mines, and hazards to miners are all recognized as part of coal's utilization. Of all the fossil fuels, coal has the most harmful immediate and long-term effects on the environment and human health (Smith, 2004). These scholars contend that the high-energy nature of fossil fuels makes them more efficient to burn, ship, and store. Besides providing for transportation, heating, and cooking, fossil fuels are used to generate electricity. Electricity is considered to be a secondary form of energy that is easier to transfer over long distances and could be applied to a variety of uses in the United States.

Robert McConnell and Daniel Abel (2008) contend that sustainable societies cannot be built on nonrenewable energy resources. Humans use energy and generate vast amounts of pollution. Thus, reducing pollution from fossil fuels requires laws to be strictly enforced. On the other hand, wind energy is the fastest growing energy source in the United States and the construction of wind farms is supported by government subsidies that partly offset subsidies for fossil fuels. In recent times, biofuel has been produced in the United States. It is often argued that biofuels cannot replace fossil fuels in transportation, but producing transport fuels from agricultural waste, and even waste cooking oil, can make a dent in expensive imports and reduce air pollution at the same time. What is important to know, however, is that pollution imposes significant and measurable costs on human and ecosystem health.

Social Impact of Waste

Seventy-seven percent of the respondents indicated that hazardous or toxic waste tends to have posed a significant threat to their health in their state and community. They stated that many substances have been declared toxic by the EPA in their state because massive daily doses administered to laboratory animals cause cancers to develop. The respondents also indicated that the problem of solid waste comes from mining, such as oil and gas production, agriculture, and industry. Waste from respondents' cities presents greater political and managerial problems. They also indicated that city waste often causes environmental impact on land and water. In some states, such as New Mexico and Indiana, nuclear waste creates special problems. These are the waste from nuclear fission reactors and nuclear weapons plants (Dye, 2011; Vig and Kraft, 2013). Radioactive waste is very dangerous to human health. It is persistent in most industrial cities' environment in the United States (Kraft, 2011). The problem of disposal of radioactive waste has been a major concern of the nuclear energy industry and the military in the country for many years. The nations waste management efforts are marked by a numerous problems. The problems include the cleaning up of superfund sites, safe disposal of hazardous and radioactive waste, and frequent local opposition to disposal sites. These problems make clear that the best solution to waste management problems in the country is to reduce the generation of waste. The respondents of this study indicated that their communities are beginning to find ways to reduce, reuse, and efficiently recycle the materials and goods that they use. The change in consumption and behavior continues to stand as a major challenge for the citizens of the United States.

Social Impact of Urbanization and Congested Cities

Fifty-three percent of the respondents indicated that urbanization and congested cities have posed a significant threat to their social and economic welfare and health. Urbanization and development have also produced unprecedented quantities of environmental pollution and waste. The pollution includes exhaust gases from the burning of fossil fuels and toxic factory emissions polluting waterways. Some of the respondents pointed out that urban pollutants such as urine and droppings have declined but are still agents of health issues in congested cities. According to Withgott and Brenna (2008) and McConnell and Abel (2008), concentrating people in cities causes waste, both human and otherwise, which if not properly treated can severely pollute rivers or the ocean. Thus, urbanization may cause epidemics. Thus, sustainable development requires economic improvements and efficiencies guided by democracy, where constant growth only occurs in human and environmental health education in the United States.

ENVIRONMENTAL JUSTICE AND SOCIAL IMPACT

According to Zachary Smith (2004) and Thomas Dye (2011), environmental protection is not a loss; it could be an investment. Most nations should not consider environmental protection to be a luxury. It should be seen as a necessity for continued economic progress and adequate quality of life. The U.S. government's role in the twenty-first century is not to let economic agents act as they please. The government must impose the collective will on industries. The government should also work with international stakeholders to establish a mechanism for regulating and protecting the environment around the world. It is important to note that ecological modernization could transform and then use the language of markets and the logic of efficiency to achieve environmental improvements goals. For the last twenty years environmentalists have argued that hazardous waste dumps are built in communities that are politically powerless, poor, and predominately minority. The impetus for the environmental justice movement began in Houston, Texas, in 1978 when African American homeowners began a long contentious battle to keep a landfill out of their neighborhood. They formed the Northeast Community Action Group (NECAG) and hired an attorney Linda McKeever Bullard. They filed a class action lawsuit to block the building of a landfill in their neighborhood. Using civil rights laws they filled the landmark case *Bean v. Southwestern Waste Management Corp*. Bullard (2007) found that from 1920 to 1978 more that 80 percent of Houston's landfills and incinerators were located in mostly Black neighborhoods. This was particularly alarming since the percentage of Blacks in Houston at the time was 25 percent.

The most significant case highlighting the most egregious abuse was the 1982 decision by a North Carolina court to dump 40,000 cubic yards or 60,000 tons of highly toxic soil contaminated with polychlorinated biphenyls (PCB) in a mostly Black county. This fueled the exposure of environmental racism in this country, and as a result, the government of the United States enacted new policies to regulate toxic soil contamination in the country. It took until 1982 to detoxify contaminated soil from this county at a state and federal cost of 17.1 million dollars. Victory came with a substantial cost (Bullard, 2007; Wright and Boorse, 2014).

In 1990 William Riley of the EPA established an Environmental Equity Work Group to investigate claims that minority and economically disadvantaged populations bear a disproportionate exposure to environmental hazards and resulting illnesses. The formation of this group was in reaction to the report by the Commission on Racial Justice of the United Church of Christ, Toxic Waste and Race in the United States (1987), and *Dumping in Dixie: Race, Class and Environmental Quality* (1990) by Robert Bullard (Sturges and Gee, 2006). Wolcott and Banks (1992) reported the following results of the investigation:

1. A significantly higher percentage of Black children compared to White children have unacceptably high blood leads levels.
2. Racial minority and low-income populations experience higher than average exposures to selected air pollutants, hazardous waste facilities, contaminated fish, and agricultural pesticides in the workplace. Exposure does not always result in an immediate or acute health effects. High exposures and the possibility of chronic effects are nevertheless a clear cause for health concerns.
3. Environmental and health data are not routinely collected and analyzed by income and race. Nor are data routinely collected on health risks posed by multiple industrial facilities, cumulative and synergistic effects, or multiple and different pathways of exposure.
4. Risk assessment and risk management procedures are not in themselves biased against certain income or racial groups. However, risk assessment and risk management procedures can be improved to better take into account equity considerations.
5. Great opportunities exist for EPA and other government agencies to improve communication about environmental problems with members of low-income and racial minority groups. The language, format, and distribution of written materials, media relations, and efforts in two-way communication all can be improved. In addition, the EPA can broaden the spectrum of groups with which it interacts. (Payne-Sturges and Gee, 2006; Wolcott and Banks, 1992).

Native Americans are a unique racial group that has a special relationship with the federal government and distinct environmental problems. Tribes often lack the physical infrastructure, institutions, trained personnel, and resources necessary to protect their members (Payne-Sturges and Gee, 2006, 4). Companies found little resistance from poor neighborhoods, which were easy targets compared to the organized homeowner associations of the more affluent who could afford counsel and organize protests. President Clinton through an executive order (12898) required all federal agencies to focus on ending the disproportionate exposures of minority and poor people to many environmental hazards. Results of such efforts include federally funded research programs on environmental hazards, initiatives to increase citizen involvement in environmental decisions, and community-based efforts to address local concerns about environmental hazards (Payne-Sturges and Gee, 2006).

Morello-Frosch and Lopez (2006) examined theoretical and methodological questions related to racial residential segregation and environmental health disparities and concluded that there were "profound racial differences in residential patterns and in environmental exposure burdens. As a result, these might imply that segregation and the resulting inequality in the toxicity of residential environments may be contributing to racial differences in morbidity and mortality" (190). Thomas Dye (2011) and Michael Kraft (2001) suggest that although much has been done since 1990, studies that address environmental health disparities are still in their infancy.

ENVIRONMENTAL POLICY DEVELOPMENT

Robert Dibie (2010) argues that ignoring environmental problems will not save revenue; however, it will become more costly in the future to attempt to solve the ecological issues. Spending a few thousand dollars today on some cleanup efforts is better than a billion-dollar lawsuit tomorrow. Cleaner environment also means happier, healthier employees (Smith, 2004). Ecological modernization movement in the United States have led to major legislation such as the Federal Water Pollution Control Act of 1948, Air Pollution Control Act of 1976 and amendments, Clean Air Act of 1963 and amendments, Water Quality Act of 1977 and amendments, Solid Waste Disposal Act of 1965 and amendments, Hazardous Materials Transportation Act of 1975, Toxic Substances Control Act of 1976, and The Comprehensive Environmental Response, Compensation, and Liability Act (CERCLA) (Superfund), Oil Pollution Act (Dye, 2011; Kraft, 2011; Peters, 2013). At the end of the 1990s environmental problems were identified as lead, radon, asbestos, ozone, water pollution, air pollution, pesticides on food, surface water contamination, global climate change, by-products of energy production, and runoff of agriculture causing soil and water contamination. According to the EPA (2010), environmental policy covers many issues but predominately the following: air quality, water quality, water pollution, waste management, ecosystem management, biodiversity protection and

Table 3.3 Environmental Variables and New Policy Goals

Environmental Variable	New Policy Goals
Education	Propose education in energy efficiency, waste recycling, and weatherization change in behavior
Energy Efficiency	Start new residential building and green residential codes; new commercial and industrial building codes.
Forestry	Invest in reforestation and tree planting
Land Use Planning	Enforce zoning ordinances to reduce auto use such as transit-oriented development ordinance, parking maximums, and so on.
Transportation	Reduce carbon content of fuels, including transit, such as biofuel standards, electric vehicles, and so on; focus on pedestrian infrastructure such as sidewalks, crossroads; emphasize alternative transportation such as discount transit passes, free bike helmet programs
Renewable Energy	Encourage buying power from green sources; encourage using renewable energy such as programs supporting solar hot water heaters and so on; generate renewable energy such as wind turbine or solar panel; require government to buy power from green sources.
Solid Waste and Recycling	Increase recycling system such as e-waste and high-technology disposal system; procure and purchase recycling content; encourage the purchase of product with minimal packaging

Source: Derived from survey in 2009 and 2013.

the protection of natural resources, and wildlife and endangered species. Table 3.3 shows new trends in environmental policies in the United States. Recent environmental policies address two sets of environmental issues: (1) the use of natural resources like forest, fisheries, oil, and land; and (2) the prevention or reduction of air, water, and land pollution.

In the United States all environmental policies are developed in a socio-political context with the goal of promoting the common good of all citizens. As a result of the basic political process of enacting environmental policies, there are many opportunities for the political system to influence environmental policies.

The trend in environmental policies in the United States shows that there have been tremendous efforts by federal, state, and local governments to adopt strategic plans. The plans include changes in the man-made and natural environment. For example, most city governments have constructed bike lanes, increased mixed-used development, and added tree canopy. State and local governments have also introduced programs to educate and reach out to citizens about environmental issues. Outreach campaign also includes expanding weatherization awareness and recycling of waste products. New building codes are also being enforced in several states across the country. It was interesting to note that decisions to address climate change through planning in state and local governments required leadership. This leadership was commonly provided by prominent elected officials such as a mayor or governor in some states.

A review of proposed environmental health indicators for climate change in the United States shows that data exist for many environmental and health measures, but more research is needed to evaluate the sensitivity and usefulness of these measures. Further attention is necessary to increase data quality and availability and to develop new environmental monitoring and surveillance databases, especially for climate-sensitive morbidity (Kraft and Furlong, 2013). Maintaining the public health infrastructure by adequately funding environmental and chronic disease surveillance systems and a well-trained public health workforce is critical.

SUMMARY

This chapter has examined some U.S. environmental policies and the subsequent social and economic impacts. The survey findings show that many people in the United States have experienced the social and economic impact of environmental issues such as drought, flooding, heat, excessive rain, vector-borne diseases, other natural disasters, and identified adaptive behavior. The chapter argues that the purpose of environmental policies is to promote the common good. It uses the ecological modernization theory to stress the point that a major outcome of the environmental policies is (1) the improvement of citizens' welfare and (2) the protection of the ecosystem. It argues that environmental protection should go hand in hand with economic development, because ignoring environmental problems will not save money; rather it will be much more costly in the future.

The chapter also argues that the environmental policies of the United States have so far made important attempts to (1) enhance the prevention or reduction of air, water, and land pollution, and (2) foster a controlled use of the nation's natural resources like forest, fisheries, oil, and land.

The chapter also discusses the consequences of not having an effective environmental policy. It contends that although the United States has enacted several important environmental policies, its citizens and their economic activities have had the potential for doing damage to the nation's environment, and the damage has a direct impact on the present and future well-being of the citizens. It discusses the evolution of environmental issues in the country. It argues that it is necessary to understand the conflict and struggles to rebuild and redirect environmental policy in the country in other to come up with appropriate solutions in the twenty-first century. This chapter points out that although the citizens of the United States have been affected socially, economically, and health wise, pollution is no longer a national question about clean water and clean air. It has taken on a pronounced international dimension, including widespread concern about global warming, ozone depletion, transborder pollution, and decreasing biological diversity.

The outcome of the research for this chapter reveals that environmental issues' impact on citizens' welfare in the United States is mostly felt in the areas of economic productivity, health, and the ongoing ability of the natural environmental to support citizens' needs. It is therefore imperative for the government to enact more laws to protect the environment. Creating the laws is one issue; another issue, however, is the effective enforcement of the laws. Environmental laws could become luxuries to be tolerated if they do not interfere with individual freedoms or economic development. The chapter contends that laws could be regarded as part of the foundation of any well-organized, democratic nation. The government of the United States needs to provide more incentives to the citizens in order to achieve substantial, sustainable development in the future.

REFERENCES

Adami, H., Lipworth, L., Titus-Ernstoff, L., Hsieh, C., Hanberg, A., Ahlborg, U., Baron, J., and Trichopoulos, D. (1995). "Organochlorine Compounds and Estrogen-Related Cancers in Women." *Cancer Causes and Control, 6*, 551–566.

Adler N. E., and Newman K. (2002). "Socioeconomic Disparities in Health: Pathways and Policies." *Health Affairs, 21*(2), 60–76.

Adolino, J., and Blake, C. (2011). *Comparing Public Policies*. Washington DC: Congressional Press.

American Lung Association. (2009). *Executive Summary: State of the Air 2009*. Washington DC: Author.

Baumer, D. and Van Horn, C. (2014). *Politics and Public Policy*. Washington DC: Congressional Press.

Bassertt, E., and Shandas, V. (2010). "Innovation and Climate Action Planning: Perspective from Municipal Plans." *Journal of the American Planning Association, 76*(4), 435–451.

Beck, U. (1986). *Risk Society: Towards a New Modernity*. Los Angeles, CA: Sage Publications.

Boswell, M. (2010). "An Assessment of the Link between Greenhouse Gas Emissions Inventories and Climate Action Plans." *Journal of the American Planning Association, 76*(4), 451–473.

Bulkeley, H., and Moser, S. (2007). "Responding to Climate Change: Governance and Social Action beyond Kyoto." *Global Environmental Politics, 7*(2), 1–8.

Bullard, R. (2007). "Dismantling Toxic Racism." *Crisis, 114*(4), 22–25.

Carson, R. (2002). *Silent Spring*. New York, NY: Houghton Mifflin. (Originally published 1962)

Centers for Disease Control. (2007). *Environmental Health Features—United States*. Retrieved from http://www.cdc.gov/Features/EnvironmentalHealth.htm

Centers for Disease Control. (2009). *Water-Related Diseases and Contaminants in Public Water Systems*. Retrieved from http://www.cdc.gov/healthy water/drinking/public/water-diseases.html

Centers for Disease Control and Prevention. (2007). "Hazardous Substances Emergency Events Surveillance (HSEES) Program: 1990–2009." Retrieved from http://www.atsdr.cdc.gov/HS/HSEES/index.html
Colborn, T., Dumanoski, D., and Meyers, J. P. (1997). *Our Stolen Future: How We Are Threatening Our Fertility, Intelligence and Survival.* New York, NY: Dutton.
Cordy, G. E. (2001). USGS Reston: A Primer on Water Quality. FS-027-01. March, 2001. Retrieved from http://pubs.usgs.gov/fs/fs-027-01/
Dibie, R. (2010). "Social Impact of Environmental Policies in Ethiopia." *Journal of International Politics and Development, 8*(1), 1–23.
Drummond, W. (2010). "Statehouse versus Greenhouse." *Journal of the American Planning Association, 76*(4), 413–456.
Dye, T. (2011). *Understanding Public Policy* (13th ed.). New York, NY: Longman.
Eisner, J. A. (2007). "Water Vapour and Hydrogen in the Terrestrial-Planet-Forming Region of a Protoplanetary Disk," *Nature, 447*, 562–564.
Fiorino, D. J. (1995). *Making Environmental Policy.* Berkeley and Los Angeles, CA: University of California Press.
Giles, W. H., and Linurd, L. (2007). "Achieving Health Equity and Social Justice." In *Prevention Is Primary: A Renewed Approach to Community Wellbeing*, edited by C. Cohen, V. Chavez, and S. Chemimi. San Francisco, CA: Jossey-Bass, 25–40.
Graham, O. L. (Ed.). (2000). *Environmental Politics and Policy, 1960s–1990s.* University Park: Pennsylvania State University Press.
Guber, D., and Bosso, C. (2010). *Environmental Regulation in the Age of Nanotechnology. Resources for the Future.* Abingdon, Oxon, UK: Earthscan.
Hays, S. P. (2000). *A History of Environmental Politics.* Pittsburg, PA: University of Pittsburgh Press.
Harris, J. M. "Global Institutions and Sustainable Development," In *Sustainability and Environmental Policy*, edited by F. Dietz, et al.
Hudson, J., Lowe, S., Oscroft, N., and Snell, C. (2007). "Activating Policy Networks: A Case Study of Local Environmental Policy-Making in United Kingdom." *Policy Studies, 28*(1), 55–61.
Kallman, M. (2008). Air Pollution's Causes Consequences and Solutions (Monthly Update). Retrieved from http://earthtrends.wri.org/Updates/node/325
Karch, A. (2007). *Democratic Laboratories: Policy Diffusion among the American States.* Ann Arbor: University of Michigan Press.
Kraft, M. (2011). *Environmental Policy and Politics.* New York, NY: Longman.
Kraft, M., and Furlong, S. (2013). *Public Policy: Politics, Analysis, and Alternatives.* Washington DC: Congressional Press.
Lauren, G., Payne-Sturges, D., and Gee, G. (2010, March). *Environmental Justice, 3*(1), 21–26. doi:10.1089/env.2009.0019.
Mazmanian, D., and Kraft, M. (2009). *Toward Sustainable Communities: Transition and Transformation in Environmental Policy* (Second edition). Cambridge, MA: MIT Press.
McConnell, R., and Abel, D. C. (2008). *Environmental Issues: An Introduction to Sustainable Development.* Upper Saddle River, NJ: Prentice Hall.
Mol, A. (2001). *Globalization and Environmental Reform: The Ecological Modernization of the Global Economy.* Cambridge, MA: MIT Press.
Morello-Frosch, R., and Lopez, R. (2006). "The Riskscape and the Color Line: Examining the Role of Segregation in Environmental Health Disparities." *Environmental Research, 102*(2), 181–196.
Mulà, I., and Tilbury, D. (eds.). (2011). *National Journey towards Education for Sustainable Development.* Paris: UNESCO, 77–81.
Myers, N., and Spoolman, S. (2014). *Environmental Issues and Solutions.* Belmont, CA: Brooke/Cole.

Payne-Sturges, D. and Gee, G. (2006). "National Environmental Health Measures for Minority and Low-income Populations: Tracking Social Disparities in Environmental Health." *Environmental Research, 102*, 154–171.
Peters, G. (2007). *American Public Policy: Promise and Performance.* Washington DC: Congressional Quarterly Press.
Peters, G. (2013). *American Public Policy: Promise and Performance.* Washington DC: Congressional Quarterly Press.
Rosenbaum, W. (2014). *Environmental Politics and Policy.* Los Angeles, CA: Sage.
Schueler, T. (2000). "Microbes and Urban Watersheds: Concentrations, Sources, & Pathways." *Watershed Protection Techniques, 3*(1), 554–565.
Selin, S., and Pierskalla, C. (2005). "The Next Step: Strengthening the Social Science Voice in Environmental Governance." *Society and Natural Resources, 18*, 933–936.
Smith, Z. (2004). *The Environmental Policy Paradox* (4th ed.). Upper Saddle River, NJ: Prentice Hall.
Sonnenfeld, D. (2000). "Contradictions of Ecological Modernization: Pulp and Paper Manufacturing in Southeast Asia." *Environmental Politics, 9*(1), 235–256.
U.S. Environmental Protection Agency. (2010). *Office of Policy, Planning and Evaluation: Comparative Assessment of Environmental Problems: Vol. 1, Overview.* http://www.epa.gov/compliance/ej/resources/reports/annual-project-reports/reducing_risk_com_vol1.pdf
U.S. General Accounting Office. (2001). *Chemical Risk Assessment: Selected Federal Agencies, Procedures, Assumptions and Policies* (Document no. GAO 01–810). Washington, DC: Author.
Vig, N., and Kraft, M. (2013). *Environmental Policy.* Los Angeles, CA: Sage.
Withgott, J., and Brennan, S. (2008). *Environment: The Science behind the Stories.* San Francisco, CA: Pearson Books.
Wolcott, R. M., and Banks, W. A., (1992). *Environmental Equity: Reducing Risk for All Communities.* United States Environmental Protection Agency Publication, Washington D.C.: EPA 230-R-92-008.
Woods, N., and Potoski, M. (2010). "Environmental Federalism Revisited: Second-Order Devolution in Air Quality Regulation." *Review of Policy Research, 27*(6), 721–727.
World Health Organization. (2002a). *Quantifying Environmental Health Impacts.* Retrieved from http://www.who.int/quantifying_ehimpacts/national/Country profile/previousestimates/en/index.html
World Health Organization. (2002b). *The World Health Report.* Retrieved from http://www.who.int/whr/2002/en/whr02_en.pdf
Wright, R., and Boorse, D. F. (2014). *Environmental Science: Toward A Sustainable Future.* Boston, MA: Pearson Education Press.

4 Analysis of Environmental Health Issues and Policies in Western Europe

Emeka Nwagwu, PhD

INTRODUCTION

The 1979 *Surgeon General's Report on Health Promotion and Disease Prevention* was the major impetus in the attempt to improve health because it highlighted the impact of unhealthy behaviors, such as smoking, substance abuse, diet and unhealthy eating, and physical inactivity on personal and community well-being. As empiricism uncovered the mounting association between smoking, lung cancer, and cardiovascular issues during the 1970s and '80s, the popular view of health as merely "the absence of illness," and the corresponding emphasis on medical interventions to remediate disease, began to be challenged (Herman, 2010). Resultantly, public health researchers adopted the broader conceptions of health as "complete physical, emotional, and social well-being." Hence, the emphasis shifted from the remediation perspective to a prevention approach to health promotion based on behavior and lifestyle modifications. Soon thereafter, several studies documented the empirical linkages between a person's routine health practices, stressful patterns of living, and their susceptibility to disease and premature death (Stokols, 1996).

Efforts to modify own health practices are often impeded by socioeconomic and cultural constraints, including low educational status, income, time, energy, chronic exposure to neighborhood violence, and proximity to friends and family members who frequently exhibit health-threatening behaviors. These are situational factors that can derail people's best efforts and intentions to improve their health practices (Perkins, 2010). Even so, recent emphasis on the health impact of global environmental change, especially global warming, underscores the importance of developing environmentally based strategies of health promotion (Baumer and Van Horn, 2014; Myers and Spoolman, 2014). As communities have become aware of their exposure to environmental hazards, there has been mounting public outcry for actions to reduce or mitigate these exposures. Health promotion and health education theories and conceptual models can be extremely useful in developing behavior and environmental change interventions to reduce exposure to environmental hazards and illnesses (Stokols, 1995).

Researchers have identified a number of factors that determine health. These factors include socioeconomic and political factors. A cluster of these determinants are socially based and collectively labeled as the *social determinants of health*. Among the social determinants of health are environmental factors: indoor-outdoor air quality, water pollution, chemical contamination, soil quality and food contamination, housing quality, noise and sanitation, lead poisoning in children, ultraviolet and electromagnetic radiation, dysfunctional environmental design, global warming, and ozone depletion, in addition to others (Myers & Spoolman, 2014; Nutbeam, 2000). These environmental factors and other forms of environmental degradation are complemented by the circumstances in which people are born, live, learn, work, play, age, and worship: circumstances that are shaped by the distribution of income as well as the power relationships in the society. They impinge on health both individually and collectively. They have also been implicated in the inequalities and the unfair and avoidable disparities in health status within and between countries (World Health Organization [WHO], 2011a). Thus, in the last fifteen years there has been greater emphasis on health promotion and health maintenance. *Health promotion* is "any planned process employing comprehensive approaches to assess, correct, control and prevent those factors in the environment that can potentially harm the health and quality of life of present and future generations" (Porter, Taak, and Hobbs, 2004). Environmental health has gained prominence as awareness of environmental stressors and exposures increased in communities across the country and the world.

THE DEVELOPMENT OF ENVIRONMENTAL HEALTH

A study of the factors at the core of environmental health promotion must incorporate individually focused behavior and lifestyle with the environmental underpinnings of health and wellness (Perkins, 2010). That is, an understanding of factors that promote environmental health must start with an understanding of both behavioral change theories and environmental enhancements or a model that combines these. Behavioral change theories offer explanations for individual behavioral patterns (Parkes, Penelli, & Weinstein, 2003). They point to environmental, personal, and behavioral factors and characteristics as determinants of health. Some of the theories of behavioral health include social ecological, social cognitive, social learning, behavior change, and health belief models (Perkins, 2010; Simeonova & Valk, 2010). Other theories include the theory of reasoned action and the trans-theoretical model. A brief description of each theory is outlined here, followed by a description of the social ecological model and its implications for health outcomes.

SOCIAL COGNITIVE THEORY

According to Santrock (2008), social cognitive theory is a learning theory that is based on the notion that people learn by watching others and that one is always influenced by the moral precepts and value system in the environment in which one is raised and reared. That is, the individual person is just as important in determining moral development because one learns by observing others, with the environment, behavior, and cognition as the factors influencing development (Chen & Land, 1986; Crone & Horner, 2003; Sugai & Horner, 2002; Tobin & Sugai, 2005).

Social learning theory: This theory posits that learning occurs within a social context: people learn from each other, including observational learning, imitation, and modeling (Ormrod, 1999). This is consistent with Santrock's social cognitive perspective.

Behavioral change theory: Frederic Skinner's immense contribution to learning theory includes the notion that complex behavior is learned gradually through the modification of simple behaviors (Akers, Krohn, Lanza-Kaduce, & Radosevich, 1979). Imitation and reinforcement are central to these theories, which postulate that individuals learn by duplicating behaviors they observe in others and that rewards are essential to ensuring the repetition of desirable behavior (Ajzen, 1985; Skinner, 1953, 127).

Health belief model: The health belief model is associated with Irwin Rosenstock, Strecher and Becker (1988). The model states that health-related action depends on the simultaneous occurrence of three classes of factors: the existence of sufficient motivation (or health concern) to make health issues salient or relevant; the belief that one is susceptible (vulnerable) to a serious health problem; and the belief that following a particular health recommendation would be beneficial in reducing the perceived threat (Bandura, 1977; Chen & Land, 1986).

Theory of reasoned action: This theory posits that individuals consider the consequences, including the unintended consequences, of a behavior before engaging in the particular behavior. Consequently, intention is an important consideration in determining behavior and behavioral change. According to Icek Ajzen (1985), "intentions develop from an individual's perception of a behavior as positive or negative, together with the individual's impression of the way their society perceives the same behavior. Thus, personal attitude and social pressure shape intention, which is essential to performance of a behavior and consequently behavior change" (Irwin et al., 2006, 2).

Trans-theoretical model: According its proponents, "behavioral change is a five- step process." An individual must navigate these steps before complete change can be achieved. These steps are "pre-contemplation, contemplation, preparation, action, and maintenance." At the pre-contemplation stage, an individual may be unaware of a problem and, hence, has no thought of changing his behavior. From pre-contemplation to contemplation, the individual develops awareness and the desire to change a behavior. During preparation, the individual intends to change the behavior, and during the action stage begins to exhibit new behavior consistently. An individual finally enters the maintenance stage once he exhibits the new behavior consistently over an extended period of time (Prochaska, 1977, 3–4). These behavioral change theories have gained wide acceptance for their effectiveness in explaining health-related behaviors and providing insight into methods that would encourage individuals to develop and maintain healthy lifestyles.

Social ecological model: Researchers have concluded that the recent rise in obesity rates is the result of changes in the physical, social, and economic environments that have made it increasingly more difficult for individuals to get the activity and health promotion foods that they need, what Prochaska and Greeni Rossi termed "obesogenic" environment: "the sum of the influences that the surroundings, opportunities, or conditions of life have on promoting obesity in individuals or populations." According to the authors, "in an obesogenic environment, food marketing, media, federal and state policies related to food and agriculture, urban design, education, and transportation work together to constrain healthy eating and physical activity in communities, worksites, homes and schools" (Prochaska, 2003, 1). The social ecological model is, therefore, a framework for examining the multiple effects and inter-relatedness of social elements in an environment. This model allows for the integration of multiple contexts in health or physical activity context.

Operationalizing the social ecological model using obesity as the unit of analysis means that intervention at multiple levels is necessary to address the obesity epidemic. At the individual level are physical and recreational activities as well as learned activity patterns, as the higher one's fitness level, the easier it is to engage in physical activity. Other parameters are food consumption and government policies that create an environment for physical activity.

At the community level are social psychology factors: education, the media, public or societal attitudes to weight and overweight individuals, body image, and self-esteem.

The model integrates person-focused behavior modification efforts with environmental interventions to enhance the person's physical and social surroundings (Stokols, 1996). The social ecological approach is a conceptual framework for understanding the dynamic interplay between persons and their socio-physical environment. The social ecological system is comprised of personal, social, cultural, and physical dimensions that could influence a variety of health outcomes. Stokols (1996) concluded from his studies that these health outcomes include physical health status, developmental maturity, emotional well-being, and social coherence. Thus, on the individual level, the environment could influence one's knowledge, attitude, belief, and the manner in which they respond to the hazard and or protect themselves from harm. The interpersonal level refers to how the family, peers, social networks, and associations to which the individual belongs responds to

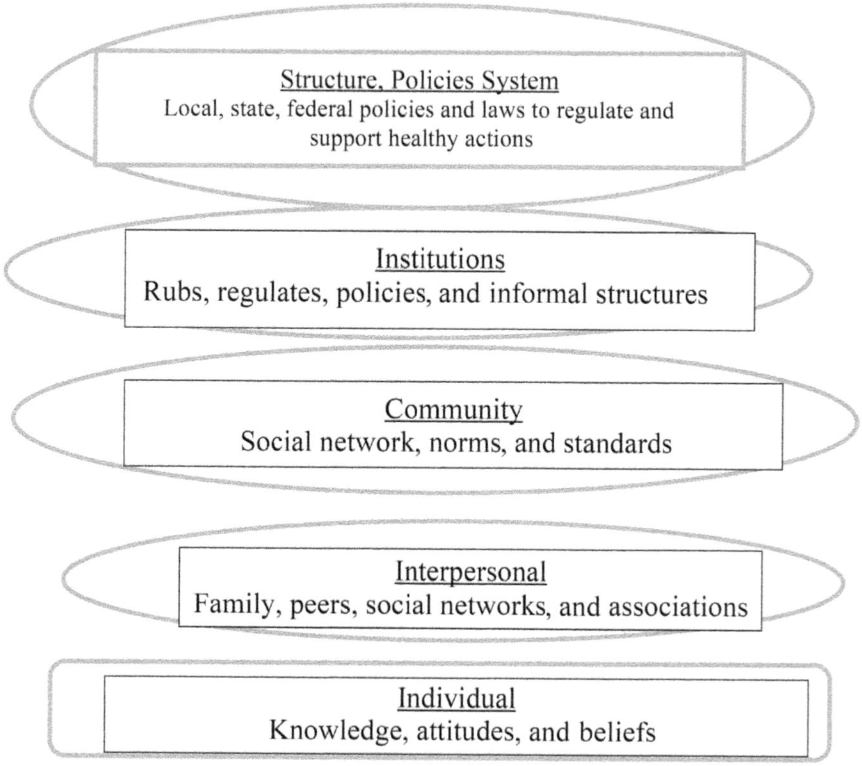

Figure 4.1 Levels of Influence in the Social Ecological Model

Source: Stokols, D. (1996). Translating social ecological theory into guidelines for community health promotion. *American Journal of Health Promotion, 10,* 282–298.

environmental concerns. For example, where people live affects their health and well-being. Although the majority of humans live in urban settings, there are close to 1 billion people who have the misfortune of living in slums characterized by poverty, disease, and deprivation. The daily conditions in which people live have a strong influence on access to good standard housing, clean water, and sanitation (WHO, 2011b). They also have implications for health equity.

The global preeminence of environmental health, spurned by the many environmental disasters of recent times (earthquakes, floods, tsunamis, nuclear explosions, etc.) compelled the WHO to establish the Commission on Social Determinants of Health (CSDH), a global network of policy makers, researchers, and civil society organizations to tackle the social causes of poor health and health inequalities. The CSDH called for

1. Greater availability of affordable housing.
2. Healthy and safe behaviors, including the promotion of physical activity.

3. The consumption of healthy foods.
4. The reduction of violence and crime (WHO, 2009).

These tasks can be achieved through environmental design and regulatory controls that include the control of alcohol sales outlets, sustained investment in rural development, economic and social policy responses to climate change and other environmental degradations, and a serious consideration of social justice by placing emphasis on health equity. Understanding how population groups experience their physical environments is fundamental to an understanding of the social determinants of health and how environmental factors affect health (Healthy People 2020, 2011). That is, reducing the societal burden of health problems that derive from environmental factors will require interventions designed to alter behavioral and environmental factors because a significant part of the total burden of disease in industrialized countries can be attributed to environmental factors. For example, the large burden of respiratory diseases and cancers is of particular concern for European countries as vulnerable groups, such as children and the elderly, are more at risk.

Also, the Commission on Social Determinants of Health recognized the complexity of environmental health, as diseases are multifactorial and addressed the issues through the integration of relevant policies and stakeholders in the determination of appropriate policy solutions. Certain environmental factors, such as exposure to polluted water, food, or air, determine health outcomes. For example, research has established that up to one sixth of all child fatalities and disease can be attributed to environmental factors. While individuals can make certain choices that affect their health, such as lifestyles, they also rely on public policies from authorized public entities for protection from health hazards. The poor, in particular, have shorter life expectancies and more illnesses than the affluent. These disparities in health outcomes draw attention to the growing need for a just and an egalitarian society where social and environmental factors that are detrimental to health are eviscerated using government policies, laws, and regulations. Peoples' lifestyles, including the conditions in which they live and work, certainly influence their health outcomes (Wilkinson & Marmot, 2003).

DIFFERENT VIEWS OF THE ENVIRONMENT

Stokols conceptualized contrasting views of the social and physical environment as a determinant of health:

1. The physical and social environment can serve as a medium of disease transmission, exemplified by waterborne and airborne diseases and the spread of contagious illness through interpersonal contact.

2. The environment can operate as a stressor, exerting detrimental effects on people's mood, performance, and physiology as the result of their exposure to uncontrollable demands such as noise, political upheaval, or interpersonal conflict.
3. The environment can function as a source of safety or danger of residing in areas that are chemically contaminated, geographically unsafe, or socially violent.
4. It can serve as an enabler of health behavior. For example, installation of safety devices in motor vehicles, proximity of physical fitness facilities to one's home or workplace, and exposure to interpersonal modeling or cultural practices can foster health promotion behavior.
5. It can also serve as a provider of health resources—community sanitation systems, public health services, legislation ensuring citizen's access to health insurance and primary care. (Stokols, 1996).

Thus, environmental conditions, especially population diversity, place of residence, and economic recession, affect health differently, depending on the individual's personality, perception of environmental controllability, health practices, and financial resources. The environment can also be a provider of resources that, intelligently used, would lead to positive health outcomes. These can be achieved through the application of social behavioral theories that facilitate behavior modifications and public policies aimed at pollution abatement, clean air and water, and urban redesign with accommodation for parks and physical and recreational facilities.

POVERTY AS AN ENVIRONMENTAL HEALTH FACTOR

Poverty is a socioeconomic factor that affects health outcomes. Poverty and health status are intimately interrelated, where the combination of both leads to a cycle—poverty leads to poor health, which, in turn leads to poverty—what, for the lack of a better term, I have chosen to label "the poverty–poor health–poverty syndrome." In essence, life's challenges associated with poverty create conditions that reduce household savings, lower earning ability, and reduce physical and emotional well-being. This combination ultimately endangers the health of the individual as well as the community.

According to research, in England 4 million children, one in three, are currently living in poverty (Hirsch & Beckhelling, 2012). This number is one of the highest rates in the industrialized world. Poverty can have a profound impact on the child, his family, and the rest of society. It often sets in motion a spiral of socioeconomic exclusion in education, employment, and mental and physical health, including social interaction. Thus, in some societies, poor children are excluded from participating fully in their rights of citizenship. Children who grow up poor are more likely to drop out of school and have reduced chances of employment, thus restricting their ability to get a good job and financially contribute to society. According to research, child poverty currently costs society £25 billion each year in England (Hirsch & Beckhelling, 2012). "Thus in its broadest sense, environmental health comprises those aspects of human health disease, and injury that are determined or influenced by factors in the environment. This includes not only the study of the direct pathological effects of various chemical, physical, and biological agents, but also the effects on health of the broad physical and social environment, which includes housing, urban development, land-use and transportation, industry, and agriculture" (Healthy People 2020, 2011).

Environmental health plays a powerful and vital role in communities and in the lives of individual living in them. Wide gaps in health disparities have been blamed on environmental factors. The gaps in life expectancy and disease rates between rich and poor, the well and the poorly educated, the manual workers and professionals, all have environmental health implications and outcomes. This has caused many communities to take affirmative actions to improve their social environment: creating better jobs, implementing policies that improve community health and safety, providing quality working environments, and increasing access to social benefits for individuals in the community (EU, 2013).

THE ROLE OF EDUCATION

An educated community is being increasingly viewed as a health community, because of the knowledge base (social and environmental factors and the conceptual framework for improving individual and public/community health). The more knowledgeable individuals are, the higher the likelihood that they will adhere to healthy living habits. Still an individual's best efforts to maintain good health may be thwarted by the surroundings, including public health policies, especially local ordinances that may inadvertently restrict certain activities. Unfortunately, the location of a community, its ethnic and socioeconomic composition, and resources availability have been found to affect the health outcomes of its inhabitants. That is exactly why community education, particularly about how to create and maintain healthy environments, prevention of the spread of diseases, and delaying the onset of chronic medical conditions are seen as critically important to health outcomes (WHO, 2011b).

Individuals with higher income tend to be able to afford better health care because they can afford to purchase health insurance. People with higher education are more gainfully employed. This means better earning and access to health insurance—which comes, primarily, through the workplace—and therefore important healthy outcomes. Most major employers provide good health insurance for their employees, and a majority of the employees take advantage of this opportunity. Worksite health promotion programs improve the health of those who have access to them. Work or place of employment is central to social status and is one of the most powerful predictors of health outcomes. Research shows that unemployment is associated with a large number of health risks. Inadequate employment has been associated with poor health outcomes.

The degree of control that employees exercise over their work situation influences health, and most employers provide avenues for employees to exercise some measure of control over their workplaces. One study indicates that every step up the career ladder has positive health consequences (WHO Regional Office for Europe, 2010). Despite its overall health benefits, work can be hazardous to one's health. The workplace influences the health of families and children. Health conditions affect work status. Income inequality affects health. An emerging social science framework that integrates labor market conditions, employment experiences, and health reached this conclusion (WHO, 2011a).

The Partnerships for Prevention, an advocacy group for better health in Europe, suggests that businesses can contribute to the health of the nation by offering employee health promotion programs, comprehensive health benefits, and safety policies as well as by partnering with community efforts to promote health and prevent disease and disability—all of which are designed for improved environmental health (Parkinson, 2003).

An educated population regularly participates in physical activities that are a major factor in staying healthy as people have become aware of the critical importance of physical activity to health and wellness. Typically these communities have built environments with recreational parks, swimming pools, exercise clubs, and other recreational facilities that are geared toward the development of an environment for healthy living and healthy lifestyles.

Community awareness and utilization of the underlying factors and issues for good health contribute to the well-being of individuals, families, and neighborhoods. The implementation of optimal environmental standards leads to a better quality of life for everyone in the community. On the other hand, disparity in health as a result of socioeconomic disadvantages, especially in inner cities, unhinges the neighborhoods. One of the objectives for Healthy People 2020 includes a healthy environment and access to health care, occupational safety and health, and public health infrastructure.

Environmental health affects the economy in almost the same way as it affects the social aspect of individual and community life. The mean income of the people living in a particular neighborhood determines the structure and social fabric of that neighborhood. Their built environment is consistent with their incomes and knowledge of healthy living. Thus the decision of a fast-food restaurant to locate in an area becomes exceedingly problematic and controversial not only because many fast-food restaurants are magnets for social deviants but also because of the reputation of such restaurants for selling unhealthy foods. Thus, gas stations and convenience stores are located only on the major roads or interstates, on the peripheries of the cities, away from individual homes and neighborhoods, especially the affluent neighborhoods. Rarely can one find a liquor store in such communities. Roads in such communities are built and maintained on a consistent basis as opposed to the inner cities where the quality of life is below the poverty level.

An affluent neighborhood maintains a sustainable environment. "Sustainable development is the development that meets the needs of the present without compromising the ability of future generations to meet their own needs" (WHO Regional Office for Europe, 2009). Sustainable development is generally thought to have three components: environment, society, and economy. These three areas are interrelated. For example, a healthy, prosperous society relies on a healthy environment to provide food and resources, safe drinking water, and clean air for its citizens.

IMPACT OF ENVIRONMENTAL HEALTH IN SELECTED EUROPEAN COUNTRIES

Air pollution, both indoors and outdoors, is a major environmental health problem affecting everyone in developed and developing countries alike. The 2005 WHO air quality guidelines (AQGs) are designed to offer global guidance on reducing the health impacts of air pollution. The guidelines first produced in 1987 and updated in 1997 had a European emphasis. The new (2005) guidelines apply worldwide and are based on expert evaluation of the most current scientific evidence (WHO, 2006). The guidelines recommend revised limits for the concentration of selected air pollutants: particulate matter (PM), ozone (O3), nitrogen dioxide (NO2), and sulfur dioxide (SO2), applicable across all WHO regions (WHO, 2006).

Denmark

Denmark thrives on trade with the EU members and globally. The Danish economy is largely driven by intensive agriculture and fisheries, which support a large agro-food industry. Unfortunately, this has been accompanied by a detrimental environmental impact. Other severe environmental pressures stem from the transportation and energy supply structures with a heavy reliance on fossil fuels. Environmental issues in Denmark also have a strong international dimension due to its association with regional economic and environmental powers, for example, the European Union and the North Sea and Baltic Sea countries, with stringent air and pollution standards. Denmark is strongly participates in global environmental issues, including environmental assistance.

The implementation of environmental policies is being further devolved to municipal authorities. Environmental policies currently focus on air pollution, the aquatic environment (nutrient discharges and groundwater contamination), biodiversity, chemical substances, environmental health, and global issues such as climate change. Land-use regulations, particularly those applicable to rural and coastal areas, are rigorous, and spatial planning contributes effectively to protection of the environment, nature, landscapes, and coastal areas.

Stringent waste incineration regulation has been a driving force for the development of cogeneration and district heating. Denmark uses economic instruments (environmental charges, environmental taxes, other economic instruments) to discourage detrimental environmental actions or to punish noncompliance with environmental requirements. The 2001 Environmental Law included the full cost-recovery principle in water management with Green taxes applying to air, water and waste management policies. The tax on sulfur emissions (1996) contributed to drastic decreases in SO_2 emissions. The wastewater tax (1997) led to a significant reduction of nitrogen, phosphorus, and organic matter in wastewater. Since the 1999 review, Denmark has introduced several new environmental taxes (e.g., CO_2 tax on gasoline, tax on ozone-depleting greenhouse gases, taxes on PVC and phthalates, tax on mineral phosphorus added to feed), and the "polluter pays" principle applies to households and, to a lesser extent, industry.

Denmark's proactive stance on protecting the environment through international cooperation has played an influential role in a number of international negotiations, particularly in the EU context. Denmark has met its international commitments (e.g., reduced discharges of phosphorous and heavy metals into the Baltic Sea and the North Sea, reduced atmospheric emissions of SO_2 and VOCs). Denmark has phased out or reduced its emissions of ozone-depleting substances ahead of internationally agreed deadlines. Environmental management is mainstreamed into a substantial part of bilateral and multilateral regional assistance programs.

England

The germ theory origins of environmental health are well illustrated by the cause and effect approach adopted by early practitioners of public health, such as John Snow, who was able to abort the nineteenth century cholera epidemic in London by removing the handle from the Broad Street pump—the major source of water contamination (Parkes, Panelli & Weinstein, 2003). Consequently, the focus here is London—a sprawling metropolitan city of several million people. The British government sought extension of time from the European Union, giving her till 2011 to comply with traffic pollution standards. The pollution situation prompted the London Liberal Democratic Member of Parliament, Sarah Ludford, to say this latest breach is yet another wake-up call for the mayor of London and the government (Vidal, 2010). Research has shown that airborne pollution in London could be responsible for up to thousands of premature deaths a year: this is an invisible public health emergency (Vidal, 2010).

Poor air quality is now considered one of the biggest public health issues facing the England. A recent report by the House of Commons Environmental Audit Committee concluded that air pollution could be contributing to 50,000 deaths in the United Kingdom a year. A study commissioned by Boris Johnson, mayor of London, calculated that more than 4,300 deaths are caused by poor air quality in the city every year, costing around £2billion a year (European Public Health Alliance, 2009).

Finland

Noise pollution is major environmental concern in Finland. One sixth of the Finnish population is exposed to daytime noise levels from motorways and railways (OECD, 2009). Efforts to reduce noise include city speed limit regulations, enactment of noise emission thresholds, implementation of noise barrier apparatus, and railway maintenance (OECD, 2009). As a result of regulations, the number of inhabitants living in areas exposed to daytime noise from city traffic above 55db has been reduced from 560,000 in 1998 to around 400,000 in 2005 (OECD, 2009).

Water pollution is another environmental concern in Finland. The primary sources of water pollution are municipal and industrial wastes, with industrial waste comprising the bulk of the pollution. Pollutants from industrial activities have resulted in increased concentrations of atmospheric sulfur, which, in turn, produces acid rains that have damaged the country's water bodies, including fisheries and other aquatic life and recreational activities. For example, the Baltic Sea has been a site of environmental concern since the 1960s due to excessive pollution from surrounding countries. In recognizing the problem, Finland has enacted regulations to help control the industrial and municipal point sources of pollution in the Gulf of Finland (OECD, 2009). As a member of the European Union since 1995, Finland has implemented or is implementing EU directives and is involved in EU environmental action, particularly in the Baltic region and in cooperation with nearby Russia (OECD, 2009).

France

The European Commission has twice referred France to the European Court of Justice over persistently high levels of nitrate pollution found in drinking water in Brittany. The pollution was caused by chemical runoff from intensive farming and irresponsible agricultural practices. The safety of drinking water in Brittany had been in question since 2001, when the Court of Justice ruled that France had failed to fulfill its obligations to keep the levels of nitrates below 50 mg (United Nations World Water Assessment Programme [UN WWAP], 2006). The Court stated that thirty-seven rivers in Brittany contained concentrations exceeding this level. Since then, France has implemented measures to reduce the volume of nitrates spread over agricultural land (UN WWAP, 2006). However, those measures have been found to be insufficient for nine rivers. In March 2007 France proposed an additional action plan to ensure subsequent compliance, although the Commission considered France's action inadequate.

The Commission asked the Court to impose on France a lump-sum fine of over €28 million and a daily penalty payment of €117,882 (Goliath, 2008). The fear of drinking polluted water made many households in Brittany refuse to drink tap water, but instead avail themselves of bottled water from the supermarkets, to avoid negative health outcomes.

Greece

In the last twenty years, Greece has been undergoing major economic and social developments, as a full member of the EU (Valavinidis & Vlachogianni, 2011). Greece's major efforts toward economic and social convergence to other European countries have been partially matched by environmental control of various forms of pollution and improving the quality of life (Bandura, 1989; Zalidis, 1998). In the field of environmental legislation and regulations, Greece has been compelled into achieving remarkably noticeable progress.

Rapid economic growth and rising income levels in the last decade increased pressures on the Greek environment conditions and requirement to match EU standards. Changes in major industrial sectors—such as energy, agriculture, transportation, tourism, housing, and road construction—urbanization, and development of coastal areas have strong implications for environmental policies, technological progress, and investment in environmental infrastructure (Valavinidis & Vlachogianni, 2011).

Although Greece has still untouched natural environment and a vast and rich cultural heritage, the demands of economic growth and development have increased pressures on environmental issues, including their health outcomes. Unplanned urbanization and expansive housing construction in forested areas and coastal zones (islands) have resulted in increased air emissions for electricity generation and excessive use of water (OECD, 2009a).

Greek agriculture has experienced significant changes with profound impact on the state of the environment (excessive use of water for irrigation, soil erosion, nitrate and phosphate pollution of rivers and lakes) (Zalidis, 1998). In the last decade water quality has been undermined severely, although urban air quality has greatly improved in big cities, such as Athens, Thessaloniki, and Patras, the result of better fuel quality, promotion of public transport, and less polluting private cars, although levels of air pollutants still remain above regulated levels in some places (Valavinidis & Vlachogianni, 2011).

Nationwide measures have improved energy efficiency, but high demand makes the task very difficult. Particulate air pollution in cities and in areas with power stations remains very high (Valavinidis & Vlachogianni, 2011). Investment efforts on air pollution abatement and control have been low. The recent use of natural gas for industrial and household power generation is expected to improve air pollution (Veropoulos & Skuras, 2008).

The integration of environmental concerns in economic, industrial, agricultural, and social policies in order to achieve cost-effective environmental protection and sustainable development is facing slow progress at a time when Greece's rapid growth demands urgent and drastic actions (Valavinidis & Vlachogianni, 2011). Public awareness, information, and education about environmental protection are changing. Enforcement of national legislation is often too slow or lacking, because of weak cooperation among ministries and agencies, and lack of rigorous inspection and prosecution, which are impeded by low numbers of environmental inspectors (National Center for the Environment and Sustainable Development, 2008). The recent economic crisis and the dramatic financial problems currently underway in Greece will probably postpone some of the environmental protection measures.

Hungary

Hungary, similar to Portugal, has considerable environmental problems, especially toxic problems related to mining. In parts of Hungary, namely, at Gyöngyösoroszi (northeast Hungary), toxic elements from the lead (Pb) and zinc (Zn) mines have generated intolerable pollution with considerable risk to the ecosystem and human health (Gasper et al., 2009). During the 1990s Hungary's mining crisis was found to pose a devastating health risk to children's health, and thus a major public health concern. The Hungarian government reacted by establishing a task force forcing environmental agencies to research and to mitigate the issue.

Environmental hazards are prominent in a number of childhood diseases, although, in most cases, environmental factors have rather multiple than single effects (Pall, Paldy, Rudnai, Dura & Varmai, 2007). Nevertheless, premature birth, congenital malformations, gastrointestinal and respiratory diseases, allergies, malignant tumors, and injuries can be considered to have possibly and partly environmental causes (Pall et al., 2007).

According to Pall, et al. (2007), there are currently 1,300 places in Hungary where not all mandated requirements for drinking water quality are being met. Most of the shortcomings do not involve an increased health risk (the concerns involve the presence of ammonium, iron, and manganese; hardness; and total organic material), but primarily concern the taste, odor, and color of the water. A government decree on drinking water quality issued in 2001 gives a list of cities and villages where there were drinking water problems that had to be eliminated by a set deadline (the end of 2009).

In spite of the enormous efforts and resources commitments of the last two decades, there are still more than 1 million people (among them over 100,000 children aged birth to fourteen) who are drinking water with arsenic level over the new health limit value introduced in 2001 (Gasper et al., 2009). Previously, the most serious drinking-water-related health problem in Hungary was methaemoglobianaemia in infants. This lethal disease has declined with the application of preventive measures by the National Public Health & Medical Officers Service (NPHMOS).

Iceland

Water pollution is a major environmental concern in Iceland. Of Iceland's 170 cu km of renewable water resources, 6 percent is used for industrial purposes (Encyclopedia of Nations, ND). The major sources of environmental pollution are the "excessive use of fertilizers, industrial runoff from aluminum smelter projects, and geothermal and hydroelectric activities" (OECD, 2012). The Icelandic economy is heavily dependent on pollution-free waters, as the fishing industry provides over one-third of Iceland's income and about 80% of the country's electricity is generated using hydropower (OECD, 2012).

Iceland became a member of a European Economic Area, thus committing itself to introducing environmental legislation set forth by the European Union, such as directives in the fields of water quality, pollution control, and waste management (OECD, 2009).

Since the early 1990s and the first OECD environmental performance review of Iceland, the Ministry for the Environment has extended the scope of its responsibilities, to include pollution prevention and control, nature protection, physical planning, and meteorology. Staffing of the ministry and of the agencies operating under its auspices has increased. Regional public health inspectorates have been created to facilitate implementation of environmental policies. Legislation has been substantially enhanced, largely as a result of Iceland's participation in the European Economic Area. Significant progress has also been made in EIA and physical planning.

Iceland's economy is heavily reliant on natural resources fishing and marine resources, hydropower, and tourism. Iceland's high rate of economic growth in recent years has been achieved by the decoupling of economic growth from environmental pressures; for example, energy intensity has fallen since 1990, and $SO_{(x)}$ and $NO_{(x)}$ emissions are growing slower than the gross domestic product (GDP). Some environmentally favorable changes in consumption patterns have been induced, notably the switch from oil to geothermal energy for domestic heating. The fishery management system (individual transferable quotas coupled with better regulations) has enabled fish stocks to recover and produce good economic returns. The National Environmental Strategy, "Towards Sustainable Development," was published in 1993, followed by the National Sustainable Development Action Plan in 1997. The Ministry for the Environment has begun regular coordination meetings with several other ministries, local authorities, and other stakeholders.

Iceland is closely associated with European countries as a member of the Nordic Council. It also has close ties with North American trading partners. Its economy is highly dependent on exports of fish and development of tourism, two sectors that require a high-quality environment and a positive, "green" image. Iceland developed its environmental policy at a fairly late stage but made significant progress in the 1990s, in particular by transposing many EU directives into its legal system and giving legal status to its international commitments. It has ratified and implemented many international agreements, and it hosts the secretariats of two working groups under the Arctic Environmental Protection Strategy (Averous, 2009).

Italy

Air and water are essential ingredients of well-being and a healthy life. Unfortunately, polluted water and air are common throughout the world. The WHO states that one sixth of the world's population, approximately 1.1 billion people, do not have access to safe water and 2.4 billion lack basic sanitation (United Nations Children Fund, 2006). For most citizens of the European Union, access to clean water is taken for granted. Most citizens fail to realize the burden of water quality as well as quantity. Pollution, whether by industry, agriculture, or households, returns eventually to the environment and may cause damage to human health or the environment itself. The WHO report also added that "air pollution causes 800,000 premature deaths from lung cancer, cardiovascular and respiratory diseases worldwide, in addition to increased incidence of chronic bronchitis, acute respiratory illness, exacerbation of asthma and coronary disease, and impairment of lung function (WHO, 2006).

The sources of pollution include road transport, stationary combustion sources (domestic coal burning, industrial sources, incinerators and waste disposals, fossil fuel power plants) and noncombustion sources (construction, quarrying and mining, cement plants, and ceramic industry), and other sources, such as forest and agricultural fires. The main pollutants found in the atmosphere include particulate matter, lead, ground-level ozone, heavy metals, sulfur dioxide, benzene, carbon monoxide, and nitrogen dioxides.

In contemporary times, Italy has been remarkable for environmental disasters, triggered mainly by industrial accidents. One episode stands out: In 1976, the city of Seveso, north of Milan, Italy, gained international reputation following an explosion at a small Swedish chemical plant, which released a significant quantity of contaminated toxic by-products (dioxin and other chemical contaminants) into the atmosphere over an extensive area, resulting in the highest residential dioxin exposure that has ever been recorded until the Chernobyl nuclear catastrophe in Russia in the 1986. More than 1,000 residents were evacuated and, following the Seveso Women's Health Study (SWHS), which investigated the effects of dioxin exposure on the reproductive health of women exposed to the 1976 industrial accident, pregnant women in the affected areas were advised by the government and the medical establishment to have abortions (Eskenazi, 2006). The investigations also found that high animal mortality was observed almost immediately and over 193 cases of chlorine (an acne-like skin condition specific to dioxin exposure) were reported. The surrounding contaminated area was divided into three exposure zones, most heavily contaminated or most exposed to the least exposed, based on surface soil dioxin measurements.

Italy has been slow in confronting its environmental problems because authority over the environment and environmental issues is highly dispersed among several agencies: the Ministry for Ecology (established in 1983), the Italian Ministry of Culture and Environmental Quality, the National Council for Research, and the Ministry for Coordination of Scientific and Technological Research. Localities also have responsibility for environmental protection, although most of the burden of planning and enforcement falls on regional authorities in Italy. The principal antipollution statute is Law No. 319 of 1976—the Merli Law (Panunzio & Landoli, 1976). This law was designed to control the disposal of organic and chemical wastes. Enforcement, however, has proved difficult, if not elusive.

Air pollution is a significant problem in Italy. United Nations sources estimate that carbon monoxide emissions increased by 12 percent in the period between 1985 and 1989 (Caselli, 2010) . In the 1990s Italy had the world's tenth highest level of industrial carbon dioxide emissions, which totaled 407.7 million metric tons per year, a per capita level of 7.03 metric tons (European Environment Agency [EEA], 2011).

Water pollution is another important environmental issue in Italy. The nation's rivers and coasts have been polluted by industrial and agricultural contaminants and its lakes by acid rain. In 2001, the nation had 160 cu km of renewable water resources with 53 percent used in farming activity and 33 percent used for industrial purposes. Facilities for the treatment and disposal of industrial wastes are inadequate. Italian cities produce 19.1 million tons of solid waste annually (Organization for Economic Cooperation and Development [OECD], 2003).

Netherlands

Air and water pollution are significant environmental problems in the Netherlands. The Netherlands has one of the world's highest levels of industrial carbon dioxide emissions totaling 155 million metric tons in

1996. The pollution resulted from industrial and agricultural waste, including heavy metals, organic compounds, nitrates, and phosphates (Perkins, 2010; Simeonova & Valk 2010). The Dutch government has embarked on a sustained effort to combat the effects and lessen the burden of the environmental hazards on its citizens by controlling air pollution and reducing sulfur dioxide emissions (Netherlands, 2010). Between 1980 and 1990 air pollution was reduced from 490,000 tons to 240,000 tons. After correction for the leap year 2008 and accounting for uncertainties in the data, total global emissions essentially stabilized in 2009. Although arithmetically global emissions decreased slightly, the uncertainties estimated in the country trends and particularly in the difference of regions with growing and with decreasing emissions are such that we must conclude that the emissions essentially stabilized in 2009. This compares with the six years following 2002 with average annual growth rates of 3.5 percent. In 2008, when the impacts of the credit crunch became visible, global emissions increased by about 1.5 percent (PBL Netherlands, 2009).

In 1971, the Ministry of Health and Environment was established; a countrywide system of air pollution monitoring by the National Institute of Public Health was put in place in 1975. Since the mid-1970s, discharges of heavy metals into industrial wastewater and emissions of most major air pollutants from industrial use of fossil fuels have been substantially reduced.

An excise tax surcharge on gasoline and diesel fuel was imposed for pollution abatement in 1981. Deforestation and wood burning were forbidden. Consequently, in 2001, 5.7 percent of the country's total land area was protected. The Netherlands currently imports 94 percent of its forest product needs, which totaled $2.92 billion in 2001 (Medieval smokestacks, 2011). With output from its national forests declining yearly, domestic wood industries are becoming more dependent on imports. Besides decreasing domestic resources, government initiatives and pending legislation will bring significant changes to the production of wood products.

The Netherlands has 11 cu km of renewable water resources, of which 68 percent is used for industrial purposes. Solid waste in the nation's cities amounts to an average of 7.6 million tons yearly (Netherlands, 2010). Aggravating the situation are the southwesterly winds, which carry the pollutants inland from coastal industries, as well as rivers that carry pollution into the Netherlands from neighboring countries (Louwman, 1992).

Poland

Poland is a land of ecological disasters and environmental degradation: air, water, lead, and even overcrowded cities. In 1991 Poland designated five official ecological disaster areas. Of the five, the densely concentrated heavy industry belt of Upper Silesia was popular for having the most serious pollution problems (Curtis, 1992). In that area, public health indicators such as infant mortality, circulatory and respiratory disease, lead content in children's blood, and incidence of cancer were uniformly higher than in other parts of Poland and dramatically higher than indicators for Western Europe (Curtis, 1992).

Experts believed that the full extent of the region's environmental damage is indeterminate. The situation has been exacerbated by overcrowding, since 11 percent of Poland's population live in the industrial belt region. With 600 persons per square kilometer, Upper Silesia ranks among the most densely populated regions of Europe. In 1991 the region's concentrated industrial activity contributed 40 percent of Poland's electrical power, more than 75 percent of its hard coal, and 51 percent of its steel.

A variety of studies reflected the effects of severe environmental degradation in Upper Silesia (Curtis, 1992). In 1990 the infant mortality rate was over 30 deaths per 1,000 births, nearly five times the levels in some countries of Western Europe; some 12,000 hectares of agricultural land had been declared permanently unfit for tillage because of industrial waste deposition; and between 1921 and 1990 the average number of cloudy days per year had increased from 10 to 183 (Curtis, 1992). Average life expectancy in southern Poland was four years less than elsewhere in the country.

Water and air pollution affect the entire country. A 2009 report found that 65 percent of the water in Poland's rivers was so contaminated that it corroded industrial equipment used in industry (WHO, 2009). After absorbing contaminants from the many cities on its banks, the Vistula River became a major polluter of the Baltic Sea, thereby ensuring that the water could not be used for irrigation. The WHO report (2009) also noted that about half of Poland's lakes had been damaged by acid rain and 95 percent of the country's river water was classified as undrinkable. Because Polish forests are dominated by conifers, which are especially vulnerable to acid rain, nearly two thirds of the forestland had sustained some damage from air pollution by 1990 (Curtis, 1992). In 1989 Polish experts estimated total economic losses from environmental damage at over US$3.4 billion, including soil erosion, damage to resources and equipment from air and water pollution, and public health costs.

Portugal

The impact of climate change on human health has been global concern. Portugal is a land of inclement weather, specifically of heat waves. During the heat wave in 2003, 583 heat-related deaths were observed. Studies determining the impact of heat on subgroups in Portugal identified the elderly as the most vulnerable group with the highest heat-related mortality rates. With Portugal's life expectancy of approximately seventy-six years (as of 2009), for the increasing number of the elderly population, the heat wave has become a huge public health concern.

Thus, the European heat wave of 2003 heightened interest in weather-related health in Europe, especially in Portugal. Studies have reported that days of usually low and high ambient temperatures were associated with increases in mortality and morbidity (Almeida, Casimiro, & Calheiros, 2010). Populations living in urban environments, especially the elderly, have a high risk of mortality from ambient heat exposure (Almeida et al., 2010). Campo de Jales, a Portuguese village surrounded by an inactive gold mine, remains a grave public health concern. Jales mine is largely contaminated with heavy metals and dust from centuries of mining operations. Exploration of the gold mines left several unsafe and unhealthy conditions, such as "abandoned structures and equipment, high-walls, open pits and mining dumps tailings, lack of drainage control, acid water and toxic materials" (Mayan, Gomes, Henriques, Silva, & Begonha, 2006, 17). A study designed to evaluate the health effects from the Jales gold mine, participants from Campo de Jales and Vilar de Macada (a place adjoining Jales) found significant higher levels of blood lead and cadmium in the population of Campo de Jales (Mayan et al., 2006).

Spain

Environmental health in Spain has been characterized by deforestation, fires, and chemical pollution. Trees and grasslands are now limited to northern Spain, namely, the Pyrenees and the Asturias-Galicia, because centuries of unplanned cutting has depleted the forest. In Spain, fire eliminates 700,000 to 1,000,000 hectares of forestland each year. Government reforestation schemes are thwarted by animal farmers whose sheep and goats graze freely over large areas of the pasture. During the 1980s, an average of 92,000 hectares (227,000 acres) were deforested annually, and soil erosion has laid waste 18 percent of the total land mass of Spain (OECD, 2004).

Air pollution is also a problem in Spain. In 1995 industrial carbon dioxide emissions totaled 223.2 million metric tons (a per capita level of 5.72 metric tons), ranking Spain twentieth compared to the other nations of the world. Industrial and agricultural sources contribute to the nation's water pollution problems. Spain is also vulnerable to oil pollution from tankers that travel the shipping routes near the nation's shores. Spain's cities produce about 13.8 million tons of solid waste per year (Eurostat, 2011).

In Spain environmental responsibility is vested in the Directorate General of the Environment, within the Ministry of Public Works and Urban Affairs. In 2001, 8.4 percent of the country's total land area was protected (OECD, 2004). In the same year, nineteen of the country's mammal species, ten bird species, ten types of freshwater fish, and 822 plant species were endangered. Endangered species include the Spanish lynx, Pyrenean ibex, Mediterranean monk seal, northern bald ibis, Spanish imperial eagle, dusky large blue and Nevada blue butterflies, and on the Canary Islands, the green sea turtle and Hierro giant lizard. The Canarian black oystercatcher and the Canary mouse have become extinct.

Sweden

Environmental issues and sustainable development have remained high on Sweden's agenda, given Sweden's low population and moderate economic growth. Environmental issues in Sweden also have a strong international aspect because of regional economic and environmental interdependencies (e.g., regarding Baltic Sea pollution, trans-boundary air pollution and Nordic cooperation). Sweden entered the European Union (EU) in 1995 because of its strong commitment to global environmental issues such as climate change, persistent organic pollutants (POPs), and environmental aid.

Environmental progress in Sweden continues as Sweden has benefited from solid institutional and regulatory frameworks, extensive use of economic instruments, and significant planning and legislative reform (e.g., the Environmental Code). The influence of EU environmental legislation has increased. Today, priority environmental issues are identified in fifteen ambitious, long-term, strategic environmental quality objectives (EQOs): reduced climate impact, clean air, natural acidity only, a nontoxic environment, a protective ozone layer, a safe radiation environment, zero eutrophication, flourishing lakes and streams, good-quality groundwater, a balanced marine environment, thriving wetlands, sustainable forests, a varied agricultural landscape, a magnificent mountain landscape and a good built environment. A sixteenth EQO, on biodiversity, is under preparation. A number of these objectives have both domestic and international dimensions.

To meet the challenges represented by the EQOs, Sweden will need to (1) implement more efficient environmental policies, (2) further integrate environmental quality concerns in economic and other policies (e.g., health, energy, transport, forestry, agriculture), and (3) further strengthen its international environmental cooperation. This report evaluates Sweden's performance in meeting its domestic objectives and international commitments concerning environmental management. Building on already very effective and innovative environmental policies, Sweden has achieved a range of environmental results that accord with its domestic objectives and international commitments. The country has further strengthened its environmental regime through the use of economic instruments. This success is the result of the priority accorded to environmental protection by all sectors of the society, including the government, the parliament, civil society, enterprises, and the public.

Sweden's environmental bonfires are driven primarily by welfare concerns, environmental expertise, and economic imperatives. In 1999, Sweden reformed its environmental legislation with the adoption of the Environmental Code. This comprehensive framework law not only consolidates previously fragmented legislation but also introduced environmental courts, environmental sanctions and ambient quality standards, clarifies the role of environmental impact assessment (EIA) and transposes the EU water framework directive and integrated pollution prevention and control (IPPC) directive, among others. The Environmental Code also spells out several principles relating to environmental policies.

Sweden gives high priority to sustainable development nationally, in Europe and globally. It adopted a national sustainable strategy in 2002, with a secretariat in the prime minister's office. Sweden's has made remarkable and significant improvements in emission intensity, energy intensity, and material intensity. This progress reflects, in part, institution-based instruments in a wide range of areas, including the integration of environmental concerns in energy, transport, and agriculture. The ongoing green tax reform is a logical extension of earlier use of economic instruments. Real efforts are being made to promote sustainable consumption and production, not only through economic instruments but also through policies favoring integrated product policy and green procurement (see OECD, 2009).

Ukraine

The world's greatest radioactive ecological accident occurred in April 1986 at the Chernobyl nuclear power plant in what is now Ukraine, precipitating the danger of a nuclear radioactive proliferation crisis. Although the crisis was confined to the Ukraine, the accident contaminated vast areas with radionuclide that created both short-term and long-term human health problems (Ginzburg & Reis, 1991). Given the enormity of the crisis, the government quickly developed policies designed to protect the public's health. In the years since the accident, policies have evolved and the government has developed new policies to respond to the threat to public health caused by the Chernobyl accident. The crisis conditions after the accident required immediate policy decisions and public health interventions without the benefit of a thorough and deliberative policy-making process (Page & Bobylena, 1995). The recent independence and nation-building efforts of Ukraine are political factors that are complicating health policy making and call into question Ukraine long-term ability to adequately deal with the health outcomes of an environmental disaster of radio activity of this magnitude.

In May 1991, the International Chernobyl Project released the results of its health assessment of the accident. The authors compared seven contaminated villages to six clean villages, but did not study evacuated people or a category the authors classified as "liquidators," the 100,000 to 200,000 people who fought the fire and did emergency cleanup at the nuclear power plant (Page & Babylena, 1995). The authors concluded that they found no statistically significant differences in health outcomes between the experimental and the control groups (International Atomic Energy Agency, 1991a, and 1991b). Public health officials in Ukraine, Belarus, and the Russian Federation believe that the study was flawed. Scientists disagree about the health effects of chronic, low-level exposure to radiation and radioactive materials (Harrison and Stather, 1990; Upton, 1992). Some scholars argue that chronic exposure to radiation in the environment and in the food supply is not a significant problem, while others argue that such exposure was causing widespread health problems, especially immune-deficiency problems. There is wide speculation that chronic, low-level exposure, caused by radiation releases from nuclear bomb testing and accidental discharges, has increased dramatically the mortality rates from septicemia and cancer in the elderly and AIDS in the young (Gould & Goldman, 1990). The human health effects of the Chernobyl catastrophe are controversial at present, and accurate assessments are difficult, since many cancers as well and many genetic disorders resulting from the Chernobyl accident may not manifest themselves for many years (Page & Bobylena, 1995).

The public health implications of the Chernobyl accident extend beyond the direct effects of exposure to radiation. The environmental and economic consequences of the accident produced psychological damage that is a significant health issue. In Ukraine, the phrase to describe this condition is "loss of full-blooded years of life" (Ginzburg, & Reis, 1991).

Exposure to radionuclide or other toxins from contaminated land is an involuntary exposure to a risk factor with the potential to cause adverse health effects. Exposures and risks over which people have little control cause greater stress and fear than do voluntary exposures to controllable risks of equal or even greater magnitude (Slovic, Fischoff, & Lichtenstein, 1979). High levels of stress and anxiety caused by fear of radiation (radiophobia) were so great that the fear itself produced adverse health, which became a significant problem in the Ukraine. Health professionals believe that large numbers of people are radiophobic (U.S. Nuclear Regulatory Commission, 2009). Lack of pragmatic understanding of post-traumatic stress disorder, reactive and situational depressions and familial strain resulting from the accident have impeded medical treatment in Chernobyl-contaminated regions (Ginzburg & Reis, 1991).

UNITED NATIONS COMMISSION ON THE EUROPEAN ENVIRONMENT HEALTH

The aim of the UN Commission on Environmental Health is to ensure satisfactory health outcome across the European Union through an integrated approach of relevant policies, aiming at involving all stakeholders in the policy-making process. Specific actions were taken to achieve these processes.

In June 2003, the Commission unveiled an EU Environment and Health Strategy with a particular focus on the health of children. DG Health and Consumers, DG Environment, DG Research, and the Joint Research Centre

Table 4.1 Life Expectancy Table for EU-27

	Total		Men		Women	
Country	**1994**	**2009**	**1994**	**2009**	**1994**	**2009**
Spain	78.1	81.8	74.4	78.6	81.8	84.9
France	78.0	81.6	73.8	78.0	82.2	85.0
Italy	78.1	81.9	74.8	79.1	82.2	84.5
Netherlands	77.6	880.5	74.6	78.7	80.4	82.9
United Kingdom	76.8	80.5	74.1	78.3	79.5	82.5
Portugal	75.5	79.6	72.0	76.5	79.0	82.6
Hungary	69.6	74.4	65.0	70.3	74.5	78.4
Poland	71.8	75.9	67.5	71.5	76.1	80.1
Ukraine	67.87	69.19	62.2	63.8	73.2	74.9
Greece	77.5	80.2	75.1	77.8	80.0	82.7
Denmark	75.5	79	72.8	76.9	78.2	81.1
Sweden	78.9	81.5	76.2	79.4	81.6	83.5
Norway	77.9	81	74.9	78.7	80.8	83.2
Iceland	79.3	81.8	77.1	79.8	81.5	83.8
Finland	76.7	80.1	72.9	76.6	80.3	83.5
United States	75.7	78.5	72.4	76.0	79.0	80.9

Source: European Commission (EuroStat), http://epp.eurostat.ec.europa.eu/statistics_explained

were commissioned to work together to reduce environmental pollution and to enhance the public health.

The EU Environment and Health Action Plan (2004–2010),was adopted in June 2004, to implement the EU 13 Action Point Strategy designed to improve knowledge of the linkage between diseases and environmental risk factors and to integrate environment and health monitoring and response in order to gather information and facilitate communication between authorities working at different levels. Table 4.1 show the life expectancy in selected European countries.

The European Union is made up of twenty-seven European countries. Incidentally England is not among them. Even so, England could not be excluded in this chapter. Figure 4.2 depicts life expectancy in many European countries as well as the per capita expenditure on health care. Sitting across the Atlantic is the behemoth, the United States, which spends 17.6 percent of its GDP on health care, while most European countries spend less than 10 percent of their GDP on health care (see Figure 4.3). The United States is included in this table and graph for purposes of comparison only. The

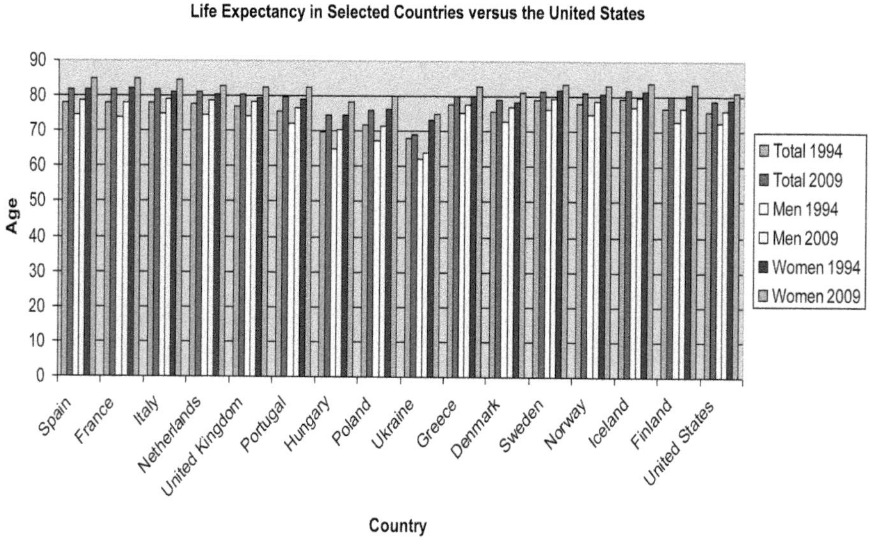

Figure 4.2 Life Expectancy in Selected European Countries and the United States
Source: European Commission (EuroStat), http://epp.eurostat.ec.europa.eu/statistics

health impacts of environmental factors in the United States have been treated elsewhere in this book. Life expectancy in England, France, Italy, Spain, and the Netherlands is higher than that of the United States, whose health care expenditure is higher than each of them. The question becomes, why? What explains the high and still-rising life expectancy in Europe and the stagnation in the United States? In the view of this author, only environmental factors can be the explanatory variable.

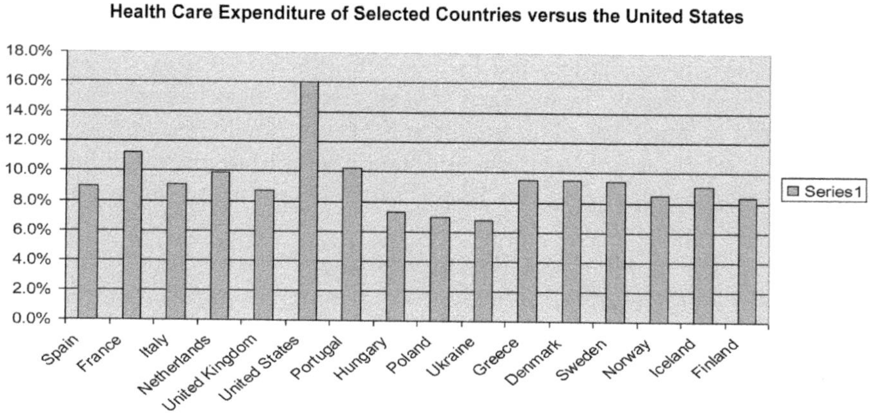

Figure 4.3 Healthcare GDP Spending—Western European Countries versus USA
Source: European Commission (EuroStat), http://epp.eurostat.ec.europa.eu/statistics_explained

In 2006, life expectancy at birth in the EU-27 countries was among the highest in the world almost seventy-six years for men and eighty-two years for women (Myers and Spoolman, 2014). The degradation of the environment, through air pollution, noise, chemicals, poor quality water, and loss of natural areas, combined with lifestyle changes, may be contributing to substantial increases in rates of obesity, diabetes, diseases of the cardiovascular and nervous systems, and cancer, all of which are becoming major public health problems for Europe's population. Reproductive and mental health problems are also on the rise. Asthma, allergies, and some types of cancer related to environmental pressures are of particular concern for children (Shrestha, 2006).

WHO estimates the environmental burden of disease in the pan-European region at between 15 and 20 percent of total deaths, and 18 to 20 percent of disability-adjusted life years. Preliminary results of a study conducted in Belgium, Finland, France, Germany, Italy, and the Netherlands, indicate that 6 to 12 percent of the total burden of disease could be attributed to nine selected environmental factors, out of which particulate matter, noise, radon, and environmental tobacco smoke were the leading factors (EEA, 2010).

The significant differences in the quality of the environment across Europe derived from urbanization, pollution, and natural resource use. Exposures and associated health risks, as well as the benefits of pollution reduction and of a natural environment, are not uniformly distributed within populations. Studies show that poor environmental conditions affect vulnerable groups especially and the elderly.

An understanding of differences in the social distribution of environmental quality will aid policy making, since specific population groups, such as those with low income, children, and the elderly, may be more vulnerable, mostly due to their health, economic, and educational status, access to health care, and lifestyle factors that affect their adaptation and coping mechanisms (EEA, 2010).

The various behavioral and social factors in these countries contribute to the change in life expectancy. Despite a substantial investment in health expenditures in the United States, limited access to care, the rate of uninsured populations, and risky lifestyle behaviors continue to reduce the span of life expectancy (Environmental Protection Agency, 2011). As of 2009, the rate of life expectancy for the United States was seventy-eight, as compared to other countries—France (eighty-one), Finland (eighty), and Japan (weighty-three) (CNN, 2009). Health economists are divided over whether the United States spends too much on health care. Researchers calculated that the seven years of additional life gained between 1960 and 2000 cost an average of $19,000 per year of life gained and concluded that the "increases in medical spending since 1960 have provided reasonable value" (Peterson & Burton, 2007, 87). It remains to be seen if the United States will adjust its total health expenditures, taking into account the current health outcome of increased life expectancy.

SUMMARY

The past fifteen years have seen tremendous growth in environmental health research and the application of the research findings to improve health outcomes. Among the factors central to this research are air and water quality; deforestation and reckless and irresponsible utilization of land and other natural resources; and conditions brought along by environmental disasters, such as the nuclear plant explosions and other natural and man-made disasters as in Italy, Poland, and the Ukraine.

Environmental health research uses a mix of person-focused and environment-focused models to study and analyze the factors that have been deterministic of environmental health promotion and education. That is, the air we breathe, the water we drink, and the food we eat, including our place of habitation, influence our health outcomes. Therefore, health promotion has moved from an emphasis on medical interventionism to prevention by the focus on individual behavioral change and social ecological applications (community, institutions, and the political systems and structures.

Intriguing to researchers is the lack of positive correlation between years of life expectancy and health expenditures (GDP). While most of the OECD, EU, and EEA countries spend less than 10 percent of their GDP on health care, their life expectancies are higher than those of the United States, which routinely spends more than 17 percent of its GDP on health care.

In this chapter we hypothesize that the only explanatory variables for this phenomenon are entirely environmental. A cursory review of the environmental policies of the selected European countries in this chapter will point to two conclusions: first, Europe is a land of environmental disasters, and second, the political leadership, with the assistance of international bodies (the United Nations and European regional organizations), recognized these problems early and took timely actions to ameliorate or reduce their impacts on their societies and economies. Some European countries have gone above and beyond the call of duty to offer environmental assistance to others, especially the developing countries, while at the same time ensuring that their environmental commitments at home are achieved through a mix of economic policy incentives, including taxation, and regulations.

REFERENCES

AgBioWorld. (2002). *Iowans who fed the world—Norman Borlaug: Geneticist.* Retrieved from http://www.agbioworld.org/newsletter_wm/index.php?caseid=archive&newsid=1400

Ajzen, I. (1985). From intentions to actions: A theory of planned behavior. In J. Kuhl & J. Beckman (Eds.), *Action-control: From cognition to behavior* (pp. 11–39). Heidelberg, Germany: Springer. Retrieved November 1, 2007, from Umass.edu Icek Ajzen Selected Publications.

Akers, R. L., Krohn, M. D., Lanza-Kaduce, L., & Radosevich, M. (1979), August). Social learning and deviant behavior: A specific test of a general theory. *American Sociological Review, 44*(4), 636–655.

Almeida, S., Casimiro, E., & Calheiros, J. (2010). Effects of apparent temperatures in daily mortality in Lisbon and Oporto, Portugal. *Environmental Health, 9*, 12.

Averous, C. (2009). *The OECD environment programme: Environmental performance of Iceland.* Paris: Author.

Bandura, A. (1977). Self-efficacy: Toward a unifying theory of behavioral change. *Psychological Review, 84*, 191–215.

Bandura, A. (1989). Social cognitive theory. In R. Vasta (Ed.), *Annals of child development: Vol. 6. Six theories of child development* (pp. 1–60). Greenwich, CT: JAI Press.

Baumer, D. and Van Horn, C. (2014). *Politics and Public Policy*. Washington DC: Congressional Press.

Caselli, I. (2010). Why Italy's trash wars are still smoldering. *Christian Science Monitor, 11*, 321–347.

Chen, M., & Land, K. C. (1986). Testing the Health Belief Model: LISREL Analysis of alternative models of causal relationships between health beliefs and preventive dental behavior. *Social Psychology Quarterly, 49*(1), 45–60.

CNN. (2009). The facts about worldwide healthcare spending. *CNN IReport.* Retrieved from http://ireport.cnn.com.

Crone, D. A., & Horner, R. H. (2003). *Building positive behavior support systems in schools: Functional behavioral assessment.* New York: Guildford.

Curtis, G. E. (Ed.). (1992). *Poland: A country study.* Washington DC: GPO for the Library of Congress.

De Decker, K. (2011). Medieval smokestacks: Fossil fuel in pre-industrial times. *Low-Tech Magazine.* http://www.resilience.org/stories/2011-09-29/

Environmental Protection Agency. (2011). *Environmental justice.* Retrieved from http://www.epa.gov/environmentaljustice/

European Environment Agency. (2010). Chapter 5: Environment, health, and quality of life. In *The European environment—state and outlook 2010: Synthesis.* Retrieved from http://www.eea.europa.eu/soer/synthesis/synthesis/chapter5.xhtml

European Environmental Agency. (2011). *Industrial air pollution cost Europe up to 169 billion in 2009, EEA reveals.* Copenhagen: EFA.

European Public Health Alliance. (2009). *Health in the European: Trends and analysis reports.* World Health Organization, Geneva, Switzerland.

European Union. (2013). *Public health: Improving health for all EU citizens.* European Commission. Brussels, Belgium: Directorate-General for Communication Publications 1049.

Eurostat: European Commission. (2011). *Air pollution statistics.* Retrieved from http://epp.eurostat.ec.europa.eu/statistics_explained/index.php/Air_pollution_statistics

Gasper, G., Emese, S., Tibor, S., Takas, T., Mathew, P., & Anton. A. (2009). Environmental impact of soil pollution with toxic elements from the lead and zinc mine at Gyongyosoroszi (Hungary). *Communications in Soil and Plant Analysis, 401*, 294–302.

Ginzburg, H. M., & Reis, E. (1991). Consequences of the nuclear power plant accident at Chernobyl. *Public Health Reports, 106*(1), 32–40.

Goliath, A. (2008). Environment: France avoids fines over water pollution in Brittany. *Europe Agriculture,15*, 97–123.

Gould, J. M., & Goldman, B. A. (1990). *Deadly deceit: Low-level radiation, high-level cover-up*. New York, NY: Four Walls Eight Windows.

Harrison J. D., & Stather, J. W. (1990). Values proposed by a nuclear energy agency expert group for radionuclides ingested by members of the public. Documents of the NRPB, 1(2), 1–238.

Healthy People 2020. (2011). *Social determinants of health*. Retrieved from http://healthypeople.gov/2020/TopicsObjectives2020/pdfs/HP2020_brochure_with_LHI_508.pdf

Herman, F. (2010). Effects of glucose ingestion rates on substrate utilization during moderate-intensity shriveling. *European Journal of Applied Physiology, 108*(2), 289–300.

Hirsch, D., Sutton, L., & Beckhelling, J. (2012). *The cost of a child in the twenty-first century*. London: Child Poverty Action Group.

International Atomic Energy Agency. (1991a). International Chernobyl project: An overview assessment of radiological consequences and evaluation of protective measures. Vienna, Austria: IAEA.

International Atomic Energy Agency. (1991b). International Chernobyl project: Technical report of radiological consequences and evaluation of protective measures. Vienna, Austria: IAEA.

Louwman, I. D. (1992). Solid waste problems and their handling in the Netherlands. *Water and Environment Journal, 6*(1), 112–117.

Mayan, O., Gomes, M., Henriques, A., Silva, S., & Begonha, A. (2006). Health survey among people living near an abandoned mine: A case study: Jales Mine, Portugal. *Environmental Monitoring and Assessment, 123*, 31–40.

Myers, N, and S. Spoolman. (2014). *Environmental Issues and Solutions*. Belmont, CA: Brooke/Cole.

National Center for the Environment and Sustainable Development. (2008). "The state of the environment." Greece National Centre for Environment And Sustainable Development. Kifissia, Athens, Greece, 1–57.

Netherlands—environment. (2010). *Encyclopedia of nations*. Retrieved from http://www.nationsencyclopedia.com/Europe/Netherlands-ENVIRONMENT.html

Nutbeam, D. (2000). Health literacy as a public health goal: A challenge for contemporary health education and communication strategies into the 21st century. *Health Promotion International, 15*(3), 259–267.

Ormrod, J.E. (1999). *Human Learning* (3rd ed.) Upper Saddle River, NJ: Prentice Hall Press.

Organization for Economic Cooperation and Development. (2004). *OECD review of Spain's environmental performance*. Spanish Ministry of the Environment in Madrid.

Organization for Economic Cooperation and Development. (2009). *OECD environmental performance reviews: Finland 2009*. Paris: Author.

Organization for Economic Cooperation and Development. (2009a). *OECD environmental performance reviews: Greece 2009*. Paris: Author.

Organization for Economic Cooperation and Development. (2012). *Government at a glance: Government performance indicators from selected sectors*. Paris: Author.

Page, W., & Bobylena, A. (1995). Environmental health policy in Ukraine after the Chernobyl accident. *Policies Study Journal, 23*(1), 141–151.

Pall, G., Paldy, A., Rudnai, P., Dura, G., & Varnai, D. (2007). Paediatric environmental health in Hungary. *Science Diet, 17*, 515–519.

Panunzio, P., & Landoli, C. (1976). *Legal aspects of aquaculture in Italy*. Istituto Centrale perla Ricerca Applicata al Mare (CRAM). Retrieved from http://ressources.ciheam.org/om/pdf/c43/99600257.pdf

Parkes, M., Panelli, R., & Weinstein, P. (2003). Converging paradigms for environmental health theory and practice. *Journal of Environmental Health Perspectives, 111*(5), 669–675.

Parkinson, M. (2003). *Environmental concerns become priority for Netherlands wood trade*. Netherlands: Department of Agriculture.

PBL Netherlands Environmental Assessment Agency. (2009). *No growth in total global CO_2 emissions in 2009*. Retrieved from http://www.pbl.nl/en/publications/2010/No-growth-in-total-global-CO_2-emissions-in-2009

Perkins, J. (2010). Green revolution. *The encyclopedia of Earth*. Retrieved from http://www.eoearth.org/article/Green_Revolution
Peterson, C. L. & Burton, R. (2007). *U.S. health care spending: Comparison with other OECD countries*. Federal Publications Paper 311.
Porter A. E., Taak P., Hobbs L. W. (2004). Bone bonding to hydroxyapatite and titanium surfaces on femoral stems retrieved from human subjects at autopsy, *Biomaterials*, *25*, 5199–5208.
Prochaska, A. (1977). The practice of radicalism: Educational reform in Westminster. In John Stevenson, ed., *London in the Age of Reform, 1750–1850*. London: Longman Press, 102–116.
Prochaska, J. O. (2003). Treating entire populations for behavior risks for chronic diseases. *Homeostasis in Health and Disease, 42*, 1–12.
Rosenstock, I., Strecker, V., & Becker, M. (1988). Social learning and the health belief model. *Health Education and Behavior, 5*(2), 175–183.
Santrock, J. W. (2008). *Educational psychology*. New York, NY: McGraw Hill.
Shrestha, L. B. (2006). *Life expectancy in the United States*. Washington DC: Government of the United States.
Simeonova, V., & Valk, A. (2010). The role of an area-oriented approach in achieving environmental policy integration in the Netherlands, and its applicability in Bulgaria. *European Planning Studies, 18*(9) 30–34.
Slovic, P., Fischoff, B., & Lichtenstein, S. (1979). Rating the risks. *Environment, 21*(3), 14–20, 36–39.
Skinner, B. F. (1953). *Science and human behavior*. New York: Free Press.
Stokols, D. (1995) "The Paradox of Environmental Psychology." *American Psychologist, 50*, 821–837.
Stokols, D. (1996). Translating social ecological theory into guidelines for community health promotion. *American Journal of Health Promotion, 10*, 282–298.
Sugai, G., & Horner, R. H. (2002). The evolution of discipline practices: School-wide positive behavior supports. *Child and Family Behavior Therapy, 24*, 23–50.
Tobin, T. J., & Sugai, G. (2005). Preventing problem behaviors: Primary, secondary, and tertiary level prevention interventions for young children. *Journal of Early and Intensive Behavior Intervention*, 2(3), 125–144
United Nations Children Fund. (2006). *Meeting the MDG drinking water and sanitation target: The urban and rural challenge of the decade*. Retrieved from http://www.who.int/water-sanitation_health/monitoring /jmpfinal.pdf
United Nations World Water Assessment Programme. (2006). *The world water development report 2: Water, a shared responsibility*. Paris: UNESCO.
United States Nuclear Regulatory Commission Information Digest. (2008–2009) Agency Publisher: Nuclear Regulatory Commission. Retrieved from http://bookstore.gpo.gov/products/sku/052-020-01027-7
Upton, A. (1992). Health effects of low-level ionizing radiation. *Physics Today, 44*(8), 34–39.
Valavinidis, A., & Vlachogianni, T. (2011). *The most important and urgent environmental problems in Greece in the last decade (2000–2010)*. Athens: University of Athens, Department of Chemistry.
Veropoulos, N. and Skuras, D. (2008). Managerial and Decision. *Economics, 29*(1), 37–55.
Vidal, J. (2010). UK air pollution causes 50,000 early deaths a year, say MPs. *The Guardian*. Retrieved from http://www.guardian.co.uk/environment/2010/mar/22/air-pollution-deaths
Whalen, R. P. (1978). *Love canal: Public health time bomb*. New York State Department of Health. Retrieved from http://www.health.state.ny.us/nysdoh/lcanal/lctimbmb.htm.

Wilkinson, R., & Marmot, M. (2003). *Social determinants of health: The solid facts* (2nd ed.). Copenhengen, Denmark: World Health Organization.
World Health Organization. (2003a). Social determinants of health. Copenhengen, Denmark: World Health Organization.
World Health Organization. (2006). *WHO air quality guidelines for particulate matter, ozone, nitrogen dioxide, and sulfur dioxide*. Retrieved from http://whqlibdoc.who.int/hq/2006/WHO_SDE_PHE_OEH_06.02_eng.pdf
World Health Organization. (2009). *Europe: Environment and health performance review: Poland*. WHO, Geveva, Switzerland.
World Health Organization. (2011a). The Commission on social determinants of health—what, why and how? Retrieved from http://www.who.int/socialdeterminants/thecommission/finalreport/about_csdh/en/. WHO.
World Health Organization. (2011b). *Health statistics and health information systems: Health system statistics*. Geneva, Switzerland.
World Health Organization. (2011c). *Social determinants of health*. Retrieved from http://www.who.int/social_determinants/en/.
World Health Organization Regional Office for Europe. (2010). *Environment and health risk: A review of the influence and effects of social inequities*. Copenhagen, Denmark: World Health Organization.
Zalidis, G. (1998). Management of river water for irrigation to mitigate soil salinization on a coastal wetland. *Journal of Environmental Management, 54*, 161–167.

5 Environmental Health Policies and Issues in Canada

Emmanuel A. Iyiegbuniwe, PhD and Robert Dibie, PhD

INTRODUCTION

In a little more than three years the world's environmental health movement has transformed several nations' public health and environment in many significant ways (WHO, 1999). Perhaps the most troubling of instances, however, is the pollution of Canada's air quality. Ambient concentrations of sulfur oxides, carbon monoxide, nitrogen oxides, and particulates—all associated with serious human health disorders—have been reduced significantly, but the nation has more work to do. As a result, climate change in Canada, the consequences of rapid population growth in some cities, the spread of toxic and hazardous chemicals, the loss of biological diversity, and air and water pollution all require diverse action by the national and provincial governments. Despite some improvement from what the nation used to be three decades ago, Canada, like several other countries in the world, continues to pollute its environment with toxic chemicals, pesticides, sewage contamination, and overfishing.

According to Boyd and Genius (2008), the presence of dozens of toxic chemicals in Canadian adults and children, including pesticides from dioxins flame retardants, and other persistent organic pollutants, has increased in the past three decades. The nation has also experienced many accidents in the handling of hazardous waste. Exposure to this preventable environmental hazard has also caused the increase of the morbidity and mortality of citizens in certain parts of Canada (Boyd and Genius, 2008). Hazardous waste has thus threatened human health in Canada, as well as the quality of life, and is detrimental to human activity in certain environments. According to Masuda and colleagues (2008), uneven exposures to hazards and the consequent health outcomes across the population and places in some Canadian provinces have caused many related diseases. These environmental health problems are congruent with notions of environmental health inequities or injustices brought about by health, economic, and environmental policies and practices that discriminate against particular communities or populations on the basis of race, class, gender, age, and culture (Taylor, 2012).

Environment Canada (2010a) reported that the country's outdoor contaminants come from natural and human pollution sources. According to the report, the most commonly measured outdoor air pollutants in Canada include ground-level ozone, particulate matter, carbon monoxide, sulphur dioxide, and nitrogen oxides. These substances tend to constitute the principal ingredients, or precursors, of smog, and some also contribute to acid rain in Canada (Environment Canada, 2010a; Hanchette, 2008). As a result, air pollution is primarily associated with everyday human activities such as industrial processes and residential activities. The effects of this outdoor contamination include increased medication use, increased doctor or emergency room visits, heightened hospital admissions, and even premature deaths.

The study of health economics is becoming more important and useful in Canada. The health care sector continues to grow in size, both in absolute dollars and as a portion of the country's overall economic activity. An Environment Canada report (2010a) indicates that climate change could affect the health and well-being of Canadians in a number of ways. Further, climate change has also caused increased smog and the number of heat waves, resulting in more temperature-related illnesses and deaths. The effects of climate change in Canada include the spread of infectious diseases such as malaria, dengue, and yellow fever; climate change has caused insects carrying these diseases to migrate northward from South America. The warming climate has also caused a decrease in the quality and quantity of drinking water in some parts of Canada as water sources become threatened by drought. In addition, the temperature increase has resulted in the spread of invasive forest pests, which are already having a dire impact on some of Canada's human population (Birdsell and Omelchuck, 2007; Environment Canada, 2010a).

The healthcare market in Canada contains many unusual economic features. However, in the absence of a comprehensive understanding of the societal structure and systems involved in the production of environmental health inequality in the country, the appropriate knowledge translation strategies and tools needed to compel public opinion leaders and government decision makers to take action is difficult to develop (Masuda et al., 2008). The unusual features include (1) the extent of government involvement; (2) the dominant presence of uncertainty at all levels of health care, ranging from the randomness of individuals' illnesses to the understanding of how well and for whom medical treatments work; (3) the large differences in knowledge between doctors (and other providers) and their patients, the consumers of health care; and (4) externalities—behaviors by individuals that impose costs or that negatively impact other Canadian citizens. These problems are present in other environmental health-related areas of the economy, but not as much as in health care, and never in such a broad combination. Therefore, it could be argued that there is a considerable shortfall in Canada's commitment to documenting the prevalence, mechanism, and impacts of environmental health inequalities in many communities, as well as a concomitant reluctance to make the policy and take the actions required to solve this problem (Masuda et al., 2008; Reed and Mitchell, 2003).

According to Draper and Mitchell (2001) attempts to theorize Canada's environmental health inequality and injustice have focused more on the notion of universal citizen rights, democratic principles, transparency, and accountability than on bringing attention to the specific consequences for particular disadvantaged groups in the country. Despite the environmental health problems previously indicated, some areas in Canada are still faced with atmospheric pollution, water pollution, climate change, ozone depletion, and solid waste management, as well as toxic, chemical, and hazardous waste management issues. The nation's national government has been vigorously pursuing the environmental health agenda by collaborating with other agencies at the city, provincial, and international levels, yet it has not been able to solve all the problems.

In Canada, many acutely dangerous ambient air toxins, especially formaldehyde and lead, have been reduced drastically or virtually eliminated. Dangerous chemical and biological pollutants in major waterways have been reduced sharply. Aggressive regulatory programs have reduced significantly the number of abandoned hazardous waste sites (Health Canada, 2008b). However, it is interesting to note that Canada is not totally free from environmental health problems. As a result, great care must be taken by the national government in setting out the extent to which both the federal and provincial governments must collaborate in order to ensure that the rights of the individual to good environmental health are not usurped by government action. By the same token, the expectations of Canadian citizens must not be raised too high for fear that the government may not be able to provide all the services that are expected to ensure a healthy environment. The citizens may have to demand that the industrial and manufacturing corporations perform their social responsibilities as well.

While the Government of Canada has been able to enact pragmatic environmental health policies, most of the cities and provinces are still lagging behind in taking action to protect their ecosystems and citizens' health. Several groups of people in Canada still live in areas where air quality is considered unhealthy (Draper and Mitchell, 2001; Health Canada, 2002; WHO, 1999). Water pollution remains an equally serious problem in some parts of Canada. Further, there is a growing recognition that the quality of the nation's domestic environment and that of the global environment have become interdependent (Kraft and Furlong, 2013). Most importantly politically, these transformations are grounded in a durable national consensus that environmental protection must now be a first-order public concern (Environment Canada, 2010a; Health Canada, 2002; Masuda et al., 2008).

This chapter discusses the management framework of environmental health issues in Canada. It explores the environmental health issues and policies that the nation has experienced in its social, economic, and environmental development over the last three decades. This development is mainly due to industrial development, technological advances, and a stable political environment. The country's social and economic transformation has resulted in both positive and negative environmental effects on the health and safety of its people. This chapter examines the impact of selected environmental health policies in Canada. It provides an extensive review of the Canadian Environmental Protection Act (CEPA), which emphasizes the use of precautionary principles. The chapter presents three case studies of environmental health and their implications for (1) children's environmental health, (2) bisphenol A, and (3) the tar sands of Alberta. The research reveals ongoing challenges to human health protection and environmental risk management. It argues that to improve and protect local environments, every community should be empowered by government, through the provision of an enabling environment and the necessary tools, to identify, analyze, and solve prevailing environmental problems. The chapter recommends the development and implementation of simple exposure and hazard assessment tools to set priorities for systematically identifying the highest-priority substances on the basis of their potential to cause adverse public health and environmental effects. It also advocates the use of participatory research to resolve ongoing environmental health problems in those local communities that have a deep distrust of Canadian government and industry experts.

THEORIES AND MODELS OF ENVIRONMENTAL HEALTH

Scientists often build process models as test beds for theories about their worldviews on specific issues. Bullock (1998) reported that having deep theoretical roots provides credibility to any field of study. In the field of environmental health and policy research, several well-known theories and principles regulate environmental behavior. The theories are environmental and human health relation theory, the health behavior lifestyle theory, the focus theory of normative conduct, the principal–agent theory, the value-belief-norm theory, and the self-determination theory. In addition, two principles—the precautionary principle and the polluter-pays principle—are also discussed.

Environment and Human Health Relation Theory

The environment and human health relations theory postulates that the environment of any nation and the health of the people who live in it are inextricably linked at both the proximate and extended timescales

(Mark-Davies and Mazumder, 2003). The theory argues that the link between workers' health and the environment manifests itself in their exposure to pollution agents. For example, the management and treatment of waste generated from human activities occur in industrialization, agriculture, logging, and urbanization. All levels of government and industries that employ workers bear the responsibility for setting policies to ensure the protection of water resources and for providing instruments to execute these health safety policies.

The environment and human health relation theory presents an argument that public policy is the driving force for dealing with occupational safety, health, and environmental issues (Dibie and Offiong, 2012). Workers' health and environmental health are strongly related. Both human and environmental health should be considered in the context of a temporal scale moving from the immediate to the long term (Mark-Davies and Mazumder, 2003). Therefore, it should be a priority for any country to establish a balance between both short- and long-term policies to address occupational safety and health (OSH) issues, as opposed to having a disproportionate amount of OSH policy aimed at solving short-term work-related health issues.

Paul Samson (2002) contends that, at the public policy level, there should be a strong desire to improve the understanding of the various relationships between human health and the environment, as well as to ensure that the data and information systems required to support workers' understanding are functioning well. Therefore, public policy managers, scientists, and information users must seek to further develop their capacity to understand, analyze, and report these issues.

It is very important to note that environmental degradation and negligence can lead to both acute and chronic human health problems. Long-term human health issues are mostly related to chemicals and physical agents, such as radon. Those of particular interest are disinfection by-products (DBPs) (Mark-Davies and Mazumder, 2003). Systematic hazard identification, education, training, and prevention are essential. It is interesting to note that in environmental law, the public trust theory is the principle that certain natural resources are preserved for public use and that the Government of Canada is required to maintain those resources for the public's reasonable use. Figure 5.1 shows the relationships among OSH policy, health, and the environment.

As a result of the Canadian government's mission to respect the public trust, its administrative units, including Environment Canada and Health Canada, must play a very important role to ensure a blossoming environmental health program. In other words, pursing a sustainability agenda for Canada is an ongoing process, based on certain values that are held in common and on behaviors that are undertaken by the public and professionals in the public service of the country.

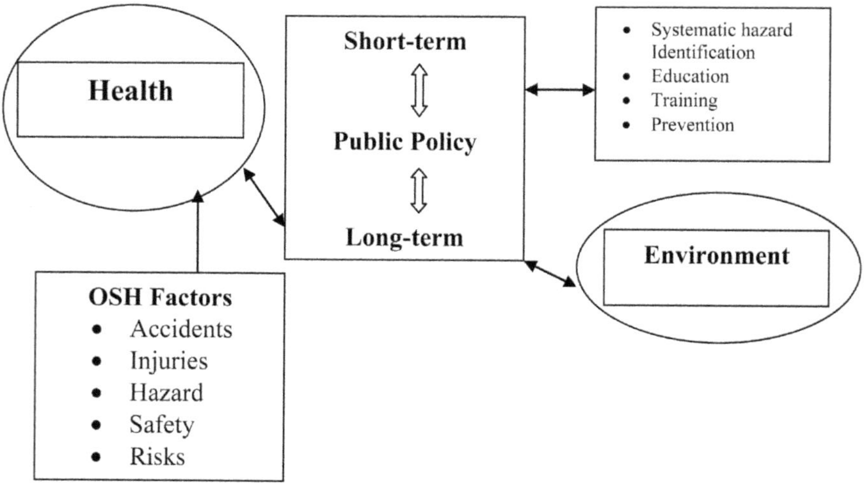

Figure 5.1 Model of the Relationship between OSH Policy, Health, and the Environment

Source: Adopted from Dibie and Offiong (2012).

Health Behavior Lifestyle Theory

The health behavior theory postulates that behavior can affect the well-being of individuals in direct and indirect ways. Novick, Morrow, and Mays (2008) contend that the behavioral effects of an individual must be considered when designing a health promotion program. Health promotion officials must consider both types of effects when developing an intervention, in order to provide for the potential indirect effects and also to include these variables in the program evaluation (Healey and Zimmerman, 2010). According to Mckenzie and colleagues (2005), very few individuals alter their behavior based on a single intervention. Rather, there is strong evidence that many different types of intervention are necessary to be successful in changing individual behaviors. Some scholars have also argued that interventions to change health behavior usually are more successful if they are employed frequently and consist of more than one activity (Mckenzie et al., 2005). Figure 5.2 shows the health behavior lifecycle model.

The theory presents an argument that people usually move through a series of steps as they attempt to change their behavior because one's behavior does not change rapidly. The first stage of behavior change is the precontemplation process because individuals do not plan to change. This is an important stage because individuals have to think things through before they take action. In this process, they also think about the various problems associated with their behavior. It is interesting to note that once they are aware of the health risks posed by their behavior, they try to avoid

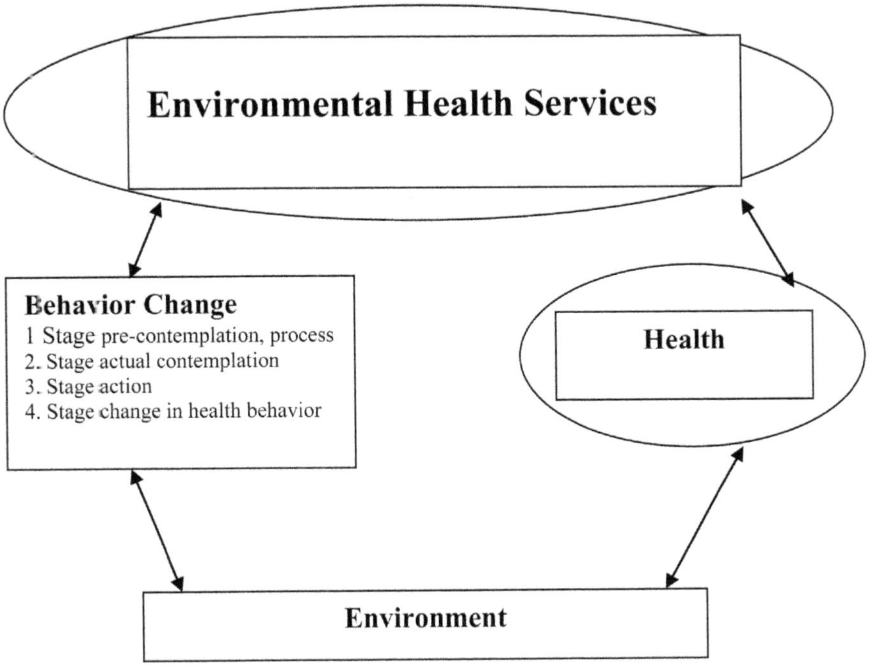

Figure 5.2 Health Behavior Lifestyle Model
Source: Adopted from Green and Kreeter (2005).

thinking or even hearing about the issue from others (Green and Kreeter, 1999; Healey and Zimmerman, 2010).

According to Mckenzie et al. (2005), the second stage of behavior change is the actual contemplation process. At this stage, the individual has become motivated to change his or her behavior but is not fully committed to making the required change. Following the second stage is the third, which involves the actual preparation to make the behavior change in the very near future. The third stage is classified as the action stage, and it is characterized by those who have changed their health behavior for less than six months. According to Anspaugh and colleagues (2000), the people at this stage of the change process encounter difficulty and often relapse into former bad behavior patterns. In most cases, people at this stage require social support, constant feedback, and reinforcement in order to maintain their previous success with behavior change. The final stage of the change in the health behavior process involves maintenance, and it is defined by individuals who have changed their health behavior for longer than six months. Health promotion officials or department of health members must understand that these individuals require assistance with coping with their new behavior, as well as with discovering alternatives to the previous bad environmental health behaviors. This theory is very important because it postulates that, for public polices to increase the chances of changing the high-risk environmental health behaviors of citizens, the government of nations around the world must pay attention to the behavioral patterns, or stages, of its citizens.

THEORIES OF ENVIRONMENTAL BEHAVIOR

The focus theory of normative conduct postulates the relationship between descriptive social norms and the likelihood of illegal antiecological behavior (Cialdini et al., 1990). The theory presents an argument that "norms do have a substantial impact on human action. However, the impact of social norms can only be properly recognized when researchers (a) separate 2 types of norms that at times act antagonistically in a situation—injunctive norms (what most others approve or disapprove) and descriptive norms (what most others do)—and (b) focus attention principally on the type of norm being studied." The theory also stresses that normative conduct has a demonstrable impact on action in a society, in a setting, and within a person. However, the impact of normative conduct is "differential depending on whether the actor is focused on norms of the culture, the situation, or the self." Studies have shown that depending on individual levels of performance relative to the norm, social norms could cause heterogeneous responses and a boomerang effect (Cialdini et al., 1990; Schwartz, 1994). Using energy conservation as an example, the boomerang effect occurs because "the moral benefit of conservation will increase for those agents, who are underperforming relative to the norm, and the opposite may occur for those who are outperforming the norm." (Cialdini et al., 1990). Subsequently, such individuals are less motivated to conserve and invariably increase their overall energy usage.

The value-belief-norm (VBN) and theory of environmentalism hypothesize that environmental attitude influences personal norms and that moral obligation is determined by beliefs in the person–environment relationship (Stern, 2000). The theory was originally founded on Schwartz's (1977) norm activation theory but has been applied to issues relating to modern environmentalism and energy conservation behavior (Ibtissem, 2010; Stern et al., 1999). It explains ecological behaviors, such as those related to ecological citizenship and political support, as well as behaviors that obtain in the private sphere (Stern et al., 1999; Stern, 2000). In applying this theory, Stern and colleagues (1999) concluded that the VBN offers the best available account for the environmental movement. The norm-activation theory was originally applied to explain social behavior but has been extended to the analysis of environmental behavior (Stern, 2000). It is based on the premise that actions occur in response to personal moral norms (Schwartz, 1977 and 1994). Stern and colleagues (1999) posit that actions are prompted in individuals who believe that environmental conditions do pose threats to public health and to the entire biosphere (awareness of consequences) and that taking action could avert potential consequences (ascription of responsibility to self).

The self-determination theory (SDT) was originally introduced more than twenty-seven years ago by two psychologists, Deci and Ryan (1985). The theory relies on the organismic theory, which is based on the assumptions of activity and structure. It recognizes that human beings actively attempt to master the forces in the environment as well as the forces of the drives and emotions in themselves. It further assumes that, in order to be effective and satisfy the full range of their needs, human beings, as active organisms, normally act on their internal and external environments (Deci and Ryan, 1985). To reflect ongoing experiences, human behavior is therefore influenced by continuously elaborated and refined internal structures. The defining characteristics of the organismic approach include internal structures with inherent principles of coherence or unity and a proactive engagement with one's environment. In the SDT, Deci and Ryan (1985) proposed that all humans have three basic needs: autonomy, competence, and relatedness. According to the authors, people often feel autonomous when they are making decisions for themselves without outside pressures, and this is viewed as operating with an internal locus of control. Conversely, people who operate with an external locus of control are not acting autonomously. Competence, on the other hand, requires that people feel as though they know what they are doing (i.e., they are capable in their pursuit). The third component of basic human needs, relatedness, requires that people feel connected to others, as though they belong to a particular group, and that the group cares about them as individuals (Ryan and Deci, 2000). It is important to note that when people make autonomous choices, they operate from an internal perceived locus of control and work toward self-determination. Deci and Ryan (2002) reported that two factors are responsible for ensuring that people reach their basic needs: their self-determination and whether not they are nurtured by social environmental factors.

Principal–Agent Model

The principal–agent theory is a formal approach that specifies how a principal (e.g., a manufacturing company) hiring an agent (e.g., equipment supply company) can design effective incentives under conditions of incomplete information about the agent's behavior (Eisenhardt, 1989; Rosness et al., 2012). The principals are the stakeholders who hire an agent, in this case the equipment company, to supply good-quality machines to enhance workers' safety and health. The principals pay the agent to provide not only good machinary but also sufficient maintenance for it (Dew-Becker, 2009). The principal–agent theory assumes that the principal and agent, while engaging in cooperative behavior, have different goals and different attitudes toward risks. It is important to note that from a legal point of view, an agency is the fiduciary relationship that arises when one person (a principal) manifests assent to another person (an agent) that the agent shall act on behalf of the principal and subject to the principal's control and when the agent manifests or otherwise consents to so act (Girasa and Ulinski, 2009). In this case, an agent is a person who acts consensually on behalf of another person known as the principal. It means that the agent must operate in the best interests of the principal. An agent has a fiduciary duty to the principal, which is more than the avoidance of carelessness. It means that the agent must operate in the best interests of the principal and avoid conflicts of interest, self-dealing, competing with the principal, accounting for moneys received and paid out, protecting confidential information, and other related responsibilities (Girasa and Ulinski, 2009; Rosness et al., 2012).

The idea is to devise a contract that changes the rules of the game so that the self-interested, rational choices of the agent coincide with the desires of the principal. For instance, the principal may invest in quality control systems to test the machines supplied by the agent or turn to outcome-based contractors to align the interests of the agents with those of the principal. Osmundsen and colleagues (2008) identified several possible challenges related to incorporating incentives into contracts between operators and contractors. If the payment rate is zero during downtimes, the principal may defer repair or maintenance work. Rewards attached to the avoidance of incidents, such as breakdowns, may promote the underreporting of the unwanted events. In addition, activities that produce easily measurable results may get disproportionately greater attention at the expense of important activities that are more difficult to monitor. Rosness and colleagues (2012) contend that the analysis of major accidents, such as the explosions at the Esso gas plant at Lonhford in 1998 and at the British Petroleum Company's Texas City refinery (2005), indicated that a rewards structure based on lost-time incidents may divert attention from personal safety and thus contribute to the occurrence of major-accidents (Dibie and Offiong, 2012). This suggests that organizations should identify and remove or mitigate the unwanted effects of current incentives systems. The principal–agent theory assumption may be threatened by the classic problem of agency capture, which occurs when an agency that regulates a single industry tends to become an advocate for the industry rather than an impartial protector of public health interests.

ENVIRONMENTAL PRINCIPLES AND CONTEXT

The foremost global cooperative approach to environmental protection was initiated during the 1972 United Nations conference on the Human Environment in Stockholm. The approach relies on the application of several important environmental principles and concepts. Environmental principles are used to resolve legal issues and other principles that courts and tribunals apply in interpretative processes. The environmental principles also serve as a basis for the negotiation and elaboration of future international environmental law (OECD, 1995). Over 900 agreements constitute the fundamental source of international environmental law with various provisions that cover one or more aspects of national, regional, subregional, and global environmental issues (Taylor, 2012). The Organization for Economic Co-operation and Development (OECD) was formed on December 14, 1960, by twenty countries that attended the Stockholm convention/conference. The Government of Canada is authorized, through CEPA, to "enter into intergovernmental environmental agreements." As one of thirty-four member countries of the OECD, Canada is entitled to apply research studies conducted in other countries, based on OECD-validated tests, as guidelines in making regulatory decisions on hazardous or toxic chemical substances (Environment Canada, 1999). CEPA (enacted in 1999) focuses on sustainable development, and the protection of the environment from pollution is rooted in two basic principles: the precautionary principle and the polluter-pays principle.

Precautionary Principle

The precautionary principle was first recognized as a policy tool in 1982 when the United Nations' General Assembly incorporated it into the World Charter for Nature. In January 1998, a diverse group of international scientists, philosophers, lawyers, and environmental activists convened at the Johnson Foundation Wingspread Conference Center in Racine, Wisconsin, USA, to discuss issues relating to this principle. The participants at the meeting, convened by the Science and Environmental Health Network, agreed on the following statement: "When an activity raises threats of harm to the environment or human health, precautionary measures should be taken even if some cause-and-effect relationships are not fully established scientifically" (SEHN, 1998). This definition is in accordance with Principle 15 of the 1992 Rio Declaration on the Precautionary Principle: "Where there are threats of serious or irreversible damage; lack of full scientific certainty shall not be used as a reason for postponing cost-effective measures to prevent environmental degradation" (OECD, 1995). Furthermore, as expressed in the Climate Change Convention (and the Rio Declaration), "The Parties should take precautionary measures to anticipate, prevent or minimize the causes of climate change and mitigate its adverse effects."

In its basic form, the precautionary principle demands that the protection of public health from potential harm from products introduced into commerce must be considered paramount, before any commercial interests. Manufacturers of new products intended to be introduced into the marketplace are considered as the juridical equivalent of guilty until proven innocent. The principle has been applied in international environmental law, particularly in the management of hazardous wastes, pollution prevention, global climate change, and protection of endangered species. Examples of international treaties where the principle has been applied are the 1992 Framework Convention on Climate Change, the 1992 Treaty on European Union (Maastricht Treaty), the 2000 Cartagena Protocol on Biosafety, and the 2001 Stockholm Convention on Persistent Organic Pollutants (POPs). For example, to ensure reaping global benefits at the lowest possible cost, regulatory authorities, when dealing with such environmental threats as climate change, must take into account adequate and cost-effective policies and measures. In 2002, the precautionary principle was widely adopted by the European Union as the centerpiece for their environmental health regulation and policy development for science and new products derived from technological innovations.

The precautionary principle also allows regulatory authorities such as Environment Canada and Health Canada to err on the side of safety when there is sufficient evidence to suggest a potential negative impact, but not enough to know with certainty. Although not completely rooted in the precautionary principle, the North American Free Trade Agreement (NAFTA) recognizes that when there is not sufficient information to conduct a risk assessment, the parties involved should act to protect human, animal, or plant life or health based on the available information. It must be recalled that NAFTA was signed in 1992 by Canada, Mexico, and the United States. Effective on January 1, 1994, NAFTA established a free-trade zone in North America by making immediate provisions that lifted tariffs on the majority of goods produced by the three signatory nations Canada, Mexico and the United States. Additionally, NAFTA required the gradual elimination of most of the remaining barriers to cross-border investment and to the movement of goods and services over a period of fifteen years.

Polluter Pays Principle

As an overarching policy goal and in the interest of policy integration, references to sustainable development have been included in various trade agreements such as NAFTA. Specifically, it was stated that one of the primary purposes of NAFTA was to:

> [c]ontribute to the harmonious development and expansion of world trade . . . in a manner consistent with environmental protection and conservation; . . . promote sustainable development; . . . [and] strengthen the development and enforcement of environmental laws and regulations.

The polluter pays principle, originally adopted in the early 1970s by the Organization for Economic Cooperation and Development (OECD) countries. The principle is based on the implementation of strict environmental regulations (OECD, 1995). The OECD provides a set of guidelines that a number of countries have adopted as statutory law, in whole or in part to protect individuals from the potential threat to public and environmental health. According to the "1972 OECD Guiding Principles on the International Economic Aspects of Environmental Policies," the Polluter Pays Principle requires that the polluter bear the expenses of carrying out pollution prevention measures or paying for damage caused by pollution (OECD, 1995). Specifically, the policy regulates toxic chemical substances, hazardous waste materials, international air and water pollution, as well as animate and inanimate products of biotechnology (Environment Canada, 1999).

The polluter pays principle is sometimes referred to as the nonsubsidization principle because it forbids governments from providing any form of subsidies for pollution control to industries. In this regard, it serves as a guide for the allocation of costs between the government and the private sector in paying for environmental pollution or protection (OECD, 1995). In March 1995, the OECD Working Group met in Ottawa, Canada, to adopt a proposal on Neurotoxicity Testing Strategies and Methods, taking into account comments from National Coordinators to the 1975 OECD and the Canadian Proposal for Neurotoxicity Testing Guidelines. An excerpt from the OECD report stated that:

> [t]he principle to be used for allocating costs of pollution prevention and control measures to encourage rational use of scarce environmental resources and to avoid distortions in international trade and investment is the so-called "Polluter Pays Principle." This principle means that the polluter should bear the expenses of carrying out the above mentioned measures decided by public authorities to ensure that the environment is in an acceptable state. In other words, the costs of these measures should be reflected in the cost of goods and services which cause pollution in production and consumption. Such measures should not be accompanied by subsidies that would create significant distortions in international trade and investment. (OECD, 1995)

The polluter-pays principle can be applied in agriculture. For example, 2, 4-dichlorophenoxy acetic acid was widely used as a herbicide in the Canadian prairies and is frequently detected as a contaminant in ground- and surface waters. The federal agency responsible for the regulation of pest control products is Health Canada's Pest Management Regulatory Agency (PMRA, 2003). Among other functions, PMRA enforces compliance with the legislation, develops pest management policies and guidelines, promotes sustainable pest management, improves the regulatory process to increase efficiency, and distributes pest management information to the general public and key stakeholders. It is expected to implement a mandatory adverse effect reporting system for all pesticides used in Canada (PMRA, 2003).

It must be noted that the concept of self-efficacy is also a very important piece of information for environmental health officials to consider. This is because, even though self-efficacy is not a theory, the process of change becomes easier to negotiate with a target audience who believe they are capable of taking the recommended action toward a desired health state, and the people in that audience are usually more successful in bringing about the behavior change toward the health of the natural environment. Successful environmental health promotion programs are usually the result of using a strong inclusive base to better understand human behavior while applying incentives to motivate the citizenry.

CANADA'S ENVIRONMENTAL POLICY

The first Canadian Environmental Protection Act (CEPA) was enacted as Bill C-74 on June 30, 1988 (Environment Canada, 2012). The 1988 CEPA replaced the Environmental Contaminants Act of 1975 and encompasses some of various existing pieces of environmental legislation including parts of the Clean Air Act, the Environmental Contaminants Act, the Canada Water Act, and the Ocean Dumping Control Act (CME, 2005; Environment Canada, 2012). CEPA also established a broad national framework for the protection of the Canadian environment. The legislation was developed as a result of a number of environmental incidents (e.g., the Love Canal chemical contamination near Niagara Falls, New York, and the St. Clair River "toxic blob" near Windsor, Ontario) that led to growing public awareness and concerns about the effect of toxic or hazardous substances in the environment (Environment Canada, 2012).

CEPA was a comprehensive environmental policy development that was enacted by the Government of Canada to manage the full life cycle of toxic substances from cradle to grave (Environment Canada, 2012). *Cradle to grave* implies that every stage in the life cycle of a toxic substance is tracked from the manufacturer or importer through its transportation, distribution, use, storage, and waste disposal. In addition, the Government of Canada is authorized to establish environmental quality objectives, guidelines, and codes of practice. It is also authorized to regulate the content of fuels, regulate the nutrient concentration in cleaning agents and water conditioners, and control ocean dumping through a permit system. CEPA empowers the Government of Canada to "regulate waste handling and disposal practices; improve, by means of guidelines and regulations, its own environmental performance and standards in relation to its operations and lands, including Indian reserves; and take action in cases of international air pollution" (Environment Canada, 2012).

CEPA was renewed and amended under Bill C-32 of 1999 as the Canadian Environmental Protection Act, 1999 (CEPA, 1999). Thereafter, CEPA 1999 became the principal federal environmental legislative tool for environmental health protection and general pollution prevention. Following a lengthy and extensive review of the original CEPA by Parliament, CEPA 1999 was finally enacted on March 31, 2000. The prevention and management of environmental risks posed by harmful substances remains the act's main mandate. Under CEPA 1999, the definition of *substance* encompasses products of biotechnology, water contaminants, anthropogenic chemicals, chemical spills, hazardous wastes, emissions (from combustion sources), and fuels. Additionally, a substance is defined as "CEPA-toxic" if the chemical contaminants can or may enter the environment in a quantity or under conditions that:

- Have or may have an immediate or long-term harmful effect on the environment or its biological diversity;
- Constitute or may constitute a danger to the environment on which life depends; or
- Constitute or may constitute a danger in Canada to human life or health. (CCME, 2005; Environment Canada, 2005)

The major intergovernmental forum for discussion and joint action on environmental issues of national concern lies within the jurisdiction of the Canadian Council of Ministers of the Environment (CCME, 2005). The fourteen-members of this intergovernmental forum work as partners in developing nationally consistent environmental standards and practices. Environment Canada, under the leadership of the Minister of the Environment, is responsible for administering CEPA. Pursuant to the 1990 Memorandum of Understanding concerning toxic substances and CEPA, signed by both Environment Canada and Health Canada, a CEPA Management Committee was established. Additionally, the overall administration of CEPA 1999 (authority and responsibility) lies within the jurisdiction of the federal Government of Canada, as is directed by section 2(1), paragraph (a). It defines the duty of the government to "exercise its powers in a manner that protects the environment and human health through the application of the precautionary principle" (Statistics Canada, 2006).

The day-to-day oversight of CEPA with regard to the regulatory process (the development and assessment of toxic substances, as well as general guidelines and codes of practice) is within the jurisdiction of both Environment Canada and Health Canada. Environment Canada was first created in 1971 with a mandate to protect the environment (through regulations and legislation), conserve the country's natural heritage, and provide weather and meteorological information to keep Canadians informed and safe (Environment Canada, 2012). Health Canada is responsible for conducting safety assessments of foods, drugs, and consumer products with a view to enabling Canadians to maintain and improve their health.

A National Advisory Committee was established as the main intergovernmental forum under CEPA 1999, section 6. This is with a view to "enabling national action and avoiding duplication in regulatory activity among governments." As a matter of priority, Environment Canada is concerned with the promotion of compliance with and enforcement of CEPA 1999 regulations. The renewed act contains many of the original important provisions and significant improvements for human health and environmental protection, including the following:

- Pollution prevention becomes the cornerstone of national efforts to reduce toxic substances in the environment;
- Processes are established to assess the risks posed by substances in commerce to human health and the environment;
- Timeframes are imposed for managing toxic substances;
- Adequate provisions are made for the use of a wide range of tools to manage toxic substances, other pollution and wastes;
- Ensures the most harmful substances are phased out or not released into the environment in any measurable quantity;
- Provisions are made to regulate vehicle, engine and equipment emissions;
- Strengthens enforcement of the Act and its regulations;
- Encourages greater citizen participation and input into decision-making; and
- Allows for more effective cooperation and partnership with other governments and Aboriginal peoples (Environment Canada, 2012).

The CEPA (1999) focuses on sustainable development and the protection of the environment from pollution. Sustainable development has been widely recognized as an emerging principle. In a publication titled *Our Common Future* (also known as the Brundtland Report) by the 1987 World Commission on Environment and Development (WHO, 2011), sustainable development was defined as "development that meets the needs of the present without compromising the ability of future generations to meet their own needs." The Brundtland Report set the stage for sustainability to be included in the global agenda and led to the first ever United Nations Conference on Environment and Development held in Rio de Janeiro, Brazil, in 1992 (also known as the Rio Declaration). A key achievement of the conference was the declaration that sustainable development is an essential part of efforts to deal with global environmental issues and to promote equitable economic development (World Commission on Environment and Development, 1987). Sustainability is based on the premise that natural resource exploitation should proceed in a way (and at a rate) that does not lead to any long-term decline of those resources and therefore guards against their future exhaustion. Environment Canada generally works collaboratively with international bodies, including the

United Nations Environment Programme (UNEP), OECD, government agencies, universities, and industry.

In 2004, the Government of Canada made a commitment to establish national indicators of air quality, freshwater quality, and greenhouse gas emissions (Statistics Canada, 2006). The overarching goal of establishing these new indicators was to provide Canadians with information regarding the state of their environment and how it is intricately linked with human activities. The development and communication of environmental management indicators are the joint responsibilities of the federal government (Environment Canada, Statistics Canada, and Health Canada) in collaboration with provincial and territorial governments. The first report from this joint effort was titled "Canadian Environmental Sustainability Indicators: Freshwater Quality Indicator: Data Sources and Methods." The report was released in 2005 as part of a series of documents published under the Canadian Environmental Sustainability Indicators (CESI) initiative (Statistics Canada, 2006).

The CESI report provided summary information and data on the status of water quality in Canada as it relates to the protection of aquatic life. The 2005 report employed the use of a water quality indicator (WQI) to provide the first national picture of Canadian water quality (CCME, 2005). Considering that the life of aquatic organisms is often influenced by the presence of numerous natural and anthropogenic water contaminants, the CESI's WQI is intended to provide a useful tool for translating water quality monitoring data into a simple overall rating (CCME, 2005). The Government of Canada anticipates that the indicators reported in this initial report "will be broadened in future reports to include other planned water uses and improvements in relation to monitoring, analysis, and surveys (CCME, 2005). As a result of their disproportionate size, volume, and the unique characteristics, the Great Lakes were treated separately in the freshwater quality indicator using data collected through Environment Canada's Great Lakes Surveillance Program (Statistics Canada, 2006). However, in a recent but related issue, Lisa P. Jackson, Administrator of the United States Environmental Protection Agency (EPA), and Peter Kent, Canadian Environment Minister, signed a "newly amended Great Lakes Water Quality Agreement" on September 7, 2012 (Environment Canada, 2012). Originally signed in 1972 and subsequently amended in 1987, the agreement represents a model of binational cooperation aimed to protect the health of the affected communities as well as the largest surface freshwater system in the world.

On March 31, 2000, Environment Canada launched its Environmental Registry Web site, as required under section 12 of CEPA 1999 (Environment Canada, 2012). The goal of the Environmental Registry is to provide all Canadians with an opportunity to have easier access to CEPA information relating to: regulations; notices and orders in the *Canada Gazette*; permits; guidelines; codes of practice; agreements; plans; policies; substances; and enforcement and compliance actions (Environment Canada, 2012). Environment Canada anticipates that by providing appropriate access to information, including judicial and administrative proceedings, the Registry will ensure public participation in environmental decision making and the overall transparency of the legislation. Although CEPA has been in existence for many years, several challenges remain with its implementation, including the categorization, prioritization, screening assessments, allocation of funds for research, and monitoring and/or surveillance for approximately 23,000 chemicals in commerce (Tyshenko et al., 2009).

SELECTED CANADIAN ENVIRONMENTAL HEALTH AND POLICY ISSUES

Published reports have shown that continuing global environmental degradation presents significant negative consequences for public health, economic productivity, and natural resources conservation (WHO, 1992). The first annual report on the state of public health in Canada by the Chief Public Health Officer David Butler-Jones provided a basic primer on personal and socioeconomic factors that impact health disparities and inequalities among Canadians (Wayne, 2008). Personal health factors include income, employment, environment, housing, education, nutrition, access to medical services, and health behaviors (e.g., smoking, alcohol drinking, drug abuse, and unsafe sex). Additionally, the report noted the need to submit government measures to appraisal for their health impact (Wayne, 2008). It also outlined a number of regional and international initiatives aimed at addressing the problems of health inequalities by referencing requirements in the provinces of British Columbia and Quebec. In the following sections, a critical analysis is undertaken of three important Canadian environmental health issues and their policy implications: children's environmental health, bisphenol A, and the tar sands of Alberta.

Children's Environmental Health

The environment has a tremendous influence on individuals' health, and children are no exception. In fact, the field of children's environmental health has emerged as an important area of research in recent years (Tyshenko et al., 2009; Wigle, 2003). Children are continuously affected by environmental threats, and the physical environment plays a very important role as a determinant of their health and well-being. In the interest of developing adequate public and environmental health policy, understanding the relationship between children's environment and their health is of utmost importance. According to the World Health Organization (2012), about 3 million children under five years of age die from environment-related diseases, and, of those children, acute respiratory infections are responsible for an estimated 1.6 million deaths each year (about 60 percent of acute respiratory infections worldwide are due to environmental conditions). Adverse environmental conditions or factors have a significant impact on children's health and disease development, and it is estimated that approximately 13 million deaths could be prevented every year by addressing such environmental problems facing children worldwide (WHO, 2012). The environmental problems include the lack of reliable and accessible safe drinking water, poor sanitation facilities, and environmental pollution (WHO, 2012). Recent estimates revealed that about 25 percent of the diseases faced by humans occur due to prolonged exposure to environmental pollution and that many of these environmental conditions often lead to or exacerbate childhood morbidity and mortality (WHO, 2012).

According to a 2006 report on *Children's Health and the Environment in North America*, considering a number of indicators, the overall health of Canadian children was reported as "very positive" by the Government of Canada. This report was the result of the June 2002 agreement by the Council of the North American Commission for Environmental Cooperation's (CEC) Cooperative Agenda to protect children from environmental risks. The CEC was made up of the environment ministers of Canada, Mexico, and the United States (Government of Canada, 2006). In the Cooperative Agenda, the three countries committed to selecting and publishing a core set of indicators of children's health and the environment for North America. This commitment was reaffirmed in the CEC Council Session in June 2003, with the adoption of Council Resolution 03–10 (Government of Canada, 2006). According to the Government of Canada's report (2006), several health indicators have improved dramatically over the past twenty years, including:

- Increased life expectancy at birth;
- Decreased perinatal, neonatal and infant mortality rates;
- Improved immunization rate for children (considered one of the best in the world); and
- Declined number of children born to teenage mothers.

Unfortunately, the report noted that Canadian children continue to face several environmental health risks. To this day, birth defects remain the single leading cause of infant deaths in Canada. For both boys and girls, unintentional injuries are the leading cause of death after the first year of life. In children one to four years of age, childhood cancer is reported as one of the top three causes of death, and the incidence rates for several types of cancer continue to increase among young Canada adults. It is believed that cancers in young adults may be associated with childhood exposures to environmental hazards. Respiratory diseases, perinatal conditions, digestive diseases, and injuries are the leading causes of infant hospitalization (i.e., children from one to four years of age).

Canada has approximately 8 million children nineteen years of age and under. The Government of Canada has made a commitment to improve the reporting of indicators of children's health and the environment, noting that a good set of indicators would allow for the translation of large data sets of complex scientific information into measures that can be understood by the general population (Government of Canada, 2006). According to Statistics Canada (2006), children below twenty years of age represent nearly a quarter, or 25 percent, of the entire population, while children under the age of four years represent 5.4 percent. Approximately 79.7 percent of the Canadian population live in urban areas; hence a greater proportion of these children are urban dwellers (Statistics Canada, 2006).

Canadian children are exposed to various harmful levels of indoor air pollutants (second-hand smoke and other contaminants) at home and in public buildings. Air pollution from heavy traffic, substandard housing, and exposure to indoor second-hand smoke constitute major environmental health risks for children from poor Canadian families. The Government of Canada (2006) estimates that in 2001, 15.6 percent of children lived in families with income levels below the low income cut-off. It is estimated that about 19 percent of Canadian children zero to seventeen years of age were regularly exposed to second-hand smoke in the home environment in 2002 (Government of Canada, 2006). However, the proportion of these children has declined in recent years.

Numerous epidemiological studies have demonstrated that children are particularly vulnerable and sensitive to air pollution. These studies have associated children's exposure to various harmful levels of ambient air pollutants with asthmatic symptoms including increased coughing and wheezing, increased use of airway medications, and increased number of hospital visits. The Government of Canada had previously published a report that presented the trends in emissions of arsenic, benzene, cadmium, chromium, dioxins and furans, hexachlorobenzene, and mercury—seven key pollutants of importance to children's health (Government of Canada, 2006). Studies have shown a continuous decline during the last decade in urban areas of several ambient air pollutant concentrations of public health importance. However, of major concern are the ever increasing levels of ground-level ozone and fine particulate matter ($PM_{2.5}$) in some regions, including the metropolitan areas of southern Ontario. The 2006 report on children's health has noted that, in view of these results, efforts aimed at "developing and portraying meaningful national measures of children's exposure to air pollution remain a challenge in Canada" (Government of Canada, 2006).

In Canada, asthma has become an epidemic, and its prevalence nationwide has increased fourfold during the past twenty years. The global burden of asthma continues to present severe challenges, and with over 235 million sufferers worldwide, the World Health Organization considers asthma to be a serious public health problem (WHO, 1999). Asthma is a "chronic inflammatory disease of the airways" characterized by wheezing, shortness of breath, tightness in the chest, and coughing. It is a lifelong disease that can limit a person's quality of life. It is estimated that over 3 million Canadians aged twelve and over (8.5 percent of the entire population) and more than one in ten children (13 percent of Canadian children) have been diagnosed with asthma (Statistics Canada, 2010). Asthma continues to be a major cause of hospitalization of children in Canada. With the exception of Canadians under twelve, twelve to nineteen, and over seventy-five years of age, females were more likely to be diagnosed with asthma than males. In general, boys were more likely to be diagnosed with asthma than girls. However, in 2010, the rate of asthma among Canadians twelve years and older was 9.8 percent for females compared with 7.1 percent for males. A number of environmental triggers or risk factors have been associated with the exacerbation of asthma, including poor indoor and outdoor air quality. Table 5.1 shows the percentage diagnosed with asthma, by sex, of household population aged twelve or older, Canada, 2001 to 2010. Table 5.1 shows the percentage diagnosed with asthma in Canada between 2001 and 2010.

Other risk factors include a family history of asthma and/or allergy (e.g., eczema, allergic rhinitis, etc.), exposure to high levels of antigen (e.g., house dust mites, cockroaches, pollen, animal dander, etc.), and exposure to tobacco smoke or other chemical irritants (Government of Canada, 2006; Statistics Canada, 2010; WHO, 1999).

The prevalence of childhood asthma has declined to its lowest level in more than a decade among Canadian children aged two to seven. Experts attribute the decline in prevalence of asthma to the reduction in exposure to cigarette smoke at home (Statistics Canada, 2010; Thomas, 2010). Other factors responsible for the recent declines in childhood ear infections and asthma are changes in the population structure, changes in diagnostic practices, the declining prevalence of respiratory allergies, improvements in air quality, and improved hygienic practices. However, among children aged two to three, the prevalence of upper respiratory infections remained

Table 5.1 Percentage Diagnosed with Asthma in Canada, 2001–2010

Year	Male (%)	Female (%)
2001	6.5	8.9
2003	6.5	8.7
2005	6.4	8.6
2007	6.2	8.7
2008	6.6	8.7
2009	6.2	8.7
2010	6.5	8.9

Source: Statistics Canada (2010a); Canadian Community Health Survey 2001–2010.

constant or declined only slightly in most regions, except in Quebec, where the prevalence has risen significantly. According to Statistics Canada, the percentage of children aged two to seven diagnosed with asthma rose from 11 percent in 1994–1995 to 13 percent in 2000–2001. However, by 2008–2009, it declined to 10 percent, according to new data released by the National Longitudinal Survey on Children and Youth (Statistics Canada, 2010). Table 5.2 shows the Canadian Federal Legislation policies for protecting children's environmental health.

According to previous studies, there are significant regional variations in the prevalence of childhood asthma, with British Columbia and the Prairie provinces showing lower rates than other provinces or regions. Data reported between 2006/2007 and 2008/2009, showed no significant regional differences in the prevalence of asthma (Statistics Canada, 2010; Thomas, 2010). The federal, provincial, and territorial governments have established several programs to prevent or reduce the morbidity and mortality rates in children due to environmental risk factors. Table 5.2 summarizes the

Table 5.2 Federal Legislation Relevant for Protecting Children's Environmental Health

Legislation	Children's Environmental Health Issue	Provisions Relevant to Children
CEPA, 1999	Exposure to chemicals	States the use of precaution in the absence of evidence establishing harm.
Canadian Environmental Assessment Act, SC 1992, c. 37	Degraded environmental quality from human activity	Regulations for potential environmental impact of various types and categories of human activity. Refers to the protection of the environmental and human health under the Act. The CEAA indirectly addresses children's environmental health.
Canada Water Act, R.S.C. 1985, c. C-11	Water quality and management	Oversees the development and implementation of water management initiatives with regulation to ensure the optimum use of those resources for the benefit of all Canadians; however, public health and children's environment health directly addressed in the statute.
Feeds Act, R.S.C. 1985, c. F-9	Harmful exposures to feeds	Regulations for the manufacture, importation, and sale of animal feeds while setting out a general prohibition against animals feeds that potentially may adversely affect animal or human health; children directly addressed.

Fertilizer Act, R.S.C. 1985, c. F-10	Harmful exposures to fertilizers	Fertilizer Regulations (Section 11) provides that they shall not contain detrimental substances injurious to vegetation. Regulation of fertilizers and supplements address health considerations in several contexts
Food and Drug Act, R.S.C. 1985	Harmful exposures to food drugs or various related products	Drugs, cosmetics, and devices are defined in the legislation, in certain cases with direct references to children's health and safety. Some regulatory details also address children's issues in relation to product rather than environmental safety, such as child-resistant packaging.
Hazardous Product Act, R.S.C. 1985, c, H-3b	Harmful products	Certain products may be prohibited if poisonous, toxic, flammable, explosive, corrosive, infectious, oxidizing, or reactive; other elements may be designated as restricted by circumstances from Part 1 of the Act.
Pest Control Products Act of 2006		The new Act improves the pesticide regulatory system by increasing transparency and strengthening health and environmental protection and postregistration control of pesticides. Pesticides will continue to require approval before they can be imported into, manufactured, sold, or used in Canada. These activities are key in protecting human health and the environment.
Radiation Emitting Device Act, R.S.C. 1985, c. 34 (1st Supp.)	EMF field exposures (e.g., from cell phone use)	Regulation is intended for the purpose of protecting persons. A radiation-emitting device may not be sold, imported, or leased if it fails to comply with prescribed standards.
Tobacco Act, 1997, c. 13a	Smoking and second-hand smoke exposure	Targets second-hand or environmental tobacco smoke. The sale of tobacco products to persons under 18 years is prohibited.

[a] Explicitly references children.
[b] Recently proposed amendments to the act include references to children.

Source: Tyshenko et al. (2009).

applicable federal legislation examined for provisions relevant to protecting children's environmental health in Canada (Tyshenko et al., 2009). For example, in its new Pest Control Products Act, Health Canada has codified special considerations of children and vulnerable other populations and conducts child-protective health risk assessments based on foods and their anticipated residues that children consume (Health Canada, 2008). The risk assessment process accounts for the unique food consumption patterns of infants and children, including breast milk, formula, and fruit juice. To ensure the protection of this vulnerable subpopulation, additional safety factors are to be incorporated when conducting risk assessments for pesticide residues in foods consumed by infants and children. The 2006 report by the Government of Canada on protecting children from environmental risks surmised that research, legislation, environmental hazards reduction programs, outreach, and the education of parents and caregivers, as well as better information to track the environmental threats to children's health, are needed to adequately protect Canadian children from environmental threats in the future.

The Case of Bisphenol (A)

Toxic chemicals are ubiquitous in our environment, and their potential exposures and effects on human health and safety, as well as their impact in the environment, continue to present serious concerns and challenges to environmental health and policy experts. Comprehensive data on toxic chemicals in the environment have only recently become available and only a few hundred of commercially available chemicals are actively regulated for unreasonable risks (Wagener, 1994). Bisphenol A (BPA), a synthetic chemical monomer widely used in the manufacture of plastics (epoxy-phenolic resins and polycarbonate), is one such toxic chemical. Epoxy resins from BPA are used worldwide in the production of food and beverage cans, mainly as interior protective liners, and polycarbonate is used to produce food containers, including milk, water, and infant bottles. A number of research studies have shown that at very low doses, exposure to BPA was associated with adverse human health effects, such as breast and prostate cancer, obesity, neurobehavioral problems, and reproductive abnormalities (Calafat et al., 2008; Health Canada, 2008a; Richter et al., 2007; Vandenberg et al., 2007; Vogel, 2009).

In 2008, the Bureau of Chemical Safety in the Food Directorate of the Health Products and Food Branch (HPFB) at Health Canada published a report summarizing their health risk assessment for BPA exposure from dietary sources (Health Canada, 2008a). Other research studies have published quantitative risk assessment results of specific chemical exposures and outcomes relating to the environment–health relationship (Baker and Nieuwenhuijsen, 2008; Merrill, 2007). Yet other studies in Canada have shown that both the physical and natural environments play key roles in understanding risk because they relate to the social context of risk (Lee et al., 2005; Lemyre et al., 2006). Previous studies have shown that small quantities of BPA can potentially leach out from bottles or containers and contaminate the water or food in them (Calafat et al., 2008; Health Canada, 2008b; Vandenberg et al., 2007; Vogel, 2009). Thus, BPA has a huge public health implication because plastic food containers may be potential sources of BPA exposure to children. The responsibility for monitoring domestic and imported foods falls within the jurisdiction of the Canadian Food Inspection Agency (CFIA). CFIA also carries out enforcement actions to prevent the sale of food containing excessive pesticide residues and enforces the specific maximum residue limits for all Canadian foods, including domestic and imported food products (Health Canada, 2008a).

Based on reported health effects, both Health Canada and Environment Canada included BPA as one of 200 chemicals in Batch 2 of the Challenge to Industry Chemicals Management Plan (Lougheed, 2009). Chemical substances included in the Chemical Management Plan can potentially cause significant public and environmental health problems, therefore manufacturers are challenged to develop adequate plans for handling and managing the affected chemical substances (Lougheed, 2009). Additionally, CEPA 1999 provisions call for the application of scientific information for conducting health risk assessments, the incorporation of "weight of evidence," and the application of the "precautionary principles."

The health risk assessment focused on potential risks posed by BPA exposure from epoxy-based products used in packaging infant formulas. Health Canada reasoned that since canned formulas are the sole source of nutrition for newborns/infants who consume these products, the risk of BPA exposure would be highest in this age group. Small amounts of BPA may also be present in the liners of metal closures for bottles or jars used to package solid infant formulas or beverages. The use of other consumer products (e.g., polycarbonate) in baby bottles and tableware was considered as an additional source of potential BPA exposure but not included in the risk assessment of prepackaged foods (Health Canada, 2008b).

In October 2007, Health Canada conducted a number of studies to measure the concentrations of BPA in canned liquid infant formulas obtained from the Canadian market. A total of twenty-one samples of various brands of canned liquid infant formula products for infants in different age groups were included in the BPA analysis (Cao et al., 2008). The results showed that BPA was detected in all twenty-one products with concentrations ranging from 2.27 to 10.2 ppb (average = 5.2 ppb). The results obtained for BPA in these studies were consistent with those previously reported by the U.S. Food and Drug Administration (Biles et al., 1997; Health Canada, 2008).

In reviewing the overall weight of evidence for BPA exposure and health risk assessment, the Food Directorate of Health Canada concluded that "the current dietary exposure to BPA through food packaging uses is not expected to pose a health risk to the general population, including newborns and young children" (Health Canada, 2008b). Furthermore, their conclusion was critical of previously published results from experimental animal studies reviewed during the assessment (Health Canada, 2008a). A number of questions were raised concerning several uncertainties related to study methodologies and the actual significance of the results and findings with respect to the potential risk to human health. The critiques noted in the Health Canada's health risk assessment report were that certain studies involved only single doses of BPA, behavioral assessments occurred at a single point in time, there was lack of a dose–response relationship, the animals used in each test group were limited in numbers, and the overall findings lacked consistency. The final report published by the Bureau of Chemical Safety, Food Directorate, Health Products and Food Branch of Health Canada recommended that the general principle of ALARA (as low as reasonably achievable) be followed in order to continuously limit BPA exposure to newborns and infants resulting from food packaging applications (Health Canada, 2008a).

In 2008, BPA continued to make headline news in major Canadian national newspapers. Toronto's *Globe and Mail* of April 15, 2008, had the caption "Canada First to Label Bisphenol A as Officially Dangerous" (Mittelstaedt, 2008). Based on the application of the precautionary principle, Canadian regulatory authorities classified BPA as a toxic substance in April 2008 (Health Canada, 2008b). Health Canada subsequently had a press release titled "Government of Canada Protects Families with Bisphenol A Regulations." Contrary to the position taken by the Canadian government, both the European Food Safety Authority and the U.S. Food and Drug Administration (FDA) maintained that BPA was "safe at estimated levels of human exposure" (Vogel, 2009). However, retailers were already removing plastic water and baby bottles made with BPA from the shelves in 2008. In early 2009, a bill was introduced in the U.S. Congress calling for the ban of BPA in children's food containers (Vogel, 2009).

The Canadian government's decision to declare BPA a toxic substance was followed by a September 2008 National Toxicology Program's Center for the Evaluation of Risks to Human Reproduction (CERHR) publication. The monograph concluded that human exposure to BPA presented possible reproductive and developmental effects (NTP–CERHR, 2008). The selection of BPA for evaluation by CERHR was based on its widespread human exposure, public concern for possible health effects from human exposures, its high production volume, and published evidence of reproductive and developmental toxicity in laboratory animal studies (Environment Canada, 2010b). Thereafter, the Canadian government classified BPA as a toxic substance and subsequently prohibited its use in any baby bottles imported or sold in Canada (Health Canada, 2008b).

BPA continued to be newsworthy, and on October 16, 2010, the Government of Canada proposed a new regulatory instrument under its

Pollution Prevention Planning Notice to address industrial releases of BPA in effluents. The announcement by Honorable Jim Prentice, the Canadian Environment Minister, was made days after the Government of Canada announced that BPA was added to Schedule 1 of CEPA 1999. The Minister noted that:

> The proposed rules will require facilities to develop and implement plans to limit releases of BPA to the environment and to submit ongoing progress reports to the Government of Canada. (Environment Canada, 2010b)

The federal government actions are targeted to the management of public health and environmental risks posed by BPA. It must be noted that following the NTP-CERHR report, the government prohibited the advertisement, importation, and sale of baby bottles containing polycarbonate plastic materials. Regulatory efforts are ongoing, and the federal government continues to collaborate with its provincial and municipal counterparts to address any potential releases of BPA during the disposal or recycling of polycarbonate and epoxy phenolic resin-based products (Environment Canada, 2010b). The first annual report on the state of public health in Canada recommended that the following five key areas be addressed to reduce disparities and inequalities in health (Wayne, 2008).

1. Social investments, particularly investments in families with children living in poverty and in early child development programs;
2. Community capacity through direction involvement in solutions, enhanced cross-sectorial co-operation, better defined stakeholder roles and increased measuring of outcomes;
3. Inter-sectorial action through integrated, coherent policies and joint actions among parties within and outside of the formal health sector at all levels;
4. Infrastructure through a better understanding of subpopulations, the pathways through which socio-economic factors interact to create health inequalities, how best practices from other jurisdictions can be adapted to improve Canadian efforts and through more advanced measurement of the outcomes of the various interventions undertaken; and
5. Leadership at the public health, community and regional health levels.

Tar Sands of Alberta

Tar sands (or oil sands) occur in the Canadian western province of Alberta and cover an estimated area of 57,000 square miles (comparable to the size of the state of Florida in the United States). The Athabascan tar sands deposits in northern Alberta are known to occur closer to the surface of the soil and are forty times larger than the Arctic National Wildlife Refuge in the United States. Tar sands are generally composed of shale and sand particles and consist of a heavy and viscous mixture of long chain hydrocarbons called bitumen. The World Energy Council has estimated that oil shales, tar sands, and other unconventional deposits typically have over ten times as much oil as liquid petroleum reserves (Cunningham and Cunningham, 2010). Canada's northern Alberta deposits constitute one of the world's largest and most accessible tar sands and are estimated to contain about 1.7 trillion barrels of synthetic crude oil. However, only 10 percent of this oil is currently economically recoverable (Rom, 2012). Almost all of the 172 billion barrels of oil reserves in Alberta occur within tar sands. The development of northern Alberta's tar sands has expanding during the last decade, and the Alberta Provincial Government estimates that by 2018 oil production will have increased from about 1.3 million barrels per day to 3 million barrels per day. It must be noted that in 2000 Canada surpassed Saudi Arabia as the largest oil supplier to the United States (Cunningham and Cunningham, 2010).

The process of mining and upgrading for tar sands involves the extraction of fuel from bitumen, a mixture of petroleum and sand or clay. Processing for tar sands involves surface mining to extract oil from sands and bitumen deposits that are typically 75 m deep. To harness the oil, hot water in large tanks is used to separate and siphon off the bitumen from the tar sands. To produce a barrel of synthetic crude oil, about 4,500 pounds of tar sands must be dug up in open pit mines, and this process is repeated several times with fresh soil-containing tar sands. The recovered bitumen must be fractionated and cleaned to make useful products. This occurs through the process of cracking, or refining, by subjecting it to a very high heat of above 900°F (Rom, 2012). The by-products are sand, water, fine clays, and minerals contained in tailing ponds. The surface mining of tar sands has huge economic value as a great source of synthetic crude oil. The production process uses vast quantities of water, and the resulting hot water is stored in trailing ponds which are often contaminated with mercury and other toxic substances. In general, tailing ponds constitute serious hazards and present risks of toxicity for migrating birds, lake trout, and pike. Runoffs from these toxic trailing ponds are increasingly seen as environmental threats that may contaminate the Athabasca River and Lake Athabasca. The extent of the degradation has proven to be very controversial.

Severe environmental and public health costs are associated with the extraction of tar sands. These costs have become important rallying points and vexing issues for several territorial governments, including the Native Cree, Chipewyan, and Metis people (Cunningham and Cunningham, 2010). Available information in the literature shows that surface mining for tar sands causes significant habitat alteration, affects wildlife, contaminates surface and groundwater, and destroys millions of hectares of boreal forests (Cunningham and Cunningham, 2010; Rom, 2012). In recent years, this has also become a rallying focus among environmental health experts and various stakeholders because the continuous extraction and refining of tar sands has a real impact on people and extensive implications for the surrounding environment. Estimates provided by the Natural Resources Defense Council showed that in the next thirty to fifty years, 6 million to 166 million birds would be lost due to disruption of the boreal forest and contamination from tailing ponds (Rom, 2012). Environmentalists have also raised policy concerns that investing billions of dollars in tar sands constitutes a continued dependence on nonrenewable fossil fuels for energy. Furthermore, many Canadians are opposed to what is presumed as "becoming an energy colony for the United States." It is estimated that generating 125,000 billion barrels of oil per day from tar sands releases 15 million m^3 of toxic sludge and 5,000 tons of greenhouse gases (Cunningham and Cunningham, 2010). Compared to conventional crude oil methods, this accounts for about 1,030 percent more greenhouse gases (Rom, 2012).

On October 5, 2009, an expert panel was appointed by the Royal Society of Canada (RSC) to assess the scientific evidence related to public health hazards and environmental issues resulting from oil production from tar sands (Weinhold, 2011). In a special report issued on December 15, 2010, the panel cited "substantial evidence that efforts to extract oil from the Alberta deposits have degraded air, land, and water quality to varying degrees." Some of the major health and environmental concerns outlined in the RSC report included (Weinhold, 2011):

- The rate of land disturbance leading to reclamation liability and its financial risks;
- The health hazards from tar sands exposure seemingly publicized more than they should be;
- Air quality based on the heavy amount of greenhouse gas emissions;
- Water quality and groundwater quantity; and
- Cumulative impact on ecosystem based on rising rate and expansion of the tar sands industry.

The RSC panel noted that there were differing interpretations of water quality data and that the establishment of a causal relationship with tar sands extraction was difficult. Furthermore, the panel concluded that there appeared to be "no significant human health threats to the general population either now or from development anticipated in the next decade" based on publicly available evidence (Weinhold, 2011). However, the panel cautioned that there were "major gaps in health and environmental data, risk assessments, government oversight, information transparency, industry efforts, and disaster preparedness." In response to the tar sands environmental health controversy, the panel made several recommendations noting that, although technologies have cut down the impact of environmental hazard to the ecosystem as well as the environment, better technologies are under development to reduce land disturbance and emissions (Weinhold, 2011). The panel also recommended the implementation of environmental practices with benefits targeted at lower-cost needs.

In another published study, surface mining operations for tar sands in northern Alberta were linked to a wide range of heavy and toxic metal (e.g., mercury, arsenic and lead) contamination of the Athabasca River and its watershed (Kelly et al., 2010). The team of researchers was led by David Schindler, an ecologist at the University of Alberta in Edmonton, Canada. The research, which was published in the Proceedings of the National Academy of Sciences, showed that "13 elements classified as priority pollutants (PPEs) by the U.S. Environmental Protection Agency were found in the Athabasca River in the province of Alberta." The authors noted that seven of the thirteen PPEs were present in concentrations that were high enough to pose significant risk to aquatic organisms. Sources of airborne emissions were bitumen upgraders and local oil sands development (Kelly et al., 2010). The results of their study contradicted widely held government and oil industry claims that natural sources (not tar sands development) accounted for the concentrations of elements found in the Athabasca River and its tributaries.

Surface water samples were obtained from the waterways upstream of the tar sands development region and compared with samples taken upstream and downstream of mining projects within the region. Additionally, snowpack samples obtained toward the end of winter from many of the same areas as the water samples were analyzed for airborne sources of PPEs. Their findings revealed higher concentrations of PPEs in the Athabasca River downstream of tailing ponds. Other tar sands deposits were found in areas downstream of watersheds stripped of soil and vegetation in preparation for mining, when compared to sites upstream of tar sands mining projects. The concentrations of seven PPEs in water and snow samples exceeded both the federal and provincial guidelines aquatic life protection. Higher concentrations of many toxic metals were found in the snowpack samples, and higher concentrations of polycyclic aromatic compounds were reported downstream of the mining activities and tar sands development. Caution must be exercised when interpreting these results because of measurement constraints and the inherent limitations of the study.

Established in 1997 to assess the state of rivers and lakes, the Regional Aquatics Monitoring Program (RAMP) is the main pollution-monitoring body in the tar sands region. The RAMP steering committee, which is funded by the oil industry, draws representatives from both federal and provincial governments, First Nations communities, and industry. The Alberta government needed to address the issues as a priority because the environmental monitoring carried out by the Cumulative Environmental Impact Assessment (CEMA) was not readily available and lacked uniformity. Six strategies were identified to improve the current conditions, including:

- Create a friendly-healthy environment for family and individuals.
- Concentrate on environmental friendliness when developing the oil sands.
- Sustain the economic growth, stability and resource optimization.
- Work towards reconciling interests with Aboriginal consultation.
- Focus on innovation and effective research to keep up the economic development.
- Use any resources out there for example, data information, measurement systems to manage the oil sands in an appropriate manner (Weingold, 2011).

On the preventive side, various areas needed to be addressed, such as the establishment of an effective integration of national and international policies to improve current environmental impacts. Some recommendations suggested by the RSC panel targeted comparing Canadian practices with international practices on health, social, economic and sustainable development components. The panel noted that any evaluation carried out must be transparent, open to the public, and updated regularly. Similarly, cumulative assessment to analyze and interpret regional data should be implemented. Further research studies on long-term effects, particularly on reclamation and groundwater, and on interactions between groundwater and surface water. With adequate funding, the research should also focus on water quality management strategies and on the monitoring of health impacts (health determinants and outcomes or end points). It is suggested that the use of participatory research be considered as a tool for resolving environmental health problems, specifically environmental contamination from the extraction of tar sands in the affected communities in Northern Alberta.

POLICY RECOMMENDATIONS AND APPROPRIATE FUTURE STRATEGIES

The health, social, and economic well-being of Canadians is fundamentally linked to the quality of their environment (Statistics Canada, 2006). There is no gainsaying the fact that governments often employ formidable power to promote their objectives for achieving sustainable environmental health policy goals. Historically, the relationships between the environment and human health have always been of keen interest in environmental health policy (ATSDR, 2002; Health Canada, 2002; Maiburg 2010). Environmental health policy encompasses a subset of policies that address the relationship between public health and the environment (Maiburg, 2010). The two core objectives of environmental health policy are environmental health (measurement, evaluation, and control of environmental stressors to human health) and ecosystem vitality (measurement and assessment of ecosystem health and natural resource management). The federal, provincial, and territorial governments in Canada have incorporated environmental attributes in their public health policies during the past several decades.

Protecting human health from environmental hazards is a much needed investment. It saves lives, prevents illness, and promotes a greener economy. Successful examples from around the globe provide leadership and inspiration for Canada. Priorities for progress in Canada should include recognizing the right of all citizens to live in a healthy environment, a right that ideally should be expressed in the appropriate policy. Developing and implementing a comprehensive pan-Canadian health and environment strategy, as well as investing in environmental health research and monitoring, are paramount. In addition, strengthening environmental laws and national standards governing air quality, drinking water, food safety, and chemicals to meet or beat the standards found in other wealthy industrialized countries should be a sustainable development goal (Maxwell, 2013).

Further, Canada's environmental health policy making should recognize a legacy of past failure and success at environmental restoration, mixed with a substantial measure of ambiguity and uncertainty about the implications for the future. To minimize the impact of environmental health disasters, accurate information and its regular delivery are important in all settings of decision making because the lives of the nation's citizens are at stake. Therefore, environmental health protection should be a work in progress in Canada.

One of the main steps that the Canadian government should adopt is the building of a more integrated research agenda on environmental health inequality in the country. This could help the nation to characterize and map out the current state of its research landscape in relation to the nation's epistemological patterns, both temporal and geographic.

The Government of Canada must realize that environmental health issues affect not just the disadvantaged and poor. Environmental health services, such as the inspection of restaurants, water quality maintenance, school immunizations, bioterrorism, and community mobilization, protect everyone regardless of social class. As a result, it is paramount that the instruments for improving the environmental health of citizens should be approached as a shared responsibility. The shared responsibility approach calls for all stakeholders to be involved in the process of ensuring good environmental health. Thus, health officials, environmentalists, health care providers, major leaders in the communities, and citizens should all be encouraged to participate and to contribute to improving the environmental health mission. It is very interesting to note that protecting the environmental health of all citizens requires the expertise of a variety of actors in the community: the private sector, churches, homeowners association, civil society organizations, and the nonprofit and business sectors, as well as the Government of Canada. These stakeholders are all needed to contribute to the well-being of individuals and populations in the country.

The process of seeking to address the fundamental social and behavioral causes of illness caused by environmental exposure requires broad thinking that goes beyond traditional disease categories, focal points of interventions, and disciplinary lines. Social, behavioral, and life scientists must collaborate to engage a biopsychosocial model. Interventions are likely to be more successful when applied in coordinated fashion across multiple levels of influences on the health and the environment in the Canada. The nation needs to embrace an attitude of collective solutions to environmental health problems and collective action to address them. It will have to join forces across disciplines.

The need for improvements in health care professionals' environmental health knowledge has been expressed by leading health institutions. The Institute of Medicine recommends the integration of environmental health concepts into all levels of medical and nursing education (Healey and Zimmerman, 2010; Shweta et al., 2012). Furthermore, the "Health Professionals and Environmental Health Education Position Statement" (Rogers, 2004), which calls on health care providers to increase their knowledge of environmental health issues and recommends the creation of faculty champions as an effective strategy for integration of environmental health knowledge into health care professionals' education and practice, is very crucial for Canada.

It is recommended that the Government of Canada should adopt appropriate indicators to track temperature: maximum temperature, minimum temperature, and apparent temperature. Apparent temperature, or the use of a heat index that combines humidity and temperature, is important in looking at mortality effects, and humidity may even become significant in drier areas of the nation because air mass moisture characteristics play a larger role during regional heat wave events (Boyd, 2012; Kraft and Furlong, 2013; Shweta et al., 2012). While we recognize that economic factors such as income inequality and poverty may also have significant impact on health inequality in Canada, it could be argued that if the government focuses on reducing poverty and tries to narrow income inequalities, the health behavior of poor citizens could improve tremendously in the country.

Further, the federal, provincial, and local governments of Canada should pay greater attention to ethical issues in public health, occupational health, environmental regulation, urban planning, and other disciplines and areas that address environmental factors in health. It is also recommended that the respective governments in Canada's federal system should increase their funding of research on environmental factors in human health. The government should also consider implementing cost-effective environmental interventions that could reduce inequalities in health, pest control, childhood education, literacy programs, healthy urban planning, disaster preparedness, and food and drug regulation (Dibie and Offiong, 2012; Resnik and Roman, 2007).

The optimist perspective, however, is that many adverse environmental exposures in Canada could be prevented through behavioral modification, strong public policy, and technological change. As evidence about environmental health and disease etiology continues to accumulate in Canada, the ongoing assessment of environmental diseases should provide a foundation for strategic improvements in environmental legislation, health policy, and public health program.

SUMMARY

This chapter has examined environmental health and related policy issues in Canada. It discussed how environmental health issues in Canada overlap in many regards. For many environmental hazards, the most serious cause of concern is the risk to human health and ecosystem. Several policy issues in the country are instrumental to both environmental and health problems. These include the extent to which individual health risks are appreciated, the extent that individuals value health risk, and how society is protected from environmental harm due to pollution or exposure to infectious diseases. The federal, provincial, and territorial governments of Canada and the various agencies (e.g., Environment Canada, Health Canada, etc.) play a critical role in developing environmental policy.

Canadian environmental regulation is grounded in the formation of CEPA in 1999. The main mandate of CEPA is the promulgation and enforcement of regulations, as well as the prevention and management of environmental risks posed by harmful substances. CEPA also focuses on sustainable development and environmental protection based on two principles: the precautionary principle and the polluter-pays principle. The precautionary principle demands that public health protection from environmental risks and products introduced into commerce must be considered before any commercial interests. Canadian children are particularly vulnerable to a variety of environmental threats, including air pollution. Asthma continues to present serious public health challenges. Based on the application of the precautionary principle, the federal government has taken far-reaching actions targeting the management of public health and environmental risks posed by BPA, a known toxic substance. The tar sands of Alberta have great economic benefits, but the extraction continues to present severe environmental public health costs, especially to the local communities and territorial governments. It is suggested that participatory research with affected communities be considered as a tool for resolving environmental health problems, specifically those communities where tar sands are extracted in Northern Alberta and where there is a deep distrust of government and industry experts.

In Canada, provincial governments have developed a substantial range of policies, regulations, strategies, and frameworks to enhance public health and safety. However, there may be significant variation in priorities and approaches to environmental health management needs and the specific circumstances of each jurisdiction. As part of its strategies on risk management and in partnership with Health Canada, Environment Canada organized a national forum through the Toxic Substances Research Initiative to support research on endocrine-disrupting substances (EDS) (Environment Canada, 1999). It involved the formation of the EDS Working Group (Endocrine Disruptor Workshop: Identifying Research Needs and Priorities). The workshop was attended by scientists and major regulatory groups within Environment Canada on October 23–24, 1997, in Niagara Falls. It created a strategy comprised of four major elements: national leadership/communication on EDS; national and international harmonization of screening and testing protocols; establishing the effects of EDS in ecosystems; and assessment/action on priority substances.

In setting the goal of any environmental policy on air and water quality issues, the issue of sustainability (maximizing available supplies and minimizing associated risks) must be considered a vital component. To be effective, environmental policy development must rely on sound science, adequate and reliable instrumentation, a careful consideration of vulnerable populations, and the participation of affected stakeholders in the regulatory process. To be involved in the improvement and protection of their local environment, every community should be empowered by government through the provision of an enabling environment and the tools necessary to identify, analyze, and solve prevailing environmental problems. There is also the need to develop and implement simple exposure and hazard tools for systematically identifying substances that have the highest priorities in terms of their potential to cause adverse health effects and environmental impacts. It is only when environmental policies are properly crafted, directed, and implemented, and all stakeholders are involved in the process that public and environmental health protection can be effectively achieved in Canada.

REFERENCES

Anspaugh, D. T., Dignan, M. B., and Anspaugh, S. L. (2000). *Developing Health Promotion Programs*. Long Grove, IL: Waveland.

ATSDR. (2002). "Healthy People in a Healthy Environment. Agency for Toxic Substances and Disease Registry." *Hazardous Substances and Public Health*, 12(2), 2–16.

Baker, D., and Nieuwenhuijsen, M.J. (Eds.). (2008). *Environmental Epidemiology*. Oxford: Oxford University Press.

Biles, J. E., McNeal, T. P., and Begley, T. H. (1997). "Determination of Bisphenol as Migrating from Epoxy Can Coatings to Infant Formula Liquid Concentrates." *Journal of Agriculture and Food Chemistry*, 45, 4697–4700.

Birdsell, J. M., and Omelchuck, K. (2007). *Building Capacity for Health Research Transfer in Western Canada: An Environmental Scan*. Edmonton: Alberta Heritage Foundation for Medical Research.

Boyd, D. R. (2012). "Environmental Hazards and Human Health." *Canadian Dimension, 46*(4), 231–256.

Boyd, D. R., and Genuis, S. (2008). "The Environmental Burden of Disease in Canada: Respiratory Disease, Cardiovascular Disease, Cancer, and Congenital Affliction." *Environmental Research 106*, 240–249.

Bullock, C. (1998). "The Leisure Ability Model: Implications for the Researcher." *Therapeutic Recreation Journal, 32*(2), 97–102.

Buzzell, M., and Jerrett, M. (2004). "Racial Gradients on Ambient Air Pollution Exposure in Hamilton, Canada." *Environment and Planning A, 36*, 1855.

Calafat, A. M., Ye, X., Wong, L. Y., Reidy, J. A., and Needham, L. L. (2008). "Exposure of the US Population to Bisphenol A and 4-tertiary-octylphenol: 2003–2004." *Environmental Health Perspectives, 116*(1), 39–44.

Cao W., Yacoub S., Shiverick K. T., Namiki K., Sakai Y., Porvasnik S., Urbanek C., Rosser C. J. (2008). "Dichloroacetate (DCA) Sensitizes Both Wild-Type and Overexpressing Bcl-2 Prostate Cancer Cells In Vitro to Radiation. *Prostate, 68*, 1223–1231.

CME. (2005). "Canadian Water Quality Guidelines for the Protection of Agricultural Water Uses: Summary Table (Canadian Council for Ministers of the Environment)." Retrieved from ceqg-rcqe.ccme.ca. Accessed September 28, 2012.

Cialdini, R. B., Reno, R. R., and Kalgren, C. A. (1990). "A Focus Theory of Normative Conduct: Recycling the Concept of Norms to Reduce Littering in Public Places." *Journal of Personality and Social Psychology, 58*, 1015–1026.

Cunningham, W. P., and Cunningham, M. A. (2010). *Environmental Science: A Global Concern* (11th ed.). New York: McGraw-Hill.

Deci, E., and Ryan, R. (Eds.). (2002). *Handbook of Self-determination Research*. Rochester, NY: University of Rochester Press.

Deci, E. L., and Ryan, R. M. (2000). "The 'What' and 'Why' of Goal Pursuits: Human Needs and Self-Determination of Behavior." *Psychological Inquiry, 11*(4), 227–268.

Deci, E. L., and Ryan, R. M. (Eds.). (1985). *Intrinsic Motivation and Self-Determination in Human Behavior*. New York, NY: Plenum.

Dew-Becker, I. (2009). "How Much Sunlight Does It Take to Disinfect a Boardroom? A Short History of Executive Compensation Regulation in America." *Economic Studies, 55*(3–4), 434–457.

Dibie, R., and Offiong J. (2012). "The Nature and Impact of Environmental Policy in Nigeria." *Journal of International Politics and Development, 10*(1–2), 113–144.

Draper, D., and Mitchell B. (2001). "Environmental Justice Considerations in Canada." *The Canadian Geographer, 45*, 93–98.

Eisenhardt, K. M. (1989). "Agency Theory: An Assessment and Review." *Academy of Management Review, 14*(1), 57–74.

Environment Canada. (2013). "Government of Canada Invests in a Community Environmental Project in Bruce County, Ontario." Retrieved from www.ec.gc.ca. Accessed March 11, 2013.

Environment Canada (2012). "Government of Canada's Response to Fall 2012 Report of the Commissioner of the Environment and Sustainable Development." Retrieved from www.ec.gc.ca. Accessed November 25, 2012.

Environment Canada. (2010a). "Minister Baird Salutes the International Year of Biodiversity 2010." Retrieved from www.ec.gc.ca. Accessed October 17, 2012.

Environment Canada. (2010b). "News Release: Government of Canada Takes Further Action on Bisphenol A." Retrieved from http://www.ec.gc.ca/default.asp. Accessed September 26, 2012.

Environment Canada. (1999). *Endocrine Disrupting Substances in the Environment*. Fact Sheet. Retrieved from www.ec.gc.ca/eds/fact/index.htm. Accessed March 10, 2013.

Frazier, L. M., Berberich, N. J., Moser, R. Jr., Cromer, J. W. Jr., Hitchcock, M. A., Monteiro, F. M., et al. (1999). "Developing Occupational and Environmental Medicine Curricula for Primary Care Residents: Project EPOCH Envi. Educating Physicians in Occupational Health and the Environment." *Journal of Occupational Environmental Medicine*, *41*(8), 706–711.

Girasa, R., and Ulinski, M. (2009). "Comparative Aspects of Executive Compensation: An Agency Perpective." *The Business Review*, *1*(2), 60–69.

Government of Canada. (2008). Federal Sustainable Development Act (S.C. 2008, c. 33). Retrieved from http://laws-lois.justice.gc.ca/eng/acts/F-8.6/. Accessed October 2012.

Government of Canada. (2006). "Children's Health and the Environment in North America: A First Report on Available Indicators and Measures. Country Report: Canada." Retrieved from www.cec.org. Accessed March 9, 2013.

Government of Canada. (2004). *Canadian Environmental Sustainability Indicators: Air Quality Indicators—Data Sources and Methods*. Retrieved from http://www.statcan.gc.ca/pub/16-254-x/16-254-x2008000-eng.htm. Accessed November 5, 2012.

Green, L. W., and Kreeter, M. W. (2005). *Health Promotion Planning: An Educational and Ecological Approach* (4th ed.). Mountain View, CA: Mayfield Publishing Company.

Hanchette. C. L. (2008). "The Political Ecology of Lead Poisoning in Eastern North Carolina." *Health and Place*, *14*, 209–216.

Healey, B., and Zimmerman, R. Jr. (2010). *The New World of Health Promotion*. Boston: Jones & Bartlett.

Health Canada. (2008a). *Health Risk Assessment of Bisphenol A from Food Packaging Applications*. Retrieved from www.hc-sc.gc.ca/fn-an/alt. Accessed September 25, 2012.

Health Canada. (2008b). "Press Release: Government of Canada Protects Families with Bisphenol A Regulations." October 17. Retrieved from http://chemicalsubstanceschimiques.gc.ca/challenge-defi/batch-lot-2/bisphenol-a/index-eng.php. Accessed October 15, 2012.

Health Canada. (2002). "Health and the Environment: Critical Pathways." *Health Policy Research Bulletin*, *4*, 3–8.

Homer-Dixon, T. (1999). *Environment, Scarcity and Violence*, Princeton, NJ: Princeton University Press.

Ibtissem, M. H. (2010). "Application of Value Beliefs Norms Theory to the Energy Conservation Behavior." *Journal of Sustainable Development*, *3*(2). 129–139.

Iyiegbuniwe, E. (2012). "Analysis of Environmental Health Policy in Belize." *Journal of International Politics and Development*, *10*(1–2), 87.

Johnson, J., and Stoskopt, C. (2010). *Comparative Health System*. Burlington, MA: Jones & Bartlett.

Kelly, E. N., Schindler, D. W., Hodson, P. V., Short, J. W., Radmanovich, R., and Nielsen, C. C. (2010). "Oil Sands Development Contributes Elements Toxic at Low Concentrations to the Athabasca River and Its Tributaries." *Proceedings of the National Academy of Science*, *107*(37), 16178–16183.

Kraft, M., and S. Furlong. (2013). *Public Policy*. Washington DC: Congressional Press.

Lee, J., et al. (2005). "Beyond The Hazard: The Role of Beliefs in Health Risk Perception." *Human and Ecological Risk Assessment*, *11*(6), 1111–1126.

Lemyre, L., et al. (2006). "The Structure of Canadians' Health Risk Perceptions: Environmental, Therapeutic and Social Health Risks." *Health, Risk and Society*, *8*(2), 185–195.

Lougheed, T. (2009). "Outside Looking in: Understanding the Role of Science in Regulation." *Environmental Health Perspectives*, *117*(3), A104–A110.

Maiburg, S. A. (2010). "Environmental Health Policy." In H. Frumkin, *Environmental Health: From Global to Local.* San Francisco: Jossey-Bass/Wiley.
Mark-Davies, J., and Mazumder, A. (2003). "Health and Environmental Policy Issues in Canada: The Role of Watershed Management in Sustaining Clean Drinking Water Quality at Surface Sources." *Journal of Environmental Management, 68,* 273–286.
Masuda, J., Zappancic, T., Poland, B., and Cole, D. (2008). "Environmental Health and Vulnerable Populations in Canada: Mapping an Integrated Equity-Focused Research Agenda." *The Canadian Geography, 52*(4), 427–450.
Maxwell, N. (2013). *Understanding Environmental Health.* Sudbury, MA: Jones and Bartlett.
Mckenzie, J. E., Neiger, B. L., and Smeltzer, J. L. (2005). *Planning Implementing and Evaluating Health Promotion Programs: A Primer* (4th ed.). San Francisco: Pearson Benjamin Cummings.
Merrill, R. M. (2007). *Environmental Epidemiology.* Sudbury, MA: Jones & Bartlett.
Mittelstaedt, M. (2008). "Canada First to Label Bisphenol A as Officially Dangerous." *Toronto Globe and Mail.* April 15. Retrieved from www.theglobeandmail.com. Accessed April 15, 2008.
Novick, L. F., Morrow, C. B., & Mays, G. P. (2008). *Public Health Administration: Principles for Population-based Management.* Sudbury, MA: Jones and Bartlett.
NTP-CERHR Monograph on the Potential Human Reproductive and Developmental Effects of Bisphenol A. (2008). National Toxicology Program, U. S. Department of Health and Human Services. Retrieved from http://cerhr.niehs.nih.gov/evals/bisphenol/bisphenol.pdf. Accessed June 17, 2012.
OECD. (1995). *Environmental Principles and Concepts.* Paris: Organization for Economic Co-operation and Development, OCDE/GD(95)124. Retrieved from www.oecd.org/trade/environmentandtrade. Accessed July 25, 2012.
Osmundsen, P., Aven, T., and Vinnem, J. E. (2008). "Safety, Economic Incentives and Insurance in the Norwegian Petroleum Industry." *Reliability Engineering and System Safety, 93,* 137–143.
PMRA. (2003). "Assessing Exposure from Pesticides in Food: A User's Guide." Science Policy Notice SPN2003–03, Pest Management Regulatory Agency, July 28.
Pope, A. M., and Rall, D. P. (Eds.). (1995). *Environmental Medicine: Integrating a Missing Element into Medical Education.* Washington, DC: Institute of Medicine, National Academy Press.
Pope, A. M., Snyder, M. A., and Mood, L. H. (Eds.). (1995). *Nursing, Health, and the Environment.* Washington, DC: Institute of Medicine, National Academy Press.
Reed, M. G., and Mitchell, B. (2003). "Gendering Environmental Geography." *The Canadian Geography, 1,* 318–337
Resnik, D., and Roman, G. (2007). "Health, Justice and the Environment." *Bioethics, 21*(4), 230–241.
Richter, C. A., Birnbaum, L. S,, Farabollini, F., et al. (2007). "In Vivo Effects of Bisphenol A in Laboratory Rodent Studies." *Reproductive Toxicology, 24*(2), 199–224.
Roberts, J. R., and Gitterman, B. A. (2003). "Pediatric Environmental Health Education: A Survey of U.S. Pediatric Residency Programs." *Ambulatory Pediatrics, 3*(1), 57.
Rogers, B. (2004). "Environmental Health Hazards and Health Care Professional Education." *American Association of Occupational Health Nurses Journal (AAOHNJ), 52,* 154–155.
Rom, W. (2012). *Environmental Policy and Public Health: Air Pollution, Global Climate Change, and Wilderness.* San Francisco: Jossey-Bass/Wiley.
Rosness, R., Blakstad, H. C., Forseth, U., Dahle, I. B., and Wiig, S. (2012). "Environmental Conditions for Safety—Theoretical Foundations." *Safety Science, 50,* 1967–1976.

Samson, P. R. (2002). "Developing Environmental Public Health Indicators: A View from Canada." *Statistical Journal of the United Nations ECE*, *19*, 105–116.

Schwartz, S. H. (1977). "Normative Influences on Altruism." In L. Berkowitz (Ed.), *Advances in Experimental Social Psychology*, Vol. 10 (pp. 221–279). New York: Academic Press.

Schwartz, S. H. (1994). "Are There Universal Aspects in the Structure and Contents of Human Values?" *Journal of Social Issues*, *50*(4), 19–46.

Shweta, P., Low, M. D., Franzini, L., and Swint, J. M. (2012). "A Review of Canadian Policy on Social Determinants of Health." *Review of European Studies*, *4*(4), 1918–7181.

Statistics Canada. (2010). "Canadian Environmental Sustainability Indicators: Freshwater Quality Indicator: Data Sources and Methods. Retrieved from www.statcan.ca/publication. Accessed November 3, 2012.

Statistics Canada. (2010a). *Canadian Community Health Survey 2001–2010*. Retrieved from www.statcan.ca/publication. Accessed October 15, 2012.

Statistics Canada. (2006). "Canadian Environmental Sustainability Indicators: Freshwater Quality Indicator: Data Sources and Methods—2005." Catalog no. 16–256-XIE. Retrieved from www.statcan.ca/publication. Accessed September 26, 2012.

Stern, P. C. (2000). "Toward a Coherent Theory of Environmentally Significant Behavior." *Journal of Social Issues*, *56*(3), 407–424.

Stern, P. C., Dietz, T., Abel, T., Guagnano, G. A., and Kalof, L. (1999). "A Value-Belief-Norm Theory of Support for Social Movements: The Case of Environmentalism." *Human Ecology Review*, *6*(2), 81–97.

Taylor, R., (2012) *Taking Sides: Sustainability*. New York, NY: McGraw Hill.

Thomas, E. M. (2010). "Recent Trends in Upper Respiratory Infections, Ear Infections and Asthma Among Young Canadian Children." Component of Statistics Canada Catalogue no. 82–003-X. Health Reports. Retrieved from http://www.statcan.gc.ca/pub/82–003-x/2010004/article/11364-eng.pdf. Accessed October 14, 2012.

Tyshenko, M. G., Turner, M. C., Craig, L., and Krewski, D. (2009). "Multi-Jurisdictional Expert Opinion for Improving Children's Environmental Health Protection." *International Journal of Environmental Health Research*, *19*(2), 81–96.

Vandenberg, L. N., Hauser, R., Marcus, M., Olea, N., and Welshons, W. V. (2007). "Human Exposure to Bisphenol A (BPA)." *Reproductive Toxicology*, *24*(2), 139–177.

Vogel, S. (2009). "The Politics of Plastics: The Making and Unmaking of Bisphenol A 'Safety.'" *American Journal of Public Health*, *99*(S3), S559–S566.

Wagener, T. P. (1994). *The Complete Guide to the Hazardous Waste Regulations* (2nd ed.). New York: Van Nostrand Reinhold.

Wayne, K. (2008). "For the Record: Public Health Primer." *Canadian Medical Association Journal*, *179*(3), 227.

Weinhold, B. (2011). "Alberta's Oil Sands: Hard Evidence, Missing Data, New Promises." *Environmental Health Perspectives*, *119*(3), A126–A131.

WHO. (2012). "Children's Environmental Health: Global Plan of Action for Children's Health and the Environment." Retrieved from http://www.who.int/ceh/en/. Accessed October 10, 2012.

WHO. (2011). "Our Planet, Our Health, Our Future. Human Health and the Rio Conventions: Biological Diversity, Climate Change and Desertification." Retrieved from www.Who.int/globalchange/publications/reports/health. Accessed March 10, 2013.

WHO. (1999). *International Consultation on Environmental Tobacco (ETS) and Child Health: Consultation Report*. Geneva: World Health Organization. Retrieved from www.smoke-free.ca/second-hand-smoke/health. Accessed July 30, 2012.

WHO. (1992). *Our Planet, Our Health: Report of the WHO Commission on Health and Environment*. Geneva, Switzerland: World Health Organization.

Wigle, D. T. (2003). *Child Health and the Environment*. New York, NY: Oxford University Press.

Wilson, K., Leonard, B., Wright, R., Graham I., Moffet, J., Pluscaukas, M., and Wilson, M. (2006). "Application of the Precautionary Principles by Senior Policy Officials: Result of a Canadian Survey." *Risk Analysis*, *26*, 981.

Wing, S. (2008). "Objectivity and Ethics in Environmental Health." *Perspectives*, *11*, 1809.

World Commission on Environment and Development (1987). *Our Common Future*. New York, NY: Oxford University Press.

Part II

Developing Countries—Africa

6 Environmental Governance in Sub-Saharan Africa

Robert Dibie, PhD

INTRODUCTION

Human-made and natural environmental disasters all over the African continent are transboundary in nature. Despite these severe environmental problems, no supranational agencies in Africa are able to take on the task of regulating the pollution of the environment in the continent. To make things even worse, no international judicial systems on the African continent have the power to enforce any environmental laws or guarantee compliance. A more complicated dilemma is that environmental problems in the continent are typically generated by many sources from different nations and have to be taken care of by voluntary agreements among African political leaders (Weatherby et al., 2009; Withgott and Brennan, 2009; Wright and Boorse, 2014).

Human-made disasters in the African continent include the problem of legitimacy of regimes, political conflicts, corruption, mismanagement of public resources, lack of resources to effectively implement environmental policies, fragmented policies or agency mandates, incoherent overall approach, a lack of participation in policy formulation and subsequent ownership of policies, and other setbacks. Natural disasters include famine, drought, flood, and tsunami (Dibie, 2008; McConnell and Abel, 2008). The human-made disasters are sometimes equated with the Marxist tradition, which contends that instability is accepted as inevitable in the progression of history (Easton, 2010; Smith, 1996; Ayittey, 2005). Thus, class conflict has become the obstacle to the sustainable development process in the African continent. The trend in Africa over the past two decades reveals that political and economic stability has become more complicated because democracy has been superseded by great political instability. Democratic pluralism has been transformed into single-party regimes and the promotion of totalitarianism in some of the nation-states (Handelman, 2009; Onimode, 2004).

The African continent, the world's second-largest landmass, encompasses many of the world's least developed countries. By global standards, urbanization is comparatively low but rising at a rapid rate. The more heavily industrialized areas at the northern and southern ends of the continent experience the major share of industrial pollution (Roorda, 2012). In other regions, the most serious environmental problems typically stem from inefficient subsistence farming methods and other forms of land degradation, which have affected an increasingly extensive area under pressure of a widely impoverished, fast growing population. Africa's distribution of natural resources is very uneven. Despite the vast availability of natural resources, some parts of the continent are at great risk of desertification, especially in the Sahel region at the edge of the Sahara but also in other dry-range areas. At the same time, the African continent harbors some of the earth's richest and most diverse biological zones (World Environmental Snapshot, 2012).

Rapid rates of resource depletion are associated with poverty and high population growth in the African continent (Theron, 2008). The average consumption pattern per capita in Africa is much higher than the rate in most developed countries (Wright and Boorse, 2014). The difference is quite clear despite developed nations' recent progress in energy efficiency and conservation. According to Niko Roorda (2012), it is impossible to sequester the global environmental challenge from related economic, social, and political challenges. Conversely, budget constraints, lagged provision of basic services to the population, and other factors such as debt service and militarization may preclude the institution of minimal environmental protection measures in the poorest countries (World Environmental Snapshot, 2012).

The welfare of the people in most countries in the African continent is dependent upon each other. Unfortunately, several governments' influence over business has become more pervasive in the past five decades. As a result of rigid government policies and overcentralized public management systems, business efforts to influence government decisions have multiplied all over Africa (Obiyan, 2005; Patel and Wilson, 2004; Weatherby et al., 2009). The outcome of centralization has caused government decisions to become a growth machine that focuses on unethical public leaders, corrupt administrators, rigged elections, and nonchalant political party behavior. Corruption and unethical practices have engulfed government and business activities, as well as strategies used by private interests. Under these circumstances, the conduct of civil servants in several African nations constitutes the greatest impediment to the attainment of sustainable development in the continent (Dibie and Gadzekpo, 2003). Ethical leadership principles have totally disappeared from the public administration system of most African nations. Most multinational corporations now seek to influence and shape the political and economic policies by affecting the actions of administrative agencies and the adjudication of issues by the judicial system (Lehne, 2010; Obiyan, 2005; Onimode, 2004).

The relationship between the private sector and several African governments that practice democracy is not an ethical one, nor do both groups share the interests of those of the citizens (Roorda, 2012). The common practice of business in most African nations rests upon values that contradict the ethos of popular rule and sustainable development. Thus, the current public administration practices in most African society could be described as one of corruption, malpractices, dishonesty, lack of integrity, and the absence of moral values and ethical conduct (Dibie, 2008; Weatherby et al., 2009). The differences in values, beliefs, religion and culture have led to different forms of administrative and accountability styles. The collapsing state of the public sector in most African nations is due largely to leaders who are not committed to the development of their societies and who lack honesty and commitment to shared governance. Although the government of a nation is expected to render social stability, regulate the market, make laws, determine the tax structure, create infrastructure, dominate key industrial sectors, and deliver goods and services to its citizens, the situation is nevertheless just not so in many African nations.

Amanda McBride and Michael Sherraden (2004) view civic service in Africa as an emerging social institution, purposefully constructed to enhance personal, societal, and economic development, as well as environmental protection. However, in several African nations the bone of contention is how they achieve sustainable development when they are confronted with problems such as (1) irregular payment of civil servants; (2) lack of office supplies and equipment to effectively function; (3) rapid turnover of governments and political systems, (4) incompatible sociocultural practices; (5) volatile economic performance; (6) rapid change in the nature, size, and complexity of tasks; (7) high rates of population growth and rapid urbanization; (8) government functions often dispersed among an excessive number of ministries, authorities, and agencies, resulting in inadequate leadership within the government for policy formulation, planning, and implementation; and (9) various public administrators and agencies who are not sufficiently competent to develop solid plans and projects for the citizens (Adamolekun, 1999; Balogun and Mutahaba, 1999; Olowu and Sako, 2003). Given the preponderant liberal democratic model inherited from past colonial regimes, political reforms and the relationships among politicians and civil servants have been allowed to degenerate.

While most political leaders or former military generals are swimming in wealth, the increased diversification of income that characterizes the household of most African household has often caused citizens to be preoccupied with scrambling to make a living in the informal sector. George Ayittey (2005), Goran Hyden (2006), and Wayne Nafziger (2005a) contend that the drastic diversification and fragmentation of economic activities continue to limit the incentives for people to form associations in ways that they used to do when their source of income was more concentrated and permanent.

The debt crisis that African nations face has also made them realize the tremendous potential of nongovernmental organizations (NGOs) as alternative instruments for providing social services to citizens. Foreign NGOs are now noted for fostering dynamic forms of economic dependence while ignoring the structural inequality of the marketplace in several African countries (Kaplan, 1994; Kamat, 2003; Weaver, Rock, and Kusterer, 1996). Foreign NGOs have also promoted new forms of gender equality and more pragmatic forms of women's liberation than those associated with the government policies of most nations in the continent (Dibie, 2008; Onimode, 2004; Thin, 2002).

This chapter explores the significance of the ties between government and business in several Africa nations and why this relationship is essential in transforming the continent into a broadly based sustainable development mechanism. The chapter contends that the economies of several nations in Africa are production and distribution systems that affect not only producers and consumers but also communities and governments. It argues that every nation in Africa needs to develop pragmatic government efforts to implement appropriate economic and environmental policies. The chapter also challenges African governments to explore a better relationship between the public, private, and NGO institutions to help transform the current precarious situation.

The chapter also examines the challenges posed by changing economic, political, and environmental circumstances and surveys the debates about various African governments' actions in a neoliberal and globalization age. Further, it stresses that public leaders in Africa must learn to effectively and ethically maneuver public institutions and agencies through the unique political, economic, ideological, and sociocultural maze in which they operate. It asserts that the mismatch between current public management practice and the needs of civil society groups has impaired the capacity of these governments to function effectively. The concluding section of the chapter provides some suggestions on the need for NGOs to collaborate with governments in Africa to establish broadly based sustainable development. It is also essential for such collaboration to promote ethical and effective administrative mechanisms, as well as an appropriate dynamic for efficiently implementing such programs in the midst of uncertainty.

SUSTAINABILITY AND URBAN LIFE FRAMEWORK

All human beings are affected in some way by exposure to environmental hazards associated with their lifestyles at work, at home, during recreation, or while travelling on the highway. It could be argued, therefore, that the environment is intimately connected with human health, illness, and mortality (Friis, 2012; McConnell and Abel, 2008). Lorraine Elliot (2007) contends that the economic theory of the voluntary provision of a pure public good predicts market failure and that the African environmental quality will not improve by means of countries in the continent acting in a decentralized manner.

One of the main theoretical dilemmas speculated by analyses of sustainable cooperation is that, in practice, the political economy of most African nations reveals that government and businesses do not exist in a vacuum. Richard Lehne (2010) contends that they are part of the broader society and that they are affected by a nation's general policies. The influences of corruption on these policies are such that they become ineffective in most African nations. Thus the status afforded foreign businesses reflects the judgment that quasi-independent businesses provide the society greater benefits than firms that are tightly controlled by government. From these African perspectives, however, this policy judgment is a complex one that reflects past events as well as current institutional developments, global trends, and societal expectations (Dibie, 2008; Lehne, 2010; Withgott and Brennan, 2009). Thus, the sustainable development goals among government, industries, and NGOs are not shared by everyone in the society, and they are certainly not permanent.

The new governance in action theory suggests that government retain considerable power to shape both the formation of new networks and the direction of their activity because it possesses valuable resources unavailable to other agencies. Rhodes (1996) contends that policy networks are neither necessary, self-organizing, nor autonomous. However, the government is in a better position to play the role of network builder given its responsibility for maintaining stability and security because of the special authority that it controls. Hudson and colleagues (2007) argued that government legitimacy arise from democratic election and financial power, which, in turn, arises from tax raising power. This argument suggests that governments are dependent upon other organizations to achieve policy goals and that they must act in an environment that is increasingly networked. On one hand, Davies (2002: 302) argues that formal partnerships have been driven by central government through inducement such as funding and are shaped by government bureaucracy and regulations, implying that such partnership are not self-organizing and are in fact, to some extent, designed. On the other hand, Kicker and Koppenjan (1997) suggest that at any one time there may be organizations and actors that are dormant members of the network. Therefore, political networks among presidents of African states may face severe difficulties and might not help to solve the environmental problems on the continent.

The concept of sustainable development explores the relationship among economic growth, social equity, and environmental quality. The concept was introduced by the World Commission on Environment and Development (WCED) in 1987. This institution's mandate is also known as the Brundtland Commission Report of 1987 (WCED, 1987). Pearce, Markandya, and Barbier (1989) contend that the major premise of the sustainable development concept calls for citizens of the world to leave to the next generation a stock of quality of life assets no less than those that they have inherited (Withgott and Brennan, 2009; Wright and Boorse, 2014). The focus of sustainable development is often on the potential for service as a social and economic development intervention that may produce a wider range of positive effects. The sustainable development concept is perceived as a dynamic process of change in which the exploitation of resources, the direction of investments, the orientation of technological development, and institutional change are made consistent with future as well as needs (Rogers, Jalal, and Boyd, 2008).

The concept calls for governments of African nations to take action in the spirit of improving living conditions for the general welfare of everyone within their respective nations. McBride and Sherraden (2004) contend that the outcome of sustainable development programs can range from peace and international understanding, to improved job skills and education, to sustained civic engagement. Pearce and Barbier (2000) and McConnel and Abel (2008) stress that sustainable development policy should include the promotion of human welfare, including the satisfaction of economic needs, the aspirations of a clean and healthy environment, and preferences in terms of social development. Figure 6.1 shows the relation between society, economy, and environment.

Figure 6.1 shows that sustainable development has three components: (1) environmental; (2) society; and (3) economic. The figure also includes four other variables: (a) making wider use of market; (b) manage links to

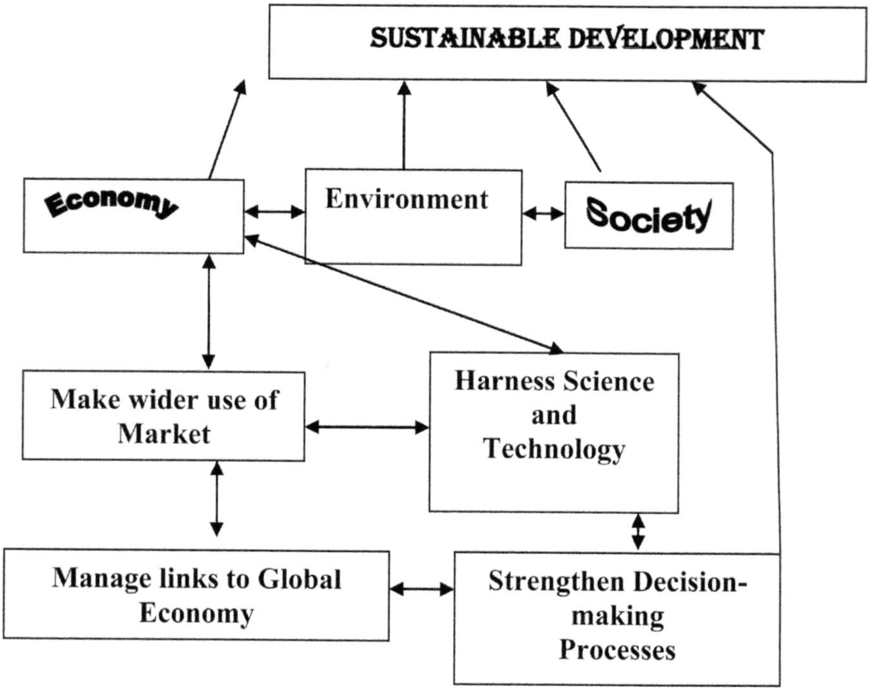

Figure 6.1 The Relationship between Sustainability and Society
Source: Dibie (2008).

global economy; (c) harness science and technology; and (d) strengthen decision-making processes. These variables were proposed by the Organization for Economic Cooperation and Development (OECD, 2001b). The OECD also argued that market oneness is not sufficient to ensure economic growth. It contends that sound macroeconomic policies, as well as institutional and social stability, are paramount to achieving sustainable development. While this framework is crucial, it is important to note that achieving global sustainable development will not be possible without the active participation of African nations (OECD, 2000a, 2001b).

Sustainable development is used in this discussion in terms of current development processes that meet the needs of the present citizens of African states without compromising the ability of future generations to meet their own needs (Dibie, 2004; Holmberg, Bass, and Timberlake, 1991; Rosenbaum, 2008). The sustainable development concept postulates that development can take place only when the per-capita utility or well-being of the African people has increased over time with free exchange or substitution between natural and human-made capital. This can be measured only when the well-being of the African people increases over time without decreasing the natural wealth of the continent. In the African context, sustainable development is the principle that will guide the various governments, NGOs, public officials, private sector, and people of the continent as they progress toward development.

Waller-Hunter and Tom Jones (2002) present an argument that increasing the scale of economic activity, trade, and investment liberalization may place more pressure on the environment. They stress that in countries (such as several African nations) where environmental policies are weak or poorly enforced, these pressures could lead to reduced environmental quality. Therefore, attempts at promoting economic liberalization in several African nations may exacerbate existing weaknesses in environmental policy regimes (Wright and Boorse, 2014). Thus, when trade and investment liberalization polices are combined with strong regulatory frameworks to protect the environment, such polices, if well enforced, could lead to a beneficial outcome on environmental conditions. The onus is therefore on policy makers in Africa to complement policies stimulating economic growth with policies intended to broaden access to income opportunities.

The pluralist theory argues that neither business nor any other single group dominates the policy-making process of any nation. As a result of countless continuous struggles to persuade public officials of the validity of their policy views, they often would be engulfed in corrupt practices. When a group fails to find support in one department or agency, it can turn to another administrative unit or political party for backing. The pluralist theory assumes that government recognizes the range of policy opinions expressed by groups in the society and then devises a policy response that balances the various viewpoints as effectively as possible (Lairson and Skidmore, 1992; Lehne, 2010).

For African nations to be sustainable, national governments on the continent have to introduce the policies needed to protect the basic human needs and inalienable rights of their people. Most African governments need policies that would promote good governance and democracy (Dia, 1993; Wunsch and Olowu, 1990). They also need to provide opportunities for all of their people to participate in the political decision-making process, as well as in the achievement of equity and social justice. African governments need to promote an environment for shared governance and the provision of self-determination and cultural diversity and, in doing so, protect their environment for future generations (Dia, 1993; Dibie, 2008). The participatory approaches to sustainable development have great potential to advance African leaders thinking about how they can best be more inclusive in the policy networks in order to make their respective nations more democratic.

DECLINE OF THE POLITICAL ECONOMY

A society or nation can be classified as civil if there is sufficient evidence of dialogue and tolerance of others participating in the discourse and political system. This concept is what pluralism and democracy prescribe. Unfortunately, that is not the case in several nations in sub-Saharan Africa. Goran Hyden (2006) noted that many international nongovernmental organizations are often offput by and tired of the ineptness of several African governments because of their unwillingness to implement good health and education policies. Thus, foreign NGOs have become prominent in the delivery of health and education in these nations. Donor funding of the voluntary sector rose from US$1.04 billion to US$2.13 billion between 1980 and 1988. This growth is almost five times higher than that for total official development assistance (ODA) to government, which actually declined quite considerably during the 1990s and between 2002 and 2005 (Hyden, 2006; Obiyan, 2005; Weatherby et al., 2009).

The level of poverty and the lack of citizen participation in any form of civic or political engagement reveal that civil society and shared governance is more easily promoted in conditions of plenty than in poverty (Dibie, 2007). In recent times, it has been very difficult in some parts of Nigeria, Rwanda, Zimbabwe, and Democratic Republic of Congo to mobilize popular support for general causes such as the promotion of human rights, the protection of the environment, and fighting HIV/AIDs. Thus the atmosphere in these countries in recent times is that of little development, minimal social mobilization, and minimal citizens' empowerment.

The *World Bank Report* (World Bank, 2007) indicates that agriculture in several African nations, where it has not been taken over by commercial and capitalist-oriented farmers, has been left in the hands of women who have little time to devote to the farming enterprise other than for subsistence.

Nafziger (2005b) contends that Africa's agriculture crisis of declining food output per person in the last forty years has been due to the neglect of agriculture and the exploitation of peasants by governing elites, whose policies include the disincentives of price ceilings and the socialist ideology in the cases of Zimbabwe, Tanzania, Ghana, and Mozambique. George Ayittey (2005) contends that these statist public policies ensured the holding of peasants in contempt because they strongly discouraged mechanized farming methods. Further, incompetent political leaders and public administrators exploited the agricultural sector for their personal again (Roorda, 2012). It is also argued that Africa is the only continent where landowners and wealthy farmers rarely invest in industry. Most landlords value consumption and real estate expenditure highly, and they lack experience in coordinating a production process with specialized work tasks and machinery and in overseeing secondary labor relations (Blakely, 1989; Nafziger, 2005b).

The political leaders in the continent have no vision. When in power, they often do not ban foreign investment, nor do they invest in infrastructure such as water supply, costal shipping, ports, harbors, new communication and technology systems postal system, bridges, river improvements, railways, electricity, gas, lighthouse, and technical research (Ayittey, 2005; Dibie, 2012; Skolnik, 2008). Instead they allow neocolonialists to continue to exploit their resources (Patel and Wilson, 2004). This practice encourages Western European governments to continue to dominate the affairs of the African political economic even after colonial periods. Thus, several postindependence African political leaders frequently fail to seek self-directed strategies and technological ideas in order take advantage of the shifts in comparative advantage from product cycle, educational policy, business assistance, financial institutions, transfer of agricultural savings to industrial, low wages policy, industrial dualistic complementarity, and foreign exchange rate policies conducive to export expansion (Nafziger, 2005). All this results in weak economies, lack of legitimacy, inefficiencies, and corruption (Dibie, 2007). With this range of predicaments, these governments are easily dominated and manipulated by wealthy domestic political leaders and powerful international interests. In some cases, the manipulation is galvanized with government subsidies, tax exemptions, or lack of regulation. This unfortunate practice often creates a situation whereby wealth is concentrated in the hands of the privileged few to the disadvantage impecunious many citizens (Hyden, 2006; Steel, Clinton, and Lovrich, 2003).

Table 6.1 shows a composite measure of three dimensions of human development: GNP per capita, life expectancy, and access to health and being educated (measured by adult literacy rate). Although these data are not in any sense a comprehensive measure of human development in the African continent, it provides a broad prism for viewing human progress and the complex relationship between income and well-being.

Table 6.1 Human Development Index in Africa

	1994		2010				
Nation	GDP per Capita	Life Expectancy	People with access to Health Services	GDP per Capita	People with access to Water (%)	Life Expectancy (years)	Adult Literacy Rate
Algeria	1,650	67.8	80	6,603	15%	71.4	59.4
Angola	896	47.2	47.270	2,180	47	41.0	44
Benin	370	54.2	18	—	—	—	35.5
Botswana	2,800	68.1	60	9,945	5	34.9	68.7
Burkina Faso	300	46	90	1,213	78	51.4	18.7
Cameroon	680	55.1	80	2,174	34	45.7	62
Chad	180	47	30	2,090	58	—	47
Cote d'Ivoire	610	52.1	60	1,551	16	45.9	39.4
CA Rep.	370	48.3	52	1,094	43.7	39.1	62
DR Congo	429	52.2	26	978	54	43.5	76.4
Egypt	2,604	68.9	60	4,211	2	70	51
Ethiopia	100	43.1	46	1,055	78	51.8	34.5
Eretria	960	50.1	50	1,109	40	43.7	25
Eq. Guinea	430	48.6	80	20,510	57	42.8	77.8
Gabon	3,880	54	68	6,623	12	54.0	65
Ghana	410	56.6	60	2,240	25	57.0	63.4
Kenya	1,404	53.6	77	1,140	39	52.1	77
Libya	—	64	80	7,570	31	73.8	60
Malawi	170	41.1	35	667	38	46.3	55.8

Mali	250	46.6	46	1,033	53.1	48.1	29.3
Mauritania	480	52.1	63	1,940	47	53.1	49
Mozambique	90	46	39	1,242	45	38.7	39.5
Namibia	1,970	55.9	55	7,586	60	51.6	85
Nigeria	280	51	51	1,128	52	43.4	56
Rwanda	80	23.1	80	1,263	26	44.2	59.2
Senegal	600	49.9	90	1,792	42.3	62.3	32.1
S. Africa	3,040	65	91	11,192	12	47.0	81.3
Sierra Leone	160	33.6	38	561	43	41.0	30.3
Tanzania	140	50.3	42	674	38	45.9	66.8
Sudan	1,084	51	70	1,943	5	56.5	44.8
Uganda	190	40	49	1,478	40	48.4	61.1
Zambia	350	42.6	50	943	40.5	37.7	77
Zimbabwe	500	50	49	2,065	19	36.6	80

Source: United Nations Development Program (UNDP) (1993–1994; 2010–2011).

The analysis of the data in Table 6.1 also reveals that poverty has been a major cause of political instability in most African nations. It shows that economic development leads to greater economic security, stability. and better education, as in the cases of South African Republic, Botswana, Gabon, and Libya. Economic development, as indicated by GNP per capita, also enlarges the middle class, whose interest is in moderating conflict, by rewarding moderate political parties and penalizing extremist ones

One serious consequence of this state of Africa's economy is the decline in the sustainable development processes. As indicated in Table 6.1, Africans have suffered in the past three decades due to human-made disasters that escalated the level of poverty. The *World Bank Report* (World Bank, 2007) indicates that the GDP per capita for most African nations in the subregion was lower in the 1990s than it was at the end of the 1960s. The report also stated that Africa's GDP per capita was the lowest in the world. To further compound the situation, life expectancy declined from the 1975 forty-six-year level, and the higher prevalence of HIV/AIDs in adults has been identified to be the cause. Table 6.1 shows the trend of GDP and other economic indicator in Africa between 1994 and 2007.

It is interesting to note that African governments do not adequately protect their citizens. In most cases, political leaders in Africa are more active in collaborating with privileged groups within their respective nations and abroad than in protecting citizens. The practice of association with the privileged is often quite at odds with the interests of pragmatic development policies that would promote sustainable development.

One of the most devastating predicaments facing the political economy of sustainable development in Africa is the 10 million estimated children orphaned due to HIV/AIDs. This number is expected to approach 40 million in twenty years' time (Steel et al., 2003; United Nations Human Development Index, 2010–2012). This is the aftermath of the widespread HIV/AIDs impact on economic and human resources in some regions of Africa.

In some regions in the continent, there are more people sixty years or older than there are people forty years or younger. The HIV/AIDs disease is beginning to destroy the most productive population segment of African stock of human capital (World Bank, 2007). The wide spread of HIV/AIDs has been explained to have been caused by a lack of pragmatic policies, lack of resources to effectively implement health policies, fragmented policies or agency mandates, incoherent overall approach, a lack of participation in policy formulation and subsequent ownership of policies, and other setbacks (Anderson, 2000; Ayittey, 2005). These problems also spill over into other economic development areas in the African continent.

In the process of promoting sustainable development agendas, most African governments have come to realize that environmental issues have to compete with many other policy issues for the finite attention of political leaders. The predicament of unchecked population growth, along with an ever increasing level of consumption, is beginning to have severe consequences on the people of Africa (United Nations Human Development Index, 2010–2012). The situation in Africa is further galvanized by deeply rooted poverty. Steel and colleagues (2003) contend that sustained poverty often leads most people in the African continent to engage in behavior that exacerbates ecological deterioration. As a result of this unfriendly environmental behavior, these African nations are overexploiting their natural resources in order to provide for pressing current needs and, in the process, are running the risk of potentially sacrificing the natural resources inheritance of future generations.

The African continent is already being affected by climate change and will be disproportionately hit by the effects of climate change, creating cumulative cycles of disadvantage that are transmitted across generations (United Nations Human Development Index, 2010–2012). The United Nations Development Program Human Development Index has identified the most highly salient public problems deserving attention in Africa as the following: (1) groundwater and surface water pollution; (2) deforestation; (3) land and soil degradation and pollution; (4) habitat loss and fragmentation; (5) air pollution; and (6) demographic change and population pressures on natural resources and in urban areas (UN Human Development Index, 2007–2008).

The UNDP 2007–2008 report further indicated that virtually all major cities in Africa suffer air quality problems due to rapid urbanization and the rising demand for electrical and petroleum-based forms of energy to support household needs and the requirement of economic activities. Steel and colleagues (2003) contend that while rapid urbanization itself need not cause serious environmental degradation, it has led to massive environmental and health problem in major cities throughout the African continent. Air, water, and soil pollution are widespread in contributing to enormous health problems such as asthma, lead poisoning, and waterborne diseases such as cholera. Table 6.2 shows the level of carbon dioxide emissions in selected African nations.

Table 6.2 also shows that uncontrolled emission due to heavy duty traffic—even in residential areas—and in some circumstances waste, effluent, and air pollution pose serious environmental health threats to the African people. Table 6.3 is derived from a survey conducted of respondents in the listed seven African nations. The table shows that some nations in African have great problems in dealing with more recent problems created by industrialization, heavy automobile traffic, large quantities of solid waste, and informal society activities

The length of time taken to collect garbage and to transport waste, the lack of adequate landfills and sanitation, and the question of who is actually responsible for garbage collection remain a major environmental problem in these nations. Thus, they cannot address their pressing development problem effectively without resolving their diffused environmental problems.

Table 6.2 Carbon Dioxide Emissions Level in Selected African Nations

	Total Emissions ($MtCO_2$)		CO_2 Emissions Annual Change (%)	CO_2 Emissions Share of World Total (%)		Population Share (%)	CO_2 Emissions per Capita (tCO_2)	
CO_2 emitters	**1990**	**2004**	**1990–2004**	**1990**	**2004**	**2004**	**1990**	**2004**
United States	4,818.3	6,045.8	1.8	21.2	20.9	4.6	19.3	20.6
China	2,398.9	5,007.1	7.8	10.6	17.3	20.2	2.1	3.8
Russian Federation	1,984.1	1,524.1	–1.9	8.8	5.3	2.2	13.4	10.6
South Africa	331.8	436.8	2.3	1.5	1.5	0.7	9.1	9.8
Nigeria	45.3	114.0	10.8	0.2	0.4	2.2	0.5	0.9
Ethiopia	3.0	8.0	12.1	0.0	0.0	1.2	0.1	0.1
Equatorial Guinea	0.1	5.4	—	0.0	0.0	0.0	0.3	10.5
Congo (Democratic Republic of the)	4.0	2.1	–3.4	0.0	0.0	0.9	0.1	0.0
Rwanda	0.5	0.6	0.6	0.0	0.0	0.1	0.1	0.1
Burundi	0.2	0.2	0.9	0.0	0.0	0.1	0.0	0.0
Global aggregates								
High-income OECD	10,055.4	12,137.5	1.5	44.3	41.9	14.3	12.0	13.2
Least developed countries	74.1	146.3	7.0	0.3	0.5	11.8	0.2	0.2
Sub-Saharan Africa	454.8	663.1	3.3	2.0	2.3	11.1	1.0	1.0
Low human development	77.6	161.7	7.7	0.3	0.6	7.8	0.3	0.3
World	22,702.5	28,982.7	2.0	100.0	100.0	100.0	4.3	4.5

Source: UNDP (1993–1994; 2010–2012).

Table 6.3 Environmental Problems due to Human Activities

Environmental Problems	Ghana	South Africa	Nigeria	Kenya	Botswana	Democratic Republic of Congo	Egypt
Garbage collection Problem	X	X	X	X	X	X	X
Inadequate landfills	X	X	X	X	X	X	X
Foul odors due to dumps in gullies	X	X	X	X	X	X	X
Waste dumped on the streets	X	X	X	X	X	X	X
Sewage treatment	X	X	X	X	X	X	X
Effect of informal sector	X	X	X	X	X	X	X
Improper disposal of waste engine oil and car batteries	X	X	X	X	X	X	X
Household with exclusive access to water	X	X	X	X	X	X	X
Uncontrolled emissions	X	P	X	X	X	X	X
Burning of Refuse	X	X	X	X	X	X	X
Large Household	X	X	X	X	X	X	X
Use of firewood to cook	X	X	X	X	X	X	X
Drainage problem when it rains	X	X	X	X	X	X	X

Source: Derived from an author-conducted survey in some African countries in 2012.

Key: X = Problem Exist; P = partially controlled.

Steele and colleagues (2003) contend that the environmental problems of African nations are as serious and urgent as the problems of underdevelopment. As alluded to earlier in this paper, the problem of governance in these nations is even more serious and urgent. It is difficult to recommend one specific solution to the problems of sustainable development and environmental protection until there is the will to transform conditions and achieve sustainable development.

With varying degrees of reluctance, most African nations have gradually come around to recognizing that rapid population growth exacerbates every other environmental and sustainable development problem they have been confronting. For over three decades, the government of these nations insisted that rapid population growth was highly beneficial to their development efforts (Steele et al., 2003; Thin, 2002). However, the Structural Adjustment Program (SAP) recommended by the World Bank, adopted by several African nations in the 1980s and 1990s, made them realize that population growth was a severe impediment to development.

POSSIBLE SOLUTION TO THE POLITICAL ECONOMY PROBLEMS

The African continent is estimated to have more than a billion people in poverty, and there are important variations between north and south. Evidence in the past four decades shows that economic growth and welfare associated with market liberalization are not always distributed equally among urban and rural groups on the continent (Handelman, 2009; OECD, 2001a; Rogers et al., 2008). To solve the environmental and political economy problems on the African continent, social development, economic growth, and environmental protection should be integrated into the national and international frameworks. For such frameworks and policy to be beneficial to the citizens of the continent, there is the need for effective and transparent government institutions, shared governance, accountability, and the solidly consistent rule of law (Waller-Hunter and Jones, 2002).

A number of formidable obstacles within the African continent have to be overcome for sustainable development to take place. First, the majority of the citizens must grasp the seriousness of the environmental degradation that is occurring and come to feel that it is in their own interest to reverse it (Steele et al., 2003). Second, the African people must become fully aware of the extent to which their respective governments have been catering to special interests and mobilize their efforts to neutralize those distorting foreign and domestic influences (Dibie, 2008; Steele et al., 2003).

Some development administration scholars have argued that the major instruments of governance are accountability and transparency of public leaders, electoral legitimacy, rule of law, development oriented leadership, and freedom of expression and association (Collins, 2000; Hyden, 2006; Obiyan, 2005; Olowu and Sako, 2003; Onimode, 2004). These scholars contend that a nation achieves sustainable development when it applies the necessary development administration tools to propel its public management toward shared governance.

Environmental Awareness

In several African nations, managing the environment and local resources is not a matter of solving technical problems; it is a way of looking for sustainable solutions. There are many technical solutions. The main impediments in these countries are the political and motivational will of both the government and the people. In fact, much of what happens in these nations will depend on the choices and activities of thousands of farming households and women in the rural communities.

Environmental resources constitute the major means of the rural people's food production problems and ultimately need to be managed by local, small-scale collaborative efforts that are linked to large-scale planning and regulatory functions (Kraft, 2011; Vig and Kraft, 2013). If the government wants to achieve their sustainable development goals, there must be a fit between their ecological objectives with regional and rural civil societies' concern of how to manage their household (Friis, 2012; Rogers et al., 2008).

The government and people must realize that in the twenty-first century, environmental concerns constitute a social construct through which all human beings organize their work, rights, responsibilities, and relationships. Understanding ecological issues will enable the governments to find more effective and equitable ways of managing their natural resources for building productive rural livelihood system and sustainable development. African governments must work with NGOs who are interested in the protection of the environment to effectively put pressure on the international community to treat their forests and natural resources with respect and dignity. Government and NGOs could bargain for environmentally friendly international trade policies, as well as setting enforceable monitoring and punitive measures for violators of biodiversity and environmental protection laws for the landscape, wildlife, and atmosphere on the continent (Dibie and Offiong, 2012). There is also the need to stimulate technology development and diffusion, as well as the promotion of structural changes necessary to make more efficient use of natural and environmental resources.

Building Administrative Capacity

Capacity in government is the process of identifying and developing the management skills necessary to address policy problems; to attract, absorb, and manage financial, human, and information resources; and to operate programs effectively, including evaluating program outcomes to guide future activities (Collins, 2000; Oluwo, 1999; Rondinelli, 1983; van Klinken, 2003; White, 1987). Figure 6.2 shows a description of how to reinforce capabilities.

The specific skills necessary to accomplish these varied management tasks are neither easily identified nor routinely implemented. In effect, it is important to look not only at the particular agency but also at the system within which the agency functions.

It should be noted that the identification of management skills depends upon the specific functions required of the managers; that is, the skills depend upon the job at hand. Managerial roles tend to have internal and external dimensions, such as meeting standard operating responsibilities, as

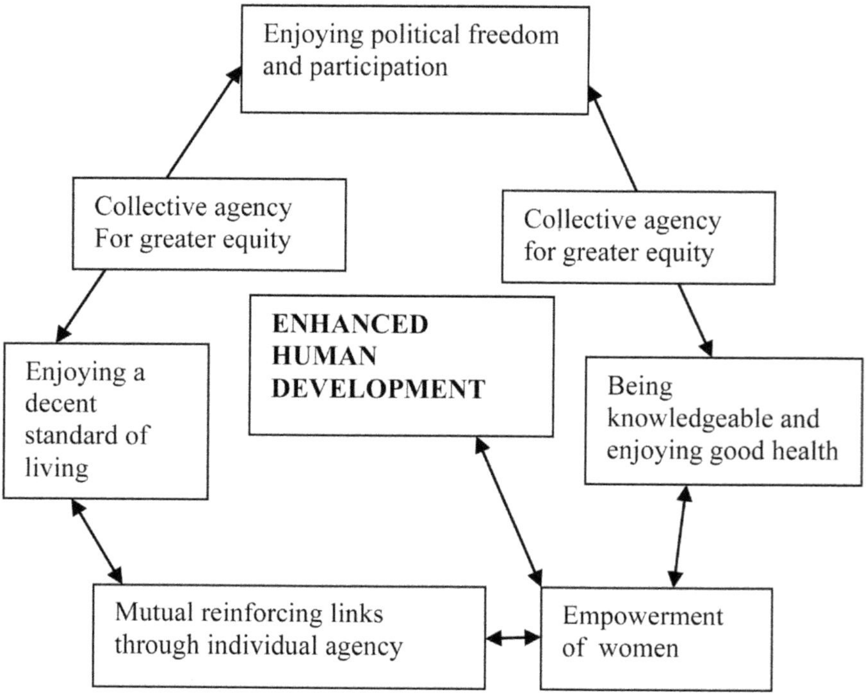

Figure 6.2 The Relationship between Human Development and Political Freedom
Source: Dibie (2007).

well as developing implementing strategies. Although the internal dimension is often the topic of interest (i.e., how are management decisions made?), the external dimension often raises the most difficult managerial questions. This can easily be seen in questions of concern to development administrators (practitioners) and scholars, such as how well developments plans are put into practice? How are communities to be mobilized? In fact, how is development to be implemented? According to Bryant and White (1982), the presence or absence of needed skills at appropriate levels, the question of capacity often explains the nonimplementability of that elusive phenomenon, development (Rogers et al., 2008; Umeh and Andranovich, 2005).

Empowerment of Women

One of the most important dimensions of development is gender. This refers to the social norms, religion, and culture that define the roles, rights, responsibilities, and obligations of men and women in the African society. Promoting gender equity in developing countries is not just a moral or human right issue. It is also recognition as fundamental to poverty reduction (Dibie, 2008; Roger et al., 2008; World Bank, 2000). It is very important for African governments to promote gender egalitarian policies that would facilitate equal access for women, girls, and boys to a range of resources including education, land, social capita, employment, equal pay, and financial capital (Dibie and Atibil, 2012). African governments should also advance the recognition of women in their contribution to the sustainable use of natural resources.

Citizen Participation

Development is defined as a process of increasing people's capacity to determine their future. This means that people need to be included in the process of sustainable development. In other words, they need to participate in the shared governance process of decision making and implementation. It is an attitude of openness to the perceptions and feelings of others. It is a concern for what difference a project makes to people's lives. It is an awareness of the contributions that others can bring to an activity (Bryant and White, 1982; Oluwo, 1999; Wright and Boorse, 2014). Popular participation is an important dimension in the administration of public services. Within the context of development administration, participation (that is, participation by the poor and deprived, not with lobbying or manipulation by well-organized and influential interests) involves the relationship of career administrators and the public interest (Denhardt and Denhardt, 2009; Denhardt, Denhardt, and Blanc, 2014). Participation in the more active sense involves exerting influence on administrative behavior and on the outputs of official action. Greater participation by the poor and deprived would mean their greater influence on decisions and programs relating to their welfare. Hence, there is a need to distinguish genuine participation, which implies real influence, from symbolic, manipulated, or controlled participation, which is intended to ratify rather than influence official behavior (Handelmen, 2009; Obiyan, 2005; Olowu and Sako, 2003; van Klinken, 2003; Weatherby, 2009).

Decentralization

Decentralization of political and administrative authority is often thought of as a necessary institutional arrangement for the maximization of public welfare, including the actual empowerment of local government to create and undertake programs for community betterment (Michie, 1989; Rondinelli, Nellis, and Cheema, 1984; Stillman, 2010; van Klinken, 2003).

Overall, decentralization can be seen as a way of increasing the effectiveness of development activities by making them more relevant and responsive to local needs and conditions, allowing greater flexibility in their implementation and providing a means of coordinating the various agencies involved at the regional or local levels (Balogun and Mutahaba, 1999; Shafritz et al., 2009).

The size of most African nations is too large to manage by weak, inefficient leaders and inadequately trained public administrators. As a result, the division of these nations into smaller, more decentralized political system is but one of the possible solutions to ethnic and religious conflict in the continent (Collins, 2000; Huxitable 1998; Stillman, 2010). Nigeria, Sudan, Ethiopia, and Democratic Republic of Congo are good candidates. The author believes that such smaller and more locally oriented politics would produce pragmatic, efficient, effective, shared governance and improved lives for the citizenry (Migdal, 1988; World Bank, 1997). It is paramount that all African nations look for better and more efficient ways to deliver goods and services to their citizens.

The more decentralized local governments are and the more regional governments can make their own policies as well as develop their growth strategies, the more relevant they can become in helping NGOs to meet their objectives (Acharya, 2002; Stillman, 2010; van Klinken, 2003). Decentralization in the governance process in the African continent could make NGOs and local governments to develop a stronger partnership as well as see each other as major stakeholders in the sustainable development process.

Social Transformation

Since market openness is not sufficient to ensure economic growth, sound macroeconomic policies, and institutional and social stability on the African continent, the sustainable development process must take into consideration local understanding of the workings of culture, society, and nature (OECD 2001b; Onimode, 2004). These factors are important because there is an inseparable relation between the material and the spiritual worlds of local people. The sustainable development process must be informed by local understanding of complex linkages among the natural, spiritual, social, cultural, political, and economic forces of society.

The emotional and spiritual well-being of the individual and the social group forms the solid rock of any development process. Social transformation is only possible if it proceeds from a development of the inner self and spiritual values of the citizens of a nation. Therefore, there is also the need to redefine both individual and corporate responsibilities to families, communities, state or regional governments, nation-state and global citizenry (Dibie and Offiong, 2012).

The practice of diverting resources for the benefit of public servants and their political patrons while taxing the public and undermining respect for government should be discouraged. On one hand, diverting the resources of government to purchase foreign properties and holding big parties constitutes a waste of resources. On the other hand, wasting public funds to purchase several cars and unnecessary properties that are of no use to the economy and people should constitute a crime against humanity (Taylor, 2012). People should be educated on how to demand transparency of government transactions, especially in the financial realm. Empowered by self-organizations, public opinion can help to hold public servants to acceptable standards of behavior (Dibie and Gadzekpo, 2003). Thus, the control of bureaucratic abuses that are endemic in several African nations requires combined pressures and sanctions from above and below. Efforts to control such abuses can be strengthened by journalism, which is free enough and courageous enough to expose and attack instances of corruption and waste of public resources (Esman, 1991; Handelman, 2009; Patel and Wilson, 2004; Thin, 2002).

Mutual interdependence is another vital linkage that should promote indigenousness as well as attest to the importance of community responsibility. This factor is also very important in the promotion of state, national, and international responsibilities (Denhardt, Denhardt, and Blanc, 2014; Kraft and Furlong, 2013). For example, national environmental policies may affect trade and investment activities at the global level, and, conversely, international trade and investment arrangements in which an African participates may affect national environmental and social policies.

Responsiveness, Transparency, and Accountability

Another major problem confronting African states is the lack of accountability and transparency. Transparency is the process of decisionmaking that is open to the public view and can be perceived as being free of interference from special interests groups or individuals at the apex of government. The lack of accountability and transparency breeds corruption in the form of abusing public office for private or personal gain. The literature on development administration suggests that the extent to which the public can affect organizational change is contingent on whether they are allies or offer votes (Dibie, 2008; Morris Maclean, 2003).

Governments in Africa should make efforts to control bureaucratic abuses that are endemic in their respective countries. Efforts to eradicate these pervasive bureaucratic abuses are serious and continuously challenging to development administration because such unethical practices entail the demoralization of public services, the failure of discipline, the waste of resources, and the consequent alienation of the public (Esman, 1991; Yukl, 2010). It should be noted that cultural and social behaviors that tend to influence or effect professional practices should not be used as a medium for encouraging unethical behavior. The fact that these unethical behaviors are inherent in the culture of these nations only confirms that nations on the continent are perpetually corrupt and cannot contribute to economic and social development. Robert Dibie and Leonard Gadzekpo (2003) contend that a conscious and principled public management with a pragmatic legal system can limit these abuses, but only if it is prepared to confront them directly.

Some of the strategies available to enforce accountability are as follows:

1. The public should have access to government information.
2. The print and electronic media should play a critical role of watchdog on the conduct of government business. One positive result should be the seriousness with which privately owned media perform the role of public watchdog, serving as citizens' allies in holding governments accountable.
3. The media should try to achieve accountability through investigative journalism.
4. Public officials should treat citizens with fairness and impartiality.
5. Respect for the rule of law and an independent judiciary constitute key mechanisms for enforcing accountability.
6. Elected assemblies such as the legislatures or parliaments should play a major role in both the enactment and enforcement of anticorruption policies.

Equity

Sustainable development includes distributional issues. To the extent that only a small segment of the population benefits from the outcome of economic development, development has not occurred. On a broader scale, development is a normative concept implying that ensuring more equity in access and benefits is a value in itself (Onimode, 2004; Rogers et al., 2008; Roorda, 2012). A normative dimension of development administration focuses on social justice demands, especially in a context of severe scarcity. It demands that resources and efforts must be concentrated on improving the life opportunities of those groups in society who are not particularly well-placed to help everybody in the nation. Accordingly, the Brandt Commission (1990) echoed Peter Guy (2013) in contending that development is more than the passage from poor to rich, from a traditional rural economy to a sophisticated urban one. It carries with it not only the idea of economic betterment, but also of greater human dignity, security, justice, and equity (Ayittey, 2005; Dibie, 2008; Hyden, 2006; Stillman, 2010).

Improved Performance in Public Management

One of the major development predicaments facing most African nations is the poor performance of the public sector. The management of the public sector is characterized by overextended responsibilities (Denhardt et al., 2014). Several African nations inherited the idea of the national and regional governments as the engine of growth from their former colonial rulers. Under these arrangements or public management styles, the domination of the economy by the national government became necessary.

A modern approach for correcting this colonial system of state domination of the economy is called "new public management." The measures are drawn largely from an international public sector modernization movement, that is, new public management (NPM). The NPM seeks to apply market principles to governmental administration with an emphasis on competition, contracting, and customer orientation (Shafritz, 2009; Stillman, 2010). It also emphasizes merit-based recruitment, promotion. and increased autonomy for managers (Denhardt et al., 2014).

Gary Yukl (2010) argues that the talents required to manage a developing economic are not equally distributed among all managers, but they could be encouraged and rewarded in several African nations. Therefore, public management education should be required by all public servants to help sensitize those who are in middle and senior positions in government in order to help them cultivate the requisite political and managerial skills. The role of African governments, like that of their NGO counterparts, is to continuously reinforce such positive and ethical behavior with appropriate incentives and rewards in order to stimulate its continuity. According to Milton Esman (1991) and Robert Dibie (2003), development administration is a social enterprise that benefits from excellent leadership in the public sector; however, this must be matched by efforts that evoke similar leadership in NGOs dispersed throughout African nations.

Public Sector Reorientation

The dynamics between weak accountability and corruption among public leaders appears to be very strong in several African nations. Corrupt practice in the public sector, in turn, has severe consequences for the entire government administration system, as well as for the structures, functions, and processes of public management.

For African nations to achieve sustainable development, corruption needs to be eradicated in the public sector (Denhardt et al., 2014; Huxitable, 1998; Olowu and Sako, 2003; Rogers et al., 2008). Government reforms or reorientation in the future should consider the following measures:

1. Reducing the role of the state in national economic management;
2. Enhancing the role of subnational governments;
3. Striking a new public–private balance with respect to the production and provision of goods and services;
4. Promoting private and public savings;
5. Striving to achieve improved performance in public management;
6. Providing adequate infrastructure (water, electricity, transportation, health care system, education, and communication) for public and private agencies;
7. Providing the necessary skills through general education and specialized training;

8. Reducing monopolies and oligopolies, as well as maintaining and expanding the social and economic infrastructure;
9. Managing new technology, as well as investigating the development of potential techniques through scientific and market research; and
10. Seeking out and assisting entrepreneurs.

Professional public administrators' cadres in African governments can, however be stimulated and be assisted by NGOs in the development and diffusion of ethical attitudes, skills, and institutions that are conducive to effective managerial behavior (Barbier, 1987; Dibie, 2008). African public administrators must be exposed to the impact of globalization and what it means to be servant leaders (Greenleaf, 1977). Public leaders must listen to subordinates and citizens, learn about their needs and aspiration, and be willing to share in their pains and frustration. Public leaders must stand for what is good and right, even when it is not in the financial interest of their government (Baumer and Van Horn, 2014; Yukl, 2006). Ethical public leaders must be willing to empower their subordinates instead of using power to dominate them. When this happens and more trust is developed, subordinates might be inspired to become ethical and service-oriented leaders themselves in the future (Denhardt et al., 2014). The outcome will be more public administrators who serve as moral agents in African society and government.

Rule of Law

The predictability of the judiciary is not yet a reality in many African states (Huxitable, 1998; Olowu, 1999; Roorda, 2012; Theron, 2008). For example, the governments and one-party states in several Africa nations do not respect judiciary decisions (especially in Nigeria, Democratic Republic of Congo, Zimbabwe, and Sudan). In several other African nations, political executives still undermine the independence of the judicial through patronage appointments. The judicial and court system are characterized by weak enforcement capacity due to a shortage of professional staff, poor physical infrastructure, and inadequate funds (Balogun and Mutahaba, 1999; Collins, 2000; Hudson et al., 2007; Stillman, 2010).

Several scholars contend that the rule of law should be underpinned by an independent court system and a predictable legal framework that helps to ensure the settlement of conflicts between the state and individuals on the one hand and among individual or groups on the other (Balogun and Mutahaba, 1999; Collins, 2000; Handelman, 2009; Kiggundu, 1989; Rogers et al., 2008). The effective implementation of the rule of law would help to ensure respect for property rights and contracts. African nations need to develop into law-based nations where the government and individuals cannot be above the law. The law should not discriminate against persons, and strengthening judicial institutions should be a priority task if African nations want to strive for sustainable development in the twenty-first century.

Economic Development Strategy

In the twenty-first century, African nations need to establish clear economic development strategy. Strategies in this respect are groups of actions that will contribute to attaining the sustainable development goals that the national and regional governments set based on the options and resources available (Bienen and Herbst, 1996; Handelman, 2009; Rogers et al., 2008). One major implication of economic development in the African continent in the future is that trade and investment flows need to be compatible with society's environmental and social policy objectives (Waller-Hunter and Jones, 2002). According to Francois Theron (2008) and Niko Roorda (2012), strategic dimensions and issues that are crucial in attaining sustainable development in several African countries must include the following:

1. *Build quality jobs for the current populations:* The major idea in economic and employment strategy is to build employment for and with the current population in both the rural and urban communities.
2. *Achieve local economic stability:* Sustainable development will be successful in African nations only if the towns and communities have a specific approach to meet all the needs of business (i.e., land, finance, labor, infrastructure, and technical assistance in addition to labor).
3. *Build a diverse economic and employment base:* Regardless of whether the nation is highly or not so technology oriented, it must develop a broad base to provide continuing employment opportunities for residents.

Freedom of Expression and Association

Freedom of expression requires independence for the press and other news media so that the people's right to participate in shared governance is backed up by the right to know. The constitutions of several African nations at independence often had a provision for freedom of expression and association. In practice, however, most African nations do not respect their citizens' freedom of expression (Onimode, 2004). The frequency of ethnic and religious conflict, the authoritarian one-party systems, and the military rule that characterized several African nations make it difficult for political leaders to obey public policies (Peters, 2013). As a result, freedom of the press has been severely curtailed, with government as sole proprietors of both the print and electronic media. In some African nations where private ownership of the media has been allowed, journalists are often harassed and publishing houses shut down from time to time (Denhardt et al., 2014; van Klinken, 2003).

Freedom of expression should be associated with democratic practice. In the twenty-first century, Africa's future, it could be argued, belongs to democracy at all levels of government. African nations must promote the advancement of ethical, legal, and political values based on dialogue and the free exchange of ideas and culture.

To achieve sustainable development, there is the need for the media, NGOs, and civil society organizations to play an active role in the social, political, and economic spheres of African nations (Dibie 2008). NGOs, community-based organizations, the private sector, and civil society organizations should become invaluable partners of government in the provision of goods, social services, the development and maintenance of local infrastructure, and the mobilization of macrocredit for financing development projects (Dibie, 2008). The media should be given the freedom to communicate this news of development processes as well as create more awareness.

Development-Oriented Leadership

The underdevelopment crisis and the frequent ethnic and political conflict phenomenon of several African nations are due in large part to poor leadership (Denhardt, Denhardt, and Blanc, 2014). Many leaders are not committed to the development of their societies, and they lack honesty and a commitment to democracy. Many scholars contend that the development crisis in Nigeria, Rwanda, Somalia, and the Democratic Republic of Congo since independence, for example, is a result of poor leadership style (Adamolekun, 1999; Huxitable, 1998; Morris Maclean, 2003; van Klinken, 2003). Rather than serving their people, leaders are served by their subjects. Leadership should not be based on a one-way social contract. An ideal and democratic leader serves the people and vice versa and should be removed from office or punished if he or she does not live up to expectations (Greenleaf, 1977; Yukl, 2010).

In the twenty-first century, African nations need development-oriented leaders. Leaders should be committed to the development of the entire society over which they rule. Development-oriented leaders should be prepared to serve the people and ensure the formulation and implementation of policies aimed at enhancing the quality of life for all citizens.

A good example of high-quality leadership and management of public resources is the avoidance of the so-called Dutch disease in the management of the windfall from a nation's major natural resources (Dale, 2003). Dutch disease is the phenomenon whereby a nation's economy is dependent on one export product. African leaders need to diversify their economy and public practices to the extent that the citizens benefit from such ventures.

Electoral Legitimacy

The frequency of ethnic and political conflicts is a major reason for the collapse of countries such as Angola, Burundi, Democratic Republic of Congo, Liberia, Sierra Leone, Nigeria, Rwanda, Somalia, Kenya, Cote d' Ivore, Sudan, among others. The frequency of conflicts is also a critical factor in the underdevelopment of the African continent. One of the solutions to civil war, civil strife, ethnic conflict, and religious unrest in these nations is electoral legitimacy.

Electoral legitimacy can be derived from periodic open, competitive, free, and fair elections that provide to the elected political executive a mandate to govern (Balogun & Mutahaba, 1999; Endres, 2004; Olowu and Sako, 2003; Stillman, 2010; Weatherby, 2009). Peace, in addition to the credibility of governments policies, should be based on electoral legitimacy. This could help African leaders to ensure that political executives are able to pay serious attention to tackling the problem of underdevelopment, including the nurturing of development administration capacity as an instrument for achieving sustainable development on the continent.

SUMMARY

This chapter has examined the predicaments facing the environmental governance and the political economy of sustainable development in Africa. It contended that African governments should embrace a dynamic process of change in which the exploitation of resources, the direction of investments, the orientation of technological development, and institutional change are made consist with future as well as present needs. Further, African nations' governments need to shape their structure of policy network, institutions, inclusiveness, and shared governance approaches. It also argued that sustainable development in African depends upon broad participation, which goes beyond established partnership networks and engages a much wider range of public institutions and individuals.

The chapter stressed the importance of fostering strong communication links between African government officials and their respective citizens. The network theory was used to emphasize the need for building trust with the different organizations that might be involved in the network for the attainment of sustainable development. At the same time, good communication and honesty from political leaders and public administrators are crucial in encouraging participation. Part of the panacea for all these public administrative problems is a lack of moral education and the effective enforcement of the rule of law. It further contended that rapid economic transformation produces social groups that find themselves left behind in the progress being made in most African countries in the past two decades. This transformation has also caused the skills, occupations, and economic activities of these groups to become less important than they were two decades earlier.

This chapter also contended that in order to reap the benefits of trade and investment liberalization, it is important that various goals of sustainable development such as growth, environmental protection, social and economic development be well integrated into national and policy goals in countries in the African continent (Dibie and Offiong, 2012). African nations should realize that developing effective burden-sharing arrangements that reflect common but differentiated responsibilities should play an important role in reconciling the trade-offs between national priorities and protecting global public goods (Waller-Hunter and Jones, 2002; Weatherby et al., 2009).

The political economy of sustainable development in Africa is experiencing a transformation from subsistence or small-scale production via the commercialization of agriculture and the introduction of new technology. However, the problems caused by environmental, social, and economic disintegration, as well as exclusion, tend to affect all nations. The onus is therefore on African nations to tap into the opportunities to benefit from participation in the growing global economic system. Some scholars argued that technologies, capital, and know-how only to nations flow where the fundamental conditions are present to attract and effectively use these flows. African nations will have to develop the policy and institutional framework needed to attract the appropriate technologies (Hudson et al., 2007; Vig and Kraft, 2013) Thus, when trade and investment policies are mutually supportive with environmental and social policies, the contribution of each of these variables to sustainable development is likely to be enhanced.

Finally, the sustainable development framework should galvanize government policies that promote broad-based growth. These policies should transform and enhance good and equitable living standard for all citizens, while promoting equitable growth for women, the poor, minority groups, and immigrants. The policies should also promote respect for human rights, good governance, and a healthy civil society of NGOs. Policies should be introduced to prevent the destruction of their environment, so that the next generation of Africans can enjoy the same or a higher living standard and level of civilization. African governments should be prepared to (1) strengthen coherence among trade, investment, environmental, and social policies by encouraging the use of environmental and social codes of conduct in the private sector, by providing a supportive regulatory and institutional framework for private sector activity, and by promoting awareness of and effective implementation of OCECD instruments dealing with multinational enterprises, corporate governance, and bribery; and (2) support opportunities and capacities for African nations to grow in a way that reinforces environmental protection and social development (Elliot, 2007; OECD, 2001a; Wright and Boorse, 2014).

REFERENCES

Acharya, B. (2002). "Networking and Collaboration Among Voluntary Organizations, Local Bodies and Government Functionaries in India: Experiences in Strengthening Local Self Governance." In *Applying Public Administration in Development: Guidepost to the Future*, edited by P. Collins. Chicester, UK: Wiley.

Adamolekun, O. (1999). "Governance Context and Reorientation of Government." In *Public Administration in Africa,* edited by O. Adamolekun. Boulder, CO: Westview.

Anderson, J. (2011). *Public Policy Making: An Introduction.* Boston, MA: Wadsworth.

Anderson, B. (2000). "AIDs Is Leading Cause of Death in Africa." *Front Lines* (April/May), 1.

Ayittey, G. (2005). *Africa Unchained: The Blueprint for Africa's Future.* New York: Palgrave Macmillan.

Balogun, J., and Gelase, M. (1999). "Redynamizing the African Civil Service for the Twenty-first Century: Prospects for a Non-Bureaucratic Structure." In *Bureaucracy and the Alternatives in World Perspective,* edited by K. Henderson and P. Dwivedi. New York: St. Martin's.

Balogun, M. J., and Mutahaba, G. (1999). "Redynamizing the Civil Service for the 21st Century: Prospects for a Non-bureaucratic Structure." *African Journal of Public Administration and Management, 11*(2), 1–31.

Barbier, E. (1987). "The Concept of Sustainable Economic Development." *Environmental Conservation, 14*(2), 101–110.

Baumer, D. and C. Van Horn. (2014). *Politics and Public Policy.* Los Angeles, CA: Sage.

Bienen, H., and Herbst, J. (1996). "The Relationship Between Political and Economic Reform in Africa." *Comparative Politics, 29*(October), 23–42.

Blakely, E. (1989). *Planning Local Economic Development: Theory and Practice.* London: Sage.

Brandt, W. (1990). *North-South: A Program for Survival: The Report of the Independent Commission on International Development Issues.* London: Penguin.

Bryant, C., and White, L. (1982). *Managing Development in the Third World.* Boulder, CO: Westview, 82.

Collins, P. (Ed.). (2000). *Applying Public Administration in Development: Guidepost to the Future.* Chicester, UK: Wiley.

Dale, R. (2003). "The Logical Framework: An Easy Escape, a Straitjacket, or a Useful Planning Tool." *Journal of Development in Practice, 13*(1), 57–70.

Davies, J. (2002). "The Governance of Urban Regeneration: A Critique of the 'Governing Without Government' Thesis." *Public Administration, 80,* 301–322.

Denhardt, R, Denhardt, J., and Blanc, T. A. (2014). *Public Administration. An Action Orientation.* Boston: Wadsworth/Cengage Learning.

Denhardt, R., and Denhardt, J. (2009). *Public Administration: An Action Orientation* (6th ed.). Belmont, CA: Thomson, 29.

Dia, M. (1993). "A Governance Approach to Civil Service Reform in Sub-Saharan Africa." Technical Paper no. 225. Washington, DC: World Bank.

Dibie, R. (2008). *Non-Governmental Organizations (NGOs) and Sustainable Development in Sub-Saharan Africa.* Lanham, MD: Lexington.

Dibie, R. (2007). "Governance and Redevelopment of Civil Society in Africa." *International Review of Politics and Development, 5*(2), 1–36.

Dibie, R. (2004). "Implication of Globalization in the Teaching of Public Administration in Africa: The Case of Nigeria." *International Journal of Public Administration, 27*(3–4), 151–169.

Dibie, R. (2003). *Public Management and Sustainable Development in Nigeria.* Aldershort, UK: Ashgate.

Dibie, R., and Atibil, C. (2012). "NGOs and the Social and Economic Empowerment of Women in Ghana and Nigeria." *Journal of International Politics and Development, 10*(1), 27–63.

Dibie, R., and Offiong, O. (2012). "The Nature and Impact of Environmental Policy in Nigeria." Journal of *International Politics and Development, 10*(1–2), 113–143.

Dibie, R., and Gadzekpo, L. (2003). "Managing Public Sector Ethics in Ghana and Nigeria." *Politics, Administration and Change, 39*, 1–24.

Easton, T. (2010). *Taking Sides: Clashing View on Environment Issues.* Boston: McGraw-Hill.

Elliott, L. (2007). "Improving the Global Environmental Policies, Principles and Institutions." *Australian Journal of International Affairs, 61*(1), 7–14.

Endres, A. (2004). "Games Theory and Global Environmental Policy," *Poiesis Prax, 3*, 123–139.

Esman, M. (1991). *Management Dimensions of Development.* Bloomfield, CT: Kumarian.

Friis, Robert (2012). *Essentials of Environmental Health.* Boston: Jones and Bartlett.

Greenleaf, R. K. (1977). *Servant Leadership: A Journey into the Nature of Legitimate Power and Greatness.* Mahwah, NJ: Paulist.

Handelman, H. (2009). *The Challenges of Third World Development* (5th ed.). Upper Saddle River, NJ: Prentice Hall.

Holmberg J., Bass, S., and Timberlake, L. (1991). *Defending the Future: A Guide to Sustainable Development.* London: Earthscan.

Hudson, J., Lowe, S., Oscroft, N., and Snell, C. (2007). "Activating Policy Networks: A Case Study of Local Environmental Policy-Making in the United Kingdom." *Policy Studies, 28*(1), 56–70.

Huxitable, P. (1998). "The African State Towards the 21st Century: Legacies of the Critical Juncture." In *The African State at a Critical Juncture,* edited by L. Villalon and P. Huxtable. Boulder, CO: Lynne Rienner.

Hyden, G. (2006). *African Politics in Comparative Perspective.* New York: Cambridge University Press.

Kaldor, M. (2003). "Civil Society and Accountability." *Journal of Human Development, 4*(1), 23–59.

Kaplan, A. (1994). "NGOs, Civil Society and Capacity-Building: Towards the Development of Strategy," 1–17. Retrieved from http://www.cdra.org.za/articles/NGOs. Accessed May 3, 2005.

Kamat, S. (2003). "NGOs and the New Democracy: The False Saviors of International Development." *Harvard International Review, 25*(1), 129–162.

Kickert, W. J. M., and Koppenjan, J. F. M. (1997). "Public Management and Networks Management: An Overview." In *Managing Complex Networks: Strategies for the Public Sector,* edited by W. J. Kickert, E. H. Klijn, and J. F. M. Koppenjan. London: Sage.

Kiggundu, M. (1989). *Managing Organizations in Developing Countries: An Operational and Strategic Approach.* West Hartford, CT: Kumarian.

Kraft, M. (2011). *Environmental Policy and Politics.* New York: Harper Collins College Press.

Kraft, M., and Furlong, S. (2013). *Public Policy: Politics, Analysis, and Alternatives.* Washington, DC: Congressional Quarterly Press.

Lairson, T., and Skidmore, D. (1992). *International Political Economy.* Belmont, CA: Wadsworth Press.

Lehne, R. (2010). *Government and Business: American Political Economy in Comparative Perspective.* Washington, DC: Congressional Quarterly Press.

McBride, A., and Sherrenden, M. (2004). "Towards a Global Research Agenda on Civic Service: Editors Introduction to This Special Issue." *Nonprofit and Voluntary Sector Quarterly,* Supplement, *33*(4), 3S–17S.

McConnell, R., and Abel, D. (2008). *Environmental Issues: An Introduction to Sustainable Development.* Upper Saddle River, NJ: Prentice Hall.

Michie, A. M. (1989). "Decentralization and Public Welfare in Rural India." Paper Prepared for the Annual Western Social Science Association Meeting, Albuquerque, April 26–29.

Migdal, I. (1988). *Strong Societies and Weak States: States–Society Relations and State Capabilities in the Third World*. Princeton, NJ: Princeton University Press.

Morris Maclean, L. (2003). "Economies of Affection in Crisis: How the Weak State Transform Informal Institutions in Africa." Paper presented at the Annual Meeting of the American Political Science Association, Philadelphia, August 28–31.

Nafziger, W. (2005a). "Meiji Japan as a Model for Africa's Economic Development." *African and Asian Studies, 4*(4), 443–464.

Nafziger, W. (2005b). *Economic Development* (4th ed.). New York: Cambridge University Press.

Obiyan, S. (2005). "A Critical Examination of the State Versus Non-Governmental Organizations (NGOs) in Public Sphere in the Global South: Will the State Die as the NGOs Thrive in Sub-Saharan Africa and Asia." *African and Asian Studies,* 2(2), 301–325.

Olowu, D., and Sako, S. (2003). *Better Governance and Public Policy: Capacity Building for Development in Africa*. Bloomfield, CT: Kumarian.

Oluwo, D. (1999). "Accountability and Transparency." In *Public Administration in Africa,* edited by L. Adamolekun. Boulder, CO: Westview.

Onimode, B. (2004). *African Development and Governance Strategies in the 21st Century*. London: Palgrave Macmillan.

Organization for Economic Cooperation and Development (OECD). (2001a). *Sustainable Development and Empowerment*. Paris: OECD.

Organization for Economic Cooperation and Development (OECD). (2001b). *Sustainable Development: Critical Issues*. Paris: OECD.

Patel, L., and Wilson, T. (2004). "Civic Service in Sub-Saharan Africa." *Nonprofit and Voluntary Sector Quarterly, 33*(4), 22–38.

Pearce, D., and Barbier, E. B. (2000). *Blueprint for a Sustainable Economy*. London: Earthscan.

Peace, D., Markandya, A., and Barbier, E. (1989). *Blueprint for a Green Economy*. London: Earthscan.

Peters, G. B. (2013). American *Public Policy: Promise and Performance*. Los Angeles, CA: Sage.

Peters, G. B. (1988). *Public Bureaucracy in Comparative Perspective*. Tuscaloosa: University of Alabama Press.

Rhodes, R. (1996). "The New Governance: Governing without Government," *Political Studies, 44*, 652–667.

Rogers, P., Jalal, K., and Boyd, J. (2008). *An Introduction to Sustainable Development*. Starling, VA: Earthscan.

Rondinelli, D. A. (1983). *Development Projects as Policy Experiments*. New York: Methuen.

Rondinelli, D., Nellis, J., and Cheema, S. (1984). "Decentralization in Developing Countries." *World Bank Staff Working Paper no. 58*. Washington, DC: World Bank.

Rcorda, N. (2012). *Fundamentals of Sustainable Development*. New York: Routledge.

Rosenbaum, W. (2008). *Environmental Politics and Policy*. Washington DC: Congressional Quarterly Press.

Rydin, Y. (1997)."Policy Networks, Local Discourse and Implementation of Sustainable Development." In *The Politics of Sustainable Development: Theory, Policy and Practice Within the European Union,* edited by S. Baker. London: Routledge.

Shafritz, J., Russell, E. W., and Borick, C. P. (2009). *Introducing Public Administration* (6th ed.), New York: Longman.

Skolnik, R. (2008). *Essentials of Global Health*. Boston: Jones and Bartlett.

Smith, B. C. (1996). *Understanding Third World Politics*. Bloomington: Indiana University Press.

Steel, B., Clinton, R., and Lovrich, N. (2003). *Environmental Politics and Policy: A Comparative Approach*. Boston: McGraw Hill.
Stillman, R. (2010). *Public Administration, Concepts and Cases*. Boston: Wadsworth.
Taylor, R. (2012). *Taking Sides: Clashing Views in Sustainability*. New York: McGraw Hill.
Theron, F. (2008). *The Development Change Agent*. Pretoria, South Africa: Van Schail.
Thin, N. (2002). *Social Progress and Sustainable Development*. Bloomfield, CT: Kumarian.
Umeh, O. J., and Greg A. (2005). *Culture, Development and Public Administration in Africa*. Bloomfield, CT: Kumarian.
United Nations Development Program (UNDP). (n.d.). *Human Development Index (2010–2012)*. New York: Oxford University Press.
United Nations Development Program (UNDP), (n.d.). *Human Development Index (1993–94)*. New York: Oxford University Press
Van Klinken, G. (2003)."Operationalizing Local Governance in Kilimanjaro." *Journal of Development in Practice, 13*(1), 71–82.
Vig, N., and Kraft, M. (2013). *Environmental Policy*. Los Angeles: Sage.
Waller-Hunter, J., and Jones, T. (2002). "Globalization and Sustainable Development." *International review for Environmental Strategies, 3*(1), 53–62.
Weatherby J. et al. (2009). *The Other World: Issues and Politics of Developing World*. Boston: Pearson.
Weaver, J., Rock, M., and Kusterer, K. (1996). *Achieving Broad-Based Sustainable Development*. Bloomfield, CT: Kumarian.
WCED. (1987). *Our Common Future*. Brundtland Report. Oxford: Oxford University Press.
White, L. G. (1987). *Creating Opportunities for Change*. Boulder, CO: Lynne Reinner.
Withgott, J., and Brennan, S. (2009). *Essential of Environment* (3rd ed.). Boston: Pearson.
World Bank. (2008). *World Bank Report*. New York: Oxford University Press.
World Bank. (2007). *World Bank Report*. New York: Oxford University Press.
World Bank. (1997). *World Development Report. The State in a Changing World*. New York: Oxford University Press.
World Bank. (2000). *Can Africa Claim the 21st Century?* New York: Oxford University Press.
World Environmental Snapshot. (2012). *Samoa Country Review*, 174–185.
Wright, R., and Boorse, D. (2014). *Environmental Science: Toward a Sustainable Future*. Boston: Pearson.
Wunsch, J., and Olowu, D. (eds). (1990). *The Failure of the Centralized State: Institutions and Self-Governance in Africa*. Boulder, CO: Westview.
Yukl, G. (2010). *Leadership in Organizations*. Upper Saddle River, NJ: Prentice Hall.

7 Environmental Policies and Issues in Ghana

Albert Assibey-Mensah, PhD
and Robert Dibie, PhD

INTRODUCTION

Ghana is an English-speaking nation in West Africa. Its population in 2011 was about 24 million and growing at a rate of around 2.9 percent per annum. The United Nations Human Development Index Report (2012) indicated that the population will reach 26.4 million in 2015 and 33 million in 2020. About 42 percent of the population in Ghana lives in urban areas, and the remaining 48 percent reside in the rural regions of the country. It is anticipated that the urban population may increase to about 46 percent in 2015 (UN Human Development Index, 2012).

The major environmental problems facing Ghana are linked with ineffective legislation. Overviews of the problems that directly and indirectly affect the environment policy in Ghana include the fact that only a small fraction of those who violate environmental policies ever reach the courts, with very few cases leading to any sentencing for environmental crimes. Further, environmental impacts and consequences are not described despite being mandatory by the national legislature. Ghana tends to have less elaborate garbage collection regulation when compared with most developed countries.

According to a 2010 report from Ghana's Environmental Analysis and Management Plan (EAMP), Ghana's environmental problems include (1) urban growth disturbances to fragile habitats; (2) new forms of stress, including pollution and health problems; (3) gaps in current regulations, mainly in the environmental health sector, including pollution monitoring and waste disposal; (4) a lack of coordinated environmental law implementation; (5) a lack of any regulation to protect coastal areas from urban and industrial development; (6) the nonexistence of effective environmental impact statement (EIS) requirements for development projects at all levels and sectors; (7) a lack of policing power and resources; (8) a lack of long-term planning; and (9) a lack of qualified staff. These conditions make it very difficult to establish monitoring stations and follow-up mechanisms. Conflicts in and overlaps of regulatory activities also play a role in the current environmental problems, as well as a lack of public participation and environmental awareness (Edoho and Dibie, 2000).

Ghana, like many other developing countries, is also experiencing the problem of air, water, and oil pollution, along with damage to biota and their habitat. Farmers in the country have to spray many pesticides to prevent insect pests from destroying their cocoa and other cash crops. By spraying these pesticides, the farmers expose other citizens and themselves to organchlorine, which can disrupt the nervous system, causing convulsions and death. According to Nancy Maxwell (2014), these pesticides are very persistent in the environment; they are lipophilic, bioconcentate in fatty tissue, and ultimately biomagnify in the food chain, as well as cause serious human health problem. Other problems arise due to the deficiency encountered in the poorer regions in the country. These deficiencies are a result of a shortage of money, food, education, medical care, safe drinking water, and electricity supply in poorer and some rural regions in Ghana. An excess in bigger cities like Accra and Kumasi creates deficiency in the rural areas. In many cases, the main issue is the poor distribution of infrastructure. The influx of immigration into rich and bigger cities is due to poverty and economic insecurity in the rural regions. On the other hand, the influx of immigration to bigger cities has also created a greater urbanization crisis and exacerbated pollution, sanitation, and waste management problems. The various environmental problems influence one other and seem to be all interconnected. The only way any of these complex environmental issues could be solved is if the government were to attempt to simultaneously solve all of them; that is, if the Government of Ghana wants to solve one problem, it must address them all. Thus, due to limited resources, it is likely that these environmental problems will never be solved. Many of the solutions to current environmental problems in Ghana, should they be implemented, would undoubtedly lead to another set of problems.

In its development process since the nation was granted independence in 1957 by the British government, Ghana has experienced various types of environmental problems. The nation's ecological conditions—in particular climate change, soil degradation, and water availability—have constantly affected the crop yields as well as the sources for food for wild animals. Further, change in the ecosystem in Ghana has influenced the abundance of human pathogens, such as malaria and cholera. The World Health Organization (WHO) report reveals that malaria constitutes about 9–11 percent of the disease burden in the country (World Health Organization and UNICEF, 2010).

Despite the enactment of Ghana's environmental policy, the governance structure is characterized by inappropriate institutional and government arrangements, including the presence of corruption and weak systems of regulation and accountability. The convoluted administration of environmental policies has made it difficult for the nation to effectively accomplish all its environmental policy goals (EAMP Report, 2010; Hilson, 2004).

Many actions to slow ecosystem degradation do not address these indirect initiatives to accomplish the nation's environmental goals. As noted by Acheampong (2001) and the Centre for Environmental and Health Research and Training (CEHRT) report (2010). The forest management in Ghana is influenced more by actions of stakeholders outside the forest sector, such as trade policies, institutions' macroeconomic decisions, and policies in other sectors such as agriculture, infrastructure, energy, and mining than by those within the Ministry.

Although Ghana is fast developing, the government has underinvested in the development and diffusion of technologies that could increase the efficient use of ecosystem services and reduce the harmful impacts of various initiatives of environmental change in the country (EAMP Report, 2010). In particular, Ghana's existing institutions are not well designed to deal with the management of common pool resources as a result of the issues of ownership and access to resources, the right to participation in decision making, and the regulation of particular types of resource use or discharge of waste. This predicament has severely influenced the sustainability of ecosystem management in the country (Acheampong, 2001; Edoho and Dibie, 2000). The struggle over the fundamental determinants of who wins and who loses from changes in the nation's environment issues among the various regulatory agencies and ministries continues to undermine efforts to solve the environmental problems that the nations face.

Another very important dynamic of the environmental problems in Ghana is the challenge of reversing the degradation of the ecosystem while increasing demand for services. The action to increase food production in Ghana, which involves increasing the use of water and fertilizers as well as the expansion of the area of cultivated lands, often leads to another set of problems. Policy in Ghana often causes the degradation of other ecosystem services, including reducing the availability of water for other uses, degrading water quality, reducing biodiversity, and decreasing forest cover, resulting in the loss of forest products and the release of greenhouse gasses. At the same time, the conversion of forest to agriculture has significantly changed the frequency and magnitude of floods, although the nature of the impact depends on the characteristics of the local ecosystem and the type of change in the land cover. Further, although demand for natural resources could be met partially in some situations, there need to be changes in the policies, institutions, and practices that are not serving the interests of the common good.

This chapter examines the social and economic impact of environmental policies in Ghana. It argues that, due to the limited socioeconomic progress that has occurred in Ghana, environmental policies have become ineffective. As in other developing nations, the situation in Ghana has remained intriguing. The environmental puzzle is linked to the pervading condition of underdevelopment, in Ghana. Despite several policy mandates that have failed to address living standards and poverty, to provide employment, and to raise incomes, despite the implementation of some programs, the Government of Ghana has not changed its approach to addressing environmental issues. Institutionally, the dearth of capable programs that address environmental concerns abating low living standards has been disappointing.

The chapter recommends the reorientation of Ghana's environmental regulatory agencies as well as the greater participation of citizens, private, nongovernmental organizations (NGOs), and community-based organizations in making decisions that relate to the nation's ecosystem and the human relationship with the environment. There is also the need for Ghana to start building sustainable cities in the near future. By using incentives and other empowering policies, the Government of Ghana could induce strategies that are capable of transforming the nation into autonomous, ecofriendly, developing economy.

PRINCIPLES AND MODELS OF ENVIRONMENTAL POLICY

The explicit concern about sustainability has generated the velocity of many countries in the world to seek environmental change. According to Niko Roorda (2012), the rate of human transformation of the environment is increasing, particularly in countries undergoing rapid industrialization or deindustrialization. Humans' ability to destroy the ecosystem that supports their life is outrageous (Myers and Spoolman, 2014). Every day we witness human beings burning through natural assets and their capacity to support life and quality of human life without thought about the future and the rights and needs of current citizens (Roorda, 2012). The self-regulating concept stipulates that the current conventional human behavior toward the environment is unsustainable (Bandura, 1986 Schunk and Zimmerman, 1994). According to Burden and Francis (1978; noted also in Lehmann, 1999, 34), altruism is a subset of prosocial behavior. They hypothesize that (1) persons with a strong selfish and competitive orientation are less likely to act ecologically, and (2) people who have satisfied their personal needs are more likely to act ecologically because they have more resources (time, money, energy) to care about bigger, less personal social and proenvironmental issues. This, of course, does not mean that powerful nations limit their ecological footprint out of environmental concerns, but it does show that more affluence leads to more ecologically sound behavior. As a result, human beings would need to voluntarily regulate themselves in order to begin to achieve sustained economic and environmental growth (Myers and Spoolman, 2014). The concept calls for self-regulated behavior by individuals and public and private institutions alike. Unless humankind develops a more harmonious lifestyle, the relationship of natural system and humanity will not be mutualistic. Self-regulation strategies support the development of human thinking, action, and judgment in explaining how people relate to outcomes and react to the environment. This thinking requires the development of reflection acquired through self-regulation. For example, self-judgment of behavior may lead to the evaluation that social interaction is needed in one situation and a change in goals in another. According to Bandura (1986), a triadic interaction between the concepts is dynamic, not always equal, and presupposes reflective thoughts to determine which process is necessary for a given situation. For example, self-judgment of behavior may lead to the recycling of personal waste products. Figure 7.1 shows the reflective self-regulated learning in human relationship with the environmental.

Therefore, social interaction, such as the government's promotion of environment issues awareness and the media's advertisement of the implication of clean and dirty environments, may lead to a change in goals. Thus, interaction in one situation is needed in another situation to achieve a change in goals on environmental issues. The self-regulation model in this chapter is intended to show how the Government of Ghana made initiatives to enact the nation's environmental policy as well as the environmental action plan.

Natural resources are both a driver and a possible constraint of economic growth. The higher GDP is, the higher is the demand for natural resources;

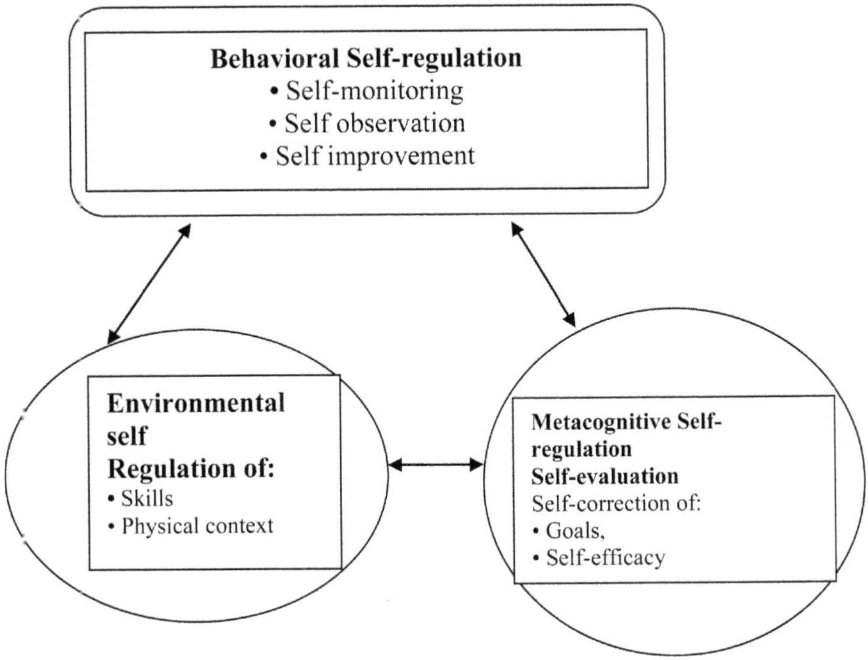

Figure 7.1 Reflective Self-Regulation towards the Environment

Source: Adopted from Schunk and Zimmerman (1994).

growing demand leads to higher production, which depletes stocks, all else being equal. The ecological footprint is a measure of humanity's demand on nature. It represents how much land and water area a human population requires to regenerate the resources it consumes and to absorb its wastes (Baumer and Horn, 2014; Vig and Kraft, 2013). According to this model, the gross national product of any country is dependent on the supply of natural resources from the environment and the demand for such natural resources. The supply of environmental product is, however, contingent on the depletion rate and the implication of fossil fuel, water, forest, and air and water pollution (Barbier, 2005). Declining stocks, on the other hand, reduce the potential medium- to longer-term production of natural resources, potentially constraining economic growth. Resource efficiency is promoted in the Ghana environmental regulation policy to reduce demand and improve the management of supply (Adelman, 1999). The rebound effect of the environment that produces the crops, biomass, or minerals is also contingent on the type of environment policy and how it is implemented in any country. Figure 7.2 illustrates the relationships between economic growth and natural resources.

The ability of the ecosystem to continue to produce should be taken into consideration because poor management of the environment can lead

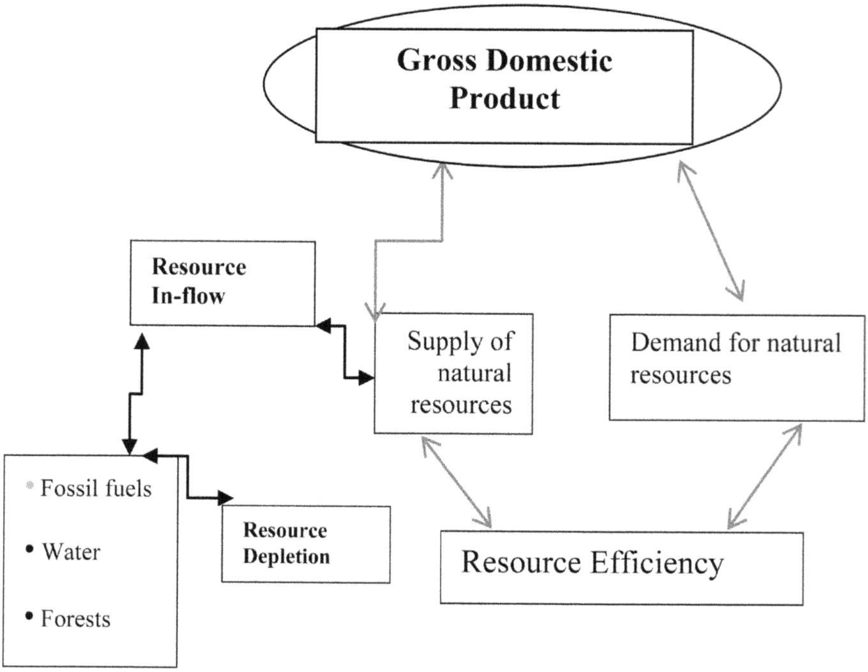

Figure 7.2 The Relations between Economic Growth and Natural Resources
Source: Adopted from Global Footprint Network (2010).

to a reduction of natural resources (Polasky and Segerson, 2009). When this happens, the intended benefits of efficiency improvements by increasing demand and gross domestic product (GDP) could never be achieved. As a result, there is a causal relationship between natural resource efficiency, economic growth, and gross domestic product.

In her seminal work, Grace Carson (1962) alerted societies to become aware of the damage caused by market forces in light of unlimited exploitations. By creating critical awareness of such exploitation, Carson specifically focused on environmental degradations, including pollution, poverty, and social inequities. Within the broader scope, the larger and more enduring outcome of Carson's admonition caused the pioneering establishment of the environmental movement. In articulating her dire concerns regarding environmental damage of the careless use of land and ecological harmful practices, Carson became critical. In arguing for responsive policies, Carson advocated projects that would accord social equity and promote stewardship. Carson felt that the twin goals of equity and stewardship would steadily tamp down the rampaging programs and moderate the wild quest for profits.

Richard Wright and Dorothy Boorse (2014) argued that society seeks sustainability when it identifies ways that regard stewardship and sound science as critical themes and when it unifies and directs long-term development, given natural capital. As defined, a natural capital consists of investment resources composed of produced assets, both renewable and nonrenewal, and of human resources (World Bank, 2008c). Progressively, given stewardship, social and economic activities based on natural capital would become instrumental in addressing poverty. As science enhances education and training, consequent knowledge becomes a useful tool, given sustainable development (Marglin, 2003; Sachs, 2005).

Environmentally, concerns about population growth are as critical as science education, using techniques fostering productivity and growth (Maxwell, 2014; Polashy and Sergerson, 2009). According to World Bank Report (2010), "Every generation should leave undiminished all the species of animals it found existing on earth." The publication suggested that environmental programs should become the first priority. In response, Robert Solow rejected the intergenerational equity position. In his disagreement, Solow, the Nobel Laureate in Economics, argued that, because scarce resources are used to address the environmental crisis, they necessarily abated institutional capacity and potential growth. Debunking Solow's analytical retort, the ecologically minded groups, under its cerebrally proactive leader, chastened and decried the declining quality of the global environment, given socially destitute communities reflecting the dark side of economic prosperity.

Besides an inconsistent management approach in public organizations, institutional capabilities coexisted with endemic corruption in light of the aid programs. Critics of the failed programs include Jeffrey Sachs, Robert

Rotbert, and Stephen Marglin (Sachs, 2005). Roth (2005) and Marglin (2003) consider Western-backed programs, sponsored largely by the Washington Consensus, the World Bank, and the IMF, as cultural imperialism.

In light of proactive development, a reasonable accommodation was the issue of debt that has dogged most developing nations, since independence (Krugman and Obstfeld, 2009; Stiglitz, 2006). Several developing nations have accumulated debt burdens. Over time, their budget deficits have risen because of tariffs imposed by the developed nations (Stiglitz, 2003). Further, decreases in export revenues the resulting from unfavorable terms in the world commodity markets have caused existing programs and the growing number of public goods to become underfunded. Because disasters have become frequent, deficits have risen continually. As emergency programs became necessary to forestall social rancor and potential disasters, they contributed enormously to increases in the debt burden. In light of social responsibility, the wave of debt forgiveness was integrated with good governance programs.

There has over time been a crucial need for Ghana to come up with measures to address the perilous environmental threats. From the vantage point of society, the environment crisis initially gained credence with the work of Rachel Carson (2002) in her book *The Silent Spring*. Carson unabashedly awakened the public consciousness of the environmental shortcomings of the market system given pollution, threats of pollution, and the misuse of natural resources and toxic chemicals that all too well have reflected untold potential dangers to the socioeconomic health and safety of societies. Over time, environmental concerns have focused on, among other issues, clean water and erosion, food supply and energy resources, climate change, air and water quality, population and poverty, health, and food safety. Carson's unprecedented caution was inconsistent with the private sector's strategic development programs and strategies of profitability that motivate firms. Without so-called social responsibility, firms are inclined to pursue their private interests to the detriment of the environments that engender those benefits. Specifically, the public bureaus directly charged with regulating the environment and governments whose social welfare role has embraced proactive policies have either become indifferent about the effective measures necessarily required to address environmental concerns or have become complacent about their power, given the material comforts arising from their disturbingly cozy relationship with the private sector.

Strategically, the expectation has been that the authority entrusted to governments ends up strategically placing them in leadership positions that empower their implementations of the rule of law through regulations and the monitoring of private sector activities in ways favorable to the environment. In addition to Carson's advocating government intervention, the work of David Bower and Barry Commoner contributed to ushering in the era of modern environmentalism, given their advocacy for not only stakeholders but also for policy actions, regulatory reforms, and activist agenda (Cunningham, 2008).Clearly, the ensuing strategies fostered by several green movements with regard to their environmental concerns were directed toward maintaining healthy, safe environmental standards as positive capacity tools that improve social welfare. Given society's welfare and economic development outcomes, the interplay of activities and the responsibilities of social institutions with respect to natural resources utilization bear tactically on the expectations of environmental development and potential progress. When enterprises strategically foster dynamic development, they tend to cobble together a networking relationship between natural resource utilization and social necessary institutions grounded in government policy (Baumer and Van Horn, 2014; Hilson, 2004). Consequently, in forging dynamic development, social welfare and firms' profit-motivated investments need to facilitate not only favorable proactive policies but also the technology and institutional needs that impact society's education and skill levels, capacity, and standards of living.

SOCIAL IMPACT OF ENVIRONMENTAL POLICIES

The enactment of Ghana's environmental policy is the result of a series of government actions initiated in March 1988. The policy was considered necessary to simultaneously address environmental issues and economy development as the nation explored its sustainable development agenda. As a result of this national goal, the Environmental Action Plan (EAP) was created to identify specific actions to be implemented so as to protect the nation's environment and to ensure better management of its natural resources. Ghana's Environmental Analysis and Management Plan (EAMP) for the Sustainable Land and Water Management Project was developed to explore sustainable development issues as defined by the World Commission on Environment and Development in 1987 and the United Nations 1992 Earth conference in Brazil. The plan provides a broad framework for the integration of environmental issues into development strategies and actions (EAMP, 2010).

The Government of Ghana's environmental plan proposed initiatives to ensure that the provision of environmental sanitation is placed highest on the national and regional agendas for achieving sustainable urban development. Other aspects of the environmental plan include the recognition of the importance of behavioral change and help bring it about through education and public awareness of environmental problems (Carroll, 2008; Issahaku, 2000; Ofori, 1991). The policy was enacted with the intention that it will help Ghana to achieve demand-driven, community-led programs of improved urban environments, as well as the sustainable use and maintenance of ecosystem in the country. Recycled liquid and solid wastes from households and establishments for the production of energy, fertilizer, and other materials used in the country were required to be regulated. However, the lofty goals have not yet been accomplished. The EAMP for the Sustainable Land and Water Management Project also required the government to conduct studies to identify suitable landfill, as well as waste management process techniques, to ensure appropriate domestic and hazardous waste disposal (EAMP, 2010).

With the EAMP as a policy instrument, the implementation process authorized the newly created Environmental Protection Agency (EPA) with the mandate of directing and steering Ghana's environmental development program and its expected progress forward. Overall, EPA's responsibility reflected an effective policy with a focus on established national priorities. The EPA's responsibilities are based not only on the focus but also on the proactive activities. The EPA is also required to address trends in the environment, economy, and society (Center for Environmental Health Research and Training, 2010; WHO, 2010). In principle, Ghana's environmental policy is expected to galvanize the skills and expertise of managers or a team of management in charge of socially efficient institutions. The expectation is that management is undeniably well-informed about crucial strategies, including the technology of private firms and the ability to exploit the country's natural resources (Schunck and Zinmerman, 1994; World Bank, 2010).

A corollary expectation was that environmental progress and growth will be systemic, based on a credible maintenance program that accords, reasonably and continually, a decent level of safety and health for the current generation. In theory, the strategic goal of the EAMP appears to be excellent action plan that is credibly oriented toward the potential future expectations of the country. In reality, however, loop holes in the implementation process of the environmental policy over time have left much to be desired. In particular, corruption and managerial ineptitude results from the inability of the monitoring personnel to act in ways consistent with the plan's mandates (Ferrell, Faedrich, and Ferrel, 2013). The dearth of proper experience, skills, and knowledge required for the plan's tactical execution has flawed the implementation process immensely. According to Edoho and Dibie (2000), there's been a lack of the framework required to implement environmental protection. Further, the authors noted that Ghana (as well as Nigeria) has merely placed such policies on paper without a clear understanding of what should be done to address the pervasive environmental crisis. They noted that such behaviors typically reflect not only the lack of knowledge but also the inability to claim ownership of the environmental principles imposed by the Western nations.

Hilson (2004) and Dibie and Gadzekpo (2003) noted that corruption has become endemic. Because corruption has pervaded the socioeconomic milieu impacting the implementation process, the lack of oversight effectively provides an illustration of the dearth of required, legitimate management thrust. Steadily and forcefully, managerial skills and know-how could, in effect, make significant strides and abate environmental stress. All the same, environmental policies that circumscribe native strategies, limiting their participation, have entailed the failure to incorporate or embrace the strategic orientation of an indigenous social and environmental thrust (Dibie and Offiong, 2012). This situation is associated with the risks of taking a backseat to somewhat irrelevant and hollow U.S. and European models that may not be compatible with Ghana's local situation.

Further, in Ghana, resource depletion continues to occur not only because of the existence of nonrenewable natural resources but also due to the irreparable pollution that has pervaded both mining and agriculture (EAMP, 2010; Klane, 2010). On several strips of farmland, as in mining regions, inadequate knowledge about geographical landmass and the lack of experience have prevented maintaining the ecosystem and ecological process essential for the functioning of the biosphere. In rural areas dotted by small farmlands, small-scale farmers have been unable to come up with the investment funds they need to cultivate crops. As a result, generations of poor people have suffered severely and have experienced tremendous difficulties in scraping up even paltry amounts of funds for farm investments (Carroll, 2008). Small farm owners don't have the resources that are operationally necessary for farm investments. The Agricultural Development Banks are concentrated in the urban areas, such as Kumasi, Accra, and other large cities, far from the rural locations of farms, and the farmers are located in the rural regions.

Farm cooperatives exist in the few rural areas but have been, for the most part, unable to benefit a significant portion of the small farmers. Without available microfinancing programs, the impoverished small farmers have had to use obsolete tools and equipment. In contrast to systems of farm mechanization and modern agricultural practices, socioeconomically poor farmers in several rural regions employ the slash-and-burn techniques to hasten the clearing of the vegetation cover. As a result, soils experience further deterioration, and the denuded open fields lose their required soil nutrients, given the slash-and-burn methods, and the chemical nutrients that crops need for quality production.

Strategically, operational approaches in agriculture, lumber, and mining that employ the burning of vegetation, as well as the removal and dumping of soil, have failed to furnish adequate protection for humans, animals, and plants and their habitat against harmful impacts. Indeed, not only are such cultures destructive in practice; they also fail to preserve biodiversity. The mining industry during the structural adjustment process in the 1980s is a prominent example. Even though Ghana has long been recognized as the world's leading nation in its production of numerous industrial and precious metals, according to Hilson (2004), such minerals contribute insignificantly to the country's gross domestic product.

Even though Ghana has long been recognized as the world leader in the production of numerous industrial and precious minerals, the unvarnished fact is that the claims of considerable gains from mining have been exaggerated (Acheanpong, 2001; Ofori, 1991). The insignificant macroeconomic gains are clearly offset by the continually expanding mining activities that eventually end up ruining the landscape, the biosphere, and the irreplaceable ecology. In particular, the literature is replete with unabashed claims of illegal exploitations by various firms and individuals, including the displacement of numerous persons whose survival crucially depends on subsistent income, and destruction of a wide range of cultural resources. Several other instances of environmental damage include excessive land degradations, contamination, and chemical pollution (WCED, 1987; World Bank, 2009).

Rural locations, in contrast to urban locations, have remained the dominant scenes for farming activities besides providing residency for almost 60 percent of Ghana's nearly 25 million people. As the breadbasket of the country, Ghana's rural regions represent the best and most productive of the landmass. Unlike the few urban locations in regions such as Accra, Kumasi, Sekondi-Takoradi, and a few relatively smaller cities in the northern and Brong-Ahafo regions, Ghana's rural locations produce staple foods, cash crops, and timber. The mainstay of cash crops is cocoa, produced mostly in Brong-Ahafo, Ashanti, and the western regions.

The growing population is exacerbating the demand on resources and services, thus increasing the generation of waste (Edoho and Dibie, 2000; Klare, 2010) by the "throw-away society," and this is accelerating the dynamics of overconsumption-poverty-inequality-environment (Agyeman and Swaine, 2008; Hill, 2005; World Bank, 2008a). The Ghana government estimated that 35 percent of urban waste generated is never collected, and reports also show that about 10 percent of the population actually does not use a municipal waste collection system (United Nations Habitat, 2008). The uncollected waste on the streets is the common cause of blocked drainages, which exacerbate the risk of flooding and vector-borne diseases and also reduce the aesthetic value of urban walkways and pavements. Worse still, the so-called green areas and river systems have been turned into SW dumping sites (Country Profiles, 2012). Generally, the urban administrations of Ghana consider waste collection as a low-priority matter, though large urban centers, such as Accra and Kumasi, are relatively better. Poor people generally depend on natural resources; hence the negligence of waste sinks the poor deeper into poverty, thereby undermining the possibility of the economic expansion of Ghana. The natural environment is crucial to the economic and social systems in Ghana. The continued advocacy for sustainable development in the past years still has a gray area on environmental conservation (Hill, 2005) , in which Ghana's waste management systems play a great role in maintaining a pollution-free environment.

The partial public relations campaign by the Government of Ghana on the revitalization of the nation's environment in the past few years has involved policy statements on sanitation. However, this institutional effort did not invite the community or the formal and informal stakeholders in waste management to participate in the decision-making process.

The ministries governing waste management under the banner of environmental protection are the Environmental Protection Agency, the Ministry of Water Resources, the Ministry of Labor and Social Affairs, the Ghana Quality and Standards Authority, and the municipalities (EAMP Report, 2010). These agencies are known to have misunderstandings about the implementation of policies and standards (Acheampong, 2001; Barbier, 2005; World Development Report, 2005).

Table 7.1 shows some essential environmental policies that have been enacted by the Government of Ghana and the national agencies that have been established to regulate each aspect of the environment. The government has also assigned responsibilities to some of its ministries and has initiated many environmental policies to reform the energy sector of the country. These initiatives were considered very crucial in attracting private sector participation as well as in solving the electricity shortages and other sustainable development problems. The goal has been to eliminate inadequate power supply as a constraint on economic development in general and to develop a well-managed, well regulated, and sustainable energy sector. The Ghana transformation process culminated in the establishment of the Public Utility Regulatory Commission (PURC) and the Energy Commission (EC) in 1997 (Issahaku, 2000; Ofori, 1991).

According to Abdul-Nasiru Issahaku (2000), the Environmental Assessment Regulations, LI 1652, policy was enacted in 1999 to give complete legal status to the Ghana Environmental Impact Assessment procedures. The public policy required that all developmental activities likely to impact adversely on the environment must be subject to environmental assessment. The objective of LI 1652 is to ensure that such development activities are carried out in an environmentally sound and sustainable manner. The key objectives of their task included:

1. Defining the screening criteria for environmental assessment for energy sector investments.
2. Determining the scope of the environmental impact assessment (EIA) for the sector.
3. Providing systematic procedures on EIS preparations for the sector.
4. Providing guidelines on common potential impacts and mitigation measures.

Ghana's National Environmental Sanitation Policy (ESP) dates from 1999. It sets out the roles of a number of ministries and institutions but

Table 7.1 List of Environmental Policies, Agencies, and Ministries

Environmental Policy	Policy Goal	Ministry or Agency in Charge of Regulation
Ghanaian Environmental Protection Council (EPC), created in 1973	The EPC was the main governmental body that decided whether an environmental impact certificate or permit should be issued to potential contractors.	Report to the president
The National Environmental Policy (NEP), 1991	Takes actions to be carried out to protect the environment and ensure better management of natural resources.	
National Environmental Action Plan (NEAP)	The aim of the policy is to ensure the sound management of the environment and the avoidance of exploitation of resources in ways that may result in irreparable damage to the environment.	Ministry of Energy
PURC Act (Act 538) of 1997	Regulates and oversees the provision of utility services by public utilities to consumers.	Public Utility Regulatory Commission (PURC) and the Energy Commission (EC)
Environmental Assessment Regulations, LI 1652, 1999.	Gives complete legal status to the Ghana Environmental Impact Assessment procedures.	
National Environmental Sanitation Policy (ESP), 1999	The policy mentions the role of the Ministry of Health in supporting hygiene education activities and contributing to regulation and standard setting for environmental sanitation services.	Ministry of Health
National Water Policy, 1998	Regulates the quality of drinking water in Ghana.	
Minerals and Mining Act, 2006, Act 703	Minerals Commission and EPA continues to offer specific capacity development programs for small-scale mining (SSM) relating to their operations.	Environmental Protection Agency

Source: EAMP (2010).

needs to be updated to include reference to the roles of Ministry of Works and Housing (MWH). The policy mentions the role of the Ministry of Health in supporting hygiene education activities and in contributing to the regulation and standard setting for environmental sanitation services.

In the past, the several techniques employed in the exploitation of natural resources have fallen woefully short. Costly farm applications have been disappointing in their implementation. Poor farmers can ill afford using alien technologies without appropriate know-how, and they are likely to face difficulties and endure irreparable negative effects due to costly changes because they verge on poverty. According to the National Environmental Action Plan (NEAP), sustainability is crucial only as long as such a strategy ensures the implementation of good management practice. Through sound management, the robust program can facilitate techniques that avoid the exploitation of resources that causes irreparable damage.

In Ghana's manufacturing sector, the Akoben was established by the nation's Environmental Protection Agency (EPA) to assess environmental performance ratings of mining and manufacturing operations in the country. The agency adopted a five-color rating scheme. Table 7.2 shows the negative impact, for example, of liquid biofuel plantations and their mitigation measures. The four colors—blue, green, gold, and orange—indicate environmental performance ranging from excellent to poor. These ratings measure the environmental performance of companies based on their day-to-day operations once they have approved environmental impact assessments (EIA) and have obtained their environmental permit to operate.

Table 7.2 Impact of Liquid Biofuel Plantations and Their Mitigation Measures

	Potential Negative Impacts Issues	Mitigating Measures
Liquid biofuels	• Ethanol and land use for energy plantations conflict with demand for food. Could face ethical issues in developing countries because most have unsustainable agricultural and food problems. • Processing of sugarcane into ethanol consumes a lot of water; (about 1 tons alcohol to over 10 tons water) and releases a lot of gaseous, liquid, and solid pollutants.	• Biofuels in most cases become desirable when crude oil prices go beyond US$30 per barrel. • Using cleaner production techniques could reduce the water consumption significantly to about 1 tons of alcohol to 4 tons of water. Scrubbing techniques tackles the gaseous emissions. The liquid and solid discharges can be processed and used as farm fertilizer. • Encourage cultivation of plants that can survive in harsh and semiarid conditions and that also do not compee with food, such as a physic nut (*Jatropha*).

Source EAMP (2010).

These ratings indicate how well companies have met the commitments they made in their EIAs in the planning stage (Environmental Protection Agency [EPA] Report, 2012).

The Chemicals Control and Management Centre (CCMC), a division of the EPA, provides Ghana with the management and control of chemicals and other related activities in the country. The CCMC has three main departments: (1) pesticides, (2) industrial consumer, and (3) national ozone unit. The CCMC has been very active in regulating the use of pesticides for agriculture, horticulture, forestry, gardening, public health, and other areas. The CCMC division has also been involved in monitoring the use of pesticides, takes regulatory actions in the event of illegal use, and provides policy advice to the government, as well as taking the lead on pesticide pollution issues in Ghana (EAMP, 2010).

According to Nancy Maxwell (2014), liquid biofuels, have the potential to supplement or to reduce total dependence on a fuel import. Sugarcane for ethanol yields sugar for local consumption to reduce sugar imports, and liquid biofuels eliminate the use of lead antiknock additives.

Ghana has been using ponds and storage reservoirs to deny downstream dwellers access to adequate water. These practices often dry up aquifers or lower groundwater tables when streams are in a recharging source (Abrahamson, 1993), and sedimentation problems can reduce the life span of the dam significantly.

The farmers in some regions of Ghana are highly dependent on rainfall. As a result, during drought periods, crops suffer and there is low production of cash crop. There are also frequent conflicts over the use of dams for other purposes, like drinking water and irrigation (Burden and Francis, 1978; Issahaku, 2000; World Bank, 2008c). Human-made lakes generally alter not only stream flow but also the water balance and hydrological cycle, consequently disturbing or altering the local microclimate (Edoho and Dibie, 2000). The filling of the lake imposes new stresses on the earth's crust, which, in turn, generate seismic movements and in some cases lead to earthquakes. The second aspect is geochemical and biogeochemical changes in water quality between the inflow and what is stored behind the reservoir. It is estimated that 23 metric tons of carbon in the form of standing biomass was submerged under the Volta Dam and that the mass in shallow soils is undergoing anaerobic respiration, emitting about 60 metric tons of methane a year (Issahaku, 2000). Also, the biological aspects consist of aquatic (phyto- and zooplankton, etc.) and terrestrial (soils, plants, and animals) biota. These aspects were very significant in the Volta Dam. Table 7.3 shows an example of negative impacts and mitigation measures.

Another environmental implication is that of the socioeconomic impact. The Volta Dam led to the displacement of about 80,000 people and more than 170,000 domestic animals. Small to large dams can flood national parks and arable lands in the northern regions of Ghana (Adelman, 1999; World Bank, 2008c).

Table 7.3 Negative Impacts and Their Mitigation Measures

	Potential Negative Impacts Issues	Mitigating Measures
Wind	Noise pollution, particularly due to airflow. Visual intrusion, particularly for low-flying aircraft. Interference with telecommunication transmission. Land use conflict: presence of towers, guy-wires will not allow additional use of land for mechanized farming. High towers or masts around 40 m high will be required for cost competitiveness due to the marginal wind speeds, 5–6 m/s in best locations. Increased bird mortality. High operation and maintenance cost compared to solar PV; moving parts require extensive lubrication.	Site specific; place site away from aviation routes and telecommunication facilities. Employ the right electronic noise filters. Use adequate warning signs for aircraft. Install markers to minimize risk for low-flying aircraft. Install deflectors on lines in areas with potential bird collisions.
Solar PV	Extensive land use for central systems. Major issue with disposal of lead-based storage batteries. Large wet storage battery system requires extensive consumption of distilled water.	Utilize roof for mountings and minimize use of ground. Hybrid and grid-connected systems minimize the use of storage batteries. For solar home systems, service company should take charge of battery maintenance and safety. Utilize solar still units to produce distilled water in remote and rural locations.

Source: EAMP (2010).

According to Dibie and Offiong (2012), one of the central themes for sustainable utilization entails the dynamic quest for long-term ecological stability and human progress. A policy that lasts into the future demonstrates an enduring feature and reflects not only the benefits it confers over time, but also its viability and the inclusive range of people or stakeholders, not just the few members of the privileged class.

According to the World Commission on Environment and Development (1987), plans to address sustainable development should include all sectors of the society. Based on the World Commission's recommendations, such decisions are weighty. The decision-making process should integrate participation by all sectors of society, including government, business, industry, and citizens, and it should be democratic. According to the United Nations, it is a human right for all people to enjoy a healthy and a safe environment. Consequently, the optimal approach should necessarily involve everyone; every member of the society must take part in the decision making that affects their life and economy. This what democracy and shared governance is all about.

Techniques used in obtaining gold, diamonds, and bauxite have scarred the land, destroyed the landscape, and endangered the biosphere. Not only have dangerous chemicals seeped into the groundwater, rivers, and lakes, creating pollution, but farm livelihoods have become jeopardized because of the pollutants deposited on land surfaces. The technology used in the mining of gold and other minerals all came from Europe. According to Agyeman and Swaine (2008), the low-level positions created for a few local people paled in comparison to the pervasively overwhelming cost imposed, in light of the huge number of miners with little knowledge about health and safety issues and practically no educational programs designed for the mining families and communities.

The agricultural sector, particularly in the production of cocoa, has undergone a significant overhaul in the last dozen years. Prior to that, Ghana lost its premier position as the largest cocoa producer for the Ivory Coast. This happened because of the structure of the Ghana Cocoa Marketing Board (GCMB) and the role that it played in handling the major cash of cocoa on the world commodity market. As a major agricultural industry crop subject to the vicissitudes of weather, the crop's export earnings provided 60–70 percent of Ghana's annual budget but declined to about 40–45 percent over time. Because of its premier role in the country's budget, farmers, most of whom tilled farms in the rural regions, got paid only a portion of the unstable world price. Because of the low payment for a bag of cocoa, farmers crossed the border to the Ivory Coast, the country adjacent to Ghana in the immediate west. The reaction of the cocoa farmers brought to light the operational inefficiency of the Ghana Cocoa Marketing Board.

In the past three decades, the unemployment rate has increased in the rural regions of Ghana. This unemployment problem has cause immigration from the rural areas to urban cities like Kumasi and Accra. The negative effects of growing unemployment has also galvanized new pressure on the Government of Ghana to come up plans for rural development and the social structures that have prevailed in the rural regions in the country. Migration, mostly of men, from the rural to urban regions, keeps growing, and unemployment in the cities increases in tandem. In the past, men in rural locations tilled the soil as heads of households, but that has been changing. The roles are being reversed, and women have become household heads with the migration of men (World Bank, 2008a, 2009, 2010). Seeing women in control of farm activities obviously runs counter to the previous tradition of men playing a leading role in agriculture. In light of such a cultural role reversal, the need for rural entrepreneurial programs with social institutions providing microfinance assistance is paramount.

Activities that nongovernmental organizations perform, including health and education services, have been integrated into the social milieu. Government-owned medical facilities, mostly located in the urban areas such as Accra, Kumasi, and Sekondi-Takoradi, provide health services as well as training and education for doctors and nurses. Rural locations do not enjoy the luxury of health facilities like those in urban areas. As a result, NGOs, including Medicine san frontier (Doctors Without Boundaries), church groups, and religious organizations, provide such services in ways consistent with the mission for evangelization.

For Ghanaians, the pathway to development seems to be via energy programs, and the oil discovered along the coast of Ghana seems to promise development. Socially, the oil enterprise has become a source of employment for people living in the western regions, particularly Takoradi and Sekondi. For the future, the educational and training institutions will facilitate a knowledge base for science programs. Along with the infrastructure, the opportunity for growth could hasten a technological transfer from the international firms, whose personnel will become purveyors of oil extraction educational programs.

A responsible government policy that ensures and respects stakeholder rights would be favorable. As such, it would seek to promote optimum and sustainable yield in resource and ecosystem utilization by providing incentives and instituting regulations. The government would be obligated to fulfill its responsibility, under the social compact, by providing pecuniary externalities, that is, providing the infrastructures required to promote and facilitate social welfare. In addition to dependable transportation and reliable communication systems, infrastructure would include electricity, potable water and clean air, and paved roads with adequate maintenance. Together, these externalities constitute a standard of living. Since resource availability comes with a cost, potential polluters should be immensely aware of their cost for polluting and for disrupting the environment. Punitive charges become necessary to prevent and eliminate pollution and environmental problems.

As a result of the social and economic impact of these environmental policies and issues, a sustainable environment and development is crucial for Ghana. The social and economic project for the nation and its larger community of current stakeholders needs to be innovative. Through NEAP advocacy, the National Youth Authority has been given a key role with the potential to alleviate environmental waste and the practice of littering. According to NEAP, youth are arguably primarily responsible for negative behaviors such as creating waste, littering, and defecation. Engaging the youth in this enterprise provides an incentive because it makes them key stakeholders in the national effort. It emphasizes that sustainable environments should consistently embrace the social compact between government and citizens, each performing its role and geared toward improving the entire welfare of society. The tools employed by performance agents are necessarily technological, social, economic, and cultural, and they eclectic, taken from all over the world, which also constitutes an important part of our environment. Besides progressive urban society, the social compact becomes a dynamic tool for stimulating proactive plans to address, promote, and facilitate rural and eco-friendly urban development.

APPROPRIATE POLICY RECOMMENDATIONS

The questions at this point are:

- How can Ghana prevent yesterday's luxuries from becoming today's necessities and tomorrow's relic?
- How can Ghana establish sustainable consumption as a norm?
- How can the new consumers in Ghana be persuaded to reduce their environmental impact and move toward sustainable consumption?

The Government of Ghana should adopt a new ethic, a new attitude toward discharging or implementing its various environmental policies. The nation's responsibility for caring for its people and ecosystem cannot be emphasized enough. Ghana must recognize that the earth has a limited capacity to provide for its people. It must also recognize its fragility and protect the land against being ravaged.

It is recommended that the Government of Ghana adopt a national strategy for sustainable development with specified goals and priorities. The sustainable development goals should harness all sectors of the nation to enable it to achieve economic, social, environmental, and security advancement. The Government of Ghana could adopt the European Union model, which specifically addresses biodiversity, climate change, renewable energy, and other major issues. The government must also establish a more inviting venue for the public to participate in and challenge decisions affecting sustainability.

One environmental initiative that could move Ghana positively toward accomplishing its already established environmental policy goals and action plans is to reduce the loss of carbon-storing ecosystems, such as the tropical forest, salt marshes, and peat lands. This would be a critical step in limiting the buildup of greenhouse gases in the nation's atmosphere. Land transformation represents the primary driving force in the loss of biological diversity in Ghana. As a matter of fact, land transformation would help the nation explore and solve other related issues in the country.

This action could also help to reduce other pressures on the nation's ecosystem as well as increasing their resilience and making them less vulnerable to those impacts of climate change that are already unavoidable. This could also encourage the Government of Ghana to continue to provide services to support citizens' livelihood as well as help them adapt to climate change.

Creating legal instruments, specific policies, and strategies to manage all the institutions involved in waste management is needed because the Environmental Protection Agency and its other divisions, particularly informal waste collectors, lack accountability. This would assist in keeping track and planning for sanitation and waste management. Waste generation is not the only problem; its collection and disposal in most cases are unaccounted for, giving the waste collectors agency and even the community the opportunity to dispose of waste in illegal ways, thereby polluting the cities, towns, roads, gullies, and rivers.

In addition, the Government of Ghana should realize that nitrogen supplies are linked to food production. Modern intensive agriculture requires large quantities of nitrogen fertilizer. As a result, controlling the growth of nitrogen fertilizer production and usage in Ghana would help the nation reduce the negative impact of the product on human beings and the ecosystem below the soil surface and better time the multiple applications to meet the needs of growing crops (Vitousek et al., 2002). The nation could cut nitrogen fertilizer use by one-third and reduce losses of nitrous oxide and nitric oxide tenfold by dissolving the fertilizer in irrigation water. The implementation of such practices should be a high priority for agronomists as well as ecologists, particularly in Ghana, because improved practices can provide an opportunity to reduce the costs of food while slowing the rate of change in the country. Other approaches to fertilizer efficiency include high-tech solutions, such as precision agriculture.

Future environmental policies in Ghana should focus on seeking the most enforceable and least bureaucratic, cheapest, most effective incentives. The Government of Ghana, nonprofit organizations, and the major corporations in the country could contribute to financial rewards for the greenest of their fellow bodies. The Government of Ghana should encourage corporations in the country to go green, which may differentiate their products from competitors'. According to Ferrell and colleagues (2013), improving a corporation's reputation for environmental stewardship may help it capture a growing market niche.

Ghana's cities and towns have formulated and implemented numerous policies on garbage collection in a bid to improve their service delivery, but flaws have been identified. The major reason is that the local governments are not responding to the rapid urbanization and increasing waste generation. This has a negative effect on an integrated approach for waste management because, since its inception in Ghana, no major changes have occurred in the collection of garbage, with the exception of Accra, the capital city. Sizable amounts of uncollected waste have been observed and are mounting, and no account of their disposal exists, which is partly the result of some waste being informally collected. This, in reality, points out a gap in the accountability and governance of Ghana. The Government needs to form and strengthen coordinating institutions to combine the separate strengths of nongovernmental organizations, international institutions, and industrial groups in order to focus their efforts on domestic and hazardous waste management in the country.

It is recommended that actions relating to policies, structures, and strategic plans relating to environmental sanitation be revised by the national Government of Ghana. The new policy should reflect the role played by CWSA, in addition to seeking convergence between the recommendations and requirements of the National Environmental Sanitation Policy and the National Water Policy. It is crucial for the Government of Ghana to clearly define the roles of the Ministries of Health and Environment in relation to sanitation and to ensure that the Ministry of Health and its staff are actively involved in an integrated approach to promoting and providing sanitation throughout the country.

The Government of Ghana continues to face environmental predicaments despite the enactment of several environmental policies. The implementation of the environmental action plan has not been very successful. It has taken a while for the Government of Ghana to realize that there might be a noncorrelation between economic development and environmental degradation. In the past ten years, the national government has taken more radical steps to rejuvenate the objectives of the National Environmental Action Plan (NEAP). There is still more for the government to do to effectively and efficiently implement the provisions of that plan. Therefore, it can be argued that what Ghana needs is a new environmental management approach that will help the nation achieve sustainable development. This new management approach must recognize the importance of the participation of local people in the environmental development process.

The citizens in both rural and urban areas of the Ghana are the major contributors to negative environmental practices, such as creating e-waste, littering, and open defecation. As a result, it would much better for the government to use incentives to build a community-based environmental program that has a chance to succeed in an increasing complex, sophisticated, and skeptical environment. The Government of Ghana must also realize that effective partnerships do not just happen. It should also design principles and strategic steps to create them. According to James Anderson (2011) and Guy Peters (2013) , no partnership can be successful without trust. Further, partnership in environmental issues in Ghana should be sponsored and promoted by committed leaders.

Environmental partnership should resist the urge to start activities until the goals, objectives, action plan, and responsibilities have been developed and agreed upon (Dibie and Offiong, 2012). In addition to the environmental policy goals for Ghana, the governance structure and principles of the partnership itself should be clearly articulated so that basic tasks can be performed effectively and efficiently. An effective partnership must establish strong leadership and determine how decisions will be made, who will provide staff service, and what the communication processes will be to keep the members informed. Actions to be addressed include (1) meeting logistics and follow-up, (2) public relations and awareness campaign, (3) coordination of activities and events including media coverage of campaign, (4) research and data collection, (5) administrative responsibility, and (6) membership recruitment and support. It is very important for the Government of Ghana to realize that a lack of structure can lead to difficulty in retaining stakeholders in the environmental protection partnership. It is paramount for the government not to undermine mutual accountability and thereby limit meaningful cooperation.

Ghana would not be the first to initiate this type of environmental or sustainable development partnership. The city of Seattle, Washington, in the United States adopted the smart growth partnership program and sought the input of its citizens to create more sustainable communities. The sustainable community planning used highly participatory processes to create liable urban areas (Vig and Kraft, 2013). Further, citizens in Pittsburgh, Pennsylvania, used a participatory sustainable community and smart growth initiatives to rejuvenate and transform the city's riverfront lands. It is therefore recommended that Ghana do the same.

Environmentally driven growth is likely to move forward by bolstering capabilities that promote sustainable development. The expectation is that, when structural adjustments become consistent with dynamic enterprises in the context of globalization, they potentially benefit those enterprises, which are crucial to the promotion of environmental growth. As Ghana implements strategic programs to become a proactive environmental protection agent and an accelerator of sustainable development, subsequent dynamics would correspondingly cultivate new institutional programs, including more effective environmental regulatory agencies, environmental policies, and reliable infrastructures, complete with the appropriate education and ecological protection training.

Increases in gains resulting from such beneficial strategies should embrace enterprises that provide the critical facilities in a safe environmental health infrastructure that resolve the silent crisis. Inevitably, such amenities should, over time, become real, capable, and adequate strategic tools that reverse the misguided compromises to environmental health and to communities exposed to threats to healthy lifestyles. The macroeconomic programs should facilitate the accessibility of both small and medium-sized firms to financial resources. In facilitating assistance, start-ups and existing programs should be cultivated and given capabilities, capital, and appropriately skilled labor. Incentives provided under these fiscal policies should promote competition because access to these financial resources should have support from public subsidies.

Based on World Bank Development Report (2009), small private cocoa farmers, as well as horticulturists, have cultivated enterprises and chalked up enormous gains, which should become critical lessons for other sectors. For strategic reasons, a favorable climate should cultivate the potential for investments and jobs. Such investments promote economic gains and abate poverty with strategies that become instrumental in fostering environmental growth. As Ghana promotes enterprises in the secondary sector, the spearheading role of agriculture calls for the integration of critical environmental protection principles that could facilitate potential firms to become ecofriendly.

In light of these strategies to promote the global agenda for Ghana, complementary aspects of fiscal and monetary policies have the potential to improve the nation's political economy and environment. Both fiscal and monetary policies should be critical, filling the pervasive gaps, given infrastructures and the public sector's performance. In promoting good environmental conditions and fostering sound governance, the availability of required infrastructures—such as facilities that educate and train people for jobs and research programs—culminates in dynamic outcomes (for example, innovation) and becomes proactive and instrumental in meeting future needs. The persistently dismal performance of Ghana's public agencies as sponsors of development means that reconstituting or restructuring the responsibilities of the public sector to affect increasingly productive outcomes is warranted.

In alignment with fiscal policy, an important aspect of the new orientation would be the creation of regional development programs that connect poor locales with mainstream markets. In essence, each region would assume, to a reasonable extent, a considerable measure of autonomy, given community development, social progress, and economic growth. Such policy freedoms would enable the autonomous entities to engage in productive activities. In general, the regional growth strategies would become consistent with effective exploitation authoritative measures in making macroeconomic decisions. For many developing nations, such as Ghana, such a strategy should necessarily address the gap in the nation's infrastructure. In particular, the shortfalls in energy, water, and sanitation have undercut strategic development.

In Ghana, poor sanitation, inadequate sanitation and drainage, and poor solid waste management have persistently become exacerbated by a rapidly growing population. Because growth in population has continued to exert negative impacts on the provision of such critically important services required for healthy lifestyles, such as sanitation and drainage and solid waste, the problem has been dubbed the silent crisis. In fact, the silent crisis has persisted, remaining among the key and very troubling urban and rural development issues.

Such newly oriented strategies should open the way for vitally necessary opportunities to protect the Ghana's environment. Not only will the regional program promote participation by stakeholders, but at the same time, as participation grows with the number of stakeholders, it should kindle the interest of other potential stakeholders, reversing the trend in the loss of opportunities. Using public service programs, Ghanaians would individually understand their role in changing their own behavior with respect to environmental pollution and personal growth, as they address poverty and sustainable development. Given the expectations of middle-income status, it could be the beginning of a development strategy that scales up foreign aid, promoting education and training, construction of infrastructure, and dynamic institutional capabilities addressing rural underdevelopment and environment growth.

SUMMARY

This chapter has examined the major environmental policies and issues in Ghana.

It evaluated the major environmental policy impact and provided specific policy recommendations concerning the human-dominated ecosystem. It suggested that the rate and extent of human alteration of the ecosystem in Ghana could affect how its citizens think about the environment. The chapter made it clear that the people of Ghana control the ecosystem in the country and that their activities will continue to affect the rest of the country. In a very real sense, the environment in Ghana is in the hands of its citizens, and how they handle it will determine how the country will accomplish its sustainable development goals.

This chapter discussed the implications of the environmental factors that influence the agricultural sectors in Ghana. Because agriculture is the major sector connecting industries and the manufacturing sector, it is crucial for the Government of Ghana to explore some form of mechanized farming. The need for the country to develop socioeconomically, politically, and environmentally remains unquestioned, given the conceptual framework for sustainable development.

The development dilemma in Ghana necessitates a pragmatic public policy paradigm shift in the search for a stable social, political, and economic milieu that institutionalizes accountability. Several lags that occur between political events and social response over time require appropriate discussion and reasoned debate.

In the past two decades, the political institutions fostering democratic outcomes in Ghana have ended up occasioning more trade and the implementation of investment policies by several international firms. The positive results of six elections have caused and furthermore attracted a growing increase in foreign direct investments (World Bank, 2009, 2010). The expectation is that optimal and healthy increases in the scale of social policy can have positive effects on environmental concerns, lessening the investment constraints in the rural regions of Ghana. Although the environmental policies are weak in Ghana, the surge of investments that embrace the conventions of the World Environmental Council and those of the indigenous local communities could foster policy ownerships and eventually lead to improvements in the environmental quality of programs.

According to Edoho and Dibie (2000), the dearth of dependable management skill and appropriate process technologies in Ghana hardly benefits the country's strategic goals. Ghana will not benefit as much as it anticipates from globalization, despite its budding democratic institutions, if it fails to implement and upgrade the environmental quality of the nation. In no small measure, the quality of Ghana's economy and projected growth will depend on the quality programs that consistently engender a socially responsible and environmentally healthy atmosphere where human health and standard of living become commensurate and direct measures of GDP.

REFERENCES

Acheampong, A. B. (2001). "Environmental Assessment of Northern Savanna Biodiversity Conservation Project (NSBCP)—Draft Report." Ministry of Lands and Forestry, Republic of Ghana.

Adelman, I. (1999). *The Role of Government in Economic Development*. Berkeley: University of California Press.

Agyeman, V. K., and Swaine, M. D. (2008). "Enhanced Tree Recruitment Following Logging in Two Forest Reserves in Ghana." *BIOTROPICA, 40*(3), 370–374.

Akhter, S. H. (2003). "Strategic Planning, Hypercompetition, and Knowledge Management." *Business Horizons* (January/February), 17–24.

Anderson, J. (2011). *Public Policy Making*. Boston, MA: Wadsworth.

Appiah-Opoku, S. (1997). "Indigenous Institutions and Environmental Assessment: The Case of Ghana." *Environment Management*, *21*(2), 159–171.

Appiah-Opoku, S. (1994). "Theoretical Orientation of Environmental Assessment in Canada: Application to the Third World." *Environments*, *22*(3), 103–110.

Apter, D., and Roseberg, C. (Eds.) (1994). *Political Development and the New Realism in Sub-Saharan Africa*. Charlottesville, VA: University of Virginia Press.

Atkinson, R. D. (2006). "Building a More-Humane Economy." *The Futurist* (May/June), 31–57.

Austin, G. (1996). "National Poverty and the 'Vampire State' in Ghana." *Journal of International Development*, *8*(4), 553–573.

Ballah, S. S. (2006). "Today's Golden Age of Poverty Reduction." *The International Economy* (Spring), 22–25.

Bandura, A. (1986). *Social Foundation of Thought and Action: A Social Cognitive Theory*. Englewood Cliffs, NJ: Prentice-Hall.

Barbier, E. B. (2005). *Natural Resources and Economic Development*. Cambridge: Cambridge University Press.

Barnard, C. (2005). "The Economy of Incentives." In *Classics of Organizational Theory* (6th ed.) (pp. 93–102). Mason, OH: Thompson-Wadsworth.

Baumer, C. and C. Horn. (2014). *Politics and Public Policy*. Washington DC: Congressional Quarterly Press.

Bell, W. (2004). "Humanity's Common Values: Seeking a Positive Future." *The Futurist* (September/October), 51–76.

Benneh, G., et al. (1990). *Land Degradation in Ghana*. Commonwealth Secretariat, University of Ghana.

Bertelsmann, E., Haltiwanger, J., and Scarpetta, S. (2005)."Microeconomic Evidence of Creative Destruction in Industrial and Developing Countries." Background paper for the WDR.

Bertelsmann, E., Scarpetta, S., and Schivardi, F. (2003). "Comparative Analysis of Firm Demographics and Survival: Micro-Level Evidence for the OECD Countries." Retrieved from http://www.oecd-ilibrary.org/docserver/download/5lgsjhvj7whf.pdf? Accessed November 19, 2012.

Burden, D. and Francis, J. L. (1978). "Who Cares About Ecology? Personality and Sex Differences in Environmental Concern." *Journal of Personality*, *46*, 190–203.
Carroll, C. (2008). "High-Tech Trash: Will Your Discarded TV or Computer End Up in a Ditch in Ghana." *National Geographic, 213* (January).
Carson, R. (2002). *Silent Spring*. Boston: Houghton Mifflin.
Centre for Environmental and Health Research and Training (CEHRT). (2010).
Cecchetti, S. G. (2005.) "Money, Banking and Financial Markets." In *The Economics of Financial Intermediation* (2nd ed.) (pp. 247–268). New York: McGraw Hill.
Chenery, H. B., (1975a). "The Structuralist Approach to Development Policy." *American Economic Review* (May), 310–316.
Chenery, H., and Syrquin, M. (1975b). *Patterns of Development 1950–1970*. New York: Oxford University Press, 122–133.
Colander, D. C. (2006). "International Financial Policy," "History of Exchange Rate Systems," Appendix A. *Economics* (6th ed.) (pp. 821–823). New York: McGraw Hill.
Coleman, I. (2006). "Women, Islam and the New Iraq." *Foreign Affairs* (January/February).
Country Profile (2012). *Ghana*. Retrieved from http://www.theafricareport.com/West/country-profile-2012-ghana.html. Accessed November 19, 2012.
Cowley, G. (2004). "Medicine Without Doctors," *Newsweek* (July 19), 23–37.
Cunningham, C. W. (2008). "Lessons Learned from Coordinating Research on the North Atlantic" (CORONA). *Ecology 89*, S1–S2.
Dibie, R. A., and Dibie, J. (2008). "The Dynamics and Nature of Civil Society." In *Comparative Perspectives on Civil Society,* edited by R. A. Dibie. Lanham, MD: Rowman & Littlefield.
Dibie, R. (2007). "Ethical Leadership, Social Responsibility and Corruption in Nigeria." *International Review of Politics and Development, 5*(1), 1–42.
Dibie, R., and Gadzekpo, L. (2003). "Managing Public Service Ethics in Ghana and Nigeria." *Politics, Administration and Change, 39* (January-June), 1–24.
Dibie, R., and Offiong, O. (2012). "The Nature and Impact of Environmental Policy in Nigeria." *Journal of International Politics and Development, 10*(1–2), 113–143.
Domar, E. D. (1947). "Capital Expansion, Rate of Growth, and Employment." *Econometrica* (April), 137–147; "The Problem of Capital Accumulation," *American Economic Review* (March), 34–55. Both reprinted in his (1957) *Essays in the Theory of Economic Growth*. New York: Oxford University Press.
Economic Co-operation and Development, Economics Department Working Paper 348.
Edoho, F., and Dibie, R. (2000). "Executing Environmental Policy and Waste Management in Ghana and Nigeria." *Journal of Sustainable Development in Africa, 2*(2), 38–70.
Esfahani, E. (2005). "Thinking Locally, Succeeding Globally." *Business 2.0* (December).
Environmental Analysis and Management Plan (EAMP). (2010). *The Sustainable Land and Water Management Project Final Report*. Accra, Ghana: Government Press.
Environmental and Social Management Framework, Social Opportunity Project—Draft Report. MLGRD, Republic of Ghana.
Environmental Protection Agency (EPA). (2012). "Ghana State of Environment Report." Accra, Ghana: EPC.
Environmental Protection Agency (EPA). (2004). "Ghana State of Environment Report." Accra, Ghana: EPC.
Environmental Protection Agency (EPA). (1995). "Environmental Impact Assessment Procedure." Accra, Ghana: EPC.
Environmental Protection Council. (1994). "Ghana Environmental Action Plan, Vol. 2." Technical background paper by six working groups. Accra, Ghana: EPC.
Federal Ministry of Agriculture and Water. (2009). "Updated Environmental and Social Management Framework (EAMP) for the Northern Savanna III/GEF-SLWM Projects, Nigeria." Abuja, Nigeria: Government Press.

Ferrell, O., Faedrich, J., and Ferrel, L. (2013). *Business Ethics*. Mason, OH: South-Western/Cengage Learning.

Fiedler, F. E. (1967). *A Theory of Leadership Effectiveness*. New York: McGraw-Hill.

Fishlow, A. (1965.) "Empty Economic Stages?" *Economic Journal, 75*(297), 112–125.

Flint, C., and Taylor, P. (2007). Geography of Imperialisms." *Political Geography: World-Economy, National-State and Locality* (5th ed.) (pp. 79–116). Upper Saddle River, NJ: Pearson-Prentice Hall.

Gerschenkron, A. (1962). *Economic Backwardness in Historical Perspective*. Cambridge, MA: Harvard University Press.

Glennerster, R., Kremer, M., and Williams, H. (2005). "The Price of Life." *Foreign Policy* (May/June).

Government of Ghana. (2004). "Environmental and Social Management Framework, Ghana Community Based Rural Development Project." Ghana.

Goldman, M. I. (2005). "Political Graft: The Russian Way," *Current History* (October).

Handelman, H. (2009). *The Challenge of Third World Development*. Upper Saddle River, NJ: Prentice Hall.

Harrod, R. F. (1948). "An Essay in Dynamic Theory," *Economic Journal, 49*(1939), 14–33; and in *Towards a Dynamic Economics*. London: Macmillan.

Hatcher, T. (2003). "Ethical Compass." *Executive Excellence* (July), 23–24.

Havens, E. (1972). "Methodological Issues in the Study of Development." *Sociologia Ruralis, 12*(3–4), 252–272.

Herrick, B., and Kindleberger, C. P. (2006). "Land and Other Natural Resources." In *Economic Development* (4th ed.) (pp. 147–164). *Economics Handbook Series*. New York: McGraw Hill.

Hill, C. W. L. (2008) "Globalization." In *Global Business Today* (pp. 4–38). New York: McGraw Hill.

Hill, J. S. (2005). "Geopolitical Analyses of Regional Markets." *World Business: Globalization, Analysis and Strategy* (1st ed.) (pp. 129–163). Mason, OH: Thompson South-Western.

Hilson, G. M. (2004). "Structural Adjustment in Ghana: Assessing the Impact of Mining-Sector Reform." *Africa Today, 51* (Winter).

Higgins, S. (2000). *Literature Review. Environmental Policies and the Livelihoods of People Living in the Forest Margins*. Reading, UK: The University of Reading.

House, R. J. (1971). "A Path-Goal Theory of Leader Effectiveness." *Administrative Science Quarterly, 16*, 321–338.

Hunt, S., and Posa, C. (2001). "Women Waging Peace," *Foreign Policy* (May/June).

International Union for the Conservation of Nature and Natural Resources. (1980). The World Conservation Strategy. IUCN.

Issahaku, A-N. (2000). "The Political Economy of Economic Reform in Ghana: Implications for Sustainable Development." *Journal for Sustainable Development, 2*(1), 13–27.

Johnson, D. (2002). "Progress Against Corruption." *The Futurist* (March/April), 23–41.

Jacobson, M. Z., and Delucchi, M. A. (2009). "A Path to Sustainable Energy," *Science News* (November), 37–70.

Kalecki, M. (1971). *Selected Essays on the Dynamics of the Capitalist Economy.* New York: Cambridge University Press.

Kessie-Adu, K. (1971). *The Politics of Political Detention*. Tema, Ghana: Ghana Publishing Corporation.

Klaff, L. G. (2003). "Getting Happy with the Rewards King." In *Work Force, Annual Edition Human Resources* (5th/6th ed.) (pp. 76–79). New York: McGraw-Hill.

Klare, M. T. (2010). "Clean, Green, Safe and Smart," *The Nation* (August 29).

Klare, M. T. (2007). "Global Warming Battlefields: How Climate Change Threatens Security." *Current History, 106*, 97–122.

Krugman, P. R., and Obstfeld, M. (2009)."Developing Countries: Growth, Crisis, and Reform." In *International Economics* (7th ed.) (pp. 602–639). Upper Saddle River, NJ: Prentice Hall.

Kutznet, S. (1966). *Modern Economic Growth*. New Haven, CT: Yale University Press.

Lancaster, C. (2005). "Development in Africa: The Good, the Bad, the Ugly," In *Current History in the Developing World, 2006–07, Annual Edition* (pp. 188–193). New York: McGraw Hill.

Lehmann, J. (1999). *Befunde Empirischer Forschung zu Umweltbildung and Unwelthewusstsein*, Opladen, Germany: Leske und Budrich.

Levine, R. (2006). "Educating Girls, Unlocking Development." *Current History* (March), 20–36.

Littlefield, E., and Rosenberg, R. (2004). " Breaking Down the Walls between Microfinance and the Formal Financial." *Finance & Development* (June), 12–29.

Lynas, M. (2007). "How to Stop Climate Change," *New Statesman* (November 8).

Mackinnon, D., and Andrew, C. (2007). "The Changing Geographies of Multinational Corporation." *An Introduction to Economic Geography Globalization, Uneven Development and Place* (pp. 135–143). Upper Saddle River, NJ: Pearson Prentice Hall.

Mankiw, G. N. (2009). *Principles of Microeconomics* (5th ed.). Mason, OH: South-Western Cengage Learning.

Marglin, S. A. (2003). "Developing as Poison," *Harvard International Review* (Spring).

Marshall, A. (1948). *Principles of Economics*. New York: Macmillan.

Martin, K., and Knapp, J. (Eds.). (1966). The *Teaching of Development Economics*. London: Frank Cass. See especially the articles by Seers, Streeten, and Myint.

Maxwell, N. (2014). *Understanding Environmental Health*. Burlington, MA: Jones & Bartlett.

Merton, R. (1985). *Social Theory and Social Structure*. New York: Free Press.

MoFA and FAO. (2007). "Draft Irrigation Policy, Strategies and Regulatory Measures." Ministry of Food and Agriculture and Food and Agricultural Organization of the United Nations, pp. 1–27.

MoFA, (2007). "Food and Agricultural Sector Development Policy (FASDEP II), First Draft-Second Revision." Ministry of Food and Agricultural, Government of Ghana, pp. 1–56.

Morrisette, J. J., and Borer, D. A. (2004–2005). "Where Oil and Water Do Mix: Environmental Scarcity and Future Conflict in the Middle East," *Parameters* (Winter).

Myers, N. and Spoolman, S. (2014). *Environmental Issues and Solution*. Belmont, CA: Brooke/Cole.

Nafisi, A. (2004). "Women's Rights—Not Just for Westerners." *American Educator* (Fall), 17–31.

NDPC/ EPA (2002). *Strategic Environmental Assessment of the Ghana Poverty Reduction Strategy. Handbook for District Development Plan Sustainability Appraisals*. Government of Ghana Press.

Ofori, S. C. (1991). "Environmental Impact Assessment in Ghana: Current Administration and Procedures—Towards an Appropriate Methodology." *The Environmentalist, 10*(2), 115–126.

Pearce, D. W., and Barbier, E. B. (2000). *Blueprint for a Sustainable Economy*. London: Earthscan.

Pearce, D. W., Markandya, A., and Barbier, E. B. (1989). *Blueprint for a Green Economy*. London: Earthscan.

Perkins, S. (2008). "Down with Carbon: Scientists Work to Put the Greenhouse Gas in Its Place." *Science News* (May 10).

Polasky, S., and Segerson, K. (2009). "Integrating Ecology and Economics in the Study of Ecosystem Services: Some Lessons Learned." *Annual Review of Resource Economics, 1*, 409–434.

Pustay, M. (2007). *International Business* (6th ed.). Upper Saddle River, NJ: Prentice Hall.

Roorda, N. (2012). *Fundamentals of Sustainable Development*. New York: Routledge.

Roth, R. I. (2005). "Strengthening Governance: Ranking Countries Would Help," *The Washington Quarterly* (Winter).

Ruttan, V. W. (1980). "How the World Feeds Itself," *Society, 17*(6), 37–64.

Sachs, J. L. (2005). "The Development Challenge," *Foreign Affairs* (March/April), 78–90.

Samuelson, P. A. (2009). "Functioning of a Mixed Capitalistic Enterprise System." In *Economics* (pp. 35–60). New York: McGraw Hill.

Schramm, C. J. (2010)."Building Entrepreneurial Economies." In *Annual Edition: Entrepreneurship* (6th ed.), edited by R. W. Price. New York: McGraw-Hill.

Schunk, D. H., and Zimmerman, B. J. (1994). *Self-Regulation of Learning and Performance Issues and Educational Application*. Hillsdale, NJ: Lawrence Erlbaum.

Simms, A. (2005). "Why We Owe So Much to Victims of Disaster," *New Statesman* (May).

Simon, H. A. (1998). *Administrative Behavior* (4th ed.). New York: Free Press.

Simpson, S. (2009). "Taming the Blue Frontier," *Conservation Magazine* (April/June).

Stiglitz, J. E. (2003). "Externalities and the Environment." In *Economics of the Public Sector* (3rd ed.) (pp. 214–232). New York: McGraw Hill.

Stiglitz, J. E. and Walsh, C. E. (2006). *Modern Economics* (4th ed) (pp. 3–20). New York: McGraw Hill.

Strategic Youth Network for Development (SYND). (2012). "Policy Brief: The National Environmental Policy of Ghana." Government of Ghana Press.

United Nations Development Program (UNDP). (n.d.). *Human Development Index (2012)*. New York: Oxford University Press.

United Nations Development Program (UNDP). (n.d.). *Human Development Index (2010)*. New York: Oxford University Press.

United Nations Habitat (2008). *Annual Report*. Retrieved from www.unhabitat.org. Accessed November 19, 2012.

Vaitheeswaran, V. (2002). "How Many Planets? A Survey of Global Environment." *The Economist* (July 6).

Vig, N., and Kraft, M. (2013). *Environmental Policy*. Washington DC: Congressional Quarterly Press.

Vitousek, P. M., et al. (2002). "Human Domination of Earth's Ecosystem." In *Sustainability*, edited by N. Smith-Sebasto. New York: McGraw Hill.

Waller-Hunter, J., and Jones, T. (2002). "Globalization and Sustainable Development." *International Review for Environmental Strategies, 5*(1), 53–62.

Weatherby, J., et al. (2009). *The Other World: Issues and Politics of Developing World*. Boston: Pearson.

Withgott, J., and Brennan, S. (2009). *Essential of Environment* (3rd ed.). Boston: Pearson.

World Bank. (2010). *Ghana: Doing Business 2009*. Washington, DC: World Bank.

World Bank. (2009). *Ghana: Doing Business 2008*. Washington, DC: World Bank.

World Bank (2008a). *Ghana: Management Report and Recommendations*. Washington, DC: World Bank.

World Bank. (2008b). *A Better Investment Climate for Everyone*. New York: Oxford University Press.

World Bank. (2008c). *Ghana: Background and Ghana's Current Economic Growth Context*. Washington, DC: World Bank.

World Business. (2010). *Globalization, Strategy and Analysis*. Mason, OH: Thompson South-Western.

World Commission on Environment and Development (WCED). (1987). *Our Common Future*. Oxford: Oxford University Press.

World Development Report. (2005). *A Better Investment Climate for Everyone*. New York: Oxford University Press.

World Health Organization and UNICEF. (2010). *Progress on Sanitation and Drinking Water:* WHO/UNICEF Joint Monitoring Program for Water Supply and Sanitation. Geneva: WHO.

Wright, R. and Boorse, D. (2014). *Environmental Science*. Boston, MA: Pearson Press.

8 Environmental Policy and Impacts in South Africa

Robert Dibie, PhD
and Kealeboga Maphunye, PhD

INTRODUCTION

Before 1994, the year in which South Africa officially joined the rest of progressive humanity to become a democratic state under Nelson Mandela, the country was in many cases perceived and treated as a pariah state owing to its racial discrimination or apartheid policies. During that period (at least from 1948 to 1993), South Africa pursued an inconsistent, albeit racially focused, approach to all policy-related matters, including those pertaining to the environment. It is in this regard that the South African government believes that "[t]here are clear links between the political history of South Africa and the social and natural environments" (DEAT, 1999b).

According to Dunn 1981 (cited in Van der Waldt, 2001: 87), public policy is "a long series of more or less related choices, including decisions not to act, made by government bodies and officials." Thomas Dye (2011) also defines public policy as what government chooses to do or not to do. It is in this regard that South Africa's environmental policies must be understood. Accordingly, in terms of past government policy, only the formerly privileged white minority benefited from environmental conservation and related policies, unlike the Black majority, who were always systematically excluded from state resources and programs (Adam and Giliomee, 1979b; ANC, 1994; Beinart and Dubow, 1995). Significantly, the white minority also had exclusive access to the most fertile, prime land, which was the environmentally stable natural habitat and residential areas; this primarily gave them disproportionate access to the country's economic and natural resources. The disadvantaged Black majority were, per the requirements of racial legislation, in most instances relegated to the periphery and to poor, inhospitable and usually environmentally unstable residential and farming lands (Beinart, 1994; Posel, 1991). In fact, the apartheid government usually ignored, circumvented or avoided enforcing international conventions on the environment because it was not a member of the United Nations and other global bodies.

During its recent and internationally publicized Polokwane policy conference, the ruling African National Congress (ANC) discussed environmental issues in South Africa. The party noted that:

> During the apartheid regime the vast majority of our people bore the brunt of poor sanitation, were located in the areas where the most polluting industries existed, and were denied the basics right to defend themselves against harmful activities (ANC, 2007).

The ANC therefore argued in favor of what it termed "transformative environmentalism based upon the idea of sustainable development, which is built upon the interconnection of environmental social and economic justice" (ANC, 2007).

The rationale for outlining the foregoing background on South Africa's former government system is to highlight and emphasize the fact that it essentially created a myriad of problems—especially those related to the environment—for the young democracy and its post-apartheid government (Posel, 1996). Arguably, the extent of the damage caused to the environment in some parts of the country, which was almost always caused by the enforcement of racial laws and policies, is incalculable, and to date the country is still reeling from its momentum.

To understand South Africa's contemporary environmental policy, this chapter is guided by the four major steps in policy making (Baumer and Van Horn, 2014; Jones, 1984; Peters, 2014). These steps are (1) initiation and definition, (2) formulation and enactment, (3) implementation, and (4) impact and evaluation. While the chapter alludes to the first three steps, its focus is on the last step, impact and evaluation. However, the impact and evaluation have to be seen in the context of the country's first nineteen or twenty years since Mandela became the country's president in 1994.

South Africa has an environmentally friendly climate, including an appealing and internationally admired landscape. Notably, the government argues that "South Africa has a vast array of resources, facilities and infrastructure to support its population although not everyone has adequate access to them" (DEAT, 1999b). The country is also one of the most sought-after international tourist destinations whose image was no doubt boosted by the FIFA Soccer World Cup that South Africa hosted in June 2010. In general, most of the tourists visiting South Africa never cease to make positive comments about the country's environment, including its weather and generally appealing natural habitat. Smith (2004: 341) cites the fact that South Africa's Constitution, in its Bill of Rights (Chapter 2, Section 24, under the heading "Environment"), states that "everyone has the right to have the environment protected for the benefit of the present and future generations."

The Department of Environmental Affairs and Tourism argues that "[t]here have also been radical changes to the Constitution, so that it now reflects the interests of all structures of the population, creates empowerment of individuals and communities, and upholds the rights of humans and the environment" (DEAT, 1999a). Naturally, such a constitutional framework is likely to create immense advantages and opportunities for the country in terms of the environment (Van Rooyen, 1999; Singer, 2001). Despite these advantages, the country faces several environmental problems, including (1) air pollution, (2) water shortages and dwindling water resources, (3) land use patterns that will not ensure sustainability, (4) recycling issues, (5) waste management challenges including illegal dumping of toxic waste, (6) unsustainable wildlife management (especially with regard to elephant and related products), and (6) building of houses and housing settlements in environmentally unsuitable locations (Sebola, 2003; Van Nicker, Van Waldt and Jonker, 2001).

Based on the country's racial legacy, the nature of most of its environmental problems is traceable to the apartheid period (Adam and Giliomee, 1976b; Haycock and Stone, 2005). For instance, many environmental problems related to air pollution, water shortages, poor land use patterns, waste management, and housing settlements in environmentally unsuitable locations usually affect Blacks and previously disadvantaged communities who usually bear the brunt of the aftermath of years of apartheid's oppressive policies (Haycock and Stone, 2005). Thus, South Africa's present government states that:

> The separate development policies and the resulting inequality [have] caused severe environmental problems through wasteful resource use and pollution among the affluent minority. As income levels improve, people consume more and produce more waste . . . (DEAT, 1999b).

Typical examples on the above include the fact that many of South Africa's so-called townships and informal settlements have traditionally been built on poor soil or environmentally risky plots set aside by policy makers to specifically house Black communities. In those communities, the usual shortage of water or inadequate sanitation often meant the use of pit latrines or the "bucket system" for sanitation as well as the use of coal and firewood as fuel, which led to bush fires, smoke, and air pollution. Some unscrupulous big businesses are even known to have taken advantage of policy failures or racial bias either to ignore or to blatantly flout environmental policies, for instance, by illegally dumping solid or hazardous waste in the vicinity of the townships and in most instances to have gotten away with such practices.

The DEAT acknowledges that "[p]olitical, economic and social inequalities of the past have resulted in high poverty rates, high levels of illiteracy, and unemployment, and distorted patterns of resource use and waste generation. Recent political reforms aim to redress these inequities, and to improve quality of life among all South Africans. These reforms take cognizance of environmental issues in terms of the capacity of the environment to meet basic needs, and adhere to sustainable development guidelines" (DEAT, 1999a).

Closely related to this acknowledgment is the fact that South Africa, like many other developing countries, also has to face the constant challenges of locating financial and relevant resources to environmental management (Haycock and Stone, 2005; Randal and Theobald, 1998). The point is that, despite the country's being relatively wealthy—at least by African standards—it nevertheless struggles to find money and other resources to boost its environmental management programs, mostly owing to competing priorities in government policy.

The chapter will examine the social impact of environmental policies in South Africa. It argues that the connection between sustainable development and environmental issues works in two ways. On the one hand, unmitigated growth in emissions has the potential to undermine sustainable development in South Africa. The chapter will contend that the projected impact of climate change may affect water, food security, the coastal system, health, and the ecosystem and thus the social and economic lives of the citizens. On the other hand, the chapter will argue that making development paths more sustainable can contribute to climate mitigation. The concluding section will recommend environmental policies that South Africa can adopt.

ENVIRONMENTAL GOVERNANCE AND THE WORLD

The literature of environmental policies often analyzes the potential for nations around the world to formulate a system of environmental governance without simultaneously sacrificing the significant advances in environmental quality that have been realized in the past few decades. The analysis is of the conceptual framework for the environmental model needed to achieve a good-quality ecosystem and of the relationship of that model with human beings, which consists of a complex set of political-institutional actors involved in the definition and implementation of environmental policy (Heineman, Blahm, Peterson and Kearney, 1990). The true test for environmental morality, however, is whether an action or individual—social or political—benefits the earth or ecosystem. Throughout the history of governance in Africa, economic and social development has followed the pattern of human capital and cultural heritance rather than the preservation of the ecosystem for the common good (Dibie, 2009). Figure 8.1 shows the connection between the biological, economic, and social systems.

The sustainable development and sustainable growth theory argues that development that does not meet the intergenerational equity criteria simply must be bad development (Myers and Spoolman, 2014; Roorda, 2012). This approach also advocates the distribution of prosperity among the

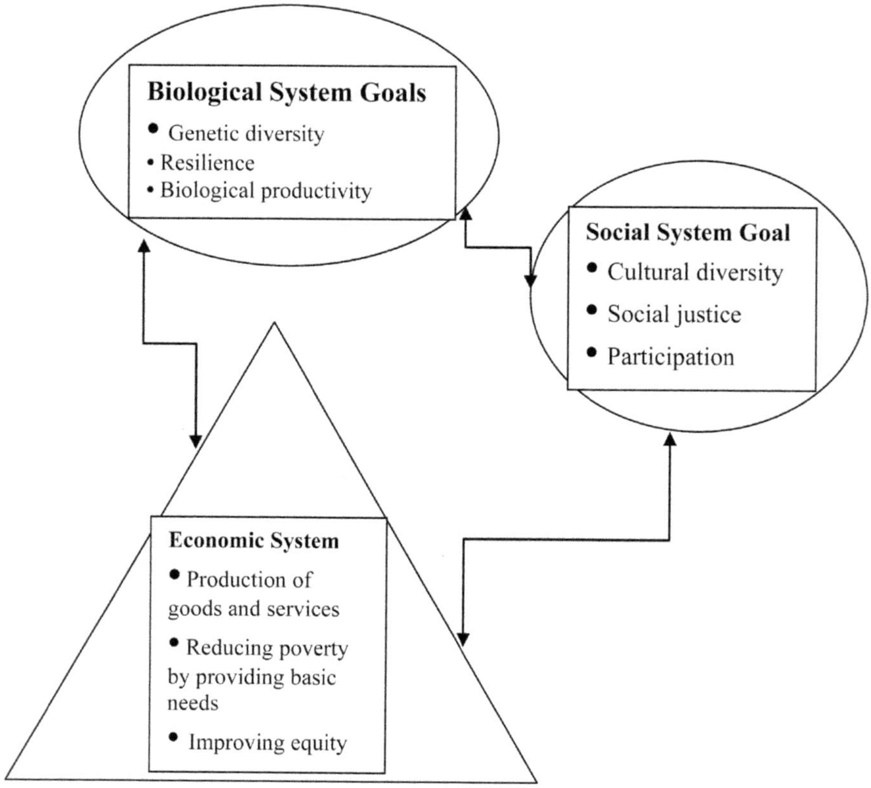

Figure 8.1 Model on the Relationship among Biological, Economic and Social Systems. Sustainable development is a driving force for dealing with environmental issues and economic growth. Both human and environmental issues should be considered in the context of a temporal scale moving from the immediate to the long term.

various parts of the current generation. That distribution among existing human beings and those of the future generation is essential to achieving sustainable development. On the other hand, sustainable growth can be achieved only if prosperity is accomplished in three categories: people, planet, and profit is accomplished any country. Similarly, Barbier (1987) identified the biological or ecological resource system, the economic system, and the social system as basic to any process of development.

According to Edward Barbier (1987) and Niko Roorda (2012), human society applies a set of goals to each system with its own hierarchy of subgoals and targets. Another approach for distinguishing between types of sustainable development involves top-down and bottom-up. Nations, governments, and multinational corporations can all focus on sustainable development by applying the top-down approach (Wright and Boorse, 2014). However, individuals can also devote themselves to the cause of sustainable development, either on their own or in groups, by applying the bottom-up and stewardship approaches (Roorda, 2012; Wright & Boorse, 2014). The stewardship approach argues that national development actions and programs should promote the management of natural resources and human well-being for the common good. This theory postulates that interaction among the three system goals change as the scale or hierarchy of the systems is extended from the local to the regional and thence to the national and even global levels. As systems theory argues, the behavior of a higher system in such a hierarchy is not readily discovered from the studying of the lower systems and vice versa. The choice of sustainable development goals, as pursued at a national level, may vary from those advocated at the local level. It is also very important to measure the progress in national development from a holistic viewpoint and not just in economic terms. In addition, justice, equity, and freedom are crucial factors for upholding sustainable development as the goal that all nations and human society including South Africa need to be moving toward.

Edward Abbey (1971) proposes a strong collaboration between eating meat and owning slaves. He contends that modern industrial people require regular exposure to the forest. Abbey argues that living outside modernity's mechanized landscape could foster independence and self-reliance. He further stipulates that, since we have agreed not to drive our automobiles into churches, cathedrals, concert halls, art museums, legislative assemblies, private bedrooms, and the other sanctums of our culture, we should do the same for selected environmental areas. According to Abbey (1971), human beings should treat our national parks with the same deference as they do cathedrals, concert halls, and legislative assemblies. While Abbey's conceptual framework may not constitute a coherent set of environmental ethics, it make sense in terms of human behavior toward the environment. Human beings tend to be more concerned about their social actualization and fail to realize that, without the environment, they cannot survive to enjoy the social and economic output.

The deep ecologists proposed a paradigm shift in which humans recognize that they are part of their ecosystems. According to Naess (1995a), the well-being and flourishing of human and nonhuman life on earth have value in and of themselves. The values of human beings would be useless in the nonhuman world. He therefore concludes that human beings should foster ecological diversity and have no right to reduce this richness and diversity, except to satisfy vital needs. Naess's conceptual framework calls for the shrinking of the number of people in the world and their economic, technological, and ideological impact. According to Naess (1995b), this change can occur only if people commit themselves to achieving a higher quality of life rather than adhering to increasingly higher standard of living. He contends that human beings must realize that bigness and greatness are not one and the same. Any nation or government that acknowledges this has a responsibility to formulate the appropriate social and environmental policies to achieve higher quality of life. The deep ecologist theory also calls for a very important realization that human beings are not really separate from the rest of the world (Eisner 2007; Peterson del Mar, 2006). This paradigm shift is necessary to turn Western civilization away from the dead end of individualism and accumulation. The paradigm shift also calls for a stop to the exploitation of other people and the environment, as well as a move toward a way of living that is at once more gentle and meaningful.

The bioregionalists contend that ecological boundaries trump political ones and that people need to become intimately familiar with and committed to the landscape, the bioregion in which they live. According to Sale (1993), a realization of humanity's place and stake in the ecosystem would be fostered by concrete interaction with the plants, creatures, and other characteristics of specific places. The message of the bioregional concept is that all human beings belong to nature and that our attempts to separate ourselves from it harm not just ecosystems but also our own lives and spirits. The bioregional theory, like the deep ecological concept, combines a reverence for and knowledge of the ecosystem with a determination to treat it more sustainably.

Figure 8.2 Nature and Animals in Cape Town, South Africa

Source: Photo by author during field research in the Republic of South Africa in 2012.

The ecofeminism theory argues that male values and institutions are primarily responsible for humanity's separation from the destruction of the environment. According to Warren (2001), men's abuse of women and nature was one and the same as the masculine fear of life and mania for domination. The ecofeminism theory calls for the reshaping of basic socio-economic relations and the underlying values of our societies around the world. Ecofeminism theory also emphasizes that women's bodies and the association of women with nature, both in fact and in men's imaginations, (Warren, 2001). The important point about the ecofeminism theory is that gender shapes how one tries to protect the environment. The bioregional, collaboration, and deep ecological theories tend to neglect the gender factor in environmental issues. Hence this paper will explore the premise that most ecological problems arise from deep-seated social problems. Therefore, the present environmental problems in South Africa cannot be clearly understood without resolutely dealing with the social problems in the society.

SOUTH AFRICA'S CURRENT ENVIRONMENTAL POLICIES AND LEGISLATION

According to Van der Merwe and Steyl (2005: 296), "South Africa has enacted a range of legislation with reference to environmental pollution. . . ." The environmental policies and laws that currently regulate South Africa's environmental situation emanate from the country's postapartheid democratic constitution. One of the key highlights of this internationally acclaimed constitution is that it enshrines progressive human rights and equality clauses that, among other things, seek to infuse public participation and citizen inputs into policy formulation and implementation.

This constitution has guided the government to grapple with environmental challenges since 1994 and has in turn spawned several laws and policies that seek to address specific environmental issues in South Africa, such as the following:

- Hazardous Substances Act (South Africa, 1973)
- Environment Conservation Act (South Africa, 1989)
- National Environmental Management Act (South Africa, 1998a)
- National Water Act (South Africa, 1998b)

Overall, the National Environmental Management Act (NEMA) has been amended several times since the late 1990s to accommodate various environmental issues and sectors. Thus, the National Environmental Management Biodiversity Act Number 10 of 2004 seeks to provide for the management and conservation of South Africa's biodiversity according to the NEMA (Number 10 of 1998). This Act covers, among other things, the following (Gubb, 2009a):

- Protection of species and ecosystems
- Sustainable use of indigenous biological resources
- Fair and equitable sharing of benefits arising from biological prospects involving indigenous biological resources
- The establishment of the South African National Biodiversity Institute

The key issues highlighted by such policies and legislation are the fact that South Africa must join the international community in addressing environmental challenges. However, it would appear that South Africa is still a long way from operating at optimum levels compared to other countries because, as some observers argue (notably Gunn):

> South African environmental legislation is largely based on the traditional "command and control" model where the threat of punishment is designed to deter any aberrant behavior. This is particularly evident from the recent National Environmental Management Amendment Bill whereby the fines and penalties associated with contraventions of environmental laws are greatly increased. Some sanctions for contraventions of our environmental laws will be accompanied by a fine of R10 million and or 10 years in jail. However strict penalties have been present in some of our environmental legislation for some time (R5 million and or 10 years in jail) and it seems to have had little effect. (Gunn, 2009)

While South Africa's Parliament eventually passed the bill, clearly this raises fundamental questions about the merit or lack of merit of South Africa's regulatory framework pertaining to the environment. Moreover, one of the country's foremost challenges in terms of such a regulatory framework is how to ensure effective or meaningful public participation in all the policy processes related to the environment so as to avoid recourse to the command–and-control model, which is a relic of the apartheid era.

According to Orkin and colleagues (1998: 483), the current environmental policy in South Africa must be viewed from the perspective of the "dominant paradigm which the state has employed, and the relative lack of formulation and influence of a counter-paradigm from civil society." They cite Cock and Koch (1991: 1), who make the following argument: "Until very recently the dominant understanding of environmental issues in South Africa was an authoritarian conservation perspective. This focused exclusively on the preservation of wilderness areas and particular species of plants and animals. Within this perspective 'overpopulation' was often identified as the main environmental problem. It was people who were perceived to be responsible for destroying trees and creating waste."

Linked to the dominant paradigm in environmental policy is the fact that the country also relies on the traditional model of policy making. According to Bryner (2003: 301), such a model is "a continual process of identifying problems, formulating governmental responses or policies, organizing administrative mechanisms for carrying out the policies, and evaluating the extent to which the policy objectives are achieved."

CIRCUMSTANCES THAT LED TO THE ENACTMENT OF VARIOUS ENVIRONMENTAL POLICIES

In the case of South Africa, the main circumstances leading to the enactment of the present environmental policies and laws were essentially the apartheid and its racial policies, which created artificial barriers to discourage interracial marriages and promoted segregated neighborhoods. However, the need to deal with environmental problems or challenges since the new government took power has also spawned new regulatory mechanisms resulting from the fact that such mechanisms have to cater for all the country's diverse population groups rather than a small minority, as was previously the case. Observers note that even before the country's democratic 1994 elections, policy documents developed by the ruling African National Congress (ANC) and its partners. South African National Civic Organisation (SANCO), Congress of South African Trade Unions (COSATU) and South African Communist Party (SACP). addressed environmental issues in the context of debates on sustainable development. This process, they argue, was part of the major contribution of the broader democratic political movement to a new model for development in South Africa (IDRC, 2009).

In the past nineteen or twenty years since the new government came to power, the main aim of enacting these policies has been to ensure that all communities and stakeholders benefit and participate equally in environmental conservation as well as in working together toward the development of environmental sustainability. However, this aside, immediately after its democratic elections, South Africa became a prominent international player not only in multilateral forums such as the United Nations General Assembly but also on other international strategic platforms, such as IBSA (India-Brazil-South Africa trilateral forum), which requires the country to take a leadership role in many issues of international significance and dispute.

Of course, in terms of global conventions and agreements, South Africa must comply (or be seen to comply) with such documents as the Millennium Development Goals (MDGs). One of these goals requires countries to integrate the principles of sustainable development into country policies and programs (Mathews and Ohadi, 2008: 754). Clearly, South Africa has developed an elaborate policy framework to address environmental issues. In fact, the South African government believes that:

> It is too soon to evaluate the effect of these policies, although it can be anticipated that their overall effect on poverty and the natural environment will be favorable in the long term if combined with strategies to reduce consumption and waste in the affluent sectors of society. (DEAT, 1999b)

In addition to these processes, one must also acknowledge the role of public participation in environmental management. According to the DEAT (1999b: 4), "South Africa has developed its national environmental policy through a process of consultation called Consultative National Environmental Policy Process (CONEPP)," which "gave all stakeholders in South Africa an opportunity to contribute ideas in the development of the environmental policy" (DEAT, 1999b: 4).

ANALYSIS OF SOCIAL AND ECONOMIC IMPACT OF ENVIRONMENTAL POLICY

The impact of environmental policy on the social welfare of South Africa's citizens can be examined from several angles. Overall, the government is convinced that "[t]here is heightened awareness about the possible negative effects of socioeconomic development on the natural environment" (DEAT, 1999b). Some of the changes are positive in a sense that the intentions of the policies and laws are to transform the previously racial regulatory framework into policy instruments that will benefit all citizens based on equity and democratic principles. Therefore, this means that, unlike during the apartheid era when some industries (e.g., cement producers) could ignore the use of protective equipment for their personnel, such practices are nowadays legally discouraged or punishable in the regulatory framework. Moreover, South Africa's vigilant and proactive civil society organizations have also gotten up the courage to challenge illegal practices, where these have been witnessed, by industries that flout available policies and legislation.

A number of the negative consequences emanating from the new policies can be outlined as follows. First, environmental policy has not always benefited all citizens equally. If anything, it has been used by previous apartheid governments as a weapon to punish previously disadvantaged communities on the basis of racial classification. Second, environmental policies have not always protected the poor and vulnerable from unscrupulous businesses owing to the fact that such businesses have in the past exploited the legal loopholes in the law to avoid prosecution or, in the rare instances where prosecution was secured, to go scot-free or with minimal sentences. Third, despite the fact that new policies and laws have been enacted to conserve the environment, many industries that usually flout them have relied on the illiteracy and ignorance of these policies and laws among the affected communities. Obviously, this makes enforcement even more difficult. Fourth, in spite of the spirit and intent of the new laws and policies, new forms of inequalities—especially based on economic power—have surfaced. For instance, Van der Merwe and Steyl (2005: 295) have noted that South Africa continues to suffer from rural–urban disparities in terms of environmental management. Thus, the "[r]ural population and landowners are generally not served by public [waste] removal practices and are mostly prohibited from using existing urban disposal facilities. As a result, disposal of waste is often environmentally unsound." In fact, even the rural population is still largely racially segregated, with white farmers and landowners occupying much of the country's fertile land.

Fifth, two related challenges need to be observed regarding the dilemma of South Africa's policy makers in terms of the environment. The first is that environmental conservation experts have raised concern "that the new [policy] regulations appear to be aimed primarily at streamlining the EIA [environmental impact assessment] process by prescribing strict time limits within which the whole process will take place and simplifying it by furnishing an exhaustive list of activities that would trigger an EIA." The second challenge is that the process "still requires a sufficient capacity and expertise within the relevant authorities, which remains lacking . . ." (De Villiers, 2009). As Bryner (2003: 301) observes, "How well government achieved important policy goals is, to a great extent, a function of the capacity of public administration." The latter concern about the state's capacity to deliver is more significant in the previously disadvantaged or Black urban and rural areas, where the lack of sufficient capacity in the form of environmental conservation or protection experts is still a major hurdle for such areas to catch up with their more developed counterparts in the rest of South Africa.

Certainly, the lack of requisite skills and expertise in environmental matters among public officials remains a key threat to South Africa's implementation of its environmental policies. Thus, "concerns about institutional capacity at all levels of government" (DEAT, 1999b) are frequently raised in public discussions on the environment. Hence, even authorities publicly acknowledged that "[i]t will be some time before all branches of government are effectively able to implement measures to alleviate poverty, curb crime and administer the already existing legal measures to mitigate negative impacts on the natural environment" (DEAT, 1999b).

Nevertheless, the biggest challenge to the implementation of South Africa's range of environmental legislation and policies seems to be the adherence to a kind of one-size-fits-all conservation paradigm. More than a decade ago, scholars warned that there were "no effective challenges to the narrowly defined conservation paradigm" and that state enterprises and academic institutions "appear to have limited influence over public policy" (Orkin et al., 1995: 499). To date, very little seems to have changed. This dominant paradigm is still largely promoted by state functionaries and representatives of big business despite proposals for the active involvement of civil society constituencies in the policy process relating to the environment (Orkin et al., 1995: 500).

South Africa has experienced not just social change, for the impacts of social change on the environment have been felt mostly in terms of the population growth rate. In 1999, the government argued that "[t]he South African population is growing, and more people means greater demands on the natural resources and the environmental services, as well as increasing the amount of waste and pollution that is generated" (DEAT, 1999a). Among the key environmental impacts that had negative effects on the society are the following:

- Water shortages
- Uncontrolled settlement development
- Deforestation from excessive collection of wood for fuel
- Soil degradation through cultivation of unsuitable soils
- Exploitation of rare and endangered species to supplement incomes (DEAT, 1999a)

South Africa's environmental challenges might not be that different from those faced by other countries. However, the legacy of apartheid has greatly distorted the country's environmental situation including the possible policy responses to these challenges. Observers note that "[t]he purpose of public policies is to change or preserve conditions in society. Impact analysis is designed to discover what effect policies have actually had. It focuses on the policy impact . . ., The objective is to determine what difference the policy has made to societal conditions" (Van der Waldt, 2001: 103).

POLICY RECOMMENDATIONS

The government of South African has a long history of regulating the environment; however, it has not fully embraced the idea of corporate environmental policies. Thus public and corporate policies have not been created equally in the country. In the future, the government of South Africa should collaborate to position its environmental policies in three groups: (1) There should be statements of values that are designed primarily for public relations purposes by both the government and corporations. These should be printed with soy inks on recycled paper. (2) Both the government and corporations should negotiate policies that describe the corporations' commitment to regulatory compliance and that explain in detail how corporations in the country will guarantee compliance. (3) Both government and corporations should enact policies that will establish a commitment to proactive environmental stewardship. These policies should proclaim a commitment to go beyond compliance in order to achieve levels of environmental quality that are far greater than those required by law.

Further, the government of South Africa should adopt the approach of taking sustainable development policy action measures. This approach starts with the development objectives and needs of the nation. South Africa should begin by examining its development priorities and identify how these can help the nation achieve more sustainable goals, either by tightening existing policy or by implementing new measures.

South Africa should renew its commitment to formulate, implement, publish, and regularly update its national and, where appropriate, regional environmental regulatory programs containing measures to mitigate climate change by addressing the sources of anthropogenic emissions and the removal of sinks of all greenhouse gases (Winkler, 2009). This commitment is currently not quantified by the government of South Africa in the same way as it is in most industrial countries. Thus, the Government of South Africa needs to foster the commitment to effectively implement the already formulated environmental policies in order to achieve sustainable development for the country.

It is also recommended that the government of South Africa combine residential and electricity policies that could enhance the nation's pursuit of economic and social sustainability. The measure could help the nation derive both local and global benefits (Eisener, 2007). According to Richard Wright (2011), most of what is good for sustainable development in South Africa also has global benefits in terms of mitigating climate change. This model could be adopted in South Africa and could also resonate with neighboring African countries. This paper argues that there are indeed policies that meet local sustainable development goals and that could also benefit other nations. There is the need for the government of South Africa to explore other sources of renewable energy in addition to what they are currently using. The adoption of more renewable energy sources may require new sets of environmental policies.

SUMMARY

This chapter has examined the social and economic impact of environmental policy in the Republic of South Africa. It argued that the basic structure of the sustainability in South Africa should be based on the impact of human production and consumption. This can be reduced through the widespread adoption of pollution control technologies and recycling. It is important to note that, while the environment in South Africa is capable of assimilation, regeneration, and resilience, the ecosystem in the nation's ecosystem has a limited carrying capacity.

Although South Africa is blessed with a moderate climate, the existing climatic, demographic, and developmental trends may exceed the carrying capacity of the ecosystem. Stressing this capacity might have dire consequences for the quality of life of all the citizens in the country. In a nutshell, climate change may affect agriculture, farmers, fishers, the food supply, the cost of heating, the water supply, and the generation of energy. These impacts may spill over into the social and economic lives of South African citizens.

Given the nature of the social and economic impact of environmental policies in South Africa, as addressed in this paper, we argued for more incremental changes in both the production and consumption patterns of citizens. The paper also called for a change in environmental policies and in how those policies are implemented in South Africa. One can also identify some interactive process of human beings that affects the use of land, water, air, energy, and other resources. These human dimensions are major factors to consider in the nation's bid for sustainable development. This chapter therefore called for a comprehensive policies for sustainable development in South Africa. Such policies should address population growth in urban and rural areas, demographic trends and factors, the use of resources, the dissemination of appropriate technology, and development. The education of the citizenry and the creation of environmental awareness are also very important.

REFERENCES

Abbey, E. (1971). *Desert Solitaire: A Season in the Wilderness*. New York: Ballantine.

Adam, H., and Giliomee, H. (1979a). *Ethnic Power Mobilized: Can South Africa Change?* New Haven, CT: Yale University Press.

Adam, H., and Giliomee, H. (1979b). *The Rise and Crisis of Afrikaner Power*. Cape Town: David Phillip.

African National Congress (ANC). (2007). 52nd National Conference, Polokwane, December 16–20, 2007. Retrieved from http://www.anc.org.za/ancdocs/history/conf/. Accessed on November 24, 2012.

African National Congress (ANC). (1994). "The Reconstruction and Development Programme." Johannesburg: Umanyano.

Barbier, E. (1987). "The Concept of Sustainable Economic Development." *Environmental Conservation, 14*(2), 101–110.

Baumer, D., and Van Horn, C. (2014). *Politics and Public Policy: Strategic Actors and Policy Domains*. Los Angeles, CA: Sage.

Beinart, W. (1994). *Twentieth-Century South Africa*. Oxford and New York: Oxford University Press.

Beinart, W., and Dubow, S. (Eds.). (1995). *Segregation and Apartheid in Twentieth Century South Africa*. London: Routledge.

Bryner, G. C. (2003), "Public Organizations and Public Policies." In B. G. Peters and J. Pierre (Eds.), *Handbook of Public Administration* (pp. 300–309). London: SAGE.

Department of Environment Affairs and Tourism (DEAT). (1999a). "Social Environment." National State of the Environment Report—South Africa. Department of Environmental Affairs and Tourism. Retrieved from http://www.ngo.grida.no/soesa/nsoer/issues/social/index.htm. Accessed on November 13, 2012.

Department of Environment Affairs and Tourism (DEAT). (1999b). "Social Environment: Data Issues." National State of the Environment Report—South Africa. Department of Environmental Affairs and Tourism. Retrieved from http://www.ext.grida.no/soesa/nsoer/issues/social/data.htm. Accessed on November 11, 2011.

De Villiers, L. (2009). "Environmental Impact Assessment—Where Are We Going?" Wildlife and Environmental Society of South Africa. Retrieved from http://www.enviropaedia.com/topic/default.php?topic_id. Accessed on November 19, 2011.

Dibie, R. (2009). "Environmental Governance and the Political Economy of Sustainable Development in Sub-Saharan Africa." *Journal of International Politics and Development*, 7, 1–28.

Dunn, W. N. (1981). *Public Policy Analysis*. Englewood Cliffs, NJ: Prentice Hall.

Dye, T. (2011). *Understanding Public Policy* (13th ed.). Upper Saddle River, NJ: Prentice Hall/Pearson.

Eisner, M. (2007). *Governing the Environment*. Boulder; CO: Lynne Rienner.

Gubb, A. (2009a). "National Environmental Management Air Quality Act (39 of 2004)." Wildlife and Environmental Society of South Africa. Retrieved from http://www.enviropaedia.com/topic/default.php?topic/164. Accessed on December 11, 2012.

Gubb, A. (2009b). "National Environmental Management Biodiversity Act (10 of 2004)". Wildlife and Environmental Society of South Africa. Retrieved from http://www.enviropaedia.com/topic/default.php/id=165. Accessed on November 11, 2012.

Gunn, A. (2009). "Are South Africa's Environmental Policies and Laws on the Right Track?" Retrieved from http://www.hg.org/article.asp?id=5883. Accessed on December 11, 2013.

Haycock, E., and Stone, A. B. (2005). "Commercialisation of Urban Environmental Maintenance Services by South African Municipalities." *Journal of Public Administration*, *40*(1), 62–72.

Heineman, R.A., Bluhm, W. T., Peterson, S. A., and Kearny, E. N. (1990). *The World of the Policy Analyst*. London: Chatham House.

International Development Research Centre (IDRC). (2009). "Chapter 2: Environmental Policy in the New South Africa." Retrieved from: http://www.idrc.ca/en/ev-82676–201–1-do_topic.html. Accessed on February 11, 2013.

Jones, C. (1984). *An Introduction to the Study of Public Policy* (Third Edition). Monterey, CA: Brooks/Cole.

Mathews, S., and Ohadi, N. (2008). "Transformative Role of Local Government to Achieve the UN Millennium Development Goals." *Journal of Public Administration*, *43*(4), 749–766.

Myers, N. and Spoolman, S. (2014). *Environmental Issues and Solutions*. Belmont, CA: Brooks/Cole.

Naess. A. (1995a). "The Deep Ecological Movement: Some Philosophical Aspects." In G. Sessions (Ed.). *Deep Ecology for the Twenty-First Century*. Boston: Shambhala.

Naess. A. (1995b). "The Shallow and the Deep, Long-Range Ecology Movements: A Summary." In G. Sessions (Ed.). *Deep Ecology for the Twenty-First Century*. Boston: Shambhala.

Orkin, M., Tshandu, Z., and Dugard, J, (1995). "Policy Research and Formulation: Two Case Studies." In P. Fitzgerald, A. MacLennan, and B. Munslow (Eds.). *Managing Sustainable Development in South Africa*. Cape Town: Oxford University Press, 469–508.

Peters, G. (2014). *American Public Policy: Promise and Performance* (Ninth Edition). Los Angeles, CA: Sage.

Peters, B. G., and Pierre, J. (Eds.). (2003.) *Handbook of Public Administration*. London: Sage.

Peterson del Mar, D. (2006). *Environmentalism*. New York: Longman/Pearson.

Posel, D. (1996). "Modernity and Measurement: Further Thoughts on the Apartheid State." *Institute for Advanced Social Research*. Seminar paper 407.

Posel, D. (1991). *The Making of Apartheid 1948–1961: Conflict and Compromise*. Oxford: Clarendon.

Randall, V., and Theobald, R. (1998). *Political Change and Underdevelopment* (2nd ed.). London: Macmillan.

Roorda, N. (2012). *Fundamentals of Sustainable Development*. New York: Routledge.

Sale, K. (1993). *The Green Revolution: The American Environmental Movement, 1962–1992*. New York: Hill and Wang.

Sebola, M. P. (2003). "The Communal Nature Reserves: A Concern for Development in Limpopo Province." *Journal of Public Administration*, *38*(4), 476–485.

Singer, P. (2001). "All Animals Are Equal." In M. E. Zimmerman (Ed.), *Environmental Philosophy: From Animal Rights to Radical Ecology* (3rd ed.). Upper Saddle River, NJ: Prentice Hall, 31–32.

Smith, F. H. (2004). "Ecologically Sustainable Development in the Utilization of Natural Resources: The Cederberg Wilderness Area Experience." *Journal of Public Administration*, *39*(2), 341–362.

South Africa. (1998a). National Environmental Management Act. Retrieved from www.acts.co.za. Accessed on November 11, 2011.

South Africa. (1998b). National Water Act. Retrieved from www.acts.co.za. Accessed on January 11, 2013.

South Africa. (1989). Environment Conservation Act. Retrieved from www.acts.co.za. Accessed on October 11, 2012.

South Africa. (1973). Hazardous Substances Act. Retrieved from www.acts.co.za. Accessed on October 11, 2012.

Van der Merwe, J. H., and Steyl, I. (2005). "Rural Solid Waste Management: A Planning Strategy for Higher Density Agricultural Regions." *Journal of Public Administration*, *40*(3), 295–313.

Van der Waldt, G. (2001). "Public Policy and Policy Analysis." In G. Van Niekerk, G. Van der Waldt, and A. Jonker. *Governance, Politics, and Policy in South Africa*. Cape Town: Oxford University Press, 86–114.

Van Niekerk, G., Van der Waldt, G., and Jonker, A. (2001). *Governance, Politics, and Policy in South Africa*. Cape Town: Oxford University Press.

Van Rooyen, E. (1999). "Development and Environmental Management in SADC: The Maputo Development Corridor." *Journal of Public Administration*, *34*(3), 235–236.

Warren, K.J. (2001). "Ecofeminism." In M. E. Zimmerman (Ed.). *Environmental Philosophy: From Animal Rights to Radical Ecology* (3rd ed.). Upper Saddle River, NJ: Prentice Hall, 253.

Winkler, H. (2009). *Cleaner Energy Cooler Climate*. Cape Town: Human Science Research Council.

Wright, R. (2011). *Environmental Science* (11th ed.). Upper Saddle River, NJ: Prentice Hall/Pearson.

Wright, R., and Boorse, D. (2014). *Environmental Science: Towards a Sustainable Future*. Boston: Pearson Learning.

9 Environmental Policy and Issues in Ethiopia

Robert Dibie, PhD

INTRODUCTION

Ethiopia is located in northeast Africa. The nation has a population of about 60.1 million people. Its population growth rate is about 2.1 percent, and the life expectancy at birth is 50 years (United Nations Human Development Index, 2008). The environmental problems in Ethiopia arise either due to human-made exploration of biodiversity or due to natural disaster. The human-made problems are further compounded by global pollution and the emission of greenhouse pollutants. In Ethiopia, as in many other African countries, the demographic growth has been very high in the past two decades (Handleman, 2009; Hyden, 2006). Apart from these factors, the environmental problems in Ethiopia are closely related to its great geographical diversity (Asefa, 2003; Shibiru, 2000; Zewdu, 2002). The nation is mountainous with high and rugged hills, flat-topped plateaus, and deep gorges incised by river valleys and rolling plains. Gedion Getahun (2009a) contends that this great geographical diversity has made the nation vulnerable to ecological problems, such as soil erosion and land degradation. Other environmental problems that currently face Ethiopia are desertification, deforestation, and the emission of industrial pollutants (Dibie, 2007). The inadequate availability of wildlife also means less meat for the citizens living in both the rural and urban areas of the nation.

Ethiopia, like most countries on the African continent, has been plagued with the problem of forest depletion (Environmental News, 2000). Barely 5 percent of the land is covered with trees, probably because the majority of the nation's rural population continue to steadfastly exploit the trees for firewood for cooking, building houses, and bridges. Forest has often been cleared for the purpose of converting it to agricultural land. The economic aftermath has been exchanging forest for farming areas to plant crops (Getahun, 2009; Tekelemichael, 2009). Severe climate change often has had adverse impacts on the social, economic, and general living conditions of the citizens of Ethiopia (Hoben, 1972). Climate change has also led to economic problems such as reduced forests, water resources, biodiversity, agriculture, human health, and wildlife (EPA, 2004; UNEP, 2000).

The United Nations report (2008) pointed out that in the past five decades Ethiopia has lost a large percentage of biomass cover due to land degradation, soil erosion, and desertification. The report also argued that the causes of the drought and famine in Ethiopia may be associated with convoluted political and economic policies. Such policies have further exacerbated the ecological mismanagement and inadequate funding of governments agencies that were supposed to implement environmental policies (United Nations, 2007; World Bank, 1997b). Politics has also negatively affected the ability of the government to aggressively seek access to environmentally friendly technologies.

Ethiopia became more popular in the 1980s due to its famine and drought disaster, when several international nongovernmental organizations (NGOs) poured in aid to help famine victims. Due to its severe environmental problems and the foreign aid given to Ethiopia since the 1980s, the government has adopted voluntary resettlement programs in regions hard-hit by drought. A United Nations report (2003a) stated that the people in the Oromiya and Amhara regions of Ethiopia have been suffering from severe environmental conditions. The report also indicated that over 13 million people in Ethiopia have faced starvation due to drought in the past twenty years. In addition, the United Nations Emergencies Unit (EUE) for Ethiopia (2003b) contends that resettlement programs initiated in the nation could cause more devastating environmental problems and threaten the living standards of people living in the rural areas.

The majority of the people in Ethiopia live in rural areas where land cultivation and agriculture are the main means of subsistence. Despite this pattern of living, there have not been enough economic progress and technological development programs to alleviate poverty in those areas. Ethiopia continues to face the problem of reconciling the demands of modern development—particularly good roads, water supply, and waste disposal systems—with the constraints of the fragile ecosystem found in the mountains (Rahmato et al., 2008). In most cases, the nation has to trade off the need for urban development and an expanding tourist industry the need to conserve natural resources.

Ethiopia is also vulnerable to climate change. The nature of its geographical diversity has often led to adverse impacts on socioeconomic activities in times of sudden climate change. For example, a shortage of rainfall often affects agriculture, water supply resources, forests, human health, biodiversity, and wildlife survival (Dibie, 2008a). As a result of inadequate national coverage of meteorological and climatological stations for effective atmospheric monitoring, the low level of awareness of and the general inattentiveness to climate change have resulted in an awful and nonchalant mistreatment of the country's biodiversity (Getahun, 2009b). This imprudent behavior of the government has negatively affected the wealth of Ethiopia.

It is also interesting to note that for several years of its existence as a sovereign nation, Ethiopia's modernization activities did not consider environmental issues in its development plan. From the outset, environmental health and development were not included in the nation's development pan. The nation enacted its first set of environmental policies between 1993 and 1999, and environmental impact assessment (EIA) legislation was not enacted until several years later. The Environmental Protection Authority of Ethiopia was enacted in September 1995.

Changes in the population all over the nation, especially the migration of people from the rural to urban areas, have also created major environmental problems in Ethiopia. The United Nations Human Development report (2008) indicated that the most of the people in Ethiopia live in traditional types of unplanned housing with low levels of services. "Low level of services" refers to urban infrastructure and services and translates to crowded housing areas and lack of basic facilities such as environmental cleanness, sewage, and sanitation systems. The UN report (2007) also pointed out that most Ethiopians have poor or inadequate access roads and drainage systems. For instance, some urban and rural households use pit latrine toilets, and a very small number of people use flush toilets. In the rural areas, the majority of the citizens use the forest and fields as their toilet, and others use pit latrines. When it rains, the latrines are often washed into the nearby rivers, lakes, unprotected wells, and springs.

Poor sanitation, bad land management, poverty, and high rate of unemployment often expose the citizens to diseases (Dibie, 2007). Thus, the rapidly expanding population in the urban areas can be viewed from an environmental perspective as burdensome and could cause a dangerous health plague if not properly handled. The enactment of the Ethiopian National Population Policy in 1993 has not helped in any way to deal with the growth of population, and officials of the National Population agency who were interviewed confirmed that the population growth in Ethiopia is due to the high fertility rate in the nation (Ethiopian Environmental Agency, 2005). Thus, unchecked population growth in Ethiopia may constitute a severe impediment to sustainable development.

This chapter examines and analyzes the economic and social impact of environmental policies in Ethiopia. It explores the local dynamics of urbanization and its effect on citizens' daily life. It also examines the complexity of the environmental management processes and the implementation of conventional and nonconventional policies in Ethiopia. It argues that increased diversity and partnership with environmental stakeholders could contribute to Ethiopian environmental policies having a positive economic and social impact. It evaluates how nongovernmental organization (NGO) programs have helped to prevent environmental degradation in the rural areas of Ethiopia. The analysis of environmental policies and population growth control programs in Ethiopia reveals that they have not been effective. The concluding section of the paper recommends some policies that could help alleviate poverty, raise environmental protection awareness, and engage citizens in appropriately contributing to solving the nation's sustainable development problems. It contends that the choice of a community-based approach and related options could help to positively improve the social impact of environmental policy in Ethiopia.

SUSTAINABILITY AND ENVIRONMENTAL PROTECTION

Sustainable development, in this discussion, refers to a means of finding alternatives to the development techniques that are currently depleting their natural resources (Easton, 2003; WCED, 1987; World Bank, 2000). The broad sentiment associated with the term is greater sensitivity to nature and natural resources. However, beyond this general association, the formulation of more rigorous environmental policies in Ethiopia has ignored the input of indigenous people who are supposed to help implement the environmental policies in their rural land. Thus, environmental problems in Ethiopia fall broadly into two categories: (1) the problems arising out of poverty or the inadequacy of development itself and (2) the problems that arise out of the very process of enacting and effectively implementing appropriate policies (De Waal, 1997; Rogers, Jalal and Boyd, 2008).

The analysis of environmental policies and programs in Ethiopia reveals that some of these policies are poorly conceived, inadequately planned, insufficiently financed, and lacking the input of the indigenous people (Rahmato et al., 2008; Tesfaye, 2009). This failing in the policy formulation process has caused the lack of popular participation in implementation, warranting a call for community-based environmental policy that can enable indigenous organizations and communities to discover and participate in policy making and implementation. To be sustainable, environmental policies must take a long-term view of all stakeholders and consider the durability of the proposed solutions. The mobilization of communities and all stakeholders is one strategy to achieve both environmental and economic growth in Ethiopia. The nation's environmental policy could be more durable if it gave priority not only to reducing costs but also promoting social and local economic benefits (Munasinghe and Swart, 2005).

Figure 9.1 shows the framework for achieving sustainable development. It illustrates how sustainable development can be achieved in three important dimensions: (1) environmental and biodiversity protection; (2) people and the reduction of poverty; and (3) social and economic transformation. From this perspective, social and economic transformations are related to efficiency, stability, and growth for all in the society, nation, or the world. Such a transformation could also lead to the empowerment of the citizens. Munasinghe (2002) contends that sustainable development should permit continuing improvements in the present quality of life at a lower intensity of resources use, while leaving enhanced stocks of assets for future generations. Such assets should include manufactured, social capital and natural

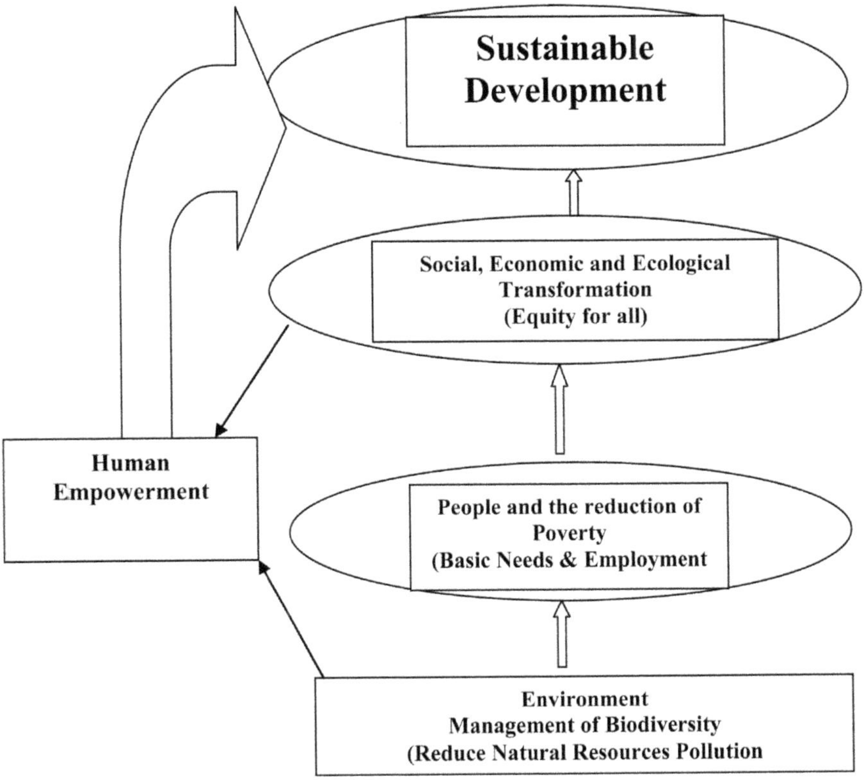

Figure 9.1 The Framework for Achieving Sustainable Development. An efficient relationship among environmental and biodiversity protection, people and the reduction of poverty, and social and economic transformation could lead to the empowerment of the citizens and sustainable development.

resources that provide opportunity for improving their quality of life (Munasinghe and Swart, 2005).

One of the social impacts of environmental conditions in Ethiopia is poverty. As indicated in Figure 9.1, an important step toward the realization of any long-term poverty reduction and meaningful economic development should be a pragmatic fight against poverty. Though Ethiopia may be capable of regeneration, assimilation, and resilience, the ecosystem in the nation has a limited ability to carry the demand.

Human consumption and waste production can be reduced through the widespread adoption of pollution control technologies and recycling. A more sustainable path would be economic reorganization on a regional rather than a global basis (Eisner, 2007). This strategy would include a good flow of food aid, adequate provision of health services for the rural poor, and the equitable provision of social infrastructure, such as water supply, schools, high rates of literacy, good technical skills, and financial resources. Further, the appropriate mixture of NGOs' expertise, traditional methods, and the technical skills necessary to achieve the desired environmental sustainable development goals should be determined (Dibie, 2001; Onimode, 2004). It is only through this kind of human empowerment approach that environmental policies can achieve their mission and objectives in Ethiopia. Tewoldeberhan Gebre-Egziabher (2009), the general manager of the Environmental Protection Authority of Ethiopia, and Jeffrey Clark (2000) contend that the social and natural calamities that continue to confront Ethiopia, if not addressed soon, might contribute to a pervasive level of poverty and perpetual underdevelopment in the nation.

NGOs have also been influenced by insights flowing from a new field of research known as earth systems science, or global change, which are composed of geology, ecology, oceanography, chemistry, paleobiology, meteorology, and earth science. This approach is necessary to sustain the practices and to monitor and react to the environment's responses to the new technology introduced by foreign NGOs in Africa. The approach also allows new techniques to develop out of local knowledge and experience in Ethiopia (MOFED, 2005; Zewdu, 2000).

The dependency theory pointed out that the development and wealth of Ethiopia are a function of the developing world (Dibie, 2008a). Thus, environmental destruction in Ethiopia may be rooted in Western imperialism and natural causes. Patterns of resource exploitation emerged that enabled Western imperialism to accumulate capital and assume political hegemony over the subordinate sub-Saharan African nations (Hyden, 2006; Waller-Hunter and Jones, 2002). Therefore, social and economic exploitation in Ethiopia could lead to a situation in which the nation's resources are used in a manner incongruent with local needs. At the same time, these patterns of resource exploitation incorporated new products and technologies from other areas of the globe (Fortune, 2000; World Bank, 2000). Ethiopia has become more dependent on overseas development assistance (ODA) and on credits from international financial institutions such as the World Bank, the International Monetary Fund, the African Development Bank, the United States Agency for International Development (USAID), and several NGOs. The United Nations country profile series report (2002) indicated that Ethiopia received development assistance averaging 15.4 percent of its GNP in the late 1990s. Between 2000 and 2008, the external assistance covered almost 9 percent of Ethiopia's gross domestic product (GDP) (World Bank, 2009).

One of the major contentions of this paper is that the government of Ethiopia may have to balance a strong relationship between economic development, poverty, and environmental quality (Steel, Clinton and Lovrich, 2003; Thin, 2002; United Nations, 2003b). The need for community-based participation was subsequently added to these criteria (Clark, 2000). A stronger collaboration among the government, NGOs, and the private sector in Ethiopia might lead to a successful set of natural resources management environmental programs in the country.

ANALYSIS OF ENVIRONMENTAL PROBLEMS IN ETHIOPIA

For any nation to develop, its need to effectively protect its environmental resources. The major bone of contention is that environmental resources protection is paramount to the social and economic development of Ethiopia because the resources from the ecosystem are instrumental to the production of goods and services. Adequate and marketable goods and services could be exchanged for financial capital that would in turn lead to the reduction of poverty and hence economic growth. The analysis of Ethiopian history reveals that the nation's environmental resources have been poorly managed, and underutilization has been. For the past four decades, a major impediment to the nation's overall development.

Ethiopia's landmass is characterized by flat-topped plateaus, rugged mountains, river canyons, lowlands, steep valleys, and rolling plains. It is estimated that 86 percent of the citizens of the nation live in rural areas (Rahmato et al., 2008). Most parts of the Ethiopia's landmass are made up of folded and fractured crystalline rocks capped by sedimentary limestone and sandstone and by thick layers of volcanic lava. Table 9.1 shows the human development index of Ethiopia with selected African countries.

The rocks, rugged mountains, and rolling plains also cause soil erosion, which is one of the major environmental problems in Ethiopia (Enrico et al., 2002a; Rahmato et al., 2007). The transition of people to more sustainable livelihoods has been very difficult in some regions in Ethiopia due to the problems of deforestation, overgrazing, and poor land management. This problem is further accelerated by the rate of erosion during the 1970s and 1980s.

Table 9.1 shows that only 46 percent of the population in Ethiopia has access to good health care. While 78 percent of the population has access to clean water, the life expectancy is approximately 52 years. Ethiopia's literacy rate is still below 50 percent at 34.5 percent (United Nations Development Program, 2008). The table also shows that economic development led to greater economic security and stability and to better GDP in 2008. Economic development, as indicated by GNP per capita, also enlarges the middle class, whose interest is in moderating conflict by rewarding moderate political parties and penalizing extremists in Ethiopia. The integration of environmental policies in the past decade in the nation has somewhat transformed traditional environmental regulation into a system of environmental governance, standard-setting organizations, NGOs, and corporations with greater coordination to address a recognized set of biodiversity problems in Ethiopia.

Table 9.1 Human Development Index in Ethiopia

	1994		2006				
Nation	GDP per Capita	Life Expectancy	People with access to Health Services	GDP per Capita	People with access to Water (%)	Life Expectancy (years)	Adult Literacy Rate
Angola	896	47.2	47.27	2,180	47	41.0	44
Botswana	2,800	68.1	60	9,945	5	34.9	68.7
Cote d'Ivoire	610	52.1	60	1,551	16	45.9	39.4
DR Congo	429	52.2	26	978	54	43.5	76.4
Egypt	2,604	68.9	60	4,211	2	70	51
Ethiopia	**100**	**43.1**	**46**	**1,055**	**78**	**51.8**	**34.5**
Eretria	960	50.1	50	1,109	40	43.7	25
Gabon	3,880	54	68	6,623	12	54.0	65
Ghana	410	56.6	60	2,240	25	57.0	63.4
Kenya	1,404	53.6	77	1,140	39	52.1	77
Malawi	170	41.1	35	667	38	46.3	55.8
Namibia	1,970	55.9	55	7,586	60	51.6	85
Nigeria	280	51	51	1,128	52	43.4	56
Rwanda	80	23.1	80	1,263	26	44.2	59.2
Senegal	600	49.9	90	1,792	42.3	62.3	32.1
South Africa	3,040	65	91	11,192	12	47.0	81.3
Tanzania	140	50.3	42	674	38	45.9	66.8
Sudan	1,084	51	70	1,943	5	56.5	44.8
Uganda	190	40	49	1,478	40	48.4	61.1
Zambia	350	42.6	50	943	40.5	37.7	77
Zimbabwe	500	50	49	2,065	19	36.6	80

Source: United Nations Development Program (UNDP) (1994, 2008).

The Environmental Protection Authority (EPA) report (Environmental Protection Authority of Ethiopia, 2005) indicated that farmers in the nation's highlands cultivate sloped or hilly land, causing topsoil to wash away during the torrential downpours of the rainy season. The effect of the rain has also been the leaching of highland soils of much fertility. Sharp sand and soils overlying crystalline rocks are often washed away, making crop growth difficult. Despite the rain activity, the volcanic soils of the highland are less readily leached and therefore are more fertile (Environmental Protection Authority of Ethiopia, 2005; Sutcliffe, 2006). Sonneveld and Keyzer (2002) estimated that nearly 2.3 billion tons of soil are washed away from agricultural land every year. The EPA has argued that if the sheet erosion problem in the nation is not checked, it may render crop production impossible in the next forty to sixty years. The presence of mosquitoes carrying malaria has kept many farmers from developing parts of Ethiopia's potentially productive lowlands. Thus, deforestation and desertification are worsened by the widespread use of traditional fuels such as firewood, which represent 96 percent of total energy consumption (Rahmato et al., 2007; Sonneveld and Keyzer, 2002).

One high positive aspect of the Ethiopian environmental ecosystem is the fact that the adverse climatic conditions in the nation over several decades has produced a habitat that harbors some unique animals and plant species. The United Nations report (2002) describes Ethiopia as having one of the highest levels of biodiversity in the world. In the past three decades, Ethiopia's environmental projects have focused on wildlife, biodiversity, forest conservation control, and waste management (Rahmato et al., 2008). The degraded areas in Trigray, the Central Rift Valley, Wollo, and Northern Shewa have been rehabilitated to a large extent through the cooperation of international NGOs and the Ethiopia government.

It is also worth noting that the population of Ethiopia is concentrated mostly in the high- and intermediate-altitude areas. The highland areas have an altitude of about 180 m above sea level. This is because 37 percent of the land in Ethiopia is covered with mountainous topography. . Zerihun Woldu (2008) and Getahun (2009b) argue that the recurrent disruption by rainfall, as well as its infiltration and retention, leading to prolonged floods and droughts, affects the ability of plants to absorb carbon and the necessary nutrients, causing the leaching of the fertility from the soil (see Figure 9.2).

From a social and economic point of view, the loss of ecosystem in Ethiopia has an adverse effect on the agricultural productivity and on the creation of wealth in the nation. These problems also affect the rearing of animals and lead to lowered biomass, further negatively affecting the livelihood of farmers and pastoralists. When faced with a bad production year, farmers find it difficult to cultivate crops, access water, acquire fuel, and find construction materials. If these conditions continue for a long time, it often leads to devastatingly low standards of living and serious financial crisis (Feoli et al., 2002a; Woldu, 2008).

The Environmental Protection Authority of Ethiopia report of 2004 describes how environmental degradation is considered as a silent and insidious encroachment on life and livelihood. Zerihun Woldu (2008) contends that environmental encroachment in Ethiopia is increasingly affecting the social, economic, and national vulnerability to severe health problems. Most scholars have also argued that the magnitude of the recurrent drought, deforestation, and incremental degradation has not been accurately reported by the Environmental Protection Authority of Ethiopia (Getahun, 2009b; Tesfaye, 2009; Woldu, 2008).

Figure 9.2 Dump Site by the Side of a Street in Ethiopia
Source: Photo by the author during field research in Ethiopia 2012

During the severe famine period of the 1980s, President Mengistu's absolute intolerance for dissidence or opposition earned him many enemies within Ethiopia and in neighboring nations. Within Ethiopia, ethnic groups such as Tigray built resistance to Mengistu's regime, and its neighbor Eritrea aggressively demanded independence (Country Watch, 2003; Herweg, 1994; Rahmato et al., 2008). At the same time, the nation's neighbor to the south, Somalia, clashed frequently with it. Toward the end of the 1980s, the resistance, chaos, and civil war that pervaded the government of Ethiopia galvanized Mengistu to loosen his regime's socialist reform policies. The aftermath of these political problems was the hardening of environmental problems. Asefa (2003) contends that policies implemented by Mengistu's regime prevented the private sector to own land. His administration abolished private ownership of land in favor of privatization so that farmers and entrepreneurs could use the land as collateral in receiving loans that could be used in developing and solving environmental problems in the nation.

Since the 1980s, almost all the forest in the nation has been destroyed. According to Tedla Shibiru (2000), less than 3 percent of Ethiopia's landscape is now covered with trees. The implication of fewer trees growing in the nation might be environmental disaster. Although Ethiopia has been suffering from drought, its government policies, citizens' interference, and economic concerns are the most important reasons for the fast depletion and serious degradation of natural resources in the nation. Medhin Zewdu

(2002) points out that the government and the people, especially those living in the rural areas, have not been able to pay appropriate attention to the efficient protection of its natural forest resources. In recent times, Ethiopia's natural resources have faced a variety of environmental crises, including drought since the 1980s and in 2001–2003 (Rahmato et al., 2008). The average temperature in the country has already increased because of the lack of trees, placing a greater burden on the fragile environment and the communities living in the rural areas. Erosion problems in some rural areas are now so severe that many foreign NGOs have been funding projects to terrace the terrain.

The landscape in the northern part of the nation has a dissected and sloping terrain, with fragile soil, and is subject to highly erosive rainstorms during the main agricultural season. Low in organic matter and with little natural tree cover, Ethiopia is subject to severe soil erosion (Hoben 1995; Sonneveld and Keyzer, 2002). The majority of the forest in Ethiopia now lies in the southwest region of the nation.

On the central and southern highlands, where rainfall is higher and distributed over most of the year, soils are generally higher in organic matter (Rahmato et al., 2008). The land is less rugged, and there is more natural vegetative cover. In the past five decades, however, the natural forest, which was extensive, has been reduced and grasslands brought under the plow by a combination of conquest, spontaneous small farmers, migration, government-sponsored resettlement, expanding commercial and state farms, and private and state exploitation of the forests. Soil erosion appears to be less severe than in the north (Hoben, 1995; Sutcliffe, 2006).

Before the 1974 revolution, the indigenous environmental management system was such that northern Ethiopians had long managed their landscapes and practiced some form of conservation. The farming techniques mostly used in the rural areas is a plow-based mixed farming system, which contributes to soil erosion through fine tilling, monocropping, and a lack of vegetation covering the soil during the heavy raining season (Rahmato et al., 2008). Historically, however, political, institutional, and economic conditions gives land rights to neither elites nor peasants. The strong incentive to invest in agricultural improvement and labor-intensive conservation measures was simply to provide food for the nation, but the intention was also to prevent the further depletion of its natural resources.

Apart from deforestation, soil erosion, population growth, rural-to-urban migration, lack of basic infrastructure, and land degradation, Ethiopia is also faced with other environmental problems. Major environmental problems in Ethiopia are emissions and industrial pollution. Gedion Getahun (2009) contends that the flower farming and export business has been seriously polluting the air in Ethiopia. Although Ethiopia has benefited immensely from flower exports to the European Union markets, it is paying the pollution price in that the flower industry uses various types of pesticides and chemicals. Wudineh Zenebe (2009) contends that the flower industry imports a large number of chemicals and pesticides and applies them to cultivate lands where they grow their flowers. From an environmental science perspective, these chemical are considered hazardous, they pollute the ecosystem, and they are carcinogenic to human beings (Rahmato et al., 2008). For example, the World Health Organization (WHO) and United Nations Environmental Programs (UNEP) report (2000) indicated that these chemicals were indeed harmful to human beings and animals. Teguam (2009) also confirmed that organic pollutants (POPs) like Aldrin, chlordane, DDT, and Dieldrin have been used for a long period in the country, and their use may continue until environmentally friendly biological control agents can be introduced to replace them.

One of Ethiopia's nightmares is its poor performance in the management of all types of waste (United Nations Development Program, 2008). Most of the industrial cities in Ethiopia are also highly polluted. For example, it has been reported that in Wonji and Methara, industrial pollutants continue to negatively affect the health of workers and people living in these cities. Asbestos and industrial smoke also fill the air of Wonji and Methara (Mequanint, 2009). The environmental pollution in cities such as Akaki Beseka and Wonji Shoa is another challenge for the Ethiopian Environmental Protection Authority (Mequanint, 2009). The textile and metal industries in these two cities have also been reported to have dumped hazardous waste into the Akaki River. Toxicologists have also found hazardous waste, metals, and chemicals such as chromium, nickel, arsenic, zinc, cadmium cobalt, copper, manganese, and other toxic organic substances. in the Akaki River (Alemu, 2009). Despite the establishment of the National Radiation Protection Authority (NRPA), Ethiopia has not been able to effectively dispose of its hazardous solid wastes. As of 2009, solid waste in Addis Ababa (the nation's capital city) was still untreated (Mequanint, 2009).

Another important area that needs attention is the Ethiopian wildlife. It should be noted that, although the nation has severe land degradation, Ethiopia is endowed with vast land potential for agricultural development and wildlife. The nation's wildlife has been reported to be incrementally decreasing due to the following reasons: (1) professional hunting of some African safari groups that are mostly from overseas; (2) the requirement of farmers in most rural areas of land to be cultivated for crops and food (Farmers tend to deforest areas in order to plough them for agriculture, bringing on the decimation of flora and fauna in the surrounding regions.); (3) extensive hunting by the local people residing near the national park area or close to the wildlife area; and (4) the farmers' livestock chasing away the wild animals. The livestock density results in the degradation of natural resources and minimizes the number of the wild animals (Getahun, 2009b).

Thus the government of Ethiopia and the EPA have not been able to grasp the notion of using incentives to promote conservation and the sustainable use of natural resources without negatively altering the ecosystem. The heavy-handed top-down approach in enforcing environmental policies has not been very effective.

NONGOVERNMENTAL ORGANIZATIONS (NGOs) AND ENVIRONMENT ISSUES

NGOs activities in Ethiopia are not new. Indigenous community-based organizations (CBOs) have been in the nation for several centuries. These community-based organizations are sometimes referred to in the literature as traditional self-help systems (Woldu, 2008; Zewdu, 2002). The Ethiopian EPA report (2005) pointed out that NGOs in the nation have tirelessly assisted the government and people in establishing community infrastructure such as health posts, water points, food warehouses, health centers, food station rotations, medical drugs, equipment and utensils, and transportation. NGOs have coordinated the facilitation of household needs, such as the construction of shelter, the provision of oxen, and the distribution of seeds. Foreign NGOs have promoted the awareness of HIV/AIDs, gender-related, and epidemic diseases (Rahmato et al., 2008). They have assisted farmers in land preparation for planting seasons. It could therefore be argued that NGOs in Ethiopia have been instrumental in capacity building at all levels. It is also worth noting that several NGOs outside Ethiopia have frequently donated funds to the government and to local civil society groups to help the funding some of these projects.

In the past five decades, some of these indigenous community-based organizations (CBOs) have been registered as functional local NGOs with the government of Ethiopia. These local NGOs are sometimes referred to by the indigenous people as *Debo* and *Afarsata* or as *Ekub* and *Edir*. CBOs are also famous for rotating saving and credit among their members. There are over 2,500 NGOs operating in the development processes in Ethiopia, and several of them are organized in associations. The Christian Relief Development Association (CRDA) is the largest organization, representing an alliance of over 350, as well as local international NGOs (Clark, 2000; Woldu, 2008). The Society for Participatory Development (SPADE) and the Development and Interchurch Aid Commission (DIDAC) are other NGO coalition groups that were also formed. The laws of Ethiopia require all NGOs operating in the nation to register with the Ministry of Justice. NGOs are also required to obtain a certificate as well as sign an operational agreement with the Federal Disaster Prevention and Preparedness Commission (DPPC) of Ethiopia (Clark, 2000; EPA Report, 2005; Zewdu, 2002).

During the 1960s, several foreign religious NGOs established locations in Ethiopian cities. Other humanitarian foreign and local NGOs were established during this period. The Ethiopian Red Cross and Swedish

Save the Children are among the first well established humanitarian NGOs in the late 1960s. The famines of 1973 and later of 1984 contributed to the increase in the number of international NGOs in Ethiopia (Rahmato et al., 2008). The 1980s are considered the beginning of severe famine in Ethiopia.

The 1980s also marked the period when several international environmental NGOs flocked to northern and eastern Ethiopia to help famine victims. Despite the aid that NGOs provided in the 1980s, the government of Ethiopia has continuously been faced with a fundamental change in the country's environmental conditions. President Mengistu's government in the late 1970s and early 1980s attempted to tighten control of the activities of NGOs because of the belief that they represented the Western values and the perspective of social and economic development (Rahmato et al., 2008). These highly regulatory efforts to control NGOs started failing in 1984–1985 because of their expanded operations in providing humanitarian aid to the people of Ethiopia. Another reason for the failure was the fact that the NGOs' presence was instrumental in securing international donations and assistance (Asefa, 2003; Clark, 2000; Dibie, 2008a). The general effort to stimulate interest in environmental protection in Ethiopia has focused on foreign aid from NGOs rather than the need to introduce sustainable development policies.

After the change of government in 1990, Ethiopia's political climate became more conducive for encouraging the growth of NGOs (Rahmato et al., 2008). It is interesting to note that more local NGOs flourished in the 1990 and that, since the late 1990s, about 89 percent of the NGOs operating in the country are local. It is also very important to note that in 1995, the government of Ethiopia started providing guidelines for NGO operations in the nation. This was because of the belief that NGOs were preventing development assistance funds, pledged by international sources, from getting directly to the national government. Jeffrey Clark (2000) contends that the NGO community perceived the guidelines to be more of an effort by the Ethiopian government to control their activities.

Foreign NGOs are now encouraging indigenous Ethiopians to plant trees. Despite the tree planting projects, more than 34 million people in Ethiopia still depend on wood for cooking (Zewdu, 2002). Thus the problems of drought, bush fire, overstocking, uncontrolled population growth, and an increasing demand for firewood and building materials have caused most of the environmental damage in Ethiopia. Failure by the government to address these issues has led to the degradation of land.

In the wake of the 1984–1985 famine and the catastrophic death of over a million citizens of Ethiopia, the government launched an ambitious environmental reclamation program that was supported by international donors and several NGOs. This reclamation project was also backed by the largest food-for-work program in Africa. Over the preceding five years, peasants constructed more than one million kilometers of soil and stone bunds on agricultural land and built almost half a million kilometers of hillside terrace. The government of Ethiopia, with the help of NGOs, closed off more than 80,000 hectares of hillside to most forms of use in order to foster the regeneration of naturally occurring plant species. NGOs also planted 300,000 hectares of trees, much of it community woodcuts (Dibie, 2008a; Rahmato et al., 2008). The Ethiopian Environmental Protection Authority Report (2005) indicated that several NGOs in the nation are involved in environmental protection and natural resource management. Some NGOs have integrated their efforts to establish the Ethiopian Coordination Committee for Combating Desertification (ENCCD).

As a result of the outbreak of famine in the mid-1980s and 2001–2003, the Ethiopian government extended its reclamation programs. Activities on peasant lands were organized by the Community Forestry and Soil Conservation Development Department of the Ministry of Agriculture, which is responsible for all food-for-work programs. The donations of grain and edible oils by the World Food Program's Project 2488, the European Economic Committee (EEC), and the United States supported the effort. Other donors provided technical equipment and tools. These funds and equipment have been used by NGOs for health (including HIV/AID prevention and treatment), child welfare and protection, education, and water and sanitation (Rahmato et al., 2008). In addition to these roles, one important area of NGOs' engagement in Ethiopia that is often ignored, mainly due to the difficulty of allocating a monetary or resource value to it, is institution building and empowerment. Despite the important roles of NGOs in Ethiopia, a lack of consensus and disagreement on priorities and different styles of operations have limited their ability to be successful in all areas. Further, a lack of appropriate policies and program guidelines and the inability to effectively mobilize stakeholders and communities by government agency officials have sometimes created unexpected impediments to addressing environmental emergencies. Table 9.2 shows the indicators of NGOs' roles and the social and economic impact of environmental issues by the flow of resources and investment across projects and sectors in Ethiopia.

Table 9.3 shows the distribution pattern of NGO projects around Ethiopia across eleven regions in the country. Tables 9.2 and 9.3 also provide information on what NGOs have invested around Ethiopia. It is also interesting to note from Tables 9.1 and 9.2 that the majority of NGOs' resources were being spent on human development such as agriculture, environmental issues, education, child welfare, food security, water and sanitation, women's and girls' empowerment, and health. In general, NGOs have played a major humanitarian role by complementing the functions of the Environmental Protection Authority of Ethiopia in the implementation of relief policies.

Most NGOs are sometimes assigned to a particular geographic location. It is also of interest to note that NGOs are now playing the role of public

Table 9.2 Projected Budgets of NGOs Working in Southern Nations Nationalities and People Region (SNNPR) in 2006–2008

Sector	Budget (Birr)	Beneficiaries in Ethiopia
Environment	22,908,165	389,631
Relief	2,361,700	175,000
Urban development	2,899,060	2,960
Capacity building	3,186,701	1,020
Food security	20,799,003	423,424
Agriculture	34,680,652	362,149
Health	105, 136,192	1,305,779
Water	70,241,461	417,486
Social welfare	51,563,340	78,760
HIV/AID	6,757,848	434,647
Education	49,465,500	112,625
Good governance	2,622,346	147,238
Conflict resolution	3,693,385	66,806
Integration development	85,100,911	281,010
Total	**461,416,264**	**4,198,535**

Source: Dessalegn et al., 2008.

Table 9.3 Actual and Planned Project Expenditures, 2004–2008

Region	Number of Projects	Total Cost (Birr)
Dire Dawa	24	36,721,184
Afar	57	637,085,008
Addis Ababa	281	1,114,223,785
Amhara	317	1,881,878,827
Benishangul Grumuz	51	153,889,665
Oromia	884	4,434,215,156
Harari	18	54,603,118
Tigray	66	492,069,831
Gambella	19	31,350,656
SNNPR	312	1,004,498,134
Somali	76	135,875,027
Total	**2,065**	**9,976,410,395**

Source: Dessalegn, 2008.

policy advocates. This new role is due to the increase in communication and collaboration between NGOs and government officials.

Clark (2000) and Woldu (2008) contend that the evolution of NGOs into implementers of long-term development projects has definitely increased their acceptance and appreciation by the government of Ethiopia. Several NGOs have also been involved in the food security program that is comprised of environmental rehabilitation schemes such as area closure, water harvesting, small-scale irrigation, and tree planting. Thus, NGOs have been very important in promoting the political and economic transformation of the people of Ethiopia. NGOs have provided the forum for involving self-motivated groups and skilled individuals in the nation-building and societal development that Ethiopia sought during difficult environmental situations. These NGOs' efforts have been supported by the international donor community.

ANALYSIS OF CURRENT ENVIRONMENTAL POLICIES

The people of Ethiopia like other countries in the African continent belief that their own wellbeing is more important than any other creature or plants. Most of the people in the rural areas of Ethiopia tend to resist environmental policies because of this belief. The indigenous people strongly believe that the surrounding biodiversity or ecosystem in Ethiopia was created for them to use. Therefore it has been a fairly difficult task for the government of Ethiopia to enforced most of its environmental policies. Table 9.4 shows Ethiopia's major environmental policies.

Until the early 1990s, Ethiopia development activities did not take environmental policies into account (Getahun, 2009b; Rahmato et al., 2008). Since 1993, the nation has, however, realized this major oversight and has adopted several environmental policies and has enacted national policies in an effort to solve the environmental problems in Ethiopia. Environmental policy measures have been taken in the past fifteen to twenty years to construct terraces, soil bunds, check dams, and diversion drains and to plant trees on hillsides and in catchment areas.

The Ethiopian government also instituted a reclamation policy and has appealed to a variety of Western donors—the World Food Program (WFP), the European Economic Community (EEC), and the United States. These foreign governments and NGOs have not only supplied food to the Ethiopian people but have also introduced massive food-for-work programs. They have helped to address the long-term, underlying cause of famine, rather than merely donating food.

The Ethiopian government environmental policy initiative also enabled the foreign NGOs to defend the criticism that they were keeping people alive only so that they could die in large numbers the next time the rains failed. The foreign NGOs operating in the nation refuted the argument that the food aid in the form of handouts would make people lazy. In addition, foreign NGOs brought in large quantities of food after the famine was over, not only to the people in the north, who were directly affected by the famine, but also to areas that had not been affected by the drought or famine (Dibie, 2009;Woldu, 2008). The extent to which the food-for-work approach has helped the people affected by drought is questionable.

Table 9.5 shows related environmental policies that are enforced by agencies other than the Environmental Protection Authority.

The major government agency responsible for the enforcement of environmental policies is the Ethiopian Environmental Protection Authority (EPA). The EPA is a fairly new agency, established in 2002. The nation had been, however, organizing conservation efforts in rural areas during the 1970s, encouraging farmers to combat erosion by building terraces and planting tree seedlings long before the EPA was established. One of the government efforts to ensure biodiversity has been to close some hilly areas to agricultural development. The EPA 2004 report indicated that one of the environmental policies is to ensure that 5.5 percent of Ethiopia's land is officially protected, although the country's system of national parks and reserves suffers from poaching and illegal logging (EPA Report 2005). The nation's environmental policies have also initiated efforts to protect the sixty animal species that are identified as threatened (Getahun, 2009b).

The Ethiopian EPA has the mission to improve and enhance the health and quality of life of all citizens and to promote sustainable social and economic development through the sound management and use of natural, human-made, and cultural resources. The EPA is committed to protecting the environment as a whole so as to meet the needs of the present generation without compromising the ability of the future generations to meet their own needs (Dibie, 2009).

Apart from the activities of the EPA, Ethiopia has ratified international agreements intended to protect biodiversity, endangered species, and the ozone layer. Ethiopia has also signed treaties limiting nuclear testing and the use of chemical and biological weapons. The nation is a party to the World Heritage Convention (Woldu, 2008).

Tables 9.4 and 9.5 show that Ethiopia has very good environmental policies, proclamations, and legislation. The major problem that the nation is confronting, however, is the ineffective implementation of a number of environmental policies. Fikremariam Tesfaye (2009) contends that the effort to protect the environment using policy and legislative measures would become effective only by directing the socioeconomic development activities and aspirations in conformity with the laws governing the natural system and thereby elevating society to a better life (Tesfaye, 2009). Further, the legal tools that are meant for environmental management and regulations are yet to be fully adopted in Ethiopia. Also, environmentally related proclamations are enforced by several other government agencies. Sometimes, however, these proclamations are duplicated at the national and regional levels, causing confusion as to who is responsible for their implementation. For example, the Ethiopian National Meteorological Services Agency (1980) is entrusted by proclamation to monitor the nation's atmosphere (United Nations, 2002). Despite Ethiopia's good intention to establish meteorological service agencies, the nation has not been able to

Table 9.4 Major Ethiopian Environmental Policies and Institutions

No	Environmental Policies	Year of Enacted
1.	Environmental Policy of Ethiopia and Conservation Strategy of Ethiopia	1997
2.	National Capacity Building program (NCBP)	1998
3.	Ethiopian Water Sector Strategy (EWSS)	2001
3.	Pollution Control Proclamation	2002
4.	Environmental Organs Establishment Proclamation—repealed the Proclamation for the establishment of the Environmental Protection Authority (EPA) and the Environmental Protection Council (EPC)	2002
5.	Environmental Impact Assessment Proclamation (EIA)	2002
6.	Rural and Agriculture Development Policies and Strategies of Ethiopia (RADPS)	2002
7.	Sustainable and Poverty Reduction Strategy Program (SDPRP)	2002
8.	National Capacity Building program	2002
9.	Ethiopian Water Sector Development Program	2002
10.	Nile Trans-Boundary Environmental Action Program	2002
	Environmental Institutions	
1.	Environmental Protection Authority (EPA)	2002
2.	Ministry of Agriculture and Rural Development	2004
3.	Ministry of Capacity Building	2002
4.	Ministry of Infrastructure development (MOID)	2002
5.	Ministry of Education (MOE)	1997
6.	Disaster Prevention and Preparedness Commission	2001
7.	Food Security Coordination Bureau	2002
8.	Ethiopia Agriculture research Organization (EARO)	2002
9.	Institute of Biodiversity Conservation (IBC)	2004
10.	National Meteorological Agencies	2002
11.	Regional Environmental Agencies (REA)	2002
12.	Ethiopian Rural Energy Development and Promotion Center (EREDPC)	2002

Source: United Nations (2002); Dessalegn et al. (2008).

Table 9.5 Other Environmental Policies Enforced Outside the EPA

	Other Environmental Policies	Year Enacted
1.	National Drug Policy	1994
2.	National Population Policy of Ethiopia	1993
3.	National Agriculture Research Policy and Strategy	1993
4.	National Science and Technology Policy	1993
5.	Health Policy	1993
6.	Energy Policy	1993
7.	National health, Science and technology Policy	1994
8.	National Policy on Disaster Prevention and Management	1997
9.	National Policy on Biodiversity Conservation and Research	1998
10.	Ethiopian Water Resources management Policy	1999
11.	National Fertilizer Policy	1999
12.	Proclamation on Institutional Arrangement for Environmental Protection	2002

Source: United Nations (2002).

adequately cover its meteorological and climatological stations. The agency does not have enough environmentally friendly technology and is bogged down by weak data generation, gathering, and analyzing ability, as well as by insufficient training and technical experts. At the time of writing, the government of Ethiopia has not been able to effectively address many chronic forest management problems.

On the one hand, the government of Ethiopia has a unique responsibility to promote and protect the environmental and health of the people based on the policies that have been enacted. On the other hand, the nation has not been able to effectively address the problem of industrialization and urbanization in big cities such Addis Ababa. Most of the urban cities have complex water, sewage and sanitation, and overcrowding problems. These cities are also characterized by unplanned and unmanaged industrialization, highly populations, high-density housing and unplanned settlement, crowded markets, low infrastructure, and poor waste disposal systems (Rahmato et al., 2008; Woldu, 2008).

The Ethiopian government, with all its many policies and environmental agencies, does not seem to be able to guarantee the nation's biodiversity and ecosystem's health. The EPA has the mandate to implement policy, strategies, laws, and standards to enhance social and economic development in a manner that fosters the people's well-being and the nation's environmental sustainability, but it has not been able to adequately control pollution.

Some of the argument that has been put forward by the EPA is that (1) the agencies resources are finite, and environmental issues are just one of many priorities; (2) there is a growing recognition that individuals, NGOs, communities, and various private institutions can form powerful collaborative relationships to improve environmental conditions that the government alone cannot replicate; (3) environmental issues are a primary public good because many aspects of human welfare, such as social and economic relationships, are contingent on it. Among other factors, environmental issues are shaped by laws, policies, employment, and income for farmers, in addition to social norms. Thus, biodiversity is very important for the people's well-being in the nation. In view of the value of biodiversity to employment, business, communities, and society in general, creating the conditions for the people of Ethiopia to become engaged in environmental policy implementation should be a shared social responsibility. The special role of the Ethiopian government must be allied with the contributions of NGOs and other sectors in the society.

Further, a fairly high population growth in the past twenty years continues to be a major environmental problem in Ethiopia. Population growth indirectly exacerbates the ecological problems that are related to land use and management, as well as putting more pressure on the ecosystem in general. One of the problems that the Ethiopian government faces is the logistical difficulties in conducting regional and local assessments. Peasant input into the process was nominal at best. In the recent past, the federal government secretariat was pressured to complete the National Conservation Strategy for an international seminar before any of the regions could complete their strategies. There was little discussion of indigenous techniques of soil preparation. Neither was there any substantial discussion or acknowledgment of indigenous terracing of runoff ponds and irrigation. Densely settled areas in the southwest were said to be at environmental risk because of population pressure with no investigation into the distinctive farming system that was used until recently and that sustained such densities for centuries.

The ecological knowledge among members of indigenous communities and economic institutions is not limited to medicinal plants. For instance, the National Conservation Strategy (NCS) survey in Ethiopia revealed that indigenous farmers were able to use the color, texture, and appearance of their crops to distinguish between poor and fertile soils (Woldu, 2008). Indigenous farmers also know local climate characteristics and when to prepare their farms for the rains or to prevent too much soil moisture using simple farm implements such as the hoe, axe, and cutlass, which minimize exposure of the subsoil to the tropical climate (Rahmato et al., 2008). Similarly, indigenous hunters have expert knowledge of the location of certain animals' plants waterbeds, forests, and other elements of the local environment. The appropriate and continuous use of these indigenous talents is one of the reasons why this discussion advocates a community-based natural resources management model as the best instrument to implement environmental policies in Ethiopia.

APPROPRIATE ENVIRONMENTAL POLICIES

Ethiopia has come a long way in adopting several environmental policies to ensure the protection of its ecosystem and citizens. These measures have been very necessary because environmental resources are the bases of any economic and social development that takes place in the nation. It is also true that environmental resources are fundamental to the production of goods and services as well as to the reduction of poverty. Therefore, any efforts to achieve sustainable development must take into consideration measures that would prevent the mismanagement of environmental resources. The health of the citizens is also at risk when social, economic, and other environmental conditions undermine health. These conditions may include toxic water, air, and poverty, unequal access to health care, pollution, poor hazardous waste disposal, and congested housing. The human approach to sustainable development theory calls for government's partners, NGOs, potential actors, and all stakeholders in the environmental policy arena to contribute to assuring a healthy environment for all the citizens of Ethiopia. They can do so by using the ecosystem for the production of goods and services for the present, at the same time considering the needs of the next generation.

To address the present and future challenges faced by the Ethiopian ecosystem and citizens, including potential actors in the private and NGOs sectors, this chapter proposes some policies and changes to be considered by all stakeholders to ensure the environmental protection and sustainable development.

Poverty Reduction and Improvement of Living Standards

The government should adopt a poverty elevation strategy that considers the multiple determinants of how to raise the poor living standards of citizens in both the rural and urban areas of Ethiopia. The nation should be attempting to improve the levels of living for its people, considering the current and anticipated levels of population growth. The government should educate the citizens and make them aware that the rapid population increase makes it more difficult to provide essential social services including housing, transport, sanitation, and security for everyone.

Partnership

The Ethiopian government should build a new generation of intersectoral partnerships that draw on the perspectives and resources of diverse communities and that actively engage them in environmental actions. The government could also promote partnership by using incentive measures to encourage conservation and the sustainable use of natural resources. Partnerships with NGOs and other diverse stakeholders could also enhance the protection of Ethiopia's natural resources and culture.

Environment and Social Framework

The government of Ethiopia should establish an environmentally friendly approach that emphasizes durability rather than insisting on strict economic growth. Promoting an equitable distribution of energy services is important for social sustainability. Thus, promoting a productive safety net program for the improvement of rural environment conditions would help the social well-being of the people of the nation. More could be achieved if some measures of economic optimality are enforced for gender equality. Also crucial is the trade-off of benefit in other areas for sustainable livelihood through the preparation, implementation, monitoring, and evaluation of current and weak environmental management plans.

Waste Management

The government of Ethiopia should prepare a waste management and pollution reduction plan for healthy lives and ecological integrity throughout the nation. The government should also establish strict measures for the use and disposal of pesticides in the country. The appropriate disposal of hazardous waste in Ethiopia is essential if human and environmental health is to be protected from hazardous chemicals. Government agencies will have to work closely with farmers and industrial groups to ensure that chemicals are used appropriately. The government should hold the industries accountable for cleaning their environment, for recycling used water and other solutions and solvents, for searching contaminated areas and assessing environmental remediation, and for industrial decommissioning, when necessary. To improve the handling and use of chemicals, the national policy and legislation of the country should be enforced to the maximum. The federal and regional governments should place sanctions on any group that fails to abide by the law.

Communication of Environmental Issues

The government should enhance and facilitate communication within all the environmental agencies and citizens. All levels of public agencies (federal and regional) that are responsible for the enforcement of the various environmental policies should be encouraged to hold meetings regularly and to share information with one other. The government should also work closely with the mass media to increase public awareness of environmental information. Distributing printed leaflets to schools and communities in both the rural and urban areas is also crucial. Further, the government of Ethiopia should organize community meetings and other participatory processes in order to involve all stakeholders in decision making on water supply, wastewater disposal, solid waste management systems, and housing development for their communities.

Clarification of National and Regional Agencies' Responsibilities

Although national and regional governments have established environmental agencies or have assigned environmental responsibilities to existing agencies, the levels of responsibilities are not clear enough. All federal, state, and local government environmental agencies should develop strategies to ensure that public environmental workers who are involved in the provision of essential environmental services demonstrate mastery of the core environmental competencies appropriate to the jobs.

Capacity Building for Public Environmental Professionals

The government of Ethiopia should institute regular leadership training for all public environmental professionals. Just as business capacity needs to be built to grow markets, institutional capacity is needed in government for environmental policies. Capacity in government is needed to enforce environmental policies, and dedicated agencies might provide an important focal point. Support and career development programs should be a high priority for the government agencies. Partnership with NGOs and other organizations to build human capacity to address environmental policy enforcement problems should be considered seriously. University programs in environmental policy, health, and occupational safety should be established in the country.

Reforestation Program

The Government of Ethiopia should also establish projects in addition to reforestation and terracing the terrain. To solve the problem of excessive interference with the natural resources in Ethiopia, the government should establish a more vigorous reforestation program. The program should strongly encourage the planting of indigenous plants and trees. Eucalyptus trees, which are good for energy and building, should also be planted all over the country. Irrigation system should be designed around eucalyptus trees, which can be planted in dry areas of the country because of the huge amount of water it requires to grow. Reforestation can also help to solve the soil erosion problem, which results from desertification, drought, degradation of natural resources, and climate change (global warming). Erosion is dangerous and could amplify the cause of serious negative economic and social development impacts.

Good Roads for Vehicles

Whenever there is wildfire in the Southern zone of Bale and Borona, it is often difficult for emergency agencies to respond because of lack of accessible roads to reach the region. The bad road situation often prevents firefighters and other emergency vehicles from reaching the region when there is bushfire. . It is recommended that the government construct and build all-weather roads. Good roads would enable the access into wildlife and forest areas to put out fires whenever it occurs. Environmentally friendly technologies should be purchased by the government to aid in the extinguishing these fires.

Programs to Stop the Depletion of Forest

A formal national dialogue should be initiated to address the issue of technology and indigenous energy in Ethiopia. It is crucial that the technology needed to make use of local energy resources, such as natural gas and biomass, be developed. This dialogue should also include the possibility of development and the implementation of energy-saving stove and fuel systems for cooking and heating purposes and so on. For this purpose, alcohols like ethanol and methanol could be used. Ethanol can be prepared from sugarcane residues and methanol from the natural gas extraction process. Such development has to combine both biomass and other hydrocarbon resources to use as energy for heating, cooking, possibly generating electricity for refrigeration and other household uses (Kraft, 2004; Tefetro, 2000; Withgott and Brenna, 2008). It is recommended that the government should explore the hydropower potential of the nation to be used to generate high-energy electricity for the country. If such programs can be established, they can help solve some of the nation's problems related to deforestation, erosion, desertification, and climate change.

SUMMARY

This chapter has examined and analyzed the economic and social impact of environmental policies in Ethiopia. It explored the local dynamics of urbanization and its effect on citizens' daily lives. It also examined the complexity of the environmental management processes and the implementation of conventional and nonconventional policies in Ethiopia. The chapter argued that because environmental resources are the foundation of economic, social, and technological development, as well as the sources of the production of goods and services, the government of Ethiopia has to effectively enforce all its environmental policies.

The chapter used the human approach to sustainable development to present an argument that Ethiopia needs to seek appropriate methods to use the environmental resources available in the country to alleviate poverty. It contends that the mismanagement and underutilization of the environmental resources has thus far reduced the overall sustainable development success story in the country.

Ethiopia has been struggling with adverse environmental conditions such as drought, famine, soil erosion, deforestation, and land degradation for quite some time. These environmental predicaments threaten the physical and economic survival of the citizenry and biodiversity. Catastrophic climate change has also reduced the environment's ability to produce household energy and biomass for food and wealth. The chapter further contended that Ethiopia needs to confront its poverty reduction problem vigorously in order to achieve sustainable development. Addressing environmental degradation and poverty reduction problems simultaneously is crucial to the achievement of any type of sustainable development in Ethiopia. Although the Ethiopian government has enacted several environmental policies, the nation's public sector still severely lacks the skills and environmentally friendly technology to take on the huge ecosystem problems that it faces.

The analysis of environmental policies and population growth control programs in Ethiopia revealed that they have not been effective. Despite the fact that the government made the eradication of hunger and reduction of poverty its prime focus by enacting the Ethiopia Poverty Reduction Strategy (PRS), many citizens in the rural areas of the nation have not benefited from the program. Further, population growth in the urban areas is another problem that has spilled over into waste disposal in the nation. The government of Ethiopia has to adopt an innovative management and planning technique for urban settlements. This measure, if effectively executed, could improve the quality of life of citizens in the cities and curtail the adverse impact of large concentrations of people on the natural environment. Improved electricity services can indirectly contribute to social sustainability. This could also enhance education, giving household members the ability to study at night and thus have a positive impact on empowerment and participation of literate civil society groups.

The analysis of the political history of Ethiopia revealed that poverty and hunger have often incited and exacerbated conflict, in addition to causing other social and economic problems. Notwithstanding these problems, it seems worthwhile for the government of Ethiopia to collaborate with grassroots support organizations, community-based organizations, foreign NGOs, and local nonprofit organizations in the future. Such collaboration could facilitate the nation's desire to achieve sustainable economic development and environmental protection. The chapter suggested that the choice of community-based approach and related options could help to positively increase the social and economic impact of environmental policy in Ethiopia.

REFERENCES

"African Countries Suffering from Freshwater Shortage." (2000). *Environmental News* (February 29).

Alemu, H. (2009). "Akaki River: The Analysis of the Pollution Level of the Akaki River and Finding Possible Ways, to Reduce the Environment Pollution." Retrieved from geocities.com/akababi/gedion. Accessed March 22, 2013.

Asefa, S. (2003). "Perspective on Reforming the Current Constitution for a Better Future in Ethiopia." *Addis Ababa Tribune* (January 16, 2013).

Clark, Jeffrey. (2000). *Civil Society, NGOs and Development in Ethiopia*. Washington, DC: World Bank.

"Death of a Continent; Africa Will Never Be the Same." (2000). *Fortune* (November 13).

De Waal, Alex. (1997). *Famine Crimes: Politics and the Disaster Relief Industry in Africa*. Bloomington: Indiana University Press.

Dibie, R. (2009). "Environmental Governance and the Political Economy of Sustainable Development in Africa." *Journal of International Politics and Development, 7*(1), 1–18.

Dibie, R. (2008a). *Comparative Perspective on Civil Society*. Lanham, MD: Lexington.

Dibie, R. (2008b). *Non-Governmental Organizations (NGOs) and Sustainable Development in Sub-Saharan Africa*. Lanham, MD: Lexington.

Dibie, R. (2007). "Governance and Redevelopment of Civil Society in Africa." *International Review of Politics and Development, 5*(2), 1–36.

Dibie, R. (2001). "Indigenous Participation in the Environmental Policy of Ethiopia and Botswana." *Journal of Sustainable Development in Africa,* 2 (Winter/Spring), 1–10.

Easton, T. (2003). *Taking Sides: On Environmental Issues* (13th ed.). New York: McGraw Hill.

Eisner, M. (2007). *Governing the Environment: The Transformation of Environmental Regulation*. Boulder, CO: Lynne Rienner.

Environmental Protection Authority, Federal Democratic Republic of Ethiopia. (2004). "The Third National Report on the Implementation of the UNCCD/NAP in Ethiopia." (February). An EPA Report, Addis Ababa.

Environmental Protection Authority of Ethiopia. (2005). "Ecosystem Department Soil Fertility Team, Sustainable land Management Best Practices: A Compilation of Field Experiences of Some Government, NGOs, Bilateral and Multilateral agencies and Community Based Organizations in Ethiopia." An EPA Report, Addis Ababa.

Feoli, E., Galzia Veurich, L., and Zerihun, W. (2002a). "Evaluation of Environmental Degradation in Northern Ethiopia Using GIS to Integrate Vegetation, Geomorphology Erosion and Socio-Economic Factors." *Agriculture, Ecosystem and Environment, 1815*, 1–13.

Feoli, E., Galzia Veurich, L., and Zerihun, W. (2002b). "Processes of Environmental Degradation and Opportunities for Rehabilitation in Adwa, Northern Ethiopia." *Landscape Ecology, 17*(4), 315–325.

Gebre-Egziabher, T. (2009). "WIC Interview on Environmental Policy in Ethiopia." General Manager's Perspective (October 18). Retrieved from geocities.com/akababi. Accessed September 23, 2012.

Getahun, G. (2009a). "Environmental Issues: Problem and Solutions." Retrieved from geocities.com/akababi/gedion. Accessed September 4, 2012.

Getahun, G. (2009b). "The Environmental Strategy of Ethiopia Present and Beyond 2000." Retrieved from geocities.com/akababi/gedion. Accessed September 18, 2011.

Handelman, H. (2009). *The Challenge of Third World Development*. Upper Saddle River, NJ: Prentice Hall.

Herweg, K. (1994). "Problems of Acceptance and Adaption of Soil Conservation in Ethiopia." Unpublished paper. Soil Conservation Research Project. Addis Ababa.

Hoben, A. (1995). "Paradigms and Politics: The Cultural Construction of Environmental Policy in Ethiopia." *World Development*, *23*(6), 1007–1021.

Hoben, A. (1972). *Land, Tenure among the Ambara of Ethiopia*. Chicago: University of Chicago Press.

Hyden, G. (2006). *African Politics in Comparative Perspective*. New York: Cambridge University Press.

Kraft, M. (2004). *Environmental Policy and Politics* (3rd ed.). Boston: Longman.

Mequanint, A. (2009). "Pollution Concern in Wonji, Wonji/Shoa and Methara Since the 1950s." Retrieved from www.geocities.com/akababi/wonji. Accessed September 21, 2012.

Ministry of Finance and Economic Development (MOFED). (2005). "Ethiopia Building on Progress: A Plan for Accelerated and Sustained Development to End Poverty (PASDEP)." Government of Ethiopia Report, 2005/2006–2009/2010. Addis Ababa.

Munasinghe, M. (2002). "The Sustainable Trans-Disciplinary Meta-Framework for Making Development More Sustainable: Application to Energy Issues." *International Journal for Sustainable Development*, *5*(1–2), 135–182.

Munasinghe, M., and Swart, R. (2005). *Primer on Climate Change and Sustainable Development: Facts, Policy Analysis, and Applications*. Cambridge: Cambridge University Press.

Onimode, B. (2004). *African Development and Governance Strategies in the 21st Century*. London: Palgrave Macmillan.

"Population Crisis in Africa." (2000). *Environmental News* (February 29).

Rahmato, D., Bantirgu, A., and Endeshaw, Y. (2008). "CSO/NGOs in Ethiopia. Southern Nations Nationalities and People Region (SNNPR): A Report Prepared for the Ad Hoc CSO/NGO Task Force Addis Ababa." Government of Ethiopia Press.

Rahmato, D., Admassie, Y., and Mekonnen, Y. (2007). "Population, Health and Environmental Integration in Ethiopia: Exploring the Opportunities and Challenges: A Report Prepared for the Population Reference Bureau, Addis Ababa." (January).

"Rainwater Harvesting Could End Much of Africa's Water Shortage." Environment Business and Energy Information Services. Retrieved from http://www.environbusiness.com/ebnnews/archivednews1/rainwater1. Accessed November 19, 2012.

Rogers, P., Jalal, K., and Boyd, J. (2008). *An Introduction to Sustainable Development*. Sterling, VA: Earthscan.

Steel, B., Clinton, R., and Lovrich, N. (2003). *Environmental Politics and Policy: A Comparative Approach*. Boston, MA: McGraw Hill.

Shibiru, T. (2000). "Community Based Natural Resources Management in Ethiopia. An Overview." Unpublished report.

Sonneveld, B. G. J., and M. A. Keyzer. (2002). "Land Under Pressure: Soil Conservation Concern and Opportunity for Ethiopia in Land Degradation and Development." Wiley InterScience. Retrieved from www.interscience.wiley.com. Accessed September 23, 2011.

Sutcliffe. J. P. (2006). "The National Biomass Planning in Ethiopia: The Woody Biomass Project." *Recommend*, *3*(4), 1–4.

Tefetro, K. (1999). "Environmental Strategy of Ethiopia." *The Quarterly News Newsletter of the CSE Projects*, *1*(7). A Publication of the Environmental Protection Authority of Ethiopia.

Teguam, T. (2009). "Status of Pops with Special References to Pesticides in Ethiopia." Environmental Protection Authority, Addis Ababa, Ethiopia. Retrieved from geocities.com/akababi/gedion. Accessed July 4, 2012.

Tekelemichael, Y. (2009). "Current Status of the Environmental Impact Assessment System in Ethiopia." *UNEP EIA Training Resource Manual*. Accessed June 10, 2013.

Tesfaye, F. (2009). "Ethiopia: Environmental Policy Implementation Still in Difficulty–EPA." *Daily Monitor* (March 13).
Thin, N. (2002). *Social Progress and Sustainable Development*. Bloomfield, CT: Kumarian.
United Nations. (2003a). "More Aid Is Needed Says Government." UN Integration Regional Information Networks. Retrieved from http://allAfrica.com. Accessed March 21, 2013.
United Nations. (2003b). Dialogue on Border Issues Unthinkable, says Eritrea. New York: Oxford University Press.
United Nations. (2002). *Ethiopia: Country Profile Series*. New York: Oxford University Press.
United Nations Development Program (UNDP) (2008). *Human Development Index 2007–2008*. New York: Oxford University Press.
United Nations Development Program (UNDP). (1994). *Human Development Index 1993–94*. New York: Oxford University Press.
United Nations Economic and Social Council. (2007). "Development Strategies That Work. Policies to Ensure Environmental Sustainability in Ethiopia." Retrieved from un.org.nvp/fontend.policy action. Accessed September 23, 2012.
United Nations Environmental Programs (UNEP). (2000). "Shortage of Freshwater in Africa." Press release (May 16).
Waller-Hunter, J., and Jones, T. (2002). "Globalization and Sustainable Development." *International Review for Environmental Strategies*, *3*(1), 53–62.
WCED. (1987). *Our Common Future*. Brundtland Report. Oxford, England: Oxford University Press.
Withgott, J., and Brenna, S. (2008). *Environment* (3rd ed.). New York: Pearson.
World Bank. (2009). *World Development Report*. New York: Oxford University Press.
World Bank. (2000). *Can Africa Claim the 21st Century?* New York: Oxford University Press.
World Bank Report. (1997a). "Environment and Sustainability." New York: Oxford University Press, 24–28.
World Bank. (1997b). *World Development Report. The State in a Changing World*. New York: Oxford University Press.
Woldu, Z. (2008). "The Population, Health and Environmental Nexus." A Background Paper for the Establishment and Launching of the PHE in Ethiopia in May. Addis Ababa University Press.
Zenebe, W. (2009). "Toxic Truth of the Flower Business." Retrieved from geocities.com/akababi/gedion. Accessed March 21, 2013.
Zewdu, M. (2002). "Sustainable Development in Ethiopia." Addis Ababa: Heinrich Boell Foundation and Institute of Biodiversity Conservation and Research.
Zewdu, M. (2000). "Government Initiated Community Based Natural Resources Management." Paper presented at a workshop on Community Based Natural Resources Management.

10 Environmental Policy and Issues in Nigeria

Offiong J. Offiong, PhD
and Robert Dibie, PhD

INTRODUCTION

The *environment*, in the normal sense, simply refers to the physical and social conditions where people live. It could also refer to natural conditions, such as the air, water, and land in which people, animals, and plants exist. Thomas Easton (2012) and Robert Taylor (2012) contend that the environment is made up of both biophysical and socioeconomic elements, including not only areas of land, air, water, plants, and animals, but also other natural and human-modified features that constitute the totality of the ecosystems. A common characteristic of the ecosystem in Nigeria is that they are interrelated (Dibie, 2000). Any disruption in one ecosystem often results in negative effects on the other ecosystems in most regions of a country.

The institutional changes that have taken place in Nigeria's governance since 1960, followed by the subsequent industrialization and democratization between 1999 and 2011, have been substantial. The dominant trend among African nations, including Nigeria, has been neocolonialism and neoliberalization. These development trends also involve a shift toward increased reliance on market relations and mechanisms in life and governance. This shift allowed for the effective use of urban land utilization by both the industrial base and the expanding population (Ebeku, 2003; Myers and Spoolman, 2014). However, these development trends continue to have a negative effect on environmental health outcomes. Over the past three decades, environmental regulations and development policies in Nigeria have been very slow in keeping pace with the expanding use of petroleum products, gas, and other energy sources and their polluting impact.

In another perspective, *environment* refers to the aggregate of all external conditions and influences affecting the life and development of organisms (Fadamiro, 1995). It consists of biotic and biodiversity components (including humans) that interact with one another (Clifford and Edwards, 2012; Ero, 1997). The interactions also include the dominance of some species of organisms over others (Enahoro and Ehi-Ebewele, 2007: 605; Friis, 2012). A lot of attention in Nigeria is paid to the so-called oil wealth that is the reason for the lion's share of the national revenue. The predicaments of conflict around this nonrenewable perspective also define the political constellation of the country. In the past six decades, however, little consideration or importance has been attached to the natural capital that sustains the structure, function, and productivity in the Niger Delta area (Edoho and Dibie, 2000).

Environmental problems cannot be considered new in or unique to Nigeria. They exist in both developing and developed worlds. It is said that environmental problems existed in human societies as early as the first century B.C.E., when the drinking water of Rome was reported to be polluted (Ruff, 1993: 20–36). The people in the Delta region of Nigeria were amazed at the attention that the United States oil spill in the Gulf of Mexico received despite the fact they have been experiencing the same massive pollution daily, for several decades, of their environment by the oil-producing companies in the area.

According to Robert Dibie (2000), the Niger Delta region represents a highly complex ecosystem that is one of the most biodiverse in the world. Yet this extraordinary but fragile ecosystem has not been accorded the care that it deserves (Adelekan, 2009). Crude oil from multiple oil spills has polluted thousands of acres of farmland, rendering them useless for growing crops. The air in the Delta region is also polluted from the flaring of natural gas that occurs every minute of the day (Dibie and Offiong, 2012).

In Nigeria, oil spills have occurred from leaking or damaged tankers, corroded pipelines, sabotage, and oil theft (UNEP, 2006). When these oil spills have occurred, rarely is attention given to them of the type seen from the U.S. government in the Gulf of Mexico in April/May 2010. In 1980, a Texaco offshore station spilled 400,000 barrels of crude oil into the Gulf of Guinea, and a tank failure at the Shell-owned Forcados Terminal resulted in a spill of 580,000 barrels (Adelekan, 2009). Between 1960 and 1997, it is estimated that over 100 million barrels of oil were spilled in the Niger Delta, making it one of the most polluted regions anywhere in the world (Adelekan, 2009; Dibie and Offiong, 2012).

The Nigerian government has not been able to hold the oil and gas multinational companies polluting its environment accountable in the same way as the United States did the Exxon Mobil for the oil spill in Alaska in 1989 and British Petroleum for the devastation in 2010 in the Gulf of Mexico. The Nigerian government should be more demanding of the multinational companies; the devastation in the Niger Delta is much more extensive and longer lasting than the damage to Louisiana and Mississippi in the United States. Nothing short of a national environmental emergency would be appropriate to handle the problems of ecological devastation in the Niger Delta, especially given the contribution of the region to the national economy (Adelekan, 2009; Ebeku, 2003; Edoho and Dibie, 2000).

Environmental problems can be classified by characteristics, substantive and procedural, which, Olokesusi asserts, influences the range of solutions as well as planning and management strategies (1987: 55–59). According to Olokesusi (1987), substantive environmental problems relate to the intrinsic nature of the environmental system, which has linkages with the land, the intensity of its usage, demographic characteristics, and socioeconomic variables, all of which have impacts on the environment. Consequently, these impacts determine the nature and intensity of environmental degradation.

A number of factors can be summarized hereto characterize Nigeria's environmental problems. These mostly include the consequences of rapid urbanization and industrialization, which have led to an increased demand for a variety of goods and services, resulting in the construction of roads, houses, and industrial structures and leading, in one way or the other, to environmental problems. It is not uncommon for many businesses in the industrial areas of Lagos such as Apapa, Ilupeju, and Ikeja and in some other industrial areas in Nigeria, discharge their chemical waste without the standard pretreatment that could negate their effects on the environment.

Orubuloye (1983: 7–16), in a seminal paper, noted that rapid population growth and migration in Nigeria have put pressure on cultivated land, housing, transportation systems, infrastructure, and trees (used for firewood), all leading to the exacerbation of environmental pollution. As part of the characteristics of environmental problems, Cointreau (1982: 33) observed that, in most Nigerian cities, the collection and disposal of refuse are usually more frequent and efficient in government reservation areas (GRAs) and in other affluent neighborhoods than in poor neighborhoods. This observation cannot be far from the truth. The residents of neighborhoods such as those in Victoria Island, Lekki Pennisula, Ikoyi, and the like in Lagos State, for instance, are inhabited by very wealthy people with heavyweight political connections in the government, and they use their influence to ensure that their environments are kept as free as possible from refuse.

Most Nigerians who work in the petroleum industry and in other similarly hazardous industries are unaware of the environmental effects of the chemicals they are exposed to. Even when they are made aware, they hardly think leaving their jobs for fear of unemployment, and yet they are not really knowledgeable of the extent of the hazards they suffer.

In identifying the characteristics of Nigeria's environmental problem, Dibie (2000: 124) observed that the major environmental problems facing Nigeria are deforestation and desert expansion, soil and beach erosion, water and air pollution from the oil industry, destruction of coastal ecosystems, industrial effluents and chemicals in environment, pollution from municipalities, and hazardous water. The dumping of toxic wastes by European nations in Nigerian rivers is also a very crucial environmental problem that faced the nation (Dibie, 2000).

Dibie's (2000) observation on environmental issues obviously captures the fundamental problems now facing the Nigerian state, with perhaps worse to follow. It is also very important to note that the resulting decades of degradation and overexploitation of natural resources, with a focus on maximizing profits at the expense of environmental management and protection in the country, have led to the declining natural base of the Niger Delta. This decline constitutes a real and critical concern for the increasing population, currently experiencing a high rate of unemployment all over the country.

The history of the Nigerian military regime and now the elected democratic government reveals how current and previous leaders in Nigeria often paid more attention to the oil wealth, which is the source of most of the national revenue (Adunbi, 2010). The challenges and conflicts associated with this nonrenewable resource also defines the political constellation of the country, but hardly ever is any consideration or importance attached to the natural capital that sustains the structure, function, and productivity in the Niger Delta (Adunbi, 2010; Etomi, 2006; Okonta, 2008). Oil-soaked pelicans and other avian fauna, fish and shrimp, mollusks and turtles, mangroves and reed grass—all biological life is under a severe threat in the Gulf of Biafra, on the coastline of the largest nation in the African continent.

Some scholars who have written about the resources of the Niger Delta in recent times are surprisingly blindfolded to vital biological factors, the decimation of which is unimaginable in Europe, the United States, and other industrialized places (Adunbi, 2010; Etomi, 2008a). On the other hand, why should there be any concern for biodiversity conservation if Nigeria does not institute mandatory requirements for environmental management by the oil and gas companies operating in its Delta and make sure that such laws are enforced to the letter (Amadi, 2010; Dibie, 2000).

This chapter will examine the nature and impact of environmental policies in Nigeria. It will stress that the probability of the Nigerian government and nongovernmental organizations (NGOs) of succeeding in resolving environmental problems of this decade will ultimately depend on public understanding, support, and action. It will critically explore the economic and social impacts of environmental policy in the nation. This study will contend that, to be effective over the next several decades, environmentalists in Nigeria will need to hone their political skills and forge broader alliances with more citizen groups with common interests in sustainability. The chapter will call on scholars and NGOs to analyze ecological problems in the nation by reviewing evidence of the risk they present to human health and well-being, as well as to the nation's ecosystem. The data on which this paper is based will be drawn from several communities in Nigeria. The chapter will also provide an extensive analysis of government data and policy reports in discussing the nature of environmental problems and their impact in Nigeria. The chapter will provide an assessment of the progress that the Nigerian government has made in dealing with environmental problems and the implications of keeping or revising present policies. In addition, the chapter will explore several questions: To what extent do environmental problems affect the social activities and lives of citizens in Nigeria? What are the barriers to effectively implementing environmental policies in Nigeria? How can these barriers be overcome? What factors could help foster a sustainable environmental policy in the Nigeria? Some policy recommendations will be provided at the end of the chapter.

SUSTAINING THE ENVIRONMENT AND PRODUCTION VALUES

The sustainable food production model suggests that organic methods of food production can contribute substantially to feeding the current and future human population on the current agricultural land base, while maintaining soil fertility (Badgley et al., 2012). Some other scholars (Clifford and Edwards, 2012; Taylor, 2012) contend that there is the possibility that the need for agricultural land could eventually be reduced, as is the case in some parts of Nigeria, where organic production methods were employed through conventional methods.

While the sustainable food production model could be implemented in Nigeria to some extent, the millennium ecosystem model presents a very interesting viewpoint. The model suggests that organic agriculture can contribute substantially to a more sustainable system of food production (Easton, 2012). On one hand, organic agriculture, when properly implemented in the nation, could produce much of Nigeria's food. On the other hand, the model presents several challenges agroeconomically, educationally, and socioeconomically. Major large-scale research institutions that are dedicated to agroecological methods of fertility and pest management, a strong extension system, and a committed public (Badgley et al., 2012). Dibie (2000) observed that, based on the way the government of Nigeria had not been effective in implementing its environmental policies, it would take a much bigger step for the nation to be disciplined and dedicated enough to adopt a comprehensive agroecological method of fertility.

It is important to note that appropriate production methods are but one component of a sustainable food system. Nigeria's viability of farming methods, land tenure for farmers, accessibility of market, availability of water, trends in food consumption, and alleviation of poverty are essential to the assessment and promotion of a sustainable food production system in the country (Badgley et al., 2012; Easton, 2012). Robert Dibie and colleagues (2011) contend that population growth, poverty, and pollution are closely connected to environmental health; thus the Nigerian government and its environmental regulatory agencies would need to determine a way to hold multinational and domestic companies accountable to the same standards as in Western industrialized countries. Nigerians can no longer dump trash in their neighbor's yard, farmland, or rivers and expect it to have no effect on the nation's social and economic lifestyle or values.

According to Stephen Polasky (2012), putting a monetary value on ecosystem services will improve decision making by clearly illustrating the consequences of alternative choice. Polasky's concept tends to serve the purpose of highlighting the relationship between ecosystem degradation and sustainability. The monetizing ecosystem approach brings with it the idea that if monetary values could be placed on ecosystem functions, concrete incentives could be designed to promote the sustainable use of ecosystems. Opponents of putting monetary value on ecosystem present an argument that the exclusive use of the monetization of ecosystem services may only favor short-term consumption of goods and services (McCauley, 2006). Opponents argue that such a short-term view of nature could be catastrophic, as has been seen with the collapse of fisheries. Kosoy and Corbera (2010) contend that an attempt to value any one component will never be able to capture the overall true value of the entire system. A very important point to note, however, is that ecosystem are interrelated; as a result any disruption in one ecosystem could have negative effects on the other ecosystems in Nigeria. Thus, any disruption in the ecosystem in Nigeria may destabilize the biodiversity in the entire affected region. The frequent oil spillage in the Niger Delta region and its subsequent impact on fishing and farming help to explain the interrelationship between ecosystem degradation and sustainability. Some scholars, however, contend that unless some economic incentive can be presented for nature conservation, most people in Nigeria will opt to take the cheaper route, which often results in the degradation of nature (Goldman, 2010; McNeely, 2007; Polasky, 2012).

The Mertonian perspective of illegitimate means may be applied to environmental crime in the Niger Delta. The Nigerian government's public administrators and political leaders, as well as the officers of multinational corporations, are likely to abandon morality to achieve economic prosperity (Clifford and Edwards, 2012). The abandonment of ethical environmental values include failure to regulate multinational corporations' exploration activities in Nigeria, failure to effectively implement environmental policies, and failure to protect the citizens from oil pollution. Abandonment may also include failure to provide treatment to citizens infected by diseases due to air, water, and hazardous pollutions. Possible unethical behavior also involves indulging in corrupt practices that may save money but that create environmental harm to Nigerian citizens.

The self-control theory, by Gottfredson and Hirschi (1990), contends that individuals who have been subjected to poor or inconsistent parenting practices are likely to have low levels of self-control. Those with low self-control in government, multinational corporations, and civil society organizations are more likely to perform unfriendly environmental behavior that may harm both the ecosystem and citizens. The self-control theory could be used to explain environmental crime in the Niger Delta region. Individuals with low self-control may not care about anyone or the environment and perform environmental crime out of self-interest. Further, individuals with low interest may perform physical harm to the environment because they are attracted to physical behavior (Pratt and Cullen, 2000). Piquero and colleagues (2010) argue that environmentally criminal behavior may be simple and easy to perform because of differential association, imitation, reinforcement, and greed.

According to Robert Taylor (2012) poor nations in the world perceive sustainable development as an attempt by the rich nations to control material consumption and population growth as restrictions on their ability to improve their standard of living. On one hand, poor nations view issues of sustainability dealing with social equity, justice, and economy as more weighty than a pure ecological view. On the other hand, there is no doubt that a major challenge to sustainability will be the growing energy and material resources demands of emerging economies such as Brazil, China, and India (Clark and Levin, 2009; Taylor 2012). The social movement of sustainability is proposed by Clark and Levin (2009). They argue that sustainability as a social movement should develop a set of indicators to benchmark changes in the natural environment and the human condition. The approach proposed methods for implementing a set of practices that could reduce the carbon footprint, minimize waste, and lessen the reliance on finite resources. Thus, if appropriate action is taken, the ecosystem can adapt and maintain itself in Nigeria. However, if the Nigerian government and citizens do not take any action or delay action, the nation's ecosystem will exceed its threshold, causing irreparable damage to biodiversity in the country.

It is interesting to note that as the citizens of Nigeria struggle to consume more material goods, the strain on the nation's natural resources, its biological diversity, and its atmosphere will intensify (Adamantiades and Kessides, 2012). Clifford and Edwards (2012) presented an argument that increased consumerism also has the consequences of expanding both the profitability and the power of transnational corporations. A number of multinational oil corporations polluting the environment in Nigeria already command greater wealth than the nation, giving them the ability to influence both the creation and enforcement of environmental policies and implementation within the country. The bone of contention remains how far the federal government in Nigeria can go to restrict the present growth of material consumption in order to ensure a healthy environment in the future.

ANALYSIS OF ENVIRONMENTAL ISSUES IN THE NATION

The characteristics of environmental problems can be linked to the history of Nigeria. The West African region called Nigeria was known to have emerged from an agrarian heritage. Nigeria became an independent nation in 1960. The advent of petroleum exploration started in 1956 when the nation was still under the British colonial regime. Due to the discovery of petroleum in the Niger Delta, farming activities incrementally gave way to pipelines and platforms. After almost three decades of petroleum exploration, the thriving agrarian population fell into decline (Adunbi, 2010; Etomi, 2008b; Okonta, 2008). The discovery of petroleum and gas in the Niger Delta region was first perceived as an ancestral promise of wealth, but not too long after exploration started, many communities in the region began to feel a sense of deprivation (Ojo, 2002). Ronald Niezen (2003) contends that the citizens in the Niger Delta region started to perceive the Nigerian government as being allied with the multinational corporations, who continue to exploit the petroleum resources. That exploitation has led to various claims and counterclaims over ownership of land and natural resources, often in ways that generate different levels of conflict (Adunbi, 2010).

Desertification is one of the major environmental problems in Nigeria in recent times. Some states in the northern part of Nigeria are gradually being encroached on by desertification. These states include Borno, Kastina, Sokoto, Kano, Jigawa, and those that are located within the Sahel savannah region and that are experiencing severe degradation. Onyewotu (1998) contends that the extent of desertification in Nigeria is no longer confined to areas above latitude 12 degrees but has now moved to as low as latitude 11.3 degrees. Kurmar (1989) reported that the rate of desert encroachment in Nigeria is about 3 miles per annum. It has been estimated that about 1,000 miles of the national land area has suffered from desertification (Adovi, 2010; Etomi, 2008b Onyewotu, 1989). Kurmar (1998) estimated that the rate of expansion of sand dunes (a consequence of desertification) in Nigeria is about 2 miles per year. This has resulted from overcultivation and excessive tilling of the land, leading to depletion of soil nutrients. The overgrazing of the land by animals such as cows, horses, goats, and camels continues to deplete the nation's vegetation. The felling of trees by local communities for economic use such as furniture, building, and firewood for cooking is also another set of environmental problem.

As a result of the lack of rainfall in some parts of Nigeria, most farmers experience poor harvests. Farmers have had to resort to crop irrigation at increased cost to be able to successfully plant and harvest crops each year. Thus, without timely and appropriate government intervention, food insecurity and starvation are inevitable. Ezirim and Onuoha (2010) observed that, due to the lack of rainfall, crop land has become unproductive, and

Figure 10.1 Example of Human Relations with the Ecosystem in Nigeria
Source: Photo by author during field research in Nigeria 2012.

previous settlements have become unavailable. Thus, farmers are forced to migrate to new settlements and compete for scarce arable land, resulting in frequent conflict in the region. In the past decade, Nigeria has been reported to have spent $62 billion on importing food during the regime of President Obasanjo (Ezirim and Onuoha, 2010). The effects of desertification also come with health risks such as blindness caused by windblown sand dunes and sunflies, among others (Ayorinde and Omisore, 2003).

Environmental pollution problem issues in Nigeria come in different forms, such as air, water, noise, and soil. Ayorinde and Omisore (2003) observed that the sources of environmental pollution in the country include the battery manufacturing, steel, plants, plastics, chemical, fertilizer, radioactive, and textile industries.

These industrial pollutants include highly poisonous wastes like hydrogen sulfur dioxide, ammonia salt, phenols, chromium, copper, and acids. Others pollutants are heavy metals such as lead, arsenic, mercury, zinc, cyanides, phosphate, and textile dyes. Inhalation of asbestos dust from asbestos and cement factories causes asbestosis, a form of cancer, and bronchitis, a respiratory ailment (Kraft and Furlong, 2012; Dye, 2011). Employees and people who live near cement factories in cities and other environments, such as Ewekoro and Shagamu in Ogun state and Gboko in Benue state, are daily exposed to these environmental pollutants and health consequences. Food crops in the vicinity of cement factories are badly affected too. These pollutants directly or indirectly find their ways into human bodies through the food chain, inhalation, or direct contact, causing sickness and diseases, some of which are incurable (Kraft and Furlong, 2012). In major cities like Lagos, Port Harcourt, clouds of smoke and effluents from manufacturing factories are seen all over the region. Common sights are gases flaring, waste dumps, and burning refuse, all contributing major pollution to the soil and air. The acid from these wastes find their way into the soil, mingling with groundwater and thereby contaminating the source of drinking water (Dibie, 2009; Etomi, 2008b). In addition, there is pollution from vehicles all over the country, especially from old cars and trucks. Nigeria has no emission or smog test requirement policy for old vehicles.

Erosion continues to be a major environmental problem in many communities in Nigeria, especially the southeast zone such as Abia, Anambra, Ebonyi, Enugu, and Imo states (Jimoh, 2006). Different types of erosions, such as sheet, rill, and gully, are pervasive in the zone. Gully erosion, however, constitutes the most significant threat to the survival of individuals and communities in these states. In this area, both active and inactive gullied surface areas range from 0.5 square miles for Ohafia and 1.05 miles for Abiriba in Abia State. The widths of the gullies range between 2.2 miles for Abiriba and 0.5 miles for Ohafia. A minimum length of 90 miles of gullied surface has been recorded at Abriba (Adoyi, 2010; Jimoh, 2006).

The federal government of Nigeria has documented about 63 active erosion sites in Anambra state. The erosion sites are reported to be very close to 177 communities (Koko, 2010). The control or prevention of the likely spread of these erosion sites is estimated to have cost over $1.3 billion in the past two decades. One such erosion site in Umuchiana-Ekwulobia has submerged 75 buildings and displaced about 437 families (Best and Lawson, 2008). In 2008, erosion washed away the rail line linking Aba and Port Harcourt, leading to a decline in economic activities in Aba, Abia state. Soil erosion has destroyed many communities and rendered about 200,000 homeless as internally displaced persons in Abia, Enugu, and Anambra states of southern Nigeria ("Erosion disrupts . . .," 2008). Many communities are wiped out annually along the 372-mile River Niger river edge from Baro in the northern state of Niger through Anambra in the east into Warri in the south (Koko, 2010).

Urbanization issues in large cities like Lagos, Ibadan, Aba, Kaduna, Onitsha, Port Harcourt, and others entail the packaging and transportation of goods, intensive fossil fuel consumption, and the major areas of extensive pollution in Nigeria. According to Ezirim (2008), the urbanization crisis is of an unprecedented proportion in Nigeria in recent times due to (1) employment opportunities, (2) industrialization, (3) food availability, (3) lifestyle considerations, and (4) escape from political conflict. Urbanization also creates congestion and traffic jams in these cities. The implications of the rapid increase in urban population have been linked to numerous adverse health impacts, particularly for poor neighborhoods. In addition, increases in urban populations have also led to rising numbers of slum dwellers in cities such as Lagos, Port Harcourt, Onitsha, Aba, and Ibadan, to mention a few. Increased populations in urban areas have also been related to diseases and accidents (Withgott and Brennan, 2009). On the one hand, the uneven distribution of population due to urbanization in Nigeria means that some states and regions may bear far more of an environmental impact than others. On the other hand, increased populations concentrated in Nigeria's big cities may intensify the impact on the environment as more people make use of resources, take up space, and generate a lot of waste.

The rate of urban unemployment in Nigeria has been increasing alarmingly, ranging from 5 percent to over 20 percent as reported by recent studies from the Federal Office of Statistics (Federal Government of Nigeria, 1999 and 2009). The increase in the urban unemployment rate has also led to higher crime rates in the cities. With the increasing environmental deterioration in the large metropolitan centers, the issue of living standard involves not only living conditions but also the ease of circulation within the city. The traffic situation in cities like Lagos, Ibadan, and Port Harcourt, for instance, is already making these cities unlivable. Apart from gradually rising levels of air pollution, these cities also present the threat of a serious health epidemic or plague for its residents. This situation attests to the poor quality of urban planning and management in most Nigerian cities.

In addition to population growth, unemployment, and health risks, poverty levels in Nigeria have increased from 28 percent in 1980 to 46.3 percent in 1985 and rose to 65.6 percent in 1996 (Federal Government of Nigeria, 1999 and 2009). Among the 175 countries listed with rises in poverty, Nigeria ranks high among those who cannot fend for themselves in terms of having access to employment opportunities and health facilities (Adoyi, 2010; Etomi, 2008b; Jimoh, 2006). Unemployment exacerbates survival instincts, which increases crime rate. Combined with poverty is reduced life expectancy, which stood between 43 and 47 years in Nigeria (United Nations Report, 2010). There is also the problem of collapsed buildings due to poor and inadequate building materials, tree felling, and ecosystem destruction.

At a fundamental level, oil exploration and exploitation make up another environmental issue facing Nigeria. Nigerian environmental crises due to petroleum exploration are not the fault of British Petroleum, Exxon Mobil, and Shell Oil. Rather, the problem is a lack of appropriate environmental policies and the inability of the Nigerian federal government to hold these multinational oil companies accountable for their operations in the country. Gbadamosi and Omole, (2003) argued that the various activities related to the exploration and exploitation of oil in Nigeria has had considerable impact on the natural environment. While Ezirim and Onuoha (2010) observed that activities such as gas flaring and seismography cause damage to the environment, far-reaching environmental destruction results from oil spillage; soil, plants, animals, and water sources are adversely affected, usually because of the toxicity of oil. The frequent oil spills in the south have also taken away the livelihood and sustenance of the populace, a major fishing community, and has destroyed the ecosystem (Gbadamosi and Omole, 2003). This has led to armed conflict between the federal government and organized Niger Delta region militants, resulting in the loss of life, economic loss, political struggle, and instability. Gbadamosi and Omole (2003) further contend that, between 1986 and 1995 alone, Nigeria experienced 2,775 oil spill incidents.

The biodiversity of any country is part of its natural assets and critical for human survival and national development (Dye, 2011). This is essentially because biodiversity provides the reservoir for genetic materials, which can be used for pharmaceutical development, wood for fuel and furniture, and food security and a natural endowment, which any country and its people must bequeath to the next generation. New analyses suggest that about 15–37 percent of a sample of 1,103 land plants and animals will eventually become extinct as a result of climate changes that are expected by 2050 (Ayorinde and Mosore, 2003). Regrettably, Nigeria is experiencing the progressive decline of its biodiversity (Ezirim and Onuoha, 2010). Natural water bodies, such as streams, lakes, and springs, are drying up due to climate-induced changes, like drought, in the northern parts of the country and due to the worsening incidence of erosion in the south, which enables the transport of pollutants to these water bodies. The intrusion of saline water as a result of erosion into streams, lakes, and rivers has led to the reduction in freshwater supplies and fish fauna (Ezirim and Onuoha, 2010).

Nigeria is a tropical country with two seasons, the rainy and dry seasons. The climate of the rainy season often causes erosion and flooding. Environmental problems pervasive in the coastal areas of Nigeria today, especially in the south-south geo-political zone and Lagos, are associated with rain and flooding. Nigeria has a coastline of approximately 553 miles, and the Niger Delta accounts for about 405 miles of the coastal zone. Over 75 percent of the 30 million people of the Niger Delta region live along the coastal area and survive mainly on fishing and agriculture (Ezirim and Onuoha, 2010). The problem of coastal erosion and flooding due to sea level rise and storm surges constitute a significant source of threat to life, property, livelihoods, and the infrastructure in the Niger Delta region (Ezirim, 2008a). The environmental condition is made worse by the destruction of the unique mangrove forests due to oil exploitation activities.

Flooding is widespread in the Niger Delta because of low relief, the reduced hydraulic capacities of water channels, and high rainfall. In the mangrove swamp forest areas, diurnal tidal movements result in floods that are exacerbated by rising sea levels, coastal erosion, and land subsidence (UNEP, 2006).

Flooding in the Niger Delta region has frequent wiped out crops and disrupted traditional farming practices and the economic welfare of the people in the area (Best and Lawson, 2008). Worse, a United Nation report has estimated that about 30 percent of Africa's coastal infrastructure, including coastal settlements in the Gulf of Guinea, Senegal, the Gambia, and Egypt, could be inundated by 2085 due to climate change (UNEP, 2006). Although scientists generally dispute the warning that sea levels will rise due to global warming, climate change and the melting of ice in the North Pole, it is strongly estimated that a 0.2-meter rise in sea level would lead to the displacement of about 200 villages in the Niger Delta region and that a projected sea level rise of more than 1 meter could flood much of the Niger Delta and force up to 80 percent of the Delta's population to higher ground, with a consequent property damage that the IPCC estimates at $9 billion (World Bank, 1996).

On July 14, 2009, property, buildings, and vehicles worth several millions of naira were washed away by a flood occasioned by persistent rainfall that lasted a week. It was a natural disaster probably caused by climate change (Dibie, 2009; Onuoha, 2010). In the event of this natural disaster, the poor people and local communities of the Niger Delta, who probably contributed the least to global emissions, bore the greatest brunt of the climate shock. This means a double tragedy for the region. First, they are affected directly by the process by which oil is extracted, and second, they are ultimately rendered vulnerable to a climate change–induced humanitarian catastrophe (Onuoha, 2008b). Climate change–caused flood is also overtaking Lagos and the adjoining Ogun state (Onuoha, 2008b). It was reported that over 2 million people living in Lagos risk being submerged by the Atlantic Ocean (Adoyi, 2010). This is the result of five years of data collection on flooding in Lagos state by the Social and Economic Research Institute (NISER), Ibadan. The research further established that the low-lying, below-sea-level coastal city of Lagos is vulnerable to storm surges, which will present infrastructural challenges for the megacity.

According to Paehler (2007), in the event of a sudden rise of sea level by only 2 yards, it is estimated that over 740,000 people would be displaced in Nigeria. A rise of 1 yard would lead to 3.7 million internally displaced persons (IDPs), and a 2.2–yard rise would render over 10 million people homeless in the country. Apart from coastal erosion and flooding, climate change–related flooding has already had immediate impacts on food production, livelihood assets, and human survival in both the rural and urban areas of Nigeria. In 2007, heavy rainfall and perennial flooding in some states, like Adamawa, Bauchi, Borno, Kebbi, Lagos, Nassarawa, Niger, Ogun, Plateau, Sokoto, and Yobe, affected about 50,000 people and killed about 63 (Onuoha, 2008b). In Bauchi State, no fewer than 24 persons were reported dead, while 5,787 farmlands and 13,609 houses, worth over $307.3 million, were destroyed by the flood. In Gombe State, similar floods induced about a 50 percent hike in the prices of grains. Consequently, 100-pound bags of maize, beans, millet, and sorghum increased to $25, $26, and $20 respectively, compared to $11, $20, $1.5, and $1 three weeks before the flood disasters (Onuoha and Ezirim, 2010).

The extent to which climate change has affected the social and economic well-being of the citizens of Nigeria over the past three decades is crucial to the way public policies are formulated in the country. Interestingly, only a very small fraction of Nigerians are aware of the seriousness of the environmental trends that are ravaging the country (Okpara in Ezirim and Onuaha, 2010). In 2008, the late President Yar'adua revealed that Nigeria loses about $5.1 billion annually due to environmental degradation (Dibie, 2009). This is even though there are environmental laws and government agencies meant to protect the environment in Nigeria, such as the Land Use Act, Petroleum Exploration Act, Federal Environmental Protection Agency (FEPA). The Federal Environmental Protection Agency has now been renamed the Federal Ministry of Environment. Some state governments also have Ministries of Environment. Socio-economic and environmental issues have continued to undermine the development of Nigeria, even on sustainable development bases.

The area referred to as the Niger Delta region was limited to the geopolitical zone occupied mainly by the minorities of southern Nigeria, which currently comprises the six states of Akwa Ibom, Bayelsa, Cross River, Delta, Edo, and Rivers. But in recent years, the Niger Delta region was redefined and enlarged to include all nine contiguous oil-producing states. The new states of Abia, Imo, and Ondo were therefore added to the original six states (Dibie and Offiong, 2012). According to Odia Ofeimun (2000), the Niger Delta people form the largest group among the ethnic minorities spread over the south-south geographical zone of Nigeria. It has a population of over 7 million people distributed in over 1,600 communities (Adunbi, 2010; Ojo, 2002).

The region has some unique characteristics, which tend to make development difficult. It is, for instance, one of the largest wetlands in the world. It covers an area of 70,000 square kilometers and is noted for its sandy coastal ridge barriers; brackish or saline mangrove, freshwater, permanent and seasonal swamp forests; and lowland rainforest. The whole area is traversed and crisscrossed by a large number of rivers, rivulets, streams, canals, and creeks. The coastal line is buffeted throughout the year by the tides of the Atlantic Ocean, while the interior is subjected to regimes of flood by the various rivers, particularly the River Niger. The Niger Delta region is the second largest delta in the world and the largest wetland in

Africa. The delta is home to an extraordinary variety of people, mostly fishers and farmers with a common history and cultural heritage.

Apart from being Africa's largest delta, the mangrove ecosystem around the Niger River estuaries is the fourth largest contiguous of its kind on earth. A combination of marine and freshwater environments, mangroves, and swamp rainforest should be the ideal natural system for the diversity of plant and animal wildlife that sustains fisheries, farming, hunting, and the gathering of nontimber forestry products, as well as forestry and tourism (Amadi, 2010). The big and difficult question is whether all of this could coexist with the petroleum industry. No one has direct answers, but a circumvention of this problem, as in the present case, is also not an option. We are therefore left with anecdotal accounts of the past: forest elephants in the Andoni and Orashi River areas, pigmy hippopotamuses in Finima on Bonny Island, manatees off Opobo, dwarf crocodiles in Patani, gray parrots in Kula, and tons of fish and shrimp all over the place! Nigerians may be a nation of storytellers, but that is not how science works.

Olokesusi (1987) considers procedural environmental characteristics as the ways and means of tackling, planning for, and managing environmental problems. We are again going to summarize the major lapses involved in the management of environmental problems. Therefore, multinational corporations such as Shell and Chevron are engaged in particular practices that are not specific to the oil enclaves of the Niger Delta but are transnational in outlook and local in implementation. Aiding these forms of practices are nongovernmental organizations engaged by multinational corporations because of their proficiency in the human and environmental rhetoric of governance. It is this form of proficiency that is producing and transforming subject populations into oil citizens in ways that are normalized and internalized (Okonta, 2008). This form of transformation, we argue, produces different outcomes that are not mutually exclusive for the oil corporations. First, it gives a semblance of minimization of conflict in resource enclaves because of its pretentious character of redefining the ownership structure and conferring some level of "ownership" of resources on oil citizens. Second, it aids the subject population in making claims of ownership of land and resources. This is done in ways that stimulate the production of new forms of citizenship, that induce the creation of the sites of power for multinational corporations, and that are evocative of colonial values in the postindependence, military, and subsequent democratic governments. This form of neocolonialism in the postcolony period will be the subject of the next section, but first we turn to three case studies that best illustrate how NGOs are becoming sites for the production of oil citizens. The three illustrations are based on ethnographic fieldwork at the offices of Chevron Nigeria Limited in Lagos and Warn, by an NGO named Our Niger Delta, and by Pro-Natural International, an international NGO with offices in the Niger Delta.

Chevron Nigeria Limited, an arm of ChevronTexaco, with headquarters in San Francisco, California, is one of the major oil corporations in Nigeria and has its Nigerian corporate headquarters on the Lekki Peninsula, an upscale area on the outskirts of Lagos. Lekki is known for its exclusivity and is an area widely known to be inhabited by the Nigerian ruling elite and business leaders, along with other members of the upper class. The

Table 10.1 Major Oil Companies Involved in the Exploration for and Marketing of Oil in Nigeria

Consortium	Shareholders	Joint Venture Operator	Production (barrels/day, 2003 estimates)	Production (%)
Shell Petroleum Development Company of Nigeria Ltd	NNPC (Nigeria, 55%) Shell (Dutch/British, 30%) Elf (France, 10%) Agip (Italy, 5%)	Shell	950,000	42.2
Mobil Producing Nigeria Ltd	NNPC (Nigeria, 58%) Mobil (USA, 42%)	Mobil	500,000	21.2
Chevron Nigeria Ltd	NNPC (Nigeria, 58%) Chevron (USA, 42%)	Chevron	485,000	18.6
Nigeria Agip Oil Company	NNPC (Nigeria, 60%) Agip (Italy, 20%) Philips (USA, 20%)	Agip	150,000	7.5
Elf Petroleum Ltd	NNPC (Nigeria, 60%) Elf (France, 40%)	Elf	145,000	6.1
Texaco Overseas (Nigeria) Petroleum Company	NNPC (Nigeria, 60%) Texaco (Nigeria, 20%) Chevron (USA, 20%)	Texaco	55,000	2.7
Other producers	Ashland (USA) Deminex (Germany) Pan Ocean (Switzerland) British Gas (British) Sun Oil (USA) Conoco (USA) BP (British) Statoil (Norway) Conoil (Nigeria) Dubri Oil (Nigeria)	Various	35,000	1.7
Total			2,320,000	100

Source: Idemudia and Ite (2006).

area overlooks the Atlantic Ocean, close to Victoria Island, another upscale neighborhood and home to many beautiful beaches. The major oil companies operating in the Niger Delta are listed in Table 10.1.

Public transportation in the area is erratic because it exists primarily for casual workers who do cleaning work for the "big man," artisans who repair the big man's car, or restaurant workers in the fast growing eatery industry in the area. Commuting is made easy for the upper class; core employees of Chevron, however, who are also situated in Lekki have homes at a comfortable distance from their Chevron workplace.

The Niger Delta region is also the main center of oil-producing activity in Nigeria and therefore the center of Nigeria's economy, accounting for more than 90 percent of the country's foreign exchange earnings and more than 80 percent of government revenue (Dibie and Offiong, 2012). Petroleum was discovered by Shell-BP in 1956, following half a century of exploration. Oil production became important in the 1960s, but the Biafra conflict and civil war between 1967 and 1970 delayed further expansion. The main reservoirs are located in and around the Niger Delta, in both onshore mangroves and shallow offshore basins, and since 1990 exploration has increasingly moved to deep, offshore areas. Unclear boundary demarcations in the Niger Delta have caused disputes relating to several strategic areas. Wunder (2003) observed that Cameroon and Nigeria both claim zones on and off the Bakassi Peninsula, and ownership of the Zafiro oil field is disputed with Equatorial Guinea. Shell continues to be the most important company, but it has been joined by a series of other multinationals over the years (Table 10.1).

The three levels of government in Nigeria lack an integrated approach to deal with environmental problems within it environment. Part of these lapses could be due to the difficulty in predicting (with some level of precision) the environmental impacts in different areas. Olokesusi (1987: 58) once pointed out that too many agencies are involved in environmental management. For instance, in Ibadan, the Oyo state capital, solid waste disposal is handled by the State Waste Disposal Board, the Sanitary Inspection Division of the Ministry of Health, the Oyo State Environmental Task-Force, and no fewer than ten certified waste disposal contractors, as well as numerous uncertified contractors. Obviously, considering this type of complex situation, it will be difficult to have an effective, integrated, and efficient environmental management.

At another level, the use of inappropriate technological measures makes for environmental problems. For instance, Cointreau (1992: 33) observed that, in Onitsha, one of the metropolitan cities in Nigeria, apart from the irregular collection of solid waste and refuse particularly from the poor neighborhoods, the trucks meant for evacuating these wastes are inadequate and unreliable. Although some efforts have been made to pass legislation to control environmental pollution in all areas of life, unfortunately many if not all the proposed laws are either not enforced or poorly enforced. In view of the problems pointed out in this section, we are of the opinion that much still needs to be done to establish and maintain vital and sound environmental planning and management strategies that can stand the test of time, given the far-reaching negative impacts on the environment.

In view of the Nigerian government's recognition of the damaging effects on the environment of floods, droughts, forest fires, technological accidents such as oil spills, industrial chemical effluents, the dumping of toxic wastes, and the contamination of rivers, lakes, soil, air, as well as other forms of pollution (Idemudia and Ite, 2006). Consequently, the Nigerian government enacted various environmental policies between the 1930s and the 1990s. Table 10.2 shows these various enactments and the years of enactment.

There is no doubt that the Nigerian government, during the pre- and postindependence era, has attempted to enact various environmental policies geared toward the regulation and control of the activities of both the

Table 10.2 Environmental Policies Enacted between the 1930s and 1990s

S/N	Policy	Year Enacted
1.	Forestry Ordinance	1937
2.	Eastern Region Forest Law	1955
3.	Oil Pipeline Act	1958
4.	Wild Animals Preservation Law of Western Region	1959
5.	Forestry Ordinance with Amendment (Northern Nigeria)	1960
6.	Wild Animals Law (Northern Region)	1963
7.	Wild Animals Law (Eastern Region)	1965
8.	Forestry Amendment Edict (Western State)	1969
9.	Petroleum Drilling and Production Act	1969
10.	Sea Fisheries Decree	1971
11.	Sea Fisheries Regulation	1972
12.	Wild Animals Preservation (Law, Lagos State)	1972
13.	Forestry Amendment Edict (Western State)	1973
14.	Wild Animals Law Amendment Edict (North-Eastern State)	1975
15.	Wild Animals Law Amendment Edict (Kano State)	1978
16.	Exclusive Economic Zone Decree	1978
17.	Kanji Lake National Park Decree	1979
18.	Endangered Species Decree	1985
19.	National Conservation Strategy for Nigeria	1986
20.	Natural Resources Conservation Council Decree	1989
21.	Federal Environmental Protection Decree	1989
22.	National Parks Decree	1991

Source: Dibie (2000).

public and private sectors, such as the disposal of hazardous and toxic wastes that affect the health of the people directly and indirectly.

The passage of these laws clearly shows that the government has been serious about the protection, conservation, and safety of important natural resources including the plants, animals, and other living things in the river, seas, and oceans. Again, the seriousness shown by the enactment of these laws has been clearly demonstrated by the passage of many other laws, such as the Minerals Act of 1958, Mineral Oil (Safety) Regulations of 1963, Oil in Navigable Waters Act of 1968, Endangered Species Act of 1990, Quarries Act of 1990, and Sea Fisheries Act of 1990. The creation of the Environmental Protection Agency (EPA) by Decree 58 in 1988, as an overall agency with the responsibility of protecting the Nigerian environment, further demonstrates the seriousness of the Nigerian government. However, it is clear that the effectiveness of all these laws is in doubt due to poor implementation strategies and other numerous logistical problems.

CONSEQUENCES OF ENVIRONMENTAL DEGRADATION IN NIGERIA

Environmental degradation connotes the deterioration of the physical environment through the activities of man by the displacement of natural landmarks and the introduction of pollutants, which in turn foul the air, water, and land, thereby endangering the life of organisms, including human lives.

Much literature has emanated on the consequences of environmental problems, particularly as they affect the Niger Delta region due to the activities of oil exploration and exploitation. Odogbor (2005: 105–114) enumerated the effects of oil spillages and industrial waste on the cultural, religious, economic, and political lives of the people. He then pointed out the serious negative implications of these effects on the people in the affected communities in Nigeria in all the spheres of their lives.

On the other hand, another scholar (Maduka, 1998: 877) catalogued major environmental problems affecting the Niger Delta region in particular. Table 10.3 presents a compendium of major environmental problems as they cut across the entire natural environment, livelihoods, aspirations, and the entire socioeconomic spectrum of the Niger Delta region. These damages can hardly be quantified in naira and kobo, and the impacts are very serious. Today's situation may likely remain the same or even worsen.

To further strengthen the fact that the Nigerian government had made enormous progress in enacting laws to protect the general environment of Nigeria, Table 10.4 lists such laws and the years they were enacted, particularly in 1990s.

A reflection of our Table 10.1 and 10.3 data should convince many that Nigeria's efforts at controlling the environment through all these laws and regulations are commendable indeed, but the problem is inadequate implementation. Our next task is to examine the various impacts of all this legislation on the Nigerian state and people.

Table 10.3 Ranking of Major Environmental Problems, Social Issues, and Priorities

Problem Type	Problem Subset	Priority Ranking
Natural environment	Coastal/river bank erosion	Moderately high
	Flooding	Moderate
	Sedimentation/silt	Low
	Substance	Low
	Exotic (water hyacinth)	High
Development related	Land degradation/soil fertility loss	High
	Agricultural decline/shortened fallow	High
	Delta forest (mangroves)	High
	Biodiversity depletion	High
	Fisheries decline	High
	Oil spillage	Moderate
	Gas flaring	Moderate
	Sewage and wastewater	High
	Other chemicals	Moderate
Socioeconomic problems	Poverty	High
	Unemployment	High
	Community-oil company conflicts	High
	Intercommunity conflicts	High
	Intracommunity conflict	Moderate
	Conflicts over land	High
	Inadequate compensation	High
	Displacements	Moderate
	Decay in societal values	High
	Poor transportation/high cost of fuel	High
	Housing pressure/infrastructure decay/crime	High

Source: Maduka (1998).

Table 10.4 Listed Laws Relevant to Environmental Issues in Nigeria

S/No	Environmental Laws	Year Enacted
1.	Oil in Navigable Waters Act, CAP 337 LFN	1990
2.	Harmful Wastes (Special Criminal Provisions) Act, CAP 165 LFN	1990
3.	Oil Pipelines Act, CAP 338 LFN	1990
4.	Minerals Act, CAP 226 LFN	1990
5.	Hides and Skin Act, CAP 167 LFN	1990
6.	Hydrocarbon Oil Refineries Act, CAP 170 LFN	1990
7.	Builders registration and related acts	—
8.	Federal Environmental Protection Agency Act, CAP 131 LFN	1990
9.	Federal Environmental Protection Agency (FEPA) Amendment Act, CAP 86 LFN	1990
10.	Rivers State Governmental Protection Agency (RISEPA) Edict	1994
11.	Agricultural (Control of Importation Act, CAP 12 LFN	1990
12.	River Basin Development Authorities Act, CAP 353 LFN	1990
13.	Petroleum Act (and its Registration), CAP 353 LFN	1990
14.	Public Health Act	—
15.	Territorial Waters Act, CAP 428 LFN	1990
16.	Explosive Act, CAP 117 LFN	1990
17.	Forestry Law of Lagos State	—
18.	Civil Aviation (Fire and Security Measures) Act, CAP 78 LFN	1990
19.	Associated Gas Re-injection Act, CAP 12 LFN	1990
20.	Criminal Code Provisions, CAP 77	1990
21.	Criminal Justice (Miscellaneous Provision) Act, CAP78 LFN	1990
22.	Wild Animals Preservation Act	—
23.	Kainji Lake National Park Act, CAP 197 LFN	1990
24.	Factories Act, CAP LFN	1990
25.	National Atomic Energy Commission Act	—
26.	Energy Commission of Nigeria Act	—
27.	Land Use Act, CAP 226 LFN	1990
28.	Mines and Quarries (Control of Building) Act, CAP 227 LFN	1990
29.	Natural Resources Conservation Act	—
30.	Navigable Waterways (Declaration) Act, CAP 287 LFN	1990
31.	Penal Code Provisions	—
32.	Quarantine Acts, CAP 384 LFN	1990

33.	Quarries Act, CAP 385 LFN	1990
34.	Tin (Miscellaneous Provision Act, CAP 229 LFN	1990
35.	Town and Country Planning Law of Lagos State	—
36.	Nigerian Urban Regional Planning Decree No. 86	1992
37.	Territorial Sea Fisheries Act, CAP 426 LFN	1990
38.	Sea Fisheries Act, CAP 404 LFN	1990
39.	Environmental Sanitation Law of Lagos State	—
40.	Environmental Sanitation Enforcement Agency Law of Lagos State	—
41.	National Parks Decree	1991
42.	Endangered Species (Control of International Trade and Traffic Act, CAP 108 LFN	1990
43.	Building Lines Federal Trunk Roads Act	—
44.	Meat Inspection and Sales Act, CAP 396 LFN	1990
45.	Forestry Law of Rivers State	—

Source: Ojo and Gaskiya, (2003).

IMPACTS OF ENVIRONMENTAL POLICIES ON THE NIGERIAN STATE

Subsequent to the dumping of toxic wastes in Koko in the former Bendel state in 1987, the Nigerian government promulgated the Harmful Wastes Decree, which was meant to provide the legal framework for the effective control of the disposal of toxic hazardous waste into any environment within Nigeria. This was immediately followed by the creation of a regulatory body—the Federal Environmental Protection Agency (FEPA) in 1988, charged with the overall responsibility of protecting and developing the Nigerian environment. States and local government councils were also encouraged to establish their own environmental regulatory bodies for the purpose of maintaining good environmental quality as it applied to their particular areas of jurisdiction.

FEPA Decree 58 of 1988 authorizes, among other things, this agency to establish and prescribe national guidelines, criteria, and standards for water quality, air quality and atmospheric protection, noise levels, gaseous emissions and effluent limits, and so on; to monitor and control hazardous substances; and to supervise and enforce compliance. This decree also gave the agency broad enforcement powers, even without warrants, to gain entry, inspect, seize, arrest, and impose stiff penalties consisting of fines or jail terms on whoever obstructs them in the discharge of their duties or makes false declarations of compliances (Gaiya, 2008: 1192–1193).

Aside from the FEPA Decree of 1988, the federal government, through Decree 86 of 1992, promulgated another law on environmental impact assessment (EIA) that aimed at protecting the Nigerian environment by making EIA compulsory for any project that may have an adverse effects on the environment. It sought to assess the likely or potential environmental impacts of proposed activities, including their direct or indirect cumulative, short-term, and long-term effects on the environment (Nwilo and Badejo, 2008: 1227).

In addition, the EIA Decree had the responsibility of identifying the measures available to mitigate adverse environmental impacts of proposed activities and the assessment of the identified measures. EIA activities are meant to be policed by the Federal Environmental Protection Agency and by the states' environmental protection agencies, respectively (Nwilo and Badejo, 2008).

The pertinent issue at this juncture is to assess the extent to which these laws and the relevant agencies have implemented them and to point out some observable constraints that may hinder their being effectively functional.

Despite the elaborate and comprehensive laws and regulations passed by the government over the years, it is uncertain whether much has been done in way of implementation and enforcement by the appropriate agencies. This observation should not be surprising because the Nigerian government is usually good at formulating public policies but grossly falls short when it comes to implementation. The literature that we have searched has not shown us a clear record of impact assessment of these elaborate laws and regulations.

First and foremost, we consider it relevant to examine the goals and principles by which EIA anchors its assessment to its ability to create a major impact (Echefu and Akpofure, 2000). According to Dibie and Offiong, before any person or authority takes a decision to undertake or authorize the undertaking of any activity that may likely or significantly affect the environment, prior consideration of its environmental effects should first be taken. In addition, in order to promote the implementation of appropriate procedures to realize the above goal, government must recruit qualified staff. Further, attempts to seek the encouragement of the development of reciprocal procedures for notification, information exchange and consultation in activities likely to have significant trans-state boundary environmental effects, inter-state compact or agreement will need to be established (Dibie and Offiong, 2012).

In addition to this, a procedural framework has been provided for EIA. The EIA process starts with the request for approval for implementation, resulting in the issuing of an environmental impact statement (EIS) and then the certificate. The stages are as follows:

1. Determining whether the Federal Environmental Protection Agency (FEPA) laws/regulations have been triggered
2. Screening a project for potential environmental effects

3. Scoping to determine the spatial and temporary dimensions of environmental effects
4. Carrying out detailed baseline studies to determine the environmental condition prior to project implementation
5. Preparing a detailed assessment report
6. Carrying out a panel review of the EIA report (if necessary) and obtaining authorization/approval where appropriate

We can categorically state that the goals of EIA and the principles to follow in order to arrive at environmental impact assessment are quite clear and straightforward. What becomes of this assessment and its impact on the nation as a whole is the task we want to dwell on now. It is pertinent to add that, apart from the provisions meant to guard against the pollution and degradation of the Nigerian environment, the Federal Republic of Nigeria's 1999 Constitution—Section 20 clearly states that "the state shall protect and improve the environment and safeguard the water, air and land forest and wild life of Nigeria" (Federal Republic of Nigeria, 1999: A887).

Unfortunately, the paternalistic attitude of some of the Nigerian judges toward matters relating to environmental hazards created by companies has rendered the enforcement of environmental laws ineffective. Some members of the judiciary, as noted by Ebeku (2003: 199–208), have been reluctant to give orders compelling companies whose operations are damaging to the environment to ease the action complained of. Perhaps these judges consider the potential loss of income and their investments at the expense of the environmental protection. In addition to this could be the fact that Nigeria's economy depends largely on the sale of crude oil. Whichever is the case, such actions retard the implementation of environmental laws and thereby relegate these laws to the status of mere paper tigers.

Between 1990 and 2004, Dibie and Offiong (2012) asserted that there were several oil-related cases filed in the Nigerian courts by affected Nigerians, ranging from pollution from oil exploration to the loss of income, loss of property, contamination of drinking water leading to waterborne diseases, and so on. A few cases need to be mentioned: *Shell vs. Tiebo Vii*, *Shell vs. Isaiah*, *Seismograph Services vs. Mark*, *Ogiale vs. Shell*, and *Shell vs. Ambah*. In most of these cases and similar ones, the courts are said to have refrained from passing judgment on how to remedy the situation of the oil spillage claims, loss of income from fishing and farming, pollution of drinking water and crops, or damage to health as a result of waterborne diseases (Mbendi, 2002). Instead of ordering offenders to address the damages done by the physical environment of these communities, they settled for compensation to the affected complainants. Thus, the environmental laws that were meant to protect human beings and other living things are thrown to the dogs. However, all is not lost because in some circles, the hope is that the judiciary may in the future begin to decide cases of environmental problems not merely on monetary compensation awards without addressing the preservation of a healthy environment (Okorodudu, 1998: 607). This type of thinking could be regarded as mere speculation that may or not come to be. The multinational oil companies, who are normally being complained against by oil communities on gas flaring, are more likely to always win cases against them on legal grounds because they have a lot of financial capacity to fight their cases instead of obeying environmental laws meant to sanitize the environment.

Given this attitude of the judiciary, FEPA (now called the Federal Ministry of Environment) has a challenge of translating the laudable provisions of the various environmental acts into effective management of the environment. Experience has shown so far that infrastructural projects suffer from environmental impact assessment because approval for such projects is often given before the environmental impact assessment takes place (if at all) (Anago, 2001; Mbendi 2002).

The key defaulters in this exercise are the various levels of government: federal, state, and local. These levels of government are said to be involved in routinely approving projects within the mandatory study list before any kind of impact assessment is made. The Niger Delta Development Commission (NDDC) decided to dredge the Ayetoro Canal prior to any EIA as required by law. In fact, according to our sources, any EIA reports are actually "post mortem just to fulfill all righteousness" and fence off resistance and complaints from nongovernmental organizations and the affected communities (Anago, 2001; Mbendi 2002).

Many rivers in Nigeria have been altered over the past four decades, showing how the natural environment in the country has been modified to serve the purposes of the population. The struggle between the people in the Niger Delta and the oil companies polluting their land and rivers is an example of the relationship between people and the natural environment. Apart from these predicaments, rivers are canalized, forests felled, towns built, and new roads constructed all for the purposes of economic interests, safety, and welfare of the richest people in the country. Although Nigeria is beginning to advance its technology for controlling rivers and the Atlantic Ocean, there is a major concern that homes are now being built in the wetlands, sometimes blocking drainage systems and causing flood during the rainy season.

The increasing awareness of the impact of environmental pollution has led to successful public campaigns against many wrong practices. For instance, a recent campaign by school children against dumping soft drinks and water cans on the street is beginning to result in a change in public behavior. There have been similar campaigns against the excessive use of plastic and its careless disposal in other major cities in Nigeria. This increasing awareness has led to successful public campaigns against many wrong practices. In addition, the social responsibility of stakeholders is very crucial in keeping the environmental clean. The Federal Government of Nigeria could adopt a system of permits and punitive measures, for example, if industries discharge pollutants, such as effluents, from their factories into waterways or rivers. If those industries are located in areas where they are not connected to a sewer and must discharge directly into a waterway, they should be required to have a permit. The government may therefore encourage some industries to have their own treatment plants. Even industries connected to sewers may be required to pretreat their wastewater before discharging it into the sewers. This is because they may contain materials that may damage the sewers or the treatment plants, or they may pose a danger to the people who work in maintaining the sewerage system.

Based on the prevailing attitude, it can safely and certainly be deduced that, EIA reports merely serve to satisfy the paperwork requirement. To further point out that EIA reports may merely serve to "fulfill all righteousness," our sources further referred us to a particular case where the federal government awarded a contract for the dredging of the Imo River to improve vehicular assess to the nation's only aluminum smelting company at Ikot Abasi in Akwa Ibom state, but some of the affected coastal communities resisted the project on the grounds that the EIA draft report was unfavorable to them. Unknown to these communities, the final report of the EIA got the approval of the Ministry of Environment, and so the dredging project commenced despite the resistance of these communities. Worst of all, it is said that the federal government refused to pay the communities the requested compensation on the grounds that the dredging exercise has potential economic benefits to the communities (Dibie and Offiong, 2012). Where, then, do justice, fairness, and, most importantly, the ability of the EIA to be in actual control of the environment—not just statutorily? It simply demonstrates the impotency of EIA on Nigeria's environment.

According to Ako Amadi (2010), Nigerian leaders plant many ceremonial trees and get much political attention from these cosmetic events. The fact, however, is that Nigerians are simply not interested in vegetation, in spite of the dangers of desertification and impacts of global warming. There is the account of the unemployed Nigerian forester who cut down all the trees and who is angry. This propensity to cut trees and the hatred of vegetation, the fable concludes, have survived to the present day (Mbendi, 2002).

Other the other hand, one may argue that there is nothing wrong with politicians' symbolically planting a tree or two and posing as environmental champions. But what are the financial costs of these events that involve the governor's motorcade, a retinue of line ministers, dancing women, school children, the statehouse press corps, armed security—all choreographed to plant a solitary tree? Would it not have been more appropriate for a governor to approve money for agroforestry in his state and watch citizens and school children plant and take care of the trees periodically (Adunbi, 2010; Amadi, 2010; Dibie, 2000)?

Ironically, the least tree planting is executed by those states that need vegetation for survival—the arid, Sahel frontline states in the north and the erosion-prone southeast. Land use planning has definitely not been a priority in those states (Roorda, 2012). Their demands are that soil stabilization be paid for out of the Ecological and Disaster Relief Fund. The request is legitimate. However, the Land Use Decree actually vests custodianship of the environment with the state and local governments (Ezirim, 2008b). Over the years, disasters could have been nipped in the bud through adequate forestry.

There is a need to manage conflicts through the interactive conflict management (ICM) (Adunbi, 2010), as opposed to the persistent militarization approach of governments. ICM is community–focused, learning-based, and action-orientated methodological approach to resolving conflict at the local level. It is less expensive and moves beyond the structurally designed, compartmentalized blueprint approach of conflict resolution. It focuses on the continuous interaction between people and their position in the broader sociocultural setting, with governments and other social agents acting as facilitators. ICM acknowledges the knowledge, skills, and experiences of the community to manage conflict. It is a pragmatic approach, and the diversity embedded in the local community can be mobilized to resolve/manage conflict.

It is a common knowledge that anytime there is outbreak of conflicts in communities in the Niger Delta, the federal government of Nigeria responds swiftly with direct military action and without appropriate consultation with stakeholders and community leaders. This has always escalated the case and led to more harm and problems than intended. The case of the Odi direct military action is a good illustration. With an interactive conflict management approach, community stakeholders and leaders will be brought to a round table, and this has been proven to be very useful in resolving conflicts in many situations. From the testimonies of the people in the area, it was clear that the incidence of high-handedness and disregard for their views in matters affecting them only helps to worsen conflicts.

The bone of contention, however, cannot be limited to the weak environmental policies in Nigeria, which chooses to utilize the benefits and risks of oil spillage in the Niger Delta region or the constant exposure to fossil fuel all over the country. Technology that enhances Nigeria's ability to exploit petroleum, fossil fuel, old-growth forests, or fisheries generally increases impact; however, the nation has not done a good job in seeking technology to reduce smokestack emissions, harness renewable energy, or improve manufacturing efficiency to decrease social and economic impact (Withgott and Brennan, 2008).

There is a shared global impact that could cause longer-term damage to ecosystems in Nigeria and around the world. According to Nancy Maxwell (2009), the ecological footprint of the earth's surface provides resources for absorbing waste and pollution. Forest areas in the southern part of Nigeria, which are needed to absorb carbon dioxide produced by burning fossil fuels, are diminishing as the demand for lumber increases in Nigeria. Unless these natural resources are replaced at a rate greater than the rate of reduction, the nation's environmental and sustainable development problems will continue to escalate (Wright and Boorse, 2014).

According to Jeffrey Sachs (2005), there are ways that Nigeria can adapt to confront these environmental challenges without destroying prosperity. This could be done by building smarter power plants that capture and dispose of the nation's carbon emissions and by increasing the use of renewable energy sources. Finally, the government of Nigeria must target its future investment on health, education, and infrastructure in order to discontinue the cycle of poverty in the country. The nation should be prepared to face the ongoing challenge of investing in its ecosystem's sustainability in the twenty-first century without compromising the welfare of its citizens.

SUMMARY

This chapter has acknowledged that the Nigerian state has taken serious steps to develop effective environmental strategies by enacting various laws and decrees, as well as by adopting the necessary procedural guidelines and strategies to ensure that the safety of the Nigerian environment prevails in all aspects. Unfortunately, many flaws render the enacted environmental policies and EIA impotent. Ashford (2000: 68), in the same vein, questioned the effectiveness of environmental policies alone as a stimulus to green innovation and underlined the importance of other factors needed to generate firm-level technological change.

The chapter listed, among these factors, regulatory requirements, including environmental policy, possible cost savings or additions to profits, public demand for a less polluting and safer industry practices, and worker demands and pressures arising from industrial relation concerns. On the other hand, it has been observed that most developing nations, including Nigeria, rarely focus on the impact of environmental policy on the technology responses of firms, but rather they focus usually on the impact of policy on emissions or on the reduction or abatement of pollution.

The negative impact of environmental policies needs to be addressed at different levels at the same time in order to move toward sustainable development in Nigeria. Although a stricter system of licenses and permits is being developed, there is great emphasis on public awareness. The involvement of the public and community is seen as the cornerstone to success in efforts toward sustainable development. Some problems of sustainability are already urgent and require immediate action. Examples are local ecosystems where the population is encroaching and where forests and water stocks are nearly depleted. In such cases already on the decline, opportunities for correction or mitigation may have been lost, and the abandonment of exploring practices and autonomy may be necessary. The urgency of some of these problems has been overlooked either because the people most affected are physically remote from the center of power or because their voices are not heard.

This chapter contended that it will not be out of place to note the many regulations with similar responsibilities. There is perhaps the need for a harmonization and for laying out clear-cut responsibilities among different regulatory agencies while, of course, keeping EIA as the apex regulator. In addition, the administration of all these environmental policies should be supported financially. Adequate financing is important for the effective monitoring and enforcement of environmental policies in Nigeria. Where compliance with the policies is neglected by any of the stakeholders, appropriate sanctions should be enforced.

Finally, there should be periodic environmental audits to ensure that there is compliance with environmentally sound practices in the form of objective evaluations of how well environmental organization is managed. There is the crucial need for the government to provide relevant equipment and related logistics for the regulatory agencies to better do their jobs. Perhaps by so doing, environmental enactments could be meaningful and effective and not merely exist mostly in theory. Despite these measures, the continued dependency of Nigeria on petroleum, gas, and the demand for the expansion of material consumption will lead to increased transnational corporations' presence in the country. These transnational corporations will continue to control most of the capital as well as curtail Nigeria's environmental protection issues, and that will remain a source of conflict well into the foreseeable future (Figure 10.2).

Figure 10.2 Human and Industrial Waste Dumped on the Street in Ibadan, Nigeria
Source: Photo taken by author during field research in Nigeria.

REFERENCES

Abegunde A. A., and Omisore E. O. (2003). "The Role of Urban and Regional Planning in Controlling Environmental Degradation in Nigeria." In V. Adekunle, E. Okoko, and S. Adeduntan (eds.), *Challenges of Environmental Sustainability in a Democratic Government. Proceedings of a National Conference Organized by Environment and Behavior Association of Nigeria.* Lagos, Nigeria: University of Lagos Press, 65–72.

Adamantiades, A., and Kessides, I. (2012). "Nuclear Power for Sustainable Development: Current Status and Future Prospects." In *Taking Sides: Clashing Views in Sustainability*, edited by R. Taylor. New York: McGraw-Hill.

Adelekan, I. (2009). "The Nigerian Press and Environmental Information for Sustainable Development." *Local Environment, 14*(4), 297–312.

Adoyi, M. A. (2010). "That Nigeria Will Not Be Hit by a Flood-Induced Tsunami." *People's Daily*, Retrieved from www.peoplesDaily.com. Accessed May 19, 2011.

Adunbi, O. (2010). "Belong to the Soil: Multi-National Oil Corporation, NGOs and Community Conflict in Post-Colonial Nigeria." PhD dissertation, Yale University, New Haven, CT, USA.

Amadi, A. (2010). "Africa Sustainable Energy and Environment Platform." An African Business Community, July 14. Retrieved from http://sustainableenergy inafrica.ning.com/profile/AndrewAmadi. Accessed November 19, 2012.

Anago, I. (2001). "Environmental Assessment as a Tool for Sustainable Development: The Nigerian Experience." Retrieved from www.jepplaw.net/page Builder-image/Pages/Environmental%20legislation 0changes. Accessed on December 3, 2010.

Ashford, N. (2000). "An Innovation-Based Strategy for Sustainable Environment." In *Innovation-Oriented Environmental Regulation: Theoretical Approaches and Empirical Analysis*, edited by J. Hemmelskamp, K. Rennings, and F. Leone. Heidelburg, Germany: Zew Economic Studies.

Badgley, C., et al. (2012). "Can Organic Farming Feed the World?" In *Taking Sides on Environmental Issues,* edited by T. Easton. New York: Mcgraw-Hill.

Best, E., and Lawson, R. (2008). "Rich Nations Put Global Warming Burden on Africa." *Daily Trust,* 28.

Clark, W., and Levin, S. (2009). "Towards a Science of Sustainability." Princeton, NJ: Princeton Environmental Institute.

Clifford, M., and Edwards, T. (2012). *Environmental Crime.* Burlington, MA: Jones & Bartlett Press.

Cointreau, S. J. (1992). *Environmental Management of Urban Solid Waste in Developing Country: A Project Guide.* Washington, DC: World Bank.

Dibie, R., and J. Offiong. (2012). "The Nature and Impact of Environmental Policy in Nigeria." *Journal of International Politics and Development,* 10, 113–142.

Dibie, R. (2009). "Environmental Governance and the Political Economy of Sustainable Development in Sub-Saharan Africa." *Journal of International Politics and Development,* 7(1), 1–28.

Dibie, R. (2000). *Understanding Public Policy in Nigeria: A Twenty-First Century Approach.* Lagos, Nigeria: Mbeyi and Associates.

Dibie, R., Mello, W., and Benitez, G, (2011). "Social Impact of Environmental Health Policy in Brazil." *Journal of International Politics and Development,* 9(1), 1–23.

Dye, T. (2011). *Understanding Public Policy.* Upper Saddle River, NJ: Prentice Hall.

Easton, T. (Ed.). (2012). *Taking Sides on Environmental Issues*, New York: Mcgraw-Hill.

Ebeku, K. (2003). "Judicial Attitudes to Redress for Oil Related Damages in Nigeria." *RECIEL, 12*(2), 199–208.

Echefu, N., and Akpofure, E. (2000). "Environmental Impact Assessment in Nigeria: Regulatory Background and Procedural Framework." Retrieved from www.iaia.org.public/documents/EIA/casestudies. Accessed on December 3, 2010.

Enahoro, J. E., and Ehi-Ebewele, O. (2007). "Effect of the Oil Industry on the Environment: Shift in Paradigm on Financial Reporting Disclosure for Sustainable Environmental Development." Paper presented at the International Conference on the Nigeria State, the Oil Industry and Niger Delta on October 22–27, 2007 at Niger Delta University, Bayelsa State Nigeria.

Ero, F. I. (1997). "Research, Environment and Resource Development." In *Proceedings of the 25th Annual Conference of Forestry Association of Nigeria Held in Ibadan, Oyo State, Nigeria*. (September 22–26).

Ezirim, G. E. (2008a). "Petro-Politics and Environmental Criminality in the Niger Delta: Advocacy or Enforcement of Global Conventions." *University of Nigeria Journal of Political Economy*, 2(1–2).

Ezirim, G. E. (2008b). "Xenophobia and Citizens' Diplomacy: Call for Reassessment of Nigeria's Afrocentric Foreign Policy." *International Journal of Communication*, 9, 267–281.

Ezirim, G. E., and Onuoha, F. C. (2010). "Climatic Change and National Security: Exploring the Conceptual and Empirical Connections in Nigeria." *Journal of Sustainable Development in Africa*, *12*(4).

Fadamiro, J. A. (1995). "Impact of Landscape Architecture on Environmental Protection and Management." In *Proceedings of the Regional Training Workshop Held at the Federal University of Technology Akure, Nigeria*. (July 26), 240–243.

Federal Government of Nigeria. (1999 and 2009). Federal Office of Statistics. Abuja, Nigeria: Government Press.

Federal Republic of Nigeria. (1999). The Nigerian Constitution of 1999. Lagos, Nigeria: Government Press.

Friis, R. (2012). *Essentials of Environmental Health* (2nd ed.). Burlington, MA: Jones & Bartlett.

Gaiya, B. (2008). "Environmental Pollution in the Niger-Delta: An Economic Perspective." Paper presented at International Conference on the Nigerian State, the Oil Industry and Niger-Delta at the Department of Political Science, Niger-Delta University, Bayelsa State- Nigeria. March 11–13.

Gbadamosi, K. T., and Omole, F. K. (2003). "Environmental Impact of Exploration in Nigeria." Paper presented at the National Conference on Challenges of Environmental Sustainability in the Democratic Government. Retrieved from http://www.kas.de/proj/home/pub/33/2/dokument_id-11468/index.html. Accessed on July 22, 2011.

Goldman, R. (2010). "Ecosystem Services: How People Benefit from Nature." *Environment*, 17, 15–23.

Gottfredson, M., and Hirschi, T. (1990). *A General Theory of Crime*. Stanford, CA: Stanford University Press.

Idemudia, U., and Ite, U. (2006). "Corporate-Community Relations in Nigeria's Oil Industry: Challenges and Imperatives." *Corporate Society and Responsible. Environmental Management*, *13*, 194–206.

Jain, R. K., et al. (1977). *Environmental Impact Analysis: A New Dimension in Decision Making*. New York: Nostrand.

Jeje, L. K., and Adesina, F. A. (1996). *Man and Environment: An Introductory Note*. Ede, Nigeria: Research, Evaluation Resources and Development Consultancy.

Jiboye, A. D. (2003). "Urbanization and Urban Degradation in Nigeria. Strategies for Improvement." Paper presented at the National Conference on Environmental Sustainability in Democratic Government.

Jimoh, I. H. (2006). "Pattern of Environmental Degradation and Development Efforts." In *Environmental Science*, edited by H. A. Saliu and A. A. Ogunsanya. Ibadan, Nigeria: Spectrum Press.

Koko, A. N. (2010). "The Environment as an Important Policy Issue." *BusinessDay* online. Retrieved from www.businessdayonline.com. Accessed May 19, 2011.

Kosoy, N., and Corbera, E. (2010). "Payments for Ecosystem Services as Commodity Fetishism." *Ecological Economics, 69*, 1228–1236.

Kraft, M., and Furlong, S. (2012). *Public Policy: Politics, Analysis, and Alternatives*. Washington, DC: Congressional Quarterly Press.

Kurmar, G. S. (1989). "Desertification: Causes and Control." *The Democratic* (March 5), *13*(2), 37–59.

Maduka, W. O. (1998). "Evolving Vibrant Communities for Development." In *Proceedings in International Seminar on Petroleum Industry*. Abuja, Nigeria, 907–925.

Maxwell, N. (2009). *Understanding Environmental Health. How We Live in the World*. Boston: Jones & Barlett.

Mbendi, A. (2002). Cited in Wunder, S. (2003). *Oil Wealth and the Fate of the Forest: A Comparative Study of Eight Tropical Countries*. London: Routledge, 303–305.

McCauley, D. J. (2006). "Selling Out on Nature." *Nature, 444*, 27–28.

McNeely, J. A. (2007). "A Zoological Perspective on Payments for Ecosystem Services." *Integrative Zoology*, 2, 68–78.

Myers, N., and Spoolman, S. (2014). *Environmental Issues and Solutions*. Belmont, CA: Brooks/Cole.

Neumann, R. (2002). *Imposing Wilderness: Struggles over Livelihood and Nature Preservation in Africa*. Berkeley: University of California Press.

Niezen, R. (2003). *The Origins of Indigenism: Human Rights and the Politics of Identity*. Berkeley and Los Angeles: University of California Press.

Nwilo, P. C., and Badejo, T. O. (2008). "Impact and Management of Oil Spill in Nigeria Coastal Environment." Paper presented at the International Conference on the Nigeria State, the Oil Industry and the Niger Delta, at the Department of Political Science, Niger-Delta University, Bayelsa State, Nigeria (March 11–13).

Odogbor, P. (2005). "Effects of Environmental Degradation on Cultural Heritage of the Niger-Delta and the Implication on Sustainable Rural Development in Orobator Eke, Ifowodo, Feturi & Edosa, Emaruma." In *Federal State and Resources Control in Nigeria*. Benin City, Nigeria: F. Parker Publishing Company.

Ofeimun, O. (2000). "The Niger Delta and the 1999 Constitution." In *The Emperor Has No Clothes: Report of the Conference on Peoples of the Niger Delta and the 1999 Constitution*, edited by O. Doifie. Benin City, Nigeria: Environmental Rights Action/Friends of the Earth, 52–71.

Ojo, G. U. (Ed.). (2002). *Empowerment in Action: ERA's Community Intervention in the Niger Delta: A Model for Development*, Benin-City, Nigeria: Environmental Rights Action/ Friends of the Earth.

Okonta, I. (2008). *When Citizens Revolt: Nigerian Elites, Big Oil, and Ogoni Struggle for Self Determination*. Trenton, NJ: Africa World Press.

Okorodudu, M. T. (1998). "Law of Environmental Protection." Text from Ebeku, K. "Judicial Attitudes to Redress Oil Related Environmental Damage in Nigeria." *RECIEL*, *12*(2) 199–208.

Olokesusi, A. (1987). "Characteristics of Environmental Problems in Nigeria." *The Environmentalist*, 1, 57–91.

Okpara, E. E. (1993). "Environmental Watch: How Much Do Nigerians Know?" *Nigerian Forum, 13*(5–6), 156.

Omisore, E. O., and Akande, C. G. (2003). "The Roles of All Tiers of Government and Non-Governmental Organizations (NGOs) in Environmental Conservation." In V. Adekunle, E. Okoko, and S. Adeduntan (eds.), *Challenges of Environmental Sustainability in a Democratic Government. Proceedings of a National Conference Organized by Environment and Behavior Association of Nigeria*. Lagos, Nigeria: University of Lagos Press, 913.

Onuoha, F. C. (2010). "Climate Change, Population Surge and Resource Over-Use in Lake Chad: Implications for Human Security in North-East Nigeria." In *Climate Change and Natural Resource Conflicts in Africa*, edited by D. A. Mwiturubani and J. van Wyk. Pretoria: Institute for Security Studies.

Onuoha, F. C. (2008a). "Environmental Degradation, Livelihood and Conflicts: A Focus on the Implications of the Diminishing Water Resources of Lake Chad for North-Eastern Nigeria." *African Journal on Conflict Resolution, 8*(2), 35–61.

Onuoha, F. C. (2008b). Oil Exploitation, Environmental Degradation and Climate Change: Assessing the Vulnerability of the Niger Delta to Natural Disasters. *Journal of Liberal Studies, 12*(1–2), 131–150.

Onuoha, F. C. (2008c). "Oil Pipeline Sabotage in Nigeria: Dimensions, Actors and Implications for National Security." *African Security Review, 17*(3).

Onyewotu, L. O. Z. (1998). "Socio-economic Consideration in the Use of Shelterbelt and Scattered Trees in the Control of Environmental Degradation in Semi-Arid Areas of Nigeria." In *Socio-Economic Implications of Desertification in Nigeria*, edited by Idumah et al. (2003). Ibadan, Nigeria: University of Ibadan Press.

Orubuloye, I. O. (1983). "The Demographic Situation." In *Population and Development in Nigeria*, edited by I. O. Orubuloye and O.Y Oyenye. Ibadan, Nigeria: Spectrum Press.

Paehler, K. (2007). "Nigeria in the Dilemma of Climate Change." *Newswatch* (July 19), 19–21.

Polasky, S. (2012). "What's Nature Done for You Lately: Measuring the Value of Ecosystem?" In *Taking Sides: Clashing Views in Sustainability,* edited by R. Taylor. New York: Mcgraw-Hill, 228–238.

Piquero, A. R., Jennings, W. G., and Farriggton, D. P. (2010). "On the Malleability of Self-Control: Theoretical and Policy Implementation Regarding a General Theory of Crime." *Justice Quarterly,* 27, 1–32.

Pratt, T. C., and Cullen, F. T. (2000). "The Empirical Status of Gottfredson and Hirschi's General Theory of Crime: A Meta-Analysis." *Criminology, 10*, 41–78.

Roorda, N. (2012). *Fundamentals of Sustainable Development.* New York: Routledge.

Ruff, L. E. (1993). "The Economic Common Sense of Pollution." In *Economics of Environment*, edited by R. Dorfman. New York: Norton.

Sachs, J. (2005). *The End of Poverty: Economic Possibilities for Our Time.* New York: Penguin, 38–41.

Taylor, R. (2012). *Taking Sides: Clashing Views in Sustainability.* New York: Mcgraw-Hill.

United Nations Environmental Programme (UNEP). (2006). "New Report Underlines Africa's Vulnerability to Climate Change." Nairobi (November 5). Retrieved from http://new.unep.org/Documents.Multilingual/Default.asp?DocumentID = 485& ArticleID = 5409&l = en. Accessed on August 7, 2007.

United Nations Report. (2010). *Environment.* New York: Oxford University Press.

Withgott, J., and Brennan, S. (2008). *Environment: The Science Behind the Story.* New York: Pearson.

World Bank. (1996). *Toward Environmental Sustainable Development in Sub-Saharan Africa.* Washington, DC: The World Bank.

Wright, R., and Boorse, D. (2014). *Environmental Science: Towards a Sustainable Future.* Boston: Pearson Learning.

Wunder, S. (2003). *Oil Wealth and the Fate of the Forest: A Comparative Study of Eight Tropical Countries.* London: Routledge.

Part III

Developing Countries—Other Continents

11 Environmental Health Policy in Brazil

Robert Dibie, PhD, William Mello, PhD, and Galia Benitez, PhD

INTRODUCTION

One of the goals of the citizens of Brazil is the enjoyment of the highest attainable standards of health. The World Health Organization (WHO) contends that every human being, regardless of political belief, economic or social status, race, religion, or nation of origin, is entitled to the highest standard of health (Skolnik, 2008). The desire for the citizens of Brazil to attain good health has galvanized enormous progress in biomedical science and public health. The health policy measures taken by the government have truly changed the nation and have given longer life spans to millions of citizens around the country.

Brazil's environmental legislation is among the most advanced in the world (Puppim de Oliveira, 2008; Luken and Rompaey, 2007), but its implementation has proven to be challenging. Puppim de Oliveira (2008) explains that the main reasons are weak intergovernmental coordination; the lack of definition of responsibilities among the federal, state, and local authorities; and the chronic shortage of resources.

A major environmental disaster that attracted international attention to Brazil in January 2011 was excessive rainfall that resulted in mudslides. Mudslides hit an area of nearly 900 square miles in lush, forested mountains about 40 miles north of Rio de Janeiro. The deaths were centered in Teresopolis and three other towns, where many wealthy citizens of Rio maintain weekend homes. Though the disaster destroyed the homes of rich and poor alike, the deaths were overwhelmingly seen in humbler areas, where the homes are flimsier, most lacking foundations, and located in steep areas known to be at high risk of mudslides (Associated Press, 2011).

The environmental problem that attracted international attention to Brazil in the 1980s was the deforestation of the Amazon (Hudson, 1997). Most of the clearing resulted from the activities of ranchers, including large corporate operations, and a smaller portion resulted from slash-and-burn techniques used by small farmers. Another unique environmental challenge for Brazil is the natural diversity that exists in such a large country (Arceneaux, 2009). While much of the attention paid to the environment goes to the Amazon, many other biomes and local environments are not taken into consideration (Teixeira, 2010). Puppim de Oliveira (2012) argues that Brazil is one of the most biologically diverse places in the world, with more species of primates, amphibians, freshwater fish, and flowering plants than any other country.

Another environmental problem that is often ignored is the enormous Brazilian savanna that requires a different strategy than the Amazon, which is different from the coastal Atlantic forest. Further, several urban areas in Brazil need their own environmental policies that can take into account issues of human development and population density. Puppim de Oliveira (2012) contends that in areas where agriculture is more intense and developed, there are serious problems of soil erosion, siltation and sedimentation of streams and rivers, and pollution with pesticides. In parts of the savannas, where irrigated soybean production expanded in the 1980s, the water table has been affected. The expansion of pastures for cattle raising has reduced natural biodiversity in the savannas.

Desertification is another important environmental problem in Brazil (Puppim de Oliveira, 2008). In the early 1990s, it became evident that the semiarid *caatinga* ecosystem of the Northeast was losing its natural vegetation through clearing and that the zone was running the risk of becoming even more arid, as in some other regions (Hudson, 1997). In urban areas, such as Rio de Janeiro, Sao Paulo, and Santa Catharina, levels of air pollution and congestion are typical and often cause health problems. Another social and environmental health problem in Brazil is related to the lack of sanitation (Hudson, 1997; Puppim de Oliveira, 2012). These problems are sometimes worse in middle-sized and small cities than in large cities, which have more resources to deal with them. The environmental problems of cities and towns finally began to get greater attention from society and the government in the 1990s; however, the problem has not been adequately resolved (Alonso and Clemencon, 2010).

With regard to the general thinking about sanitation, Tesh and Paes-Machado (2004), using opinion polls, show that Brazilians, as well as environmentalists, think that urban sanitation is a major environmental problem. Mueller (1995) studied nine metropolitan regions in Brazil. Regarding the lack of the adequate provision of water and the disposal of human excrement and household water, this author argues that these services have improved since 2010; however, in many places the problem is still precarious. Sixty percent of the Brazilian government investment in the southeastern region is in water, and 40 percent is in sewer. The lack of sanitation is a challenge in Brazil. Heller and Oliveira (2010) explain that this problem is the result of the conflict among the federal, state, and local levels of government, as well as the conflict between the private and the public sectors. They concluded that the traditional sanitation indicators do not capture the internal environmental problems within the regions.

In the past three decades, Brazil has experienced tremendous social, economic, and environmental development. These developments, however, come with some positive and negative environmental and health effects. Several scholars have attributed the development in Brazil to technological advances, the stable political environment and industrial development (Drummond and Barros-Patiau, 2006; Guilhem and Avevdo, 2007; Olivia et al., 2003). The amount of carbon dioxide emissions in Brazil is still much lower than in most of the countries in South America (Smith et al., 2000). Brazil has undertaken a number of measures to reduce greenhouse gas emissions, including introducing cleaner fuels such as natural gas and biofuels and improving public transportation. Other measures include using sugarcane for ethanol fuel. Some scholars have also contended that the government is not doing enough (Guilhem and Avevdo, 2007).

Despite these health policy measures, the extent of the disparity in environmental health in Brazil remains one of the greatest moral and intellectual problems of the twenty-first century. Almost every day we hear stories of human exposure to environmental contamination. Nancy Maxwell (2009) contends that most contamination arises from the transportation and transformation of chemicals in the environment. In Brazil, as in most nations of the world, chemical contaminants can be found in the air, water, and soil, which is dry and fine enough to become airborne and dissolved in bodies of water. The oil in bodies of water and on surfaces is also regarded as sediment.

The antagonism between environment and development characterized the debate in Brazil during the 1970s and 1980 (Arceneaux, 2009; Arceneaux and Pion-Berlin, 2005). During these two decades, the expectation was based on the idea that the development policies implemented in Brazil by the government and supported by the international organizations, such as the International Monetary fund (IMF) and the World Bank, were improving social inclusion by increasing societal income and employment. From this perspective, natural areas were looked upon as underutilized spaces that were available for the development of industrial complexes and export industries. However, reality showed that, as social development grew, other problems emerged, such as land conflicts, exclusion of marginalized societal sectors, and the achievement of national inclusion only by devastating new territories and initiating conflicts with the native populations. The assessment of Chico Mendes, a leader from the rubber tappers, marked the height of conflict between developers and environmentalists. The cause of the rubber tappers brought attention to the notion of treating nature in a sustainable manner, as well as to the existence the people of the forest and having respect for them (Alonso and Clemencon, 2010; Arceneaux and Pion-Berlin, 2005).

This chapter examines the social impact of environmental policies and health issues in Brazil. Although several studies have explored environmental issues in Brazil, few have critically examined the social impact of environmental health policy in the nation. With the aid of 876 in-depth surveys, this study contends that, although environmental scientists have learned about the behavior of contaminants in the environment—the chemicals, their physical transformation, and their movement within or between environmental media—contaminants in the air and water continue to affect the social welfare and health of citizens. The chapter draws on data from several communities in Brazil and conducts a cross-national comparison of the social impact of environmental health policy. In addition, the study explores the following questions: To what extent does environmental health affect the social activities and lives of citizens in the major cities of Brazil? What are the barriers to effectively implementing environmental health policies in Brazil? How can these barriers be overcome? What factors could help foster a sustainable environmental health policy in the nation? Without practicing sustainable development, Brazil faces a deteriorating environment and may even be inviting ecological disaster in the near future.

ENVIRONMENTAL HEALTH AND ITS IMPEDIMENTS

The impediments to environmental health and the factors associated with occupational diseases and injuries include noise, dust, toxic heavy metals and their fumes, carbon monoxide, chemicals, ionizing radiation, microbial agents, lifting heavy weights, repetitive motion, accidents, and work-related stress (Friis, 2011). Thus, protecting the environment in Brazil means creating a region in which the air is safe to breathe, the water is safe to drink, the land is arable and free from toxins, wastes are managed effectively, infectious diseases are kept at bay, and natural areas are preserved. According to *Health People 2010*, the physical and social environments play a major role in the health of individuals and community in any nation. In the case of Brazil, the social environment includes agriculture, housing, urban development, land use, transportation, industry, violence, injury, and exposure to such factors as work-related stress.

Human health and environmental health are strongly interlinked. The World Health Organization (WHO) estimates that 25 percent of the world's preventable diseases are attributable to environmental factors. This percentage is even higher, 35 percent, if we concentrate on developing regions such as Africa (WHO, 2002). Since 1972, the WHO has recognized the importance of studying this relationship and defines environmental health as ". . . those aspects of human health, including quality of life, that are determined by physical, chemical, biological, social and psychological factors in the environment" (WHO, 2002). By this definition, the environmental health concept encompasses a wide variety of risks. David Ball, using a risk taxonomy proposed by the International Risk Governance Council (IRGC) (Renn, 2006), argues that the following risks might be incorporated under the environmental health heading: (1) physical agents (e.g., ionizing radiation, noise, kinetic energy, etc); (2) chemical agents (e.g., toxic substances, environmental pollutants, compound mixture, etc); (3) biological agents (e.g., fungi and algae, bacteria, other pathogens); (4) natural forces (wind, earthquakes, floods, etc.); (5) sociocommunicative hazards (e.g., terrorism, human violence, mass hysteria, etc.); and (6) complex hazards (e.g., food, consumer products, technologies, constructions etc.). Out of all these general risks, there are few key areas where developing nations are particular vulnerable such as poor quality water or scarcity and sanitation, vector-borne diseases, poor ambient indoor and outdoor air, toxic substances, and global environmental change (WHO, 2002).

On the one hand, it has been argued that one of the growing needs of citizens in Brazil is to balance economic growth or demands for environmental preservation. On the other hand, however, the nation is faced with significant challenges to its environmental and human health programs. Pollution and population growth are often associated with environmental health. Thus, the citizens are being affected in some way by exposure to environmental hazards associated with lifestyle, at work, at home, during recreation, or on the highways (Ioris, 2010; Jacob, 1997).

Several scholars have pointed out that the implications of the social impact of environmental health policies tend to originate from the intentions of the government. For example, despite their best intentions, the government of Brazil has not been able to fully improve the quality of public health services delivery in all the regions of the nation (Fallon and Zgodzinski, 2009; Murphy, 2005; Skolnik, 2008; Teitelbaum and Wilenski, 2007). This unequal practice has been inconsistent between and across jurisdictions in the nation. Further, capacity differs markedly between federal and regional government agencies. These differences are due not only to funding disparities but also to the organizational ability to deliver health services and programs of a consistently high quality that reflect the current state of environmental health.

The ecological theory postulates that the factors influencing environmental health behavior occur at several levels (Murphy, 2005; Skolnik, 2008). These levels include public policy, individual, intrapersonal, interpersonal, and institutional factors. The theory also contends that behavior both influences and is influenced by the social environments in which it occurs (Murphy, 2005). The health belief model on the other hand, suggests that people's health behaviors depend on their perception of the likelihood of getting the illness, the severity of the illness if they get it, the barriers to engaging in preventive behavior, and the benefits of engaging in behavior that will prevent the illness. According to Murphy (2005), people's health behavior also depends on whether they feel that it is possible to carry out the appropriate behavior if they tried it; that is, people's health behavior is dependent on their self-efficacy (Skonik, 2008; Teitelbaum and Wilensky, 2007).

The stages of change in the health behavior model contend that change in behavior is a process and that different people are at different stages of readiness for change. According to Murphy (2005), the stages of change are (1) precontemplation, (2) contemplation, (3) decision and determination, (4) action, and (5) maintenance. It is interesting to note that people who are aware of their health problem may be willing to do something about it. In this sense, they might seek help in addressing the health problem.

In an attempt to achieve success in health promotion, the literature on social impact assessment examines a variety of factors that go beyond health. According to Richard Skolnik (2008) and Mclaughlin and McLaughlin (2008), these impacts include the influences of demography, gender, culture, communities, buildings, and health. These models also suggest that assessment could be carried out in a way that builds on local processes, engages the community fully, and proactively tries to maximize the potential good that can come from the health and policy evaluation process. The advantages of this model are that it promotes community development and empowerment, builds capacity, and develops social capital (Vanclay, 2008).

Most general studies about environmental health policy are conducted in the well-developed, industrialized world. Listorti and Doumani (2001) assert that "this situation has introduced an inadvertent bias into the analysis of environmental health issues in developing countries." These scholars call for the understanding of how environmental health issues differ in developing countries to counterweight this bias.

Water quality is strongly related to the prevalence of diseases, such as cholera or diarrhea, in developing nations. The WHO estimates that 88 percent of the preventable disease burden is attributable to unsafe water supply sanitation and hygiene. Diarrheal disease amounts to an estimated 4.1 percent of the total disease burden, with 1.8 million people suffering from it every year (WHO, 2002). Esrey (1996), studying water contamination at the point of usage and comparing different types of system to reduce diarrhea, concludes that sanitation and personal hygiene are much more effective than improving water quality. Other authors who study water contamination at the storage point, such as Musa and colleagues (1999), argue that poor water storage negates the benefits of new water installations. Wright and Boorse (2014) contend that the quality of drinking water declines significantly after collection in many settings, and they add that the extent of the contamination varies considerably across a multitude of settings. Therefore, policies that aim to improve water quality and point of usage are paramount but must be also complemented with preventative policies geared toward safer household storage and treatment.

Vector-born infectious diseases, such as yellow fever, malaria, dengue, among others, and their resurgence in the last few decades have been of concern, especially in developing nations and tropical areas where these diseases occur more frequently (Gratz, 1999). Gubler (1998) calls for adequately trained personnel and academic institutions to train them, as well for governments to change their policies and approaches to diseases prevention. Dias and colleagues (2002) caution about Chagas disease in Latin America and how controlling it has become a problem because the government and authorities give priority to other diseases. They also discuss the risk associated with the decentralization of public health programs and the lack of tradition and expertise in the treatment of this disease.

Air pollution kills approximately 800,000 people each year, and this is a major issue in developing nations (WHO, 2002). While two-thirds of these deaths occurred in Asia, Latin American urban centers also face high levels of air pollution. It is widely known that outdoor air pollution is associated with chronic health effects, as attested to by voluminous epidemiology studies that research disease rates in vitro and in human populations (Bascom et al., 1996). Bella and colleagues (2005) studied the consequences of air pollution in three Latin American cities: Sao Paulo (Brazil), Ciudad de Mexico (Mexico), and Santiago (Chile). They show that an air pollution control policy would have vast health benefits for each of the three cities, averting numerous adverse health outcomes, including over 156,000 deaths, 4 million asthma attacks, 300,000 children's medical visits, and almost 48,000 cases of chronic bronchitis in the three cities over the twenty-year period. Other researchers have shown that acute respiratory infections are the most common illness and cause of death among children in the developing world. Albertini and Krupnick (2000) studied respiratory symptoms in Taiwan by comparing the cost-of-illness and willingness-to-pay estimates. They found that the estimate ratios in Taiwan were similar to those from the United States.

Indoor air pollution is also an important cause of mortality and immobility relating to acute respiratory infections, asthma, and lung cancer in developing nations. Compared to developed nations, developing nations have higher concentrations of pollution and personal exposure to biomass. Globally, almost 3 billion people use biomass and coal as their primary source of energy (Rom, 2012). In 2000, approximately 2 million deaths were attributed to exposure to these indoor air pollutants (Ezzati and Kammen, 2002). Numerous epidemiology studies on the impact of air pollution and human health have found that woman and children are at greater risk (Balakrishnan et al., 2002; Ezzati and Kammen, 2002; Smith et al., 2000).

In response to the Global Agenda 21, adopted in 1992 at the United National Conference on Environment and Development (UNCED), the concept of sustainable development has increasingly received attention of policy makers around the world. During this conference, governments unanimously committed themselves to go beyond traditional developmental goals of maximizing economic outputs to consider seriously the environment and social effects of development (Sachs and Warner, 1995). Goals 4, 5, and 6 of the Earth Summit addressed directly the importance of the symbiosis between human health and environmental health within development. However, as Myers and Spoolman (2014) explain, development studies have neglected the linkages between social and environmental variables. Munasinghe (2007) caution that within this literature no simple generalization is possible and that important opportunities have been missed for combining environmental and social impacts into win-win outcomes.

The sustainable development literature looks at macroeconomic and limited policy reforms and how these influence economic development and its anticipated environmental and social impacts. Munasinghe and Cruz (1994) point to key macroeconomic variables that are widely used by these studies such as trade deficit, food security, and employment and industrial protection. Munasinghe (2007) summarizes in a chart the most used macroeconomic variables and their expected environmental and social impacts. Regarding the trade deficit, export promotion might promote more deforestation; however, industrial jobs creation may reduce pressure on land resources. Regarding food security and unemployment, agriculture intensification reduces migration to ecologically fragile areas, but there is also a potential to overuse fertilizer and chemicals. Regarding industrial protection, openness leads to more energy efficiency or less polluting technologies and could also attract hazardous industries.

Some specific examples of this macroeconomic literature are the studies of O'Ryan and colleagues, who found that in Chile, the taxing of PM10 emission yields better environmental results than taxing SO_2 and NO_2. However, they show that macroeconomic impacts are slightly negative, while limited effects (positive and negative) can be strong. Therefore, they conclude that policy makers should consider flexible policies with adjustment periods targeting the taxation of environmentally bad impact instead of good because doing so achieves fairly good environmental results without economic and social losses. In regard to social policy, applied simulations show very low impact, highly driven by pollutants. Therefore, they suggest that environmental policies may have social impacts but not vice versa. Jacob (1997) and Puppim de Oliveira (2012) pointed out that in Brazil deforestation is a spillover from macroeconomic policies. Binswanger (1989) is another example at the limited policies level in the case of Brazil. Binswanger studied general tax policies, special tax incentives, the rules of land allocation, and the agriculture credit system and found that these policies have a strong impact on deforestation in the Amazon. Mahar (1988), however, cautions that these policies reduce the land available to the poor. In the worst case, policies introduced to improve the environmental health of the poor may introduce starvation and malnutrition in exchange for cleaner air.

This study adopts the perspective that if more public leaders are trained to be development-oriented managers, there is the possibility that accountability and performance monitoring would improve considerably in the area of environmental health in Brazil. Exposing leaders to the skills of implementing growth-oriented and market-friendly economic environmental health policies would enhance the quality of life of all their citizens. Such pragmatic policies would also correct the problem of maladministration and corruption in the public sector. Other methods that could help to improve developments in Brazil include the effective rule of law, changes in social and health behavior, occupational health, and the commitment to social responsibility by leaders.

ANALYSIS OF ENVIRONMENTAL TREND

The antagonism between environment and development characterized the debate in Brazil during the 1970s and 1980s. During those two decades, the expectation was that the government's development policies implemented in Brazil and supported by international organizations, such as the IMF and the World Bank, were bringing social inclusion by increasing societal income and employment (Arceneaux, 2001 & 2009). From this perspective, natural areas were looked upon as underutilized spaces that were available for the development of industrial complexes and export industries. However, reality showed that social dissatisfaction grew, and other problems emerged, such as land conflicts, the exclusion of marginalized societal sectors, and national inclusion at the expense of devastating new territories and introducing conflicts with the native populations. The assessment of Chico Mendes, a leader from the rubber tappers, marked the height of conflict between developers and environmentalists. The cause of the rubber tappers brought attention to the notion of treating nature in a sustainable manner and to the existence of the people of the forest and respect for them (Puppim de Oliveira, 2008 & 2012).

The United Nations Conference on the Human Environment in Stockholm in 1972 sparked a concern in Brazil for the environment. Legislation up to this point was centered on flora and fauna, a situation that changed when water and mineral resources were included. In 1973, the Special Secretariat for the Environment subordinated to the Ministry of the Interior was created in addition to other state agencies, such as the State Foundation for Engineering and Environment (FEEMA) and the Company for Environmental Technology and Sanitation (CETESB), which aimed to control industrial pollution. This led in 1975 to the environmental theme being incorporated into the Brazilian Development Plan.

In 1980, the health industrial disaster of Cubatao in Sao Paulo and the burning of the Amazon intensified Brazilian consciousness of the importance of protecting the environment (Puppim de Oliveira, 2008). As official rhetoric about the importance of considering the environment began emerging within the Brazilian government, officials began pressing for reforms (Ioris, 2010). In 1987, Brazilian legislation created the National Environmental Policy (Federal Law No. 6938), which pushed for the decentralization of environmental functions. Since then, the state and civil society have matured in their thinking. Environmentalists have become a multisector movement, and the government has created a more complete system of national and regional coordination (Leis, 2000).

In 1981, the Brazilian government created the National Environmental System (SISNAMA) and the National Environmental Council. The former was formed by government agencies' oversight policies affecting natural resources; the latter includes members of government but also NGOs, labor, and other civic organizations (Puppim de Oliveira, 2008). In the last decade, three main channels were developed: (1) the evaluation of environmental impact, which ranges from impact studies to public hearings; (2) public civil actions; and (3) Comision Nacional del Medio Ambiente (CONAMA) and other state and municipal councils. However, Leis (2000) argues that real society participation is reduced because there is no room for negotiation and for groups to generate new alternatives (Ioris, 2010). In 1989, the Institute for Environment and Renewable Natural Resources (IBAMA) was created as an agency to pull together all the agencies of different government departments responsible for any aspect of the environment. In 1992, the Ministry of Environment, which included the IBAMA, was created (Puppim de Oliveira, 2008).

According to Rex Hudson (1997), although Brazil's public policies regarding the environment are generally advanced, their implementation and the enforcement of environmental laws have been far from ideal. Laws regarding forests, water, and wildlife have been in effect since the 1930s. Brazil achieved significant institutional advances in environmental policy design and implementation after the Stockholm Conference on the Environment in 1972. Specialized environmental agencies were organized at the federal level and in some states, and many national parks and reserves were established (Arceneaux, 2009).

According to a World Bank report in 1992, the Brazilian Environmental Sustainability Agenda Technical Assistance Project supports the government on advancing the pace of reform supporting the Programmatic Reform Loan for Environmental Sustainability as defined in the Policy Letter presented by the government on July 7, 2004. According to Puppim de Oliveira (2008), there are five projects components: (1) strengthening of the environmental management system, including defining environmental indicators and bolstering the Ministry of Environmental Management capacity; (2) improving the biodiversity conservation and the sustainable use of natural resources (the Green Agenda); (3) improving management of environmental risks due to hazardous chemicals (the Brown Agenda); (4) improving water resources management (the Blue Agenda); and (5) mainstreaming environmental sustainability in selected government sectors, for example, fiscal sector, hydropower sector, tourism, etc. (World Bank, 1992). Despite favorable laws, promising institutional arrangements, and external funding, the government has not, on the whole, been effective in controlling damage to the environment. This failure is only in small measure because of the opposition of antienvironmental groups. A much greater reason can be attributed to the traditional separation between official rhetoric and actual practice in Brazil. It is also related to general problems of governance, fiscal crisis, and lingering doubts about appropriate trade-offs between the environment and development. Some of the most effective governmental action in the environmental area has occurred at the state and local levels in the most developed states and has involved nongovernmental organizations (NGOs). In 1994, the PNMA began to stress the decentralization and strengthening of state environmental agencies, a tendency that subsequently gained momentum (Hudson, 1997).

The allocation of water is key to improving social justice, the standard of leaving, and environmental conservation. Brazil in 1997 approved a water law (Law 9) that replaced the water code of 1934 and that put in place an extensive regulatory system. Despite the commitment of the government for sustainable development, the reality is that other government officials keep stressing the importance of economic development over sustainable development goals (Ioris, 2010). The main problem, as Ioris (2010) argues, is that the new decision-making forums have been dominated by the same rural oligarchy (e.g., landowners), construction companies, and industrialists who traditionally have controlled the economic and social opportunities for water use. Brazil in the last two decades has embarked on some of the largest water projects in the world, such as the construction of dams and other water-related structures.

The Global Agenda 21, which was developed in 1992 by the United Nation, is an overarching program of action for sustainable development in the twenty-first century. The agenda postulates the importance of sustainability which entails the capacity of the earth's natural system and human cultural system to survive, flourish, and adapt to changing environmental conditions into the future. What we do to the environment today should not compromise the benefit that future generations may need to survive. According to the Agenda 21, there are two very important reasons why we should care about sustainability:

1. The long term survival of our own species and culture depends on our leaning how to live more sustainably.
2. The moral and ethical obligation to the eco-system requires human beings not to crowd out or eliminate large numbers of the planet's other species and the places where we live.

According to Alonso and Clemencon (2010), Brazil has not established an effective local agenda at a regional level; the only agenda established is the Brazilian Amazonian Agenda 21. This later agenda will be used as a blueprint for the Amazonian Cooperation Treaty among Bolivia, Brazil, Colombia, Ecuador, Guyana, Peru, Surinam, and Venezuela (Alonso and Clemencon, 2010; Trinidade, 2000).

A specific example of this macroeconomic literature is the study by Alonso and Clemencon(2010) , which found, for the case of Brazil, that, in relation to environmental policies, the taxing of PM10 emission yields better environmental results than taxing SO_2 and NO_2. However, they show that the macroeconomic impacts are slightly negative and that sector effects (positive and negative) can be strong. Therefore, they conclude that policy makers should consider a flexible policy with adjustment periods in order to tax environmentally bad impacts instead of good ones because that achieves fairly good environmental results without economic and social loses. In regard to social policy, the simulation showed very low impact, highly driven by pollutants. Therefore, they suggest that environmental policies may have social impacts but not vice versa. Heller and colleagues (2010) contend that, in Brazil and the Philippines, deforestation is a spillover from macroeconomic policies. Another example at the sector policies level is the case of Brazil Binswanger (1989), who studied general tax policies, special tax incentives, the rules of land allocation, and the agriculture credit system and found that these policies have a strong impact on deforestation in the Amazon. Mahar (1988) cautions that these policies reduce the land available to the poor.

Regarding industrial protection and economic openness, the literature accepts that economic openness leads to increasing industrial pollution. This concern is based on the main assumptions that incentives for exports will lead to greater exploitation of national resources and that free trade increases pollution in developing nations, appealing to the race-to-the-bottom arguments that dirty industries will relocate from developed countries to developing countries with less environmental restrictions. Wheeler (2001) specifically tests the race-to-the-bottom model in China, Brazil, and Mexico and concludes that the most dangerous forms of air pollution have declined. However, the evidence is mixed. Kosmo (1989) demonstrated that, in Turkey, capital-intensive industries have more of a tendency to pollute. Katz (1991) showed how the fishmeal industry in Chile eliminated bacterial contamination of its products by treating the effluents. He makes the argument that, if an export-oriented industry needs to meet international standards, which are higher than the local standards, it needs to invest in new technologies or methods that lead to cleaner products and processes. Birdsall and Wheeler conducted a study of twenty-five Latin American countries over three decades—the 1960s, 1970s, and 1980s. They argued that "[o]penness encourages cleaner industries" (1993: 143).

RESEARCH METHODOLOGY

The objective of this research is to examine the social impact of environmental health policy in Brazil. The specific objectives of the study are:

1. Assess the extent to which available environmental health policy impacts the social and economic behavior of citizens in Brazil.

2. Conduct an analysis of the significance of occupational and environmental health in Brazil.
3. Assess the major agent of occupational diseases in Brazil.
4. Finally, examine the preventive mechanism being used by the government in Brazil to address occupational diseases.

The data on which this analysis will be derived are from a questionnaire that was administered in the Ceara, Fortaleza, Brasilia, and Rio de Janeiro regions of Brazil. The selection of Brazil is based on three important criteria: (1) the existence of a democratic government; (2) the existence of environmental health policies and ability to assess environmental risks, benefits, and costs, while considering the social implication; and (3) the ability to administer a questionnaire that has greater validity. Combining the in-depth interviews and questionnaire data analysis of the social impact of environmental health policy in Brazil will enable the authors conduct a systematic comparison of how the nation handles their occupational health problems, sustainable development processes, and service provisions in lieu of government.

DEMOGRAPHICS

Rio de Janeiro—The city of Rio de Janeiro is located in the southeastern region of Brazil along the country's extensive Atlantic coastline and is the commercial capital of the Brazilian state of the same name. The city was the capital of Brazil from 1760 until 1960 before the capital was transferred to Brasilia. Rio is the second largest Brazilian state, both in population and in GDP. The population of the City of Rio de Janeiro is estimated at 6,186,000 inhabitants, or 11,513,000 when considering the metropolitan area, and it covers 1,759 square miles (including metropolitan regions) (Arceneaux, 2009). The city's economic activity is dedicated to the service industry and accounts for the country's second largest GDP, being divided as follows: service industry, 65.5 percent; industrial activity, 11 percent; agricultural activity, 0.5 percent. The city's remaining income is through tax revenue (Weatherby et al., 2009). The main industrial activities are chemical, food processing, steel, ship building, textile, clothing, and furniture. The city continues to host the headquarters of many major national and international corporations.

Brasilia—Brasilia is the capital of Brazil, and it is located in the midwestern region of the country. The city is the administrative center and home of the three branches of the federal government. The city's population is 2,557,000, and, when considering the metropolitan area, the population increases to 3,599,000 (United Nations Human Development Index, 2010). Unlike most Brazilian cities, Brasilia is an entirely planned urban environment with modern architecture by Lucio Costa, designed by the world-renowned architect, Oscar Niemeyer. The city officially assumed the role of the country's capital in April 1960. Brasilia occupies an area of 2,204 square miles and continues to be one of the fastest growing cities in the world. The primary economic activity draws on its role as the seat of government, and public administration is therefore responsible for 54 percent of the city's GDP. Industry is responsible for 10 percent of Brasilia's economic activity, and services account for 28.7 percent of the local economy, commerce 6.1 percent, and agriculture 0.2 percent. Brasilia is the third richest Brazilian city after Sao Paulo and Rio de Janeiro (United Nations Human Development Index, 2010). The primary industrial activities are construction, food processing, pharmaceutical, and graphics.

Ceara—The State of Ceara is located in the northeastern region of Brazil, bordering the extensive Atlantic coastline. By population, Ceara is the eighth largest Brazilian state and seventeenth by geographic area. The state's population is estimated at 8,472,000 occupying 56,505.4 square miles with an approximate annual per capita income of $2,500 (United Nations Human Development Index, 2010). Because of its extensive coastal areas, 56 percent of the state's GDP is derived from the service sector, primarily from tourism. The industrial sector of the economy is second, occupying 37.9 percent of the GDP and derived primarily from garment, leather footwear and products, cashews, and fruits. A large part of the industrial activity is for export (Arceneaux, 2009). Agricultural activity ranks third with 5.4 percent of the state's economic activity (United Nations Human Development Index, 2010).

Fortaleza—The City of Fortaleza is the capital of the State of Ceara with a population of 2,458,545 inhabitants in the city and 3,529,000 inhabitants when considering the metropolitan area (Arceneaux, 2009). By population, Fortaleza is Brazil's fifth largest city and the second largest in the northeast region, occupying 121 square miles, making it a city with one of the highest demographic densities in the country with 7,587.68 inhabitants per square kilometer, unequally distributed throughout the city's five subregions. It has the ninth largest GDP in the country (United Nations Human Development Index, 2010). Just like the state, the city's main economic activity is the service and tourist industry, with a per capita income of approximately $4,500 per year; however, income inequality is high, with 43 percent of the population living in abject poverty (Arceneaux, 2009).

Data are not usually available in a single location or from a single source. Secondary data will be reviewed first. The first part of the project will also be an analysis of the current environmental health policy literature available in the United States. This will be following by in-depth interviews of industrial workers and other citizens in Brazil. The responses from the in-depth interviews will be used to determine the extent to which available environmental health policy impacts the social and economic behaviors of citizens in Brazil.

The second part of the research will involve administering a questionnaire to industrial workers in Brazil. Data collected through focus groups and individual interviews will be used as a vehicle for formulating analysis for analyzing the significance of occupational and environmental health in Brazil.

ANALYSIS AND DISCUSSION OF FINDINGS

The data collected from the respondents were analyzed using the SSPS computer program to find out the relationship between environmental health policies, preventive mechanism, workers' safety, and their actual contribution toward sustainable development processes in Brazil. Tables 11.1 and 11.2, along with Figure 11.1, summarize the results from 876 out of 1,200 surveys administered in Brazil.

Air Quality

Air pollution has been one the major environment health problems in Brazil. In Brasilia, 25 percent of the respondents indicated that the nation's air policy had had great impact in their lives. In Ceara, only 30 percent of the respondents stated that air pollution had affected their health. On the other hand, it is interesting to note that in Fortaleza only 2 percent and in Rio de Janeiro only 4 percent of the respondents indicated that air pollution had had any impact on their health. Generally, the Brazilian Air Pollution Index (API) shows that between 100 and 200 people with existing heart or respiratory illnesses may experience any significant aggravation of their symptoms. In some parts of the nation, there may also be widespread symptoms in the healthy population (e.g., eye irritation, wheezing, coughing, phlegm, and sore throats). Further, respondents indicated that about 5 to 10 percent of disease-adjusted life-years lost in Brazil are due to environmental causes. By far the most important factor is fine particulate-matter pollution in urban air in cities such as São Paulo and Santa Catharina.

Climate Change

A major environmental disaster that often affects the lives of the citizens of Brazil is climate change due to excessive rainfall, resulting in mudslides. In 2011, mudslides hit an area of nearly 900 square miles in lush, forested mountains about 40 miles north of Rio de Janeiro. In Brazil, 95 percent of the respondents indicated that climate change often affects agriculture, biodiversity, and energy sources. In Ceara, Fortaleza, and Rio de Janeiro, an average of 55 percent of the respondents stated that climate change has

Table 11.1 Environmental Impact on Health

Variable	Brasilia 240		Ceara 210		Fortaleza 204		Rio de Janeiro 222	
	Yes	No	Yes	No	Yes	No	Yes	No
Have you had any medical problems caused by safety negligence on the job within the last year?	15	85	20	80	47	53	8	92
Have you ever been injured on the job due to mistake of your company or employer?	10	90	7	93	20	80	8	92
Did you receive sick leave for the injuries in question?	18	82	10	90	24	76	16	84
Do you think environmental health and development are intimately interconnected in Brazil's development plan?	83	17	57	43	88	12	73	27
Do you think chemicals and pesticides polluting water and rivers in the country has affected your life style?	100	0	93	7	0	100	97	3
Do you think hazardous wastes are present in water and rivers have affected your health or someone in your family?	100	0	100	0	100	0	100	0
Do you feel that waste dumped on the streets and resulting in foul odor due to ineffective implementation of environmental policies has affected citizens' health?	98	2	100	0	100	0	89	11
Do you feel uncontrolled emission may affect the lives and health of citizens?	98	2	100	0	100	0	90	10
Do you think poorly managed sewage treatment has constituted health problems in your region?	100	0	100	0	100	0	100	0
Has poor drainage system affected your business, street, farm or home?	78	22	80	20	80	20	83	17
Have you experience any loss in biodiversity in your region?	100	0	100	0	100	0	92	8
Do you think that improper disposal of engine oil and old motor batteries could affect the health of citizens?	100	0	97	3	100	0	98	2

Source: Derived from questionnaire analysis (2011).

Table 11.2 Awareness of Environmental Policy Impact

Variable	Brasilia 240		Ceara 210		Fortaleza 204		Rio de Janeiro 222	
	Yes	No	Yes	No	Yes	No	Yes	No
Do you know any government policies in Brazil?	84	16	57	43	64	36	57	43
Do you know any environmental policies enacted by the government of Brazil?	65	35	34	66	50	50	30	70
Do you think environmental policies have affected your life style in any way?	94	6	57	43	76	24	78	22
Area of Major Environmental Polices Impact	Waste 60 Air 25	Sustain 15	Sust. 60 Air 30	Waste 10	Waste 57 Sust. 47	Air 2	Sust. 60 Waste 36	Air 4
In your opinion how seriously does your fellow co-worker talk about environmental policies and issues at work?	51	49	51	49	38	62	35	65
Do you believe sustainable development requires agriculture, environment, biodiversity, energy and other issues?	95	5	51	49	68	32	68	32
Do you notice any social impact of irregular garbage collection in your city?	100	0	80	20	100	0	89	11
Do you think the government is doing enough to effectively educate the citizens about environmental policies?	24	76	14	86	76	34	27	73
Do you think environmental policies have affected economic and social trends in your life and region?	81	19	65	35	82	18	73	27

Source: Derived from questionnaire analysis (2011).

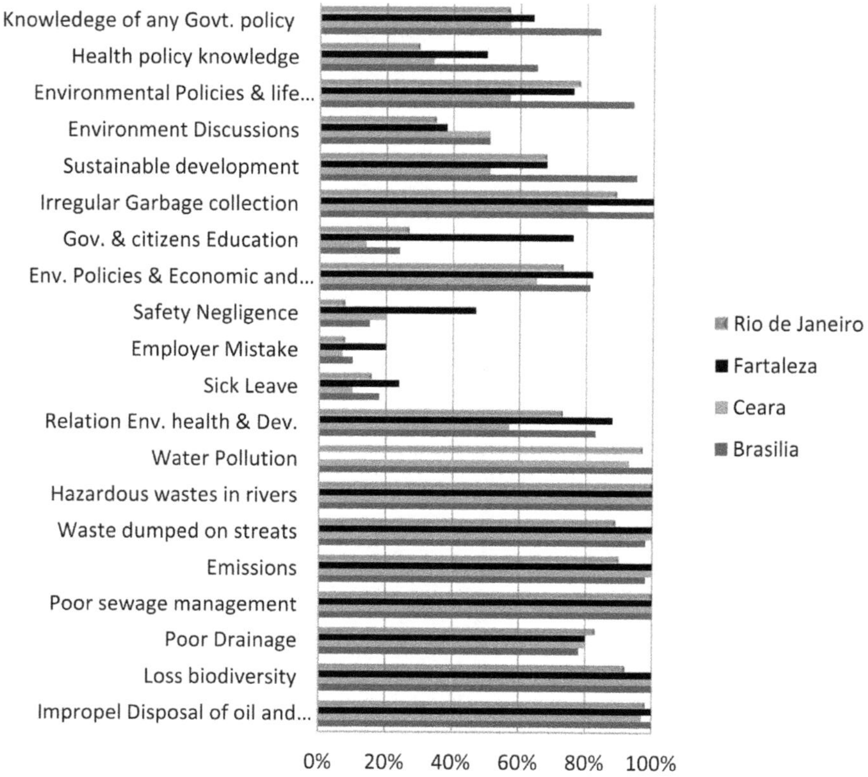

Figure 11.1 Summary of Survey Findings in Brazil

had a major impact in their region. A change in the type, distribution, and coverage of vegetation may occur given a change in climate; this much is obvious to farmers. In any given scenario, a mild change in climate may result in increased precipitation and warmth, resulting in improved plant growth and the subsequent sequestration of airborne CO_2. Larger, faster, or more radical changes, however, may well result, in certain circumstances, in vegetation stress, rapid plant loss, and desertification.

Food Safety

Eighty percent of the respondents in Brasilia, 65 percent in Ceara, 82 percent in Fortaleza, and 73 in Rio de Janeiro indicated that environmental policies haves affected economic and social trends in their lives and region. They indicated that safe food handling by farmers and department stores has minimized the health issues that would have resulted. This environmental health policy requirement includes safe procedures for all processes, such as receiving, repacking, food storage, preparation and cooking, cooling and reheating, displaying products, handling products when serving customers, packaging, cleaning and sanitizing, pest control, transport and delivery, and dealing with the causes of cross-contamination. Brazilian government regulatory agencies normally hold people responsible for serving unsafe food, and those found guilty are often liable for heavy fines. Generally, respondents are pleased that industry will be forced to take food safety seriously under this food safety legislation.

Solid Waste Management, Including Landfills and Recycling

One hundred percent of the respondent in Brasilia and Fortaleza, respectively, indicated that the improper disposal of engine oil and old motor batteries could affect the health of citizens. Further, 100 percent of respondents in Rio de Janeiro, Ceara, Fortaleza, and Brasilia, respectively, indicated that hazardous waste is present in the water and rivers of the country. They also indicated that hazardous waste has affected their health or the health of someone in their family. The same percentage of respondents indicated that they feel waste, dumped on the streets and resulting in foul odors due to ineffective implementation of environmental policies, has affected citizens' health in Brazil. Waste management is the collection, transport, processing, recycling or disposal, and monitoring of waste materials. The term usually relates to materials produced by human activity and is generally undertaken to reduce their effect on health, the environment, or aesthetics. Waste management is also carried out to recover resources from it. Waste management can involve solid, liquid, gaseous, or radioactive substances, with different methods and fields of expertise for each. Disposing of waste in a landfill involves burying the waste, and this remains a common practice in Brazil. Landfills were often established in abandoned or unused quarries, mining voids, or borrow pits. A properly designed and well managed landfill can be a hygienic and relatively inexpensive method of disposing of waste materials. Older, poorly designed or poorly managed landfills can create a number of adverse environmental impacts, such as windblown litter, the attraction of vermin, and the generation of liquid leachate. Another common by-product of landfills is gas (mostly methane and carbon dioxide), which is produced as organic waste breaks down anaerobically. This gas can create odor problems, can kill surface vegetation, and is a greenhouse gas.

Hazardous Waste

Sixty percent of respondents in Brasilia and 57 percent in Fortaleza indicated that waste management generally was a major concern. They indicated that hazardous waste were particularly dangerous to the health of citizens and wish that the government of Brazil would do more to prevent any social and economic impact of the inappropriate handling of such dangerous waste in the nation. In Brazil, hazardous wastes are defined as wastes that exhibit the following characteristics: ignitability, corrosively, reactivity, or toxicity. Many types of businesses generate hazardous waste. Some are small areas that may be located in a community. For example, dry cleaners, automobile repair shops, hospitals, exterminators, and photo processing centers all generate hazardous waste. Some hazardous waste generators are larger companies like chemical manufacturers, electroplating companies, and oil refineries.

Substandard Housing

In the four regions surveyed were a number of rundown areas of cities characterized by substandard housing and squalor and lacking in tenant security. The respondents indicated that 24 percent of urban dwellers in Rio de Janeiro and another 37 percent in São Paulo and Santa Catharina lived in slums. According to our survey respondents, the proportion of urban residents has increased from about 25 percent in 1990 to over 51 percent in 2010. The urban residents in the four regions under study living in slums are expected to decrease to about 6 to 3 percent in the near future. Although the proportion of urban residents has increased in Brasilia and Rio de Janeiro, a relatively larger growth in the number of urban residents is noticed in São Paulo and Santa Catharina. The respondents indicated that some of the factors that have led to urbanization are (1) industrialization, (2) food availability, (3) employment opportunities, (4) lifestyle consideration, and (5) escape from political conflict. The implication of rapid urban residential growth could be linked to numerous adverse health impacts, particularly for the poor. However, due to a rising population, especially in urban areas, the number of slum dwellers is increasing. Among the most important cause of morbidity and early death in urban areas in Brazil are environmentally related diseases and accidents (Ioris, 2010).

Land Use Planning

Eighty-five percent of the respondent in Brasilia and 92 percent in Rio de Janeiro indicated that adequate land planning in the country has prevented them from suffering from any medical problems. The federal, state, and local governments in Brazil have used and continue to use land planning as an important means of their social policy. Land planning is very important to ensure that land is used efficiently for the benefit of the wider economy and population, as well as to protect the environment. Architecture, urban design, urban planning, landscape architecture, and urban renewal usually address the selection of physical layout, the scale of development, aesthetics, costs of alternatives, and the selection of building materials and their impact on landscape and species.

Noise Pollution Control

Generally, respondents indicated that noise pollution is very common in São Paulo, Santa Catharina, and Rio de Janeiro. Eighty-three percent in Brasilia and 73 percent in Rio de Janeiro indicated that environmental health and development are intimately interconnected with Brazil's development. It is very interesting to note that 88 percent in Fortaleza indicated that they were affected by noise pollution. The number concerned about noise pollution in Fortaleza was higher than that of Rio de Janeiro, with its much larger population and more numerous industries. In Brazil, noise pollution refers to excessive, displeasing human-, animal-, or machine-created environmental noise that disrupts the activity or balance of human or animal life. The source of most outdoor noise in Brazil is mainly the construction and transportation systems, including motor vehicle noise, aircraft noise, and rail noise. On the one hand, poor urban planning in the old part of the cities and industrial areas may give rise to noise pollution because side-by-side industrial and residential buildings can result in noise pollution in the residential areas. On the other hand, indoor and outdoor noise pollution sources in Brazil include car alarms, emergency service sirens, mechanical equipment, fireworks, compressed air horns, grounds systems, electronic megaphones and loudspeaker equipment, barking dogs, appliances, lighting hum, and audio entertainment.

Occupational Health and Industrial Hygiene

It is interesting to note that 90 percent of the respondents in Brasilia, 93 percent in Ceara, and 92 percent in Rio de Janeiro indicated that they have never been injured on the job due to a mistake or requirement of the company they work for. An average of 15 percent of the respondents from the four regions under study indicated that they have received sick leave for the injuries that they might have sustained at work. In Brazil, occupational health and safety officers promote health and safety procedures in both public and private industrial organizations. They recognize hazards and measure health and safety risks, set suitable safety controls in place, and give recommendations on avoiding accidents to management and employees.

Vector Control

Vector-born infectious diseases in Brazil include yellow fever, malaria, dengue, among others, and their resurgence in the last few decades has been of concern, especially in the nation's tropical areas where these diseases occur more frequently (Gratz, 1999). One hundred percent of the respondent in Brasilia, Ceara, Rio Janeiro, and Fartaleza, respectively, indicated that poorly managed sewage treatment and poorly constructed and managed drainage systems have led to health problems in their respective regions. Vector control is any method to limit or eradicate the mammals, birds, and insects, such as mosquitoes, rodents, flies, cockroaches, or other arthropods, that transmit disease pathogens. Vector control in Brazil is extremely important. Because the impacts of disease and virus are devastating, the need to control their vectors is prioritized. Vector control in Brazil can have tremendous impacts because it increases mortality rates, especially among infants. In very large cities such as Rio de Janeiro, São Paulo, and Santa Catharina, due to the high movement of the population, the spread of disease is a greater concern for the government in effectively managing an outbreak. Respondents were concerned about the risk associated with the decentralization of public health programs and the lack of tradition and expertise in the treatment of these vector-born infections.

SUMMARY

This paper has examined environment health problems in Brazil. It contends that environmental health programs in Brazil seek how to increase benefits to the rural poor, shape the environmental impact of future land use trends, and mitigate the global impacts of climate change and biodiversity loss. The government of Brazil has also been promoting the incorporation of landscape-level planning into its planning and policies. To achieve these goals, the nation tends to support innovative partnerships with experienced implementing partners organized in three consortium structures. Such initiatives are organized predominantly by local organizations but include international organizations as well (Arceneaux, 2001 & 2009). Consortium activities also engage civil society in decision making for large-scale development projects.

In Brazil, environmental health issues tend to affect all human health–related aspects of both the natural environment and the built environment. In Brasilia, Ceara, Fortaleza, and Rio de Janeiro, environmental health concerns of the respondents include the lack of environmental policy implementation in a number of institutions. This research reveals that government agencies in Brazil often fail to implement environmental protection policies mainly because they lack political support, have insufficient resources, have underdeveloped institutional capacity, and tend to overlook the importance of cooperation at the local level. This study recommends that this trend can be reversed by decentralizing policy implementation into the hands of development-oriented agencies and nongovernmental organizations whose operation are focused on sustainable environmental issues. To make the process work, central authorities should offer incentives to ensure increased attention to environmental protection objectives in the major states and cities. At the same time, an independent body with oversight authority should be in place to prevent development agencies from neglecting environmental concerns.

According to recent estimates about 5–10 percent of disease-adjusted life-years lost in Brazil is due to environmental causes. By far, the most important factor is fine particulate-matter pollution in urban air (Dibie et al., 2012; Myers and Spoolman, 2014). Thus, urban areas need their own environmental policies that can take into account issues of air pollution, human development, and population density. Therefore, policies that are aimed at improving water quality at the point of usage are very important, but they need to be complemented with preventative policies gear toward safer household storage and treatment.

It is important to note that the foundation of sustainable development and environmental policy in Brazil must be biodiversity. The concerns for food production, energy security, and combating climate change all rely on a diverse array of natural resources. These national needs must be conserved, but they must also be developed in a responsible and sustainable way. Brazil is yet to develop an effective legal international framework toward this end. Thus, covering access to biodiversity and genetic resources has not been strategically mapped out to achieve these environmental goals. Brazil needs to find a middle ground on allocating the environmental benefits that accrue from the use of genetic resources found in areas such as the Amazon. These benefits must further be linked into existing political institutions and national sustainable development goals.

REFERENCES

Alberini, A., and Krupnick, A. (2000). "Cost-of-Illness and Willingness-to-Pay Estimates of the Benefits of Improved Air Quality: Evidence from Taiwan." *Land Economics*, *76*(1), 37–53.

Alonso, A. and Clemencon, R. (2010). "Environmentalism Brazil: Between Domestic Identity and Response to International Challenges." *The Journal of Environment and Development*, *19*, 247–265.

Arceneaux, C. (2009). "Latin America." In *The Other World: Issues and Politics of Developing World* (8th ed.), edited by Joseph Weatherby et al. New York: Longman.

Arceneaux, C. (2001). *Bounded Missions: Military Regimes and Democratization in the Southern Cone and Brazil.* University Park, PA: Penn State University Press.

Arceneaux, C., and Pion-Berlin, D. (2005). *Transforming Latin America: The International and Domestic Origins of Change.* Pittsburgh, PA: University of Pittsburg Press.

Balakrishnan, K., et al. (2002). "Daily Average Exposures to Respirable Particulate Matter from Combustion of Biomass Fuels in Rural Households of Southern India." *Environmental Health Perspectives*, *110*(11), 1069–1075.

Bascom, R., et al. (1996). "Health Effects of Outdoor Air Pollution." *American Journal of Respiratory Critical Care Medicine*, *153*, 477–498.

Bella, M. L., et al. (2005). "The Avoidable Health Effects of Air Pollution in Three Latin American Cities: Santiago, São Paulo, and Mexico City." *Environmental Research*, *100*(3), 431–440.

Binswanger, H. (1989). "Brazilian Policies That Encourage Deforestation in the Amazon." Environment Working Paper 16. E. Department. Washington, DC: World Bank.

Birdsall, N. and Wheeler, D. (1993). "Trade Policy and Industrial Pollution in Latin America: Where are the Pollution Havens?" *The Journal of Environment and Development*, 2(1), 137–149.

"Brazil Flood Survivors Left to Fend for Themselves." (2011). Associated Press (January, 15).

Cole, B., et al. (2004). "Prospects for Health Impact Assessment in United States: New and Improved Environmental Impact Assessment or Something Different." *Journal of Health Politics, Policy and Law*, *29*(6), 1153–1186.

Dias, J., et al. (2002). "The Impact of Chagas Disease Control in Latin America—A Review." *Memórias do Instituto Oswaldo Cruz*, *97*, 603–612.

Dibie, R., Mello, W., and Benitez, G. (2012). "Social Impact of Environmental Health Policy in Brazil." *Journal of International Politics and Development*, *9*(1), 1–24.

Drummond, J., and Barros-Patiau, A. F. (2006). "Brazilian Environmental Laws and Policies, 1934–2002: A Critical Overview." *Law and Policy*, *28*(1), 83–96.

Esrey, S. (1996). "Water, Waste and Well-Being: A Multi-Country Study." *American Journal of Epidemiology*, *143*(6), 608–623.

Ezzati, M., and Kammen, D. M. (2002). "The Health Impacts of Exposure to Indoor Air Pollution from Solid Fuels in Developing Countries: Knowledge, Gaps, and Data Needs." *Environmental Health Perspectives*, *110*(11), 1057–1068.

Fallon, F., and Zgodzingski, E. (2009). *Essentials of Public Health Management*. Sudbury, MA: Jones & Bartlett.

Friis, R. (2011). *Essentials of Environmental Health*. Sudbury, MA: Jones & Bartlett.

Gratz, N. G. (1999). "Emerging and Resurging Vector-Borne Disease." *Annual Review of Entomology*, *44*, 51–75.

Gubler, D. J. (1998). "Resurgent Vector-Borne Diseases as a Global Health Problem." *Emerging Infectious Diseases*, *4*(3).

Guilhem, D., and Avevdo, A. (2007). "Brazilian Public Policies for Reproductive Health: Family Planning, Abortion and Prenatal Care." *Journal of Compilation, 13*, 68–77.

Heller, L., et al. (2010). "Políticas Públicas de Saneamento. Por Onde Passan os Conflitos." In *Desenvolvimento e Conflictos Ambientales*, edited by A. Zhouri and C. Laschefki. Belo Horizonte, Brazil: Editora Ufmg.

Hudson, R. (Ed.). (1997). *Brazil: A Country Study*. Washington, DC: Government Printing Office for the Library of Congress.

Ioris, A. A. R. (2010). "The Political Nexus Between Water and Economics in Brazil: A Critique of Recent Policy Reforms." *Review of Radical Political Economics*, *42*(2), 231–250.

Jacob, P. (1997). "Environmental Problems in Sao Paulo, Brazil: The Challenge for Co-Responsibility and Innovation Crisis Management." *Journal of Contingencies and Crisis Management*, *5*(3), 131–137.

Katz, R. (1991). *Urban Air Pollution and Standards. The Case of Mexico City and Santiago*. Washington, DC: World Bank.

Kosmo, M. (1989). "Economic Incentives and Industrial Pollution in Developing Countries." Economic Development Workshop Paper. Washington, DC: World Bank, 25–98.

Leis, H. R. (2000). "A Politico-Communicative Model to Overcome the Impasse of the Current Political-Technical Model for Environmental Negotiation in Brazil." In *The Environment, Sustainable Development and Public Policies*, edited by C. Cavalcanti. Cheltenham and Camberley, UK: Edward Elgar.

Listorti, J. A., and Doumani, F. M. (2001). *Environmental Health: Bridging the Gaps*. Washington, DC: World Bank.

Luken, R. A., and Rompaey, F. V. (2007). *Environment and Industry in Developing Countries*. Cheltenham and Camberley, UK: Edward Elgar.

Mahar, D. (1988). "Government Policies and Deforestation in Brazil's Amazon Region." E. Department Environment Working Paper 7. Washington, DC: World Bank.

Maxwell, N. (2009). *Understanding Environmental Health*. Sudbury, MA: Jones & Bartlett.

McLaughlin, C. P., and McLaughlin, C. D. (2008). *Health Policy Analysis*. Sudbury, MA: Jones & Bartlett.

Mueller, C. C. (1995). "Environmental Problems Inherent to a Development Style: Degradation and Poverty in Brazil." *Environment and Urbanization*, *7*, 67.

Munasinghe, M. (2007). "Economy-Wide Policies and the Environment." *The Encyclopedia of Earth*. Retrieved from http://www.eoearth.org/.

Munasinghe, M., and Cruz, W. (1994). *Economywide Policies and the Environment: Lessons from Experience*. Washington, DC: World Bank.

Murphy, E. (2005). "Promoting Health Behavior." *Health Bulletin 2*. Washington, DC: Population Reference Bureau.

Musa, H. A., et al. (1999). "Water Quality and Public Health in Northern Sudan: A Study of Rural and Peri-Urban Communities." *Journal of Applied Microbiology*, *87*, 676–682.

Myers, N., and Spoolman, S. (2014). *Environmental Issues and Solutions*. Belmont, CA: Brooks/Cole.

Olivia, R., Gemal, A. L., Nobrega, A. W., and Araujos, A. C. (2003). "Pesticide Monitoring Program of the Ministry of Health of Brazil." *Food Additives and Contaminations*, *20*(8), 758–763.

Puppim de Oliveira, J. (2012). *Green Economy and Good Governance for Sustainable Development: Opportunities, Promises and Concerns*. New York, NY: United Nations University Press.

Puppim de Oliveira, J. (2008). *Implementation of Environmental Policies in Developing Countries: A Case of Protected Areas and Tourism in Brazil*. Albany, NY: State University of New York Press.

Renn, O. (2006). *Risk Governance: Towards an Integrative Approach*. Geneva: International Risk Governance Council.

Rom, W. (2012). *Environmental Policy and Public Health*. San Francisco, CA: Jossey-Bass.

Sachs, J. D., and Warner, A. M. (1995) *Economic Reform and the Process of Global Integration*. Brookings Papers on Economic Activity, Washington, D.C., 1–118.

Skolnik, R. (2008). *Essentials of Global Health*. Sudbury, MA: Jones & Bartlett.

Smith, K. R., et al. (2000). "Indoor Air Pollution in Developing Countries and Acute Lower Respiratory Infections in Children." *International Journal of Respiratory Medicine*, *55*(6), 518–532.

Tesh, S. N., and Paes-Machado, E. (2004). "Sewers, Garbage, and Environmentalism in Brazil." *The Journal of Environment & Development*, *13*, 42–72.

Teixeira, I. (2010). Brazil's Minister of the Environment Discussed the Challenges and Goals of Her Ministry at the Wilson Center (October 20). Retrieved from http://www.newsecuritybeat.org/2010/12/a-review-of-brazils-environmental. Accessed November 20, 2013.

Teitelbaum, J. B., and Wilenski, S. (2007). *Essentials of Health Policy and Law*. Sudbury, MA: Jones & Bartlett Learning.

Trinidade, S. C. (2000). "Agenda 21: A Sustainable Development Strategy Supported by Participatory Decision-Making Processes." In *The Environment, Sustainable Development and Public Policies*, edited by C. Cavalcanti. Cheltenham and Camberley, UK: Edward Elgar.

United Nations Human Development Index. (2010). New York: Oxford University Press.

Vanclay, F. (2008). "Social Impact Assessment: International Principles." Retrieved from http://www.iaia.org/members/Publications. Accessed December 17, 2012.

Weatherby, J., et al. (Eds.). (2009). *The Other World: Issues and Politics of Developing* World (8th ed.). New York: Longman.

Wheeler, D. (2001). "Racing to the Bottom? Foreign Investment and Air Pollution in Developing Countries." *The Journal of Environment & Development*, *10*(3), 225–245.

WHO. (2002). *Environment and Health in Developing Countries.* HaELP Series. Geneva, World Health Organization.

World Bank. (1992). *Development and the Environment.* New York: Oxford University Press.

Wright, R., and Boorse, D. (2014). *Environmental Science: Toward a Sustainable Future.* New York: Pearson.

12 Environmental Policies and Issues in Jamaica

Robert Dibie, PhD

INTRODUCTION

Jamaica is known for its rich heritage with regard to forest coverage and flowing streams, especially in the earliest years of its existence (Evelyn and Camirand, 2003; Ahmed, 1998). Most of Jamaica's natural vegetation has been lost after centuries of improper land use and is now suffering major environmental problems that include flooding, soil erosion, destruction of wildlife and wildlife habitat, and decreased surface flows in streams and rivers (Brown and Edwards, 2005; Evelyn and Camirand, 2003). The goal of ensuring a sustainable future in Jamaica becomes more difficult as the planet warms, rainfall patterns shift, and climate-related natural disasters become more frequent (Clifford and Edwards, 2012). Jamaica, like most developing nations, has invested in health education and infrastructure to prevent the escalation of extreme poverty (Government of Jamaica, 1987). However, continuing environmental degradation in the country tends to threaten the long-term sustainability of the entire nation's social and economic development (Sachs, 2005).

Tourism is Jamaica's largest source of income. The bauxite industry is the nation's second largest revenue earner—and a major environmental hazard because it is often a leading cause of deforestation (Evelyn and Camirand, 2003; Neufville, 2001). In its fifty years of operation, the bauxite industry has stripped 5,099 hectares of trees, including some 3,218 hectares of forest (ENACT, 2003). More than 30 percent of Jamaica is classified as forest, of which 64 percent is unprotected. Most of this forest is privately owned and crown (i.e., vacant/waste) land (Dodman, 2004; ENACT, 2003). Improper land use has resulted in the destruction of about two-thirds of the natural forests of Jamaica since the 1960s. Woodland, scrub forest, and dry limestone forests are a key component of Jamaica's ecology and economy (Ministry of Energy, Mining and Communication, 2006–2020; Ministry of Environment, 1998). Ninety-four percent of all Jamaica's forests are disturbed, and 20 percent of land within the forest reserves has been impacted by human activity (ENACT, 2003).

The Government of Jamaica, like several Caribbean countries, has enacted environmental policies over the past two decades (Gordon et al., 2010; Government of Jamaica, 1996 & 2001). A close analysis of these policies and reviews by several scholars reveals that the policies have not been very effective (Brown and Edwards, 2005; Dodman, 2004; Neufvile, 2001). Despite the nation's goals to pursue a sustainable development vision, its implementation strategies have not been sustainable in some key areas: (1) Diversification from fossil-fuel and exploration of renewable energy sources have not met with much success because there are still great demands for petroleum (Ministry of Mining, Energy and Commerce, 2009). (2) Sustained improvement in the area of conservation has proved to be difficult. (3) Only key stakeholders were asked to review and make comment on policy, resulting in a lack of public participation. (4) The public education campaign to promote energy conservation has been insufficient (some members of the population are unable to purchase the high-cost energy-saving equipment). (5) The environmental data used to inform policy and direct planning has gaps in it. (6) Though Jamaica has sufficient water to meet all the nation's demands, the supply of water to urban households and industrial users is often erratic, and the infrastructure to move the water to the areas where it is needed is inadequate in some parts of the country. Even though the country has made significant progress in providing water service to its people, the poorest 20 percent of the population has not benefited from these improvements (Dodman, 2004). One-third of the poorest households rely on standpipe for their water, 30 percent obtain their water from untreated sources such as rivers, and only 21 percent of the poorest households have flush toilets (*Jamaica Survey of Living Conditions*, 1997).

The source of water for a quarter of rural households is still from standpipes. Rural citizens travel far distances to fetch water, and 22 percent use rainwater tanks, which are usually untreated and hence unsafe for consumption (*Jamaica Survey of Living Conditions*, 1997). Additionally, septic tanks, pit latrines, and other types of on-site sanitation are sometimes improperly constructed and pose a threat to the health and quality of aquifers and surface water (Dodman, 2004).

The Water Resources Authority (WRA) has had responsibility for the regulation, control, and management of the nation's water resources since 1996 (Policy for the National System of Protected Areas, 1997). Despite some improvements in resource management and allocation, watershed management is often inadequate (Dodman, 2004). Integrated water resources planning is not always achieved in practice due to failures in institutional coordination. In many cases, water is being used inefficiently due to excessive agricultural irrigation. In most cases, the coordination of water use has negatively affected the agricultural sector of the nation.

The provision of water is capital intensive, and it is evident that the water sector is faced with the ongoing conflict between economic and environmental development for scarce budgetary allocation. The Ministry of Mining, Energy and Commerce of Jamaica (2009) reported that the main areas of concern are the deterioration and malfunction of municipal water supply. In addition, ineffective sewage treatment infrastructure, the management of industrial waste, and toxic waste often affect the nation's water resources. Further, rapidly increasing water demands in deficient areas has also had some adverse effects of deforestation. In most cases, the coordination and the use of water have caused another set of problems for agricultural irrigation.

Jamaica's municipal solid waste includes organic or compostable waste, paper or cardboard (newspaper, office paper) packaging materials, plastic (packaging, records, compact disks, toys), glass (bottles, scrap, mirrors), metal or tin (cans, scrap, nails, aluminum), wood or board (furniture, construction scrap), textiles (upholstery, clothes), and electronics e-waste (televisions, component sets, computers). Untreated municipal solid waste contributes to poor health standards, adds to greenhouse gas emissions, and is sometimes beset by spontaneous fires (Friis, 2013).

In the past decade, Jamaica's population has increased to about 2.7 million people, with an estimated 10,000 tons per week of solid waste, excluding commercial and industrial waste (Brady, 1991). Prior to 1993, it was pointed out that of the 26 sewage plants, 23 were not actually functioning (Greenfield, 1993). The impact of increased urbanization and public health concerns have been added to existing problems because, in some communities due to unscheduled pickups, people have disposed of garbage in gullies, drains, and rivers and on roadsides (Dodman, 2004). The same cry has been uttered almost a decade later by the Director of Enforcement and Compliance at the National Swage Water Management Authority (NSWMA), Ann Rodriquez (Wright, 1998), who stated that persons who dump garbage into drains "due to limited access to formal collection systems, constitute an irresponsible behavior." The location of some communities near drains and a lack of awareness of the associated dangers make for another type of environmental problem.

Funding in the public sector that is under economic strain has been an issue, where the lack of conventional solid waste disposal site equipment results in the improper and inadequate burial of solid waste. The country has no sanitary landfills, only small dumpsites, with a drastic reduction in their number over the last ten years primarily due to sanitation and environmental problems, which include leaching of toxic and hazardous substances into ground- and surface water because the disposal sites were unlined. Reportedly, workers and livestock, which rummage the garbage, have been infected. Waste sometimes involves medical disposals (Brown and Edwards, 2005; Dodman, 2004). Uncovered waste and uncontrolled burning due to build-up of methane gas and low-flash-point substances have given rise, in some cases, to an intense stench (Easton, 2012; Myers and Spoolman, 2014).

The chapter examines the social and economic impact of environmental issues in Jamaica. It uses primary and secondary data methods to discuss the variety of environmental challenges facing the country. It argues that there are close relationships among population, consumption, and technology in Jamaica's environmental issues. It further points out that, although environmental problems created by industrialization, heavy automobile traffic, large quantities of waste, and informal sector activities are now beginning to be addressed, there is still a lot the government needs to do with respect to adequate implementation. Despite the efforts of the government, various environmental problems continue to affect the social and economic affairs of the citizens. This chapter states that there is still no clear relationship between urban size, prosperity, and environmental problems. The concluding section recommends that national economic development should go hand in hand with environmental development. The chapter not only identifies the most pressing environmental problems in Jamaica, but it also shows the potential of how citizens, communities, and government can respond to them.

THE ECONOMIC VALUES OF ECOLOGY

One of the most interesting aspects of the study of environmental issues is that people differ in opinion on the seriousness of ecological problems and what should be done about them (Dibie and Bourke, 2011). A major component of environmental science is ecology, which is the biological science that studies how all organisms, or living things, interact with one another and with the environment (Miller and Spoolman, 2014). According to Paul Ehrlich and John Holden's (1970) connection model, the population size of a nation influences the consumption of resources per person. Further, technology affects the environmental impact of the populations in both developing and developed nations. Thus, while many people in affluent countries over consume, many poor people in developing countries suffer from not having enough resources (Baumer and van Horn, 2014; Miller and Spoolman, 2014). The model shows a very important relationship among population, consumption pattern, and technology in developed and developing nations in the world. Figure 12.1 shows the relationships among these three factors.

In Jamaica, the major environmental impact of population is observed in Kingston and Montego Bay. The population of Kingston of over 600,000 has led to the degradation of renewable resources because many people are involved in informal sector business to stay alive. With respect to the impact of technology in the country, the bauxite industry, a chemical factory, cement manufacturing, brewing, and petroleum refining in Jamaica have all been causing environmental problems (Dodman, 2004).

John Losey and Mace Vaughan (2012) point out that the natural system provides ecological services on which humans depend. Uncountable organisms

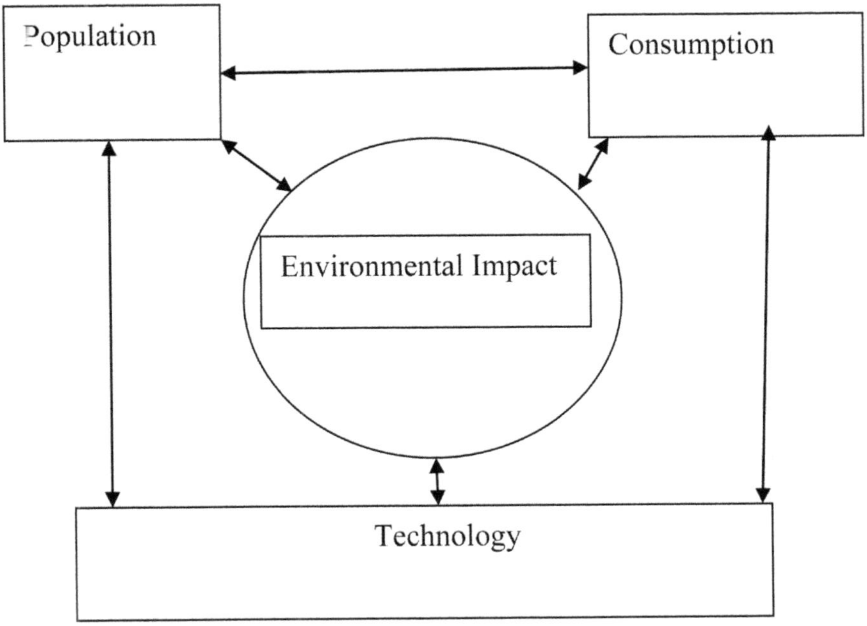

Figure 12.1 Environmental Impact Model. The model shows a very important relationship between population, consumption patterns, and technology in developed and developing nations.

are involved in the complex interactions the put food on the table and remove waste. They contend that, although human life could not persist without these services, it is difficult to assign them even an approximate economic value, which can lead to their being assigned a lower priority for funding or action than other needs for which values are more readily calculated (Losey and Vaughan, 2012). Thus, action that leads to substantial decreases in biodiversity should be avoided because of the high probability of a major disruption in essential services by countless organisms. This is interesting to know because although multicriteria analysis is currently being used for environmental impact assessment in many countries, the concept of cost-benefit analysis has again become fashionable, along with various pricing techniques associated with it. Other techniques associated with cost-benefit analysis include contingency valuation methods, hedonic prices, and costs of replacement of ecological services (Gatto and De Leo, 2012; Sachs, 2005). It is interesting to note that among these pricing techniques, the contingent valuation method is the only approach that is capable of providing an estimate of existence value and is the one biologists and environmentalists have mostly used recently.

According to Miller and Spoolman (2012) and Thomas Dye (2011), pollution cleanup and prevention are both needed in any country. However, environmental scientists and some economists will argue that nations should encourage their citizens to adopt prevention mechanisms because they work better and in the long run are cheaper than cleanup. They contend that making a shift toward sustainable living will require much greater reliance on pollution prevention to keep from upsetting the earth's chemical cycling process, in keeping with one of the three principles of sustainability (Kraft, 2011). In Jamaica, a national shift toward sustainability ultimately depends on the actions of individuals within the country (NREM, 2007).

Sustainability should begin with how and where the citizens in Jamaica live. Sustainability begins at the personal and local levels. The role of the Jamaican government and its National Environmental Protection Agency (NEPA, 2001 & 2006) is to encourage citizens with incentives to adopt sustainable lifestyles. Thus, the theoretical framework of the social and economic impact of environmental policies and issues has the potential to illuminate the complicated ways in which population, consumption, and technology interact and influence each other. The environmental impact approach reflects an awareness that understanding the dynamics of how the interaction occurs is crucial for the attainment of sustainable development in Jamaica.

RESEARCH METHODOLOGY

The research for this study was conducted by analyzing both primary and secondary data sources between 2009 and 2012. The purpose of the study was to find the social and economic impact of environmental issues and policies in Jamaica. The research method includes conducting secondary and primary research.

Secondary research includes conducting library research and a review of government documents and reports about environmental agencies, ecological assessments reports, environmental NGOs, and their subcontractors' reports in Jamaica. The secondary data analysis offers an important starting point for an investigation into environmental problems in the country. Empirical evidence of the economic and social impact of environmental policies and issues previously documented by the Government of Jamaica was analyzed.

The primary research involves the division of Jamaica into four regions:

1. Kingston to Port Maria–Port Antonio to Morant Bay
2. Spanish Town to St. Ann's Bay
3. Falmouth to Mandeville–May Pen
4. Montego Bay to Black River–Savanna, La Mar to Negril

Cities were randomly selected from a list of towns in each of the four regions.

Questionnaires were administered to at least 2,200 respondents from the selected cities. The respondents returned 1,520 completed questionnaires, for a 67 percent response rate. Interviews were also conducted (interviewees included farmers, agricultural ministry officials, managers of environmental agencies, NGOs, and other citizens).

One thousand five hundred interviews were conducted. The interviews lasted approximately thirty minutes each. The author also interviewed some staff of the National Environmental Protection Agency (NEPA) of Jamaica. While the staff of NEPA served as an important information source for the impact of environment policies in the country, they often did not show a desire to be transparent. Some of them were not helpful, and others were biased. However, it is not unusual to find those committed to environmental policy implementation to be unwilling to reveal its flaws. Consequently, the author also interviewed fifty staff members of environmental NGOs across Jamaica. While the two sets of in-depth interviews did not definitively identify the most critical deficiencies in the environmental efforts in Jamaica, they offer a systematic exploration of the social and economic impact of environmental policies contributing to the sustainable development process of the nation.

Information on waste management practices outside the so-called transaction sector was gathered through questionnaires given to respondents drawn from various Jamaican regional cities and towns. Only a few respondents returned their questionnaires. The sample of respondents was derived from the list of government regulatory agencies and NGOs operating in Kingston. The pooling of data from a multitude of independent samples was achieved within the multiple comparatives framework. Here meta-analysis was employed to pool common statistics addressing similar environmental research issues in solid waste management, agricultural issues, air quality control, water quality, energy policy issues, and biodiversity problems. More emphasis was put on effect size analysis where summary measures such as proportion were subjected to such tests as chi-square tests and the normal economic and social impacts.

The feedback of both the in-depth interviews and questionnaires was organized into an analysis matrix to allow us to summarize and develop typologies and descriptive statistics to link our findings to our research goals. An SPSS program was used to analyze the data. Comments from interviews were quantified using percentages. The validity of the research results includes comparing the questionnaire and interview outcomes.

The author evaluated the impact of environmental issues on the array of climate change and relevant environmental policies identified for adoption and justified and operationalized each of the impacts. This chapter reports the survey findings. The survey outcome area closely relates to the various social and economic impacts of environmental issues in Jamaica.

ANALYSIS AND DISCUSSION OF FINDINGS

As previously discussed, in this chapter, the most common economic and social impacts of environmental issues in Jamaica are the lack of knowledgeability and the ineffective implementation of a number of environmental policies.

This research supports each of these criticisms of the way the National Environmental Protection Agency (NEPA) staff has been enforcing policies. However, the author also found that these environmental policy deficiencies are far from an adequate explanation of why so many of the citizens have suffered some environmental consequences. The analysis that follows provides information on the nature of the economic and social impacts of environmental issues and policies in Jamaica.

Eighty-six percent of the respondents contend that environmental problems in the country were caused by the deviant behavior of citizens. It is interesting to note that 75 percent of the respondents indicated that structural issues, the pattern of governance, and governmental irresponsibility are some the causes of inadequate implementation of environmental policies.

Figure 12.2 Human Waste Dumped in the Gullies in Jamaica
Source: Photo by author during field research in Jamaica 2011.

Sixty-six percent indicated that the lack of environmental protection education was the cause of deviant behavior by the citizens, mostly outside Kingston, the capital city. This group also acknowledges that poor housing conditions were a major cause of sanitation and waste management problems.

While the informal sectors tend to provide employment and other positive economic impacts for citizens, 73 percent of the respondents indicated that the sector members often practice unsafe procedures and use unsuitable or dangerous equipment that is harmful to the neighboring citizens and environment. They stated that illegal gas stations, the improper disposal of waste engine oil, the lack of facilities for discarding car batteries, and the generation of thick smoke often cause health problems. Table 12.1 is derived from the survey conducted between 2009 and 2012.

Table 12.1 Social and Economic Impact of Environmental Issues

Variables	Social and Economic Impact	Yes (%)	No (%)
Agriculture	Farmers in the Spanish Town and Mandeville region were mostly unemployed due to pollution of their farmland and rivers. The rate of unemployment was generally high in the region. Population is growing faster than agricultural food supplies, making price of food to increase.	87	13
Air pollution	Major pollution problem due to untreated or partially treated liquid effluents from oil and petrochemical industries, breweries and alcohol distilleries, sugar refineries, and canning and packaging plants. Pesticide, fertilizer, and herbicide runoff from farms also pollute the land and rivers around Kingston, Port Maria, Morant Bay, and St. Ann's Bay.	89	11
Water pollution	Major pollution problem due to untreated or partially treated liquid effluents from oil and petrochemical industries, breweries and alcohol distilleries, sugar refineries, and canning and packaging plants. Pesticide, fertilizer, and herbicide runoff from farms also pollute the land and rivers around Kingston, Port Maria, Morant Bay, and St. Ann's.	67	33
Waste management	There is a concern about the solid waste coming from industry and urban sources comprising plastic, metal, glass, paper, vegetable, and food waste. Pollution from the bauxite industry, oil from the informal sector, and petrochemical installations make up the major problem.	69	31

(*Continued*)

Table 12.1 (Continued)

Variables	Social and Economic Impact	Yes (%)	No (%)
Deforestation	There are complaints about deforestation resulting from logging, land clearing for agriculture, quarrying, and property development along hillsides. The deforestation problem has also worsened erosion.	71	29
Natural hazard	In Jamaica the average annual rainfall ranges from 2,880 mm in the west to 1,200 mm in the east. Most of the respondents indicated that they have been affected by flood due to excessive rainfall and hurricanes. Jamaica is known for having serious hurricanes due to its geographical location. Hurricanes often result in severe health, social, environmental, and economic impacts on the citizens.	66 west coast 82 east coast	34 west 18 east
Environmental health issues	Dust, smoke, and other bad environmental practices often cause respiratory and intestinal problems, as well as outbreaks of gastroenteritis caused by poor sanitation among children in western and eastern parts of the country. The respondents indicated that a higher overall temperature often increased the range of major disease vectors, such as mosquitos.	96	4

Source: Derived from survey (2009–2012).

It is important to note that Jamaica's soil is naturally high in lead. This means that the presence of additional lead from discarded batteries can cause extremely high soil and water levels of the element (Dodman, 2004).

Table 12.2 shows the various environmental problems identified by respondents and the nature of the social and economic impact in some cities and towns in Jamaica. The evidence in the table mirrors the social and economic impacts of environmental issues observed by the respondents all over the country.

The nature of urban planning and urbanization in Jamaica has been associated with crime and violence. Fifty-seven percent of the respondents indicated that the conditions of the urban environment and poor sanitation are just some of the factors working against the health of the urban poor in the country. They contend that these factors had a positive correlation with poverty, unemployment, limited education, bad diet, and lack of access to natural resources. For one thing, the respondents argued that the problem of unemployment was not isolated from crime and violence. For another,

Table 12.2 Economic and Social Impact of Environmental Issues

Environmental Problems	Nature of Social and Economic Impact	Affected Cities/Towns
Sanitation	Siting low-income housing with septic tanks and absorption pit systems in areas where the water table is high or that are prone to flooding	Kingston, Manchester, Montego Bay, Plymouth
Atmospheric change	Building housing developments in areas that will not be able to receive curbside garbage collection services due to hilly terrain and other factors without an appropriate solidwaste management plan for the community	Kingston, Mandeville, Montego Bay
Waste management	There are concerns about the length of time taken to collect garbage, transportation of waste, and the lack of adequate landfills and holding stations. The inadequate handling of waste has caused foul odors in some streets as well as vermin and lots of flies.	Kingston, Montego Bay and Ocho- Orio, Lucea, Port Maria, Mandeville, Spanish Town, Negril, Falmouth
Dumps in gullies	There are complaints about frequent dumping of waste in the gullies. This practice often creates a habitat for disease vectors and is a cause of blocking the channels. Dangerous flooding often occurs after heavy rainfalls. Rubbish is often piled up in the gullies. Sewage often flows several feet beside residential homes.	Plymouth, Greenwich Town, Montego Bay,
Burning of refuse	Burning of refuse in the rural areas has been identified as a major contributor to poor air quality. The spillover of the smoke to urban areas often results in deteriorated air quality in the urban population centers.	Ochio Rio, St. Ann Bay, Black River, Morant Bay, Port Antonio, Savanna La Mar
Informal sector	Waste from the informal sector located near residential homes often causes air pollution and, in turn, serious environmental health threats. Informal sector activities include smelting scrap metal and illegal gas stations and mechanics garages.	Spanish Town, Kingston, Black River, Montego Bay, Mandeville, Port Antonio, Negril
Natural hazard	Jamaica is known for having serious hurricanes due to its geographical location. Hurricanes often result in severe health, social, environmental, and economic impacts on the citizens.	All over the country

Source: Derived from survey (2009–2012).

the lack education was identified as key to several environmental problems in the society.

The secondary data that were analyzed reveal that Jamaica's limestone hosts deposits of bauxite and igneous rock containing occurrences of copper and gold (Rabacheysky, 1994). The Ministry of Mining, Energy and Commerce of Jamaica (2009) indicated that one half of the nation's land mass is used for bauxite mining and alumina production. Limestone is used domestically in the production of lime, road construction, and for other purposes (Ministry of Mining, Energy and Commerce of Jamaica, 2009). The United States is Jamaica's major market for bauxite alumina production up to the current decade. The nation also traded with Trinidad and Tobago, the United Kingdom, and Canada. Apart from tourism, bauxite and alumina are the nation's second leading income. This accounts for about 22% of the nation's annual gross domestic product (GDP). It is also reported that Jamaica is the third largest producer of bauxite ore after Guinea and Australia (Ministry of Mining, Energy and Commerce of Jamaica, 2009). According to Rabacheysky (1994), the Jamaican bauxite industry is covered by a number of environmental and mining laws dating back to

Table 12.3 Jamaican Production of Mineral Products
(Metric tons unless otherwise specified)

Commodity	1992	1993	1994
Aluminum Bauxite, dry equivalent, gross weight thousand tons	11,300	11,400	11,800
Alumina	2,920	2,990	3,180
Cement, hydraulic	481	451	400
Gypsum	145,000	152,000	152,000
Iron and steel: Steel, crude	25,000	25,000	24,500
Lead, refined (secondary)	1,000	800	800
Lime	179,000	151,000	151,000
Petroleum refinery products thousand 42-gallon barrels	7,000	7,000	7,500
Salt	14,000	13,500	13,500
Silica sand	15,600	21,300	21,300
Limestone thousand tons	5,480	4,300	4,500
Marble, cut and/or polished	4,620	5,780	6,000
Sand and gravel	1,350	1,550	1,600

Source: Rabchevsky, G. A. (1994).

Table 12.4 Jamaican Structure of Mineral Industry
(Thousand metric tons unless otherwise specified)

Commodity	Major Operating Companies and Major Equity Owners	Location of Main Facilities	Annual Capacity
Alumina	Jamalco Aluminum Co. of America/ Government, 50% Clarendon Alumina Production Ltd., 50%)	Halse Hall plant at Clarendon Clarendon Parish	750
Alumina	Alumina Partners of Jamaica (Alpart) Nain, St. Elizabeth Parish 1,450 (Kaiser Aluminum Corp., 65%; Hydro Aluminum Jamaica AS, 35%)	Nain, St. Elizabeth Parish	1,450
Alumina	Alcan Jamaica Co. (JAMALCAN) Kirkvine, Manchester Parish and 548 (Alcan Aluminum Ltd. Canada, 93%; Government, 7%)	Kirkvine, Manchester Parish and Ewarton, St. Catherine Parish	548 548
Bauxite	JAMALCAN Bauxite Mines	St. Ann and St. Catherine Parishes	1,600
Bauxite	Jamaica Bauxite Mining Ltd. Lydford Mines	Lydford, St. Ann Parish	2,500
Bauxite	Jamalco Bauxite Mines Mocho Mountains,	Clarendon Parish	2,500
Bauxite	Kaiser Jamaica Bauxite Co., Ltd.	Kirkvine, Manchester Parish; Water Valley, Discovery Bay, St. Ann Parish	4,500
Cement	Cement Caribbean Cement Co. Ltd.	Rockfort. St. Andrews Parish	1,700
Petroleum products thousand 42-gallon barrels	Petrojam Ltd. (Petroleum Corp. of Jamaica, 100%)	Kingston, St. Andrews Parish	9,000

Source: Rabchevsky, G. A. (1994).

1947. In addition, the nation has enacted new environmental laws since then (Ministry of Mining, Energy and Commerce of Jamaica, 2009; Rabchevsky, 1994).

Despite several environmental laws that have been enacted in Jamaica, pollution from the bauxite industry is still a major issue to the citizens and government. Dust and suspended particulates, acid rain, and sodium pollution is a major health issue in Jamaica. Pollution levels in ground water is still very high. There have been consistent oil spills from refineries. Further, sulfur dioxide emission and oil burning in power plants as well as alumina calcining plants also contribute to air pollution in Jamaica (Ramon, 2009; Ministry of Local Government and Community Development, 2001). Over the years there has been a steadily growing mass of evidence indicating that the air quality has had a measurable impact on residents' health. Results obtained from a number of studies now confirm long-suspected fears that air pollution kills, even at levels of pollutant concentration well within the legally allowable limits established by the National Environmental Protection Agency and Ministry of Health (Friis, 2013; Maxwell 2014). Epidemiologic studies previously conducted convincingly show a link between high levels of particulate air pollution and death due to lung cancer, respiratory disease, and heart ailment. According to Nancy Maxwell (2014), the positive correlation between pollution levels and increased risk of premature death was very significant only in relation to concentrations of fine, inhalable particles, varying in levels of total suspension particulates. It is also very interesting to note that the costs of income security and health care programs have risen so rapidly that their solvency is threatened, yet benefits for poor people are barely adequate for their survival, and the health status of Jamaicans has not improved significantly.

STRATEGIES FOR IMPROVEMENT OF THE PRESENT ENVIRONMENTAL PROBLEMS

The purpose of this section is to outline the key strategies that could positively impact the natural environment in Jamaica. Identifying environmental issues and risks will give the nation's responsible public administrators the opportunities to respond with appropriate government policies for the promotion of sustainability. For many countries, including Jamaica, the collective participation of employees in making sustainable decisions can result in ecological protection and at the same time contribute to finding positive solutions to questions about the use of natural resources and the well-being of the nation's society (Ferrell et al., 2013; Harris, 1992).

For the protected areas in Jamaica to be maintained, funds need to be allocated for the upkeep and upgrading of the affected lands. These funds must be made available to the local authorities responsible for these protected areas. Pollutants, such as nitrogen in sewage, fertilizers, and other substances, often destroy protected areas (Dye, 2011). There is a need for the close monitoring of the various municipalities' treatment, collection, and sewage sites in order to prevent overflows. The adoption of such measures should be based on new information about changing patterns of temperature, precipitation, and other urban planning variables.

There is the need to adopt alternative measures to effectively address the environmental health problems associated with contaminated water. Adequate provisions need to be put in place for sanitation, drainage, and garbage collection. Relevant authorities must also ensure that passenger ships comply with the rules regarding waste disposal. Not only are companies responsible for destroying protected areas; individuals are as well. Simple factors such as the overuse of fertilizers on vegetables or lawns all contribute to the degradation of the land. Several projects have been launched related to the replanting of mangroves. These projects must be assigned to specific persons in order for them to be kept going. The authorities also need to ensure that the Fishing Industry Act is upheld.

Major projects need to be implemented to clean up hazardous waste sites. The question is: Are there enough municipal landfills open to receive waste? There needs to be an increase in the amount of waste disposal sites (Sacha-Renée, 2006). There needs to be a discontinuation of the removal of hollow trees, dead standing trees, rotting logs, and underbrush from parks just for the sake of appearances when indeed some of these objects are nesting sites for many exotic birds, bats, and bears.

Insufficient funding is the main impediment of the implementation of the forest plan (which includes the Forest Policy 2001) since its adoption in July 2001. There is clearly a need for new incentives to encourage private sector participation. The participation of the private sector is further limited by the lack of information about the financial returns of growing trees (Alba, 2008). The way forward to improve the environmental conditions facing Jamaica, with regard to the Forest Policy 2001, involves compliance with the strategies that were adopted. These include increasing environmental awareness among the Jamaican population, which will inevitably increase public participation in working toward the ultimate goal of sustainable development. The elements of shared governance are needed, whereby people are urged to participate in the decision-making and implementation processes. The public and private sectors, along with NGOs, must collaborate and bear responsibility for the Jamaican environment (Dibie, 2009).

There is also the need to identify and develop the management skills necessary for addressing environmental policy problems. Environmental managerial skills are necessary for attracting, absorbing, and managing financial, human, and information resources and for operating programs effectively, including evaluating program outcomes to guide future activities (Dibie, 2009; Vig and Kraft, 2013). The acquisition of appropriate management skills will enhance the nation's agricultural productivity and water management needs, which are required to feed the increased population while protecting already stressed ecosystems. Further, long-term, large-scale. Integrated management will help meet increased demands on natural resources while conserving biodiversity and maintaining terrestrial carbon stocks. The challenge of developing appropriate management skills, however, is not only to increase access to universities, but also to enable the University of the West Indies, Mona Campus Kingston, and the University of Technology, Kingston to function as engines of development for Jamaica.

Some of the functions of the regulatory agencies in Jamaica need to be decentralized. Decentralization could enhance the effectiveness of development activities by making them more relevant and responsive to local needs (Geohegan and Bennett, 2003). This is very important for greater flexibility in their implementation, and it provides a means for coordinating the various agencies involved at the regional or local level. Decentralization in this sense will encourage the maximization of public welfare benefits (Dibie, 2009; Kraft, 2011).

There is an immediate need for social transformation in Jamaica in terms of all policy issues, not just as it relates directly to the environmental problems, because the decision-making process of other policies are closely intertwined with social and economic lifestyle (Kraft and Furlong, 2013). The transition to modern economic growth in Jamaica will involve urbanization, changing gender roles, increased social mobility, changing family structure, and increase awareness of environmental policies (Sachs, 2005). Thus, policies enacted in these areas could affect environmental issues and the efforts to conserve and manage natural resources. This is especially important in the quest to achieve sustainable development where there needs to be a balancing act between the products and services provided by agriculture and its conservation and management. Empowered by self-organizations, public opinion can help to hold public servants to acceptable standards of behavior (Dibie and Bourke, 2011; Nadakavukaren, 2011).

It is clear that the driving force in the transformation process is having appropriate environmental policies. There are also elements of environmental concerns that are geared toward the financial impact of soaring oil prices. In many parts of the world, as in Jamaica, regulatory enforcement is weak in these situations; the true driving force for pollution prevention is the economic incentive (Cheremisinoff, 2001). While it is not that there is anything intrinsically wrong with the financial origin of environmental policy, the lack of incentive prevents policy makers from exploring the best alternative sources of energy and conservation rather than the cheapest.

It has also been observed that the availability of cheap oil and gasoline discourages conservation and investments in efficiency, in addition to making the development of alternatives, including domestic sources of oil, more difficult (Cheremisinoff, 2001; Friis, 2013). The recommendations are for the full adoption of the Kyoto Protocol of 1997. Improved research in areas related to best alternative technology (BAT) in sustainable energy resources is also recommended for Jamaica. Qualitative as well as quantitative data need to be gathered in order to fill the country's data gaps on existing trends and issues in energy practices and on the impact of economic conditions on these practices.

The government needs to review its existing policy on incentives with a view to encourage sustainable energy use (Maxwell, 2014), for example, citizens who use solar energy to supply some percentage of their electricity rather than taking it from the public grid. Industries should be encouraged to utilize energy conservation through recycling or recovery processes in their plant design. Incentives for the importation of hybrid vehicles should be encouraged. Individuals or companies that utilize biogas technology or that voluntarily invest in total quality environmental prevention (TQEP) should also be given tax breaks as incentives (Cheremisinoff, 2001; Maxwell 2014). A policy to technically analyze the impact of energy policy—for example, qualitative statistics on targets and results—should be developed and commissioned.

The circulation and subsequent evaluation of the policy stakeholders needs to be extended beyond those directly involved in energy production and sale. For example, environmental groups, farmers, NGOs, students in environmental studies, and students in primary and secondary education institutions should be exposed to practical experiences with managing the ecosystem of Jamaica. Various interest groups in Jamaica should be encouraged to participate in the environmental policy evaluation process. According to Gatto and De Leo (2012), the breadth of goals embraced by environmental impact assessment is much wider than that of cost-benefit analysis. On the one hand, environmental impact assessment provides a conceptual framework and formal procedures for comparing different alternatives to a proposed project. Environmental impact assessment could also foster interdisciplinary team formulation to investigate all possible environmental, social, and economic consequences of a proposed activity (Kraft and Furlong, 2013). On the other hand, these factors explicitly reflect the relative importance of the different environmental, social, and economic compartments and indicators (Gatto and De Leo, 2012; Nadakavukaren, 2011).

Strengthening international collaboration in tracking energy issues will also help Jamaica to address some of its environmental problems. For example, collaboration with NGOs and the World Bank could enhance the sponsorship and technical support for renewable energy. Because of the volume of goods and services procured each year, the Government of Jamaica can effectively stimulate the supply of goods and services that are environmentally friendly, thereby assisting in the achievement of national sustainable development objectives. This initiative will challenge manufacturers and suppliers to implement green production and service delivery techniques. It will also challenge them to increase their own competitive advantage, both locally and overseas. Following are some policy options to explore in solving the environmental problems in Jamaica:

1. Implement the Protected Areas System Master Plan.
2. Share decision-making authority and power with resource users. For example, develop comanagement systems within the communities. These committees would issue permits on behalf of NEPA and monitor and collect fines. These committees would be legal entities, but citizens would be at the helm. Village justice is well-known in Jamaica. The shame that a family bears when a member does wrong is still very strong in rural communities. This would assist in maintaining the desired attitude within the community. This would free NEPA so that its more expensive resources could be used more efficiently.
3. Increase and strengthen the management capacity of NEPA, so that they will be able to deal with new protected areas. There is at present no law protecting plants. The CITES convention that Jamaica signed is used to prevent endemics from leaving Jamaica. No policy exists to prevent persons from removing these plants from their habitat.
4. There has been limited success in developing financial mechanisms and a centralized leadership for a protected areas system. Several international commitments to environmental conservation have been agreed to. No additional allocations or plans to meet the obligations have been made.
5. The Government of Jamaica should conduct research to assist in sustainable development to build strong protected areas. This means that planning and guidance for the existing protected areas are based not on assumptions but rather on scientific processes.
6. A number of organizations are doing similar activities. The management of environmental policies could be better served with a little more coordination and clarity on roles so that there is no duplication of effort.
7. Economic issues and crime reduction have taken and will continue to take center stage of the Government of Jamaica's short- to midterm priorities. Consequently, the protected areas system will be unlikely to receive any additional funding for protected areas management in the near future. However, this situation provides an opportunity to determine the economic value of protected areas to the country, to capitalize on the promotion of tourism, and to place protected areas at the forefront of the Government of Jamaica's priorities.
8. Financial survival and sustainability are ranked among the highest priorities. Achieving financial sustainability has no easy solution but can be assisted by the development of financial mechanisms and better allocation and coordination of responsibilities. At present, there is little coordinated fund-raising and almost no revenue generation within the protected areas system. A huge opportunity now exists

for a more business-like approach, which has been successfully implemented by other countries.
9. Create an enabling environment for the private sector to take the lead role in the development of biodiversity and sustainable development programs.

It is very important to note that while there is unequivocal evidence that the climate is changing, there is a lot of uncertainty regarding the pace and extent of the change and the impacts on Jamaica. This uncertainty makes environmental policy decision making in the country more complex. It also magnifies the need for Jamaica to build its knowledge and analytical base and to strengthen the capacity of the government and business institutions in the areas of water resources monitoring, risk management, planning and coordination, weather forecasting, land use information, and disaster preparedness.

The continuing growth in the population of Jamaica and the increase in the demand for resources will make it difficult to propose achieving sustainable development in the country in the near future. Dibie and Bourke (2011) and Dodman (2004) contend that it can be done, but others think that for sustainable development to work, either population or resource demand must be reduced (Kraft, 2011; Miller and Spoolman, 2014). Generally, sustainability is assumed to promote two controversial ideas. The first is that the country's environmental problems can be solved only by radically restructuring how people live, work, and consume in Jamaica. The second is that achieving sustainability will require drastic reductions in the current levels of economic and political inequality between social groups and country (Baland et al., 2006; Clifford and Edwards, 2012). According to Clifford and Edwards (2006), given the wide gap between the ideals of sustainability and the actual systems of production and consumption, sustainability has become one of the most significant guiding principles of emerging efforts to achieve environmental protection at the international level, including in Jamaica (Kraft and Furlong, 2013; Vig and Kraft, 2013). The point is that Jamaica cannot continue its current production and consumption practices indefinitely. The nation has to minimize its current practices that degrade the environment, consume nonrenewable resources, and threaten its ecosystems.

SUMMARY

This chapter examined the social impact of environmental policies and issues in Jamaica. It identified the changes that may have occurred due to the enactment of various environmental policies. The chapter pointed out that the environmental problems in Jamaica are caused mostly by the relationships among population, consumption, and technology. Other likely causes of environmental problems include natural hazards, the informal sector pollution practices, and industry. Suggestions were made for possible solutions to improve the environmental problems facing the nation. However, it must be pointed out that these strategies are not new to the government agencies. The overall environmental problem is that of implementation, proper monitoring, and the adoption of appropriate policies. Solving these will allow for the achievement of policy objectives, the monitoring of managerial performance, transparency, and accountability. This could also allow for making adjustments to targets in order to respond to shortfalls, over-achievements, and emerging issues (Forestry Department, 2009–2013).

The chapter also indicated that the Government of Jamaica should encourage its citizens to adopt the principles of sustainability. It argues that prevention is better than cleanup. There is strong government support for environmental concerns in Jamaica, as is evident in numerous policies, including the energy policy. While strides have been made in the achievement of the objectives of the energy policy, there have been setbacks in sustaining alternative renewable energy sources. For example, the wind farm in Manchester is out of commission due to technical difficulties. The incentive to individuals using less than 100 KW/h per month is an example of an economic incentive for citizens (Dibie, 2010). According to Dibie (2010), incentives could encourage conservation and investment in the efficient use of the ecosystem in Jamaica. Incentives could be supplemented with strong programs to educate the public on the importance of environmental policies and how sustainability relates to the citizens' health and economic benefits. The benefits derived from the implementation of environmental policies mean a better standard of living.

Four forces, however, run counter to the protection of Jamaica's environment: (1) the continued demand for increased standards of living, (2) the growth and expansion of multinational corporations in the country, (3) the increased expansion of the free trade movement, and (4) urbanization. Using research to identify these factors will determine whether the nation can achieve sustainable development by solving social and economic issues. The prospects of introducing a modified energy policy are tied to patterns of consumption or conservation in Jamaica. There is the need for more emphasis on public awareness and for debate on the policy with a view to arriving at alternatives that are acceptable to the citizens. Finally, Jamaica must foster economic systems that spread the benefits of science, technology, and the division of labor to all parts of the country and to all its people.

REFERENCES

Ahmed, B. (1998). "Role of the Environment in the Sustainable Development of the Caribbean." *Social and Economic Studies, 47*(4), 83–97.

Alba, A. (2008). National Forestry Plan (NFP) Update. Information as of 2004, 1–9.

Baland, J. M., Bardhan, P., and Bowles, S. (2006). *Inequality, Cooperation, and Environmental Sustainability*. Princeton, NJ: Princeton University Press.

Baumer, D., and Van Horn, C. (2014). *Politics and Public Policy: Strategic Actors and Policy Domains*. Los Angeles, CA: Sage.

Brady, L. (1991). "Landfill Site Rationalization. Project Final Report. Kingston, Jamaica: PAHO/WHO." *Solid Waste Management: Critical Issues for Developing Countries*, edited by Elizabeth Thomas-Hope. Geneva: WHO.

Brown, C., and Edwards, S. (2005). *Situation Analysis of Jamaica's Protected Areas System Plan*. Washington, DC: Center for Park Management.

Cheremisinoff, N. P. (2001). *Handbook of Pollution Prevention Practices*. New York: Marcel Dekker.

Clifford, M., and Edwards, T. (2012). *Environmental Crime*. Burlington, MA: Jones & Bartlett.

Dibie, R., and Bourke, M. (2011). "Economic and Social Impacts of Environmental Issues in The United States." *Journal of International Politics and Development*, *9*(2), 33–46.

Dibie, R. (2010). "Social Impact of Environmental Policies in Ethiopia." *Journal of International Politics and Development, 8*(1), 1–23.

Dibie, R. (2009). "Environmental Governance and the Political Economy of Sustainable Development in Sub-Saharan Africa." *Journal of International Politics and Development, 7*(1), 1–23.

Dodman, D. (2004). "Community Perspective on Urban Environmental Problems in Kingston, Jamaica." *Social and Economic Studies*, *53*(3), 31–38.

Dye, T. (2011). *Understanding Public Policy* (13th ed.). New York: Longman.

Easton, T. (Ed.). (2012). *Taking Sides on Environmental Issues*. New York: McGraw-Hill.

Ehrlich P. R., Ehrlich, A. H., and Holden, J. P. *Human Ecology. Problems and Solutions*. 304 Seiten, 8, 79 Abb. 26. Cited in G. T. Miller and S. E. Spoolman. (2010). *Environmental Science*. Belmont, CA: Brooks & Cole, 23–34.

ENACT. (2003). "Policy on Strategic Environmental Assessment (Draft)." MoLE Office of the Cabinet. Kingston, Jamaica: Jamaica Printing Services.

Evelyn, O. B., and Camirand, R. (2003). "Forest Cover and Deforestation in Jamaica: An Analysis of Forest Cover Estimates over Time." *International Forestry Review*, *5*(4), 354–363.

Ferrell, O. C., Faedrich, J., and Ferrell, L. (2013). *Business Ethics*. Mason, OH: South-Western Cengage Learning.

Forestry Department. (2009–2013). "Strategic Forest Management Plan (SFMP)." Kingston, Jamaica: Jamaica Printing Services.

Friis, R. (2013) *Essentials of Environmental Health*. Boston: Jones & Bartlett.

Gatto M., and De Leo, G. (2012). "Pricing Biodiversity and Ecosystem Services: The Never-Ending Story." *Taking Sides on Environmental Issues,* edited by Thomas Easton. New York: McGraw-Hill, 51–53.

Geohegan, T., and Bennett, N. (2003). "Local Forest Management Committees: A New Approach to Forest Management in Jamaica." Caribbean Natural Resources Institute.

Gordon, L., Devon Payne-Sturges, D., and Gee, G. (2010). "Environmental Health Disparities: Select Case Studies Related to Asian and Pacific Islander Americans." *Environmental Justice*, *3*(1): 21–26.

Government of Jamaica. (2001). Forestry Policy. Forestry Department. Kingston, Jamaica: Jamaica Printing Services.

Government of Jamaica. (1996). Forest Act (1996). Forestry Department. Kingston, Jamaica: Jamaica Printing Services.

Government of Jamaica. (1987). "Jamaica: Country Environmental Profile." *International Institute for Environment and Development,* (3), 13–15.

Harris, J. M. (1992). "Global Institutions for Sustainable Development." In *Sustainable Development and Environmental Policy*, edited by F. Dietz et al. Orange Park, FL: Sigma Press, 37–51.

Jamaica Survey of Living Conditions. (2004). Statistical Institute of Jamaica; Jamaica Printing Services.

Kraft, M. (2011). *Environmental Policy and Politics*. New York: Longman.
Kraft, M., and Furlong, S. (2013). *Public Policy: Politics, Analysis, and Alternatives*. Washington, DC: Congressional Press.
Losey, J., and Vaughan, M. (2012). "The Economic Value of Ecological Services Provided by Insects." In *Taking Sides on Environmental Issues,* edited by T. Easton. New York: McGraw-Hill, 39–51.
Maxwell, N. (2014). *Understanding Environmental Health*. Burlington, MA: Jones & Bartlett.
McConnell, R., and Abel, D. C. (2008). *Environmental Issues: An Introduction to Sustainable Development*. Upper Saddle River, NJ: Prentice Hall.
Miller, T., and Spoolman, S. E. (2014). *Living in the Environment*. Belmont, CA: Brooks & Cole.
Miller, T., and Spoolman, S. E. (2010). *Environmental Science*. Belmont, CA: Brooks & Cole.
Ministry of Energy Mining and Communication. (2006–2020). "Jamaica Green Paper Policy 2006–2020." Kingston, Jamaica: Jamaica Printing Services.
Ministry of Environment. (1998). "Jamaica National Environmental Action Plan (JANEP) Status Report." Kingston, Jamaica: NRCA. Jamaica Printing Services.
Ministry of Local Government and Community Development. (2001). "National Solid Waste Management Policy." Kingston, Jamaica: Jamaica Printing Services.
Ministry of Mining, Energy and Commerce of Jamaica. (2009). Retrieved from http://www.scribd.com/doc/92045092/Jamaica-s-National-Energy-Policy-2009-2030. Accessed July 28, 2011.
Myers, N., and Spoolman, S. (2014). *Environmental Issues and Solutions*. Belmont, CA: Brooks & Cole.
Nadakavukaren, A. (2011). *Our Global Environment: A Health Perspective*. Long Grove, IL: Waveland Press.
National Environmental Protection Agency (NEPA). (2006). "Jamaica's National Environment Action Plan (JaNEAP) 2006–2009." Kingston, Jamaica: Jamaica Printing Services.
National Environmental Protection Agency. (2001). "Jamaica's Environment." Annual Report. Kingston, Jamaica. Jamaica Printing Services.
National Policy and Systems of Protected Areas. (1997). Kingston, Jamaica: Jamaica Printing Services.
Natural Resources and Environment Management (NREM). (2007). "Vision 2030 Jamaica: National Development Plan." NREM and HRRC Change.
Neufville, Z. (2001). "Environment—Jamaica: Bauxite Blamed for Deforestation." Annual Report. Kingston, Jamaica. Jamaica Printing Services.
Rabchevsky, G. A. (1994). "The Mineral Industry in Jamaica." Retrieved from http://minerals.usgs.gov/minerals/pubs/country/1994/9515094.pdf. Accessed June 16, 2012.
Ramon, E. (2009). "Prospects for the Oil Importing Countries of the Caribbean." Inter-American Development Bank CCB working paper CSI-114.
Sachs, J. (2005). *The End of Poverty*. New York: Penguin Press.
Sacha-Renée, T. (2006). "A Framework to Manage Jamaica's Protected Areas." Kingston, Jamaica: Jamaica Printing Services.
Vig, N., and Kraft, M. (2013). *Environmental Policy: New Direction for the 21st Century*. Los Angeles, CA: Sage Press.
Wright, E. D. (1998). "Riverton City Dump: Evaluation of a Jamaican Landfill Site." In *Solid Waste Management: Critical Issue for Developing Countries*, edited by E. Thomas-Hope. 1998, 189–234.

13 Environmental Policies and Issues in Trinidad and Tobago

Robert Dibie, PhD

INTRODUCTION

Of all the economic development that has taken place in Trinidad and Tobago, half has occurred over the past four decades with devastating effect on biodiversity in the nation's ecosystem (Kenny, 2008). The effects also include the rapid extinction of species, the loss of forest and wetland, and harm to human health. In an attempt to solve the environmental problems in the country, the government of Trinidad and Tobago established its National Environmental Policy in 1990 and revised the Act in 2005. The goal for the establishment of this crucial policy was the need to develop a framework to address the environmental problems facing the nation. Since 2005, the formulation of other environmental policies has not been transparent and has not involved local stakeholders. Political leaders tend to rely more on technology rather than involving the local community in decision making that relates to the environment (Dhillion et al., 2010; Gaskell et al., 2005; Johnson, 2007).

An analysis of the economic development trends in Trinidad and Tobago reveals that the country continues to have the fastest development in the Eastern Caribbean region. Such rapid development trends come with all types of environmental problems. For example, the pollution of both marine and land resources by oil companies has been a major environmental concern in the past three decades (Government of Trinidad and Tobago. 2010). The bauxite mining industry has also been creating major environmental health problems (Coke, Weir, and Hill, 1992). In addition, the oil and gas companies have been responsible for over 80 percent of the industrial pollution in the country (Caribbean Environmental Health Institute Report, 2006). Other sources of pollution reported by the Caribbean Environmental Health Institute (2006) include petrochemical installations, breweries and alcohol distilleries, sugar refineries, and canning and packaging plants. It was also reported that food processing plants and two rum distilleries were emitting their waste into the adjacent water courses. Some sugar refineries have also been emitting phosphate-based industrial waste into the lower reaches of the River Caroni basin (Tikka et al., 2000).

The agricultural sector has also been reported to have polluted the environment through pesticide, herbicide, and fertilizer runoff from farmlands, animals, and other organic waste. Detergents, sanitary waste, and incompletely processed sewage have also been reported to flow into the River Caroni and the ocean (National Environmental Policy Report, 2009).

Trinidad and Tobago have also experienced environmental problems with land reclamation and conversion projects. In the past two decades, wetland has been cleared and buildings constructed. A good example of such projects is the Movie Town Complex at Invaders Bay in Port of Spain. Mangrove forests were completely destroyed and the land reclaimed that was supposed to be a mangrove restoration project area (Kenny, 2008; Trinidad and Tobago Biodiversity, 2009). However, the government has not been able to implement the project because once an ecosystem is destroyed, it is virtually impossible to restore the biodiversity fully. Thus, all over the country, wetland unsuccessfully competes with economic development. Another environmental problem associated with the mangrove wetland is illegal hunting, poaching, and fires, particularly in the Caroni Swamp. Hunters have been reported to enter the protected swamp and kill animals, such as the scarlet Ibis (the endangered national bird), some species of snakes, and wildfowl, and take some exotic species of orchids indigenous only to Trinidad and Tobago (Spencer, 2006). It is interesting to point out that these illegal hunting practices take place regardless of hunting seasons (Government of Trinidad and Tobago, 2002).

Further, harmful emissions and effluents by industries in the Point of Lisas industrial estate, as well as indiscriminate slash–and-burn practices by citizens, are other environmental problems that have plagued the nation. Deforestation on the island of Trinidad and Tobago due to clearing of land, including hillsides, for development, in addition to excessive unauthorized quarrying, has caused environmental health problems (Spencer, 2006). During the past two decades, there has been a major concern about the destruction of the Bucco Reef because of sewerage, sand, and people. In addition, the intensive agricultural venture in Nariva Swamp has led to extensive pesticide pollution that entered the catchment area from rainfall runoff. The runoff has poisoned aquatic life and rendered the swamp lifeless. In the Bucco Reef and Bon Accord Complex, the fertilizer and pesticide runoff contributed to high nutrient levels, leading to eutrophication (Tikka et al., 2000). The government and industries have been very effective in addressing the Nariva Swamp problem and the pollution of the nation's waters and aquifers (National Environmental Policy for Trinidad and Tobago, 2008). There has been significant progress in conserving critical ecosystems and avoiding development projects that tend to cause coastal and marine degradations, with a major environmental policy shift toward coastal protection through zone management. However, the government of Trinidad and Tobago has not been very effective in the implementation of the National Environmental Policy.

Although some level of recycling is practiced in Trinidad and Tobago, solid waste from industries and urban sources such as metal, glass, plastic, and food has been identified as a major cause of environmental and health problems in the country. In Trinidad and Tobago, unlike other developing countries, hazardous waste is still being disposed of in the same way as domestic solid waste. It is reported that some of the poor citizens of the country, birds, and other animals still often sift the content of landfills all over the country for items of value. These citizens and animals are routinely exposed to hospital waste, heavy metal, noxious and poisonous chemicals, and vectored diseases.

According to the National Environmental Policy Report (2009) of Trinidad and Tobago, polluters of the environment do not care about the consequences or pay the fines. Public infrastructure and other resources are not adequate for effective environmental management and protection. The behavior of some of the citizens of the country continues to be unacceptable due to lack of enforcement, lack of alternative approaches, and lack of incentives to motivate citizens to comply with public policies.

Many of the environmental problems that have existed in Trinidad and Tobago in the past two decades find their roots in the attitudes and behavior patterns of the citizens of the country (Tikka et al., 2000). The government's attempts to mitigate these environmental problems and improve sustainability, therefore, rest with the development and effective implementation of environmental education to empower its citizens to become more responsible. This is one area that the government of Trinidad has not been able to fully accomplish.

The environmental problems in the country are compounded by unprofessional and poor agricultural activities. Indiscriminate clearing of land by slash-and-burn, deforestation activities in the northern range, bushfires, and shafting cultivation have all contributed to the high soil erosion levels and sedimentation overload. Vegetable cultivation in Aranguaez and crops planted on steep terrain cause soil erosion and high siltation during rainfalls. In the Nariva Swamp, large-scale watermelon and rice cultivation and illegal cultivation of marijuana are persistent problems (Tikka et al., 2000).

In addition to these problems, there is the great problem of traffic congestion on all the roads in the country, especially those leading to the major cities. Almost every household in Trinidad and Tobago owns a car and in some cases two or more. There are also a number of public and private vehicles for hire to any point in the country. The congestion of vehicles is usually worse during peak periods of the workweek and during public holidays. Most of the vehicles operating in Trinidad and Tobago use fossil fuel as their energy source. The main pollutants from the combustion of fossil fuel are carbon monoxide (CO), nitrogen oxide, sulphur dioxide, ground-level ozone, volatile organic compounds (VOCs), and lead. Several health issues or diseases are associated with these pollutants, including irritation of the lungs, respiratory infections like influenza, skin allergies, and reproductive defects (Kraft, 2011). In addition, lead is reported to hamper fetus development and can cause retardation in children after chronic exposure (Dye, 2011).

Trinidad and Tobago sometimes experience a short period of slight to moderate haze from local industries and transboundary pollution from neighboring countries (National Environmental Policy Report, 2009). The air quality statuses on the two islands differ according to their geographical locations, the industrial and commercial activities, populated areas, and the traffic conditions, especially in Trinidad compared to Tobago. During the past two decades, Port of Spain has tended to be more prone to air pollution than any other areas, and the air quality was good only 47 percent of the time, moderate 50 percent of the time, and unhealthy for the remaining 3 percent of the time (Caribbean Environmental Health Institute Report, 2006).

This chapter examines sustainable environmental issues in Trinidad and Tobago. It contends that the lack of rigid sustainable environmental policies in the country has exacerbated partially ecofriendly human activities. It explores how environmental changes have been one of the most complex challenges facing Trinidad and Tobago. The chapter uses interviews and survey questionnaire data to present an argument that Trinidad and Tobago will bear the brunt of the effect of environmental change, even as the country strives to overcome poverty and advance economic growth. Inadequate sustainable environmental policies are inextricably related to modern civilization in the nation. The chapter has two objectives. The first is to investigate the economic and social impact of environmental issues and policies in Trinidad and Tobago. The second objective is to evaluate the nature of sustainable environmental policies on the islands. The chapter concludes with a discussion of the survey results, limitations, and directions for future sustainable environmental research in the country. The future challenging goal for Trinidad and Tobago is the search for environmental pollution solutions and comprehensive new policies. Sustainable transformation in the future would require the industrial sector, governments, nongovernmental organizations (NGOs), and the citizens to act now, act differently, and act together. Sustainable environmental change cannot be attainable without these stakeholders cooperating at the national level to improve energy efficiencies, develop and deploy clean technologies, and expand the desire to grow green by absorbing gases.

ECONOMIC GROWTH AND ECOLOGICAL SUSTAINABILITY

Economic growth and ecological sustainability are closely related to the means of securing livelihoods, basic needs, political participation, and citizen's rights (not working against them) (Dye, 2011; Peters, 2013).

Healthy ecosystems provide invaluable functions and services that are essential to the marine environment and human well-being. Unfortunately, some human activities in Trinidad and Tobago have had an adverse impact on the structure and functioning of marine ecosystems (United Nations, 2011). Therefore, environmental sustainability must be conceptualized so as to support and sustain livelihoods and basic needs, and not in ways that automatically counterpoise nature against the survival needs of the most vulnerable. According to Robert Dibie (2010), there might be trade-offs among these different goals. The costs and burdens must not fall on the poor citizens who are most vulnerable. The citizens must have a voice in negotiating resolution through open and genuinely participatory political processes (Alba, 2008; Clifford and Edwards, 2012). Further, environmental strategies to enhance livelihoods and fulfill needs can likely help lay the basis for reduced rates of mortality, fertility, and good life.

The governments of all countries have a crucial role to play in environmental protection (Kraft, 2011; Vig and Kraft, 2013). In promoting sustainable consumption practices among the people of Trinidad and Tobago, the government should publicize various strategies to implement sustainable consumption and development. This is necessary in order to educate and foster environmental awareness and concern among the citizens of the country. The government should also enhance its social advertising campaign (Dibie, 2009). The most significant factor that may affect the ecosystem in the country is not the official government policy but the public awareness of the environmental problems (Dye, 2011). Many people in Trinidad and Tobago tend to have high ecological concerns but also have the sentiment that the preservation of the environment is a major responsibility of the government (Miller and Spoolman, 2014).

According to Andrew Basiago (1995), the future of sustainable growth, equity, global environmentalism, and biodiversity must guide decision making about environmental challenges. Sustainability can be related to the natural capital perspective. This perspective stipulates that the stock of natural capital consists of natural ecosystems that produce a flow of valuable goods and services. In addition, natural ecosystems consist of mineral resources, such as nitrogen, phosphorous, water, carbon, and oxygen, and biogeochemical cycles that are critical for life (Easton, 2012). Thus, the ecosystem provides services that could consistently produce a flow of valuable goods and services. The services provided by the ecosystem include (1) controlling climate and disease, (2) nutrient cycles and crop pollination, (3) production of food and water, and (4) spiritual and recreational benefits (Myers and Spoolman, 2014; Vig and Kraft, 2013).

A model of the unsustainable use of renewable resources contends that supplying people with renewable resources and producing the resulting waste and pollution can have a large environmental impact (Miller and Spoolman, 2014). The ecological footprint is the amount of biologically productive land and water needed to supply the people in a particular country or area on an ongoing basis. It is also very important to produce renewable resources as well as absorb and recycle the waste and pollution produced by such resources use. Thus if a country's total ecological footprint is larger than its biological capacity to replenish its renewable resources and absorb the resulting waste and pollution, that nation is said to have an ecological deficit. From this perspective, the country might be living unsustainably by depleting its natural capital instead of living on the income provided by such capital.

Empirical research and practice methodologies that are ground in the ecological perspective enable the government to access the biopsychosocial forces of the environmental culture of any nation or society (McConnell and Abel, 2008; Smith 2004). According to Withgott and Brennan (2008), the environmental culture perspective also allows for consideration of the interchange between various ecosystems, including plants, animals, forests, fishes, sun, and human beings. Such biodiversity is germane to the function and life cycle of the people of Trinidad and Tobago.

Biodiversity, as used in this chapter, refers to a variety of earth's species, the genes they contain, the ecosystems in which they live, and the processes of energy flow and nutrient cycling that sustain all life. Wood and Potoski (2010) observed that an ecological theoretical perspective enables the development of environmental culture and sensitive methodologies. Such methodologies allow for more relevant conjecture that can then be assessed within its native space in a country such as Trinidad and Tobago. The outcome could facilitate a conduit for more objective analysis of environmental pollution. Thus the conceptual and theoretical frameworks suggested herewith are critical to the interpretation of the social and economic impact of environmental policies and issues in Trinidad and Tobago. It is important to note that to stay alive, grow, and reproduce, human beings must use some of the chemical energy stored in the biomass that they take for their own respiration.

The natural capital degradation model shows the major ways in which human beings have altered natural system to meet their growth population's resource needs and wants. The model also shows a very important relationship between population, consumption pattern, and technology in developed and developing nations. Figure 13.1 shows the economic and social impact of environmental issues. It is interesting to note that recent analysis estimates that it takes at least 4–5 hectares of productive land to provide water, energy settlement area, and food for one person living in a developed country's city (United Nations, 2011). So if 10 million people were to live in Trinidad and Tobago, then we would need 50 billion hectares of productive land. However, that amount is seven times the total of all the productive land in the Caribbean Island countries. Figure 13.1 shows the natural capital degradation model.

According to Thomas Dye (2011), Zachary Smith (2004), and Guy Peters (2013), the limits in productive land are among the main reasons

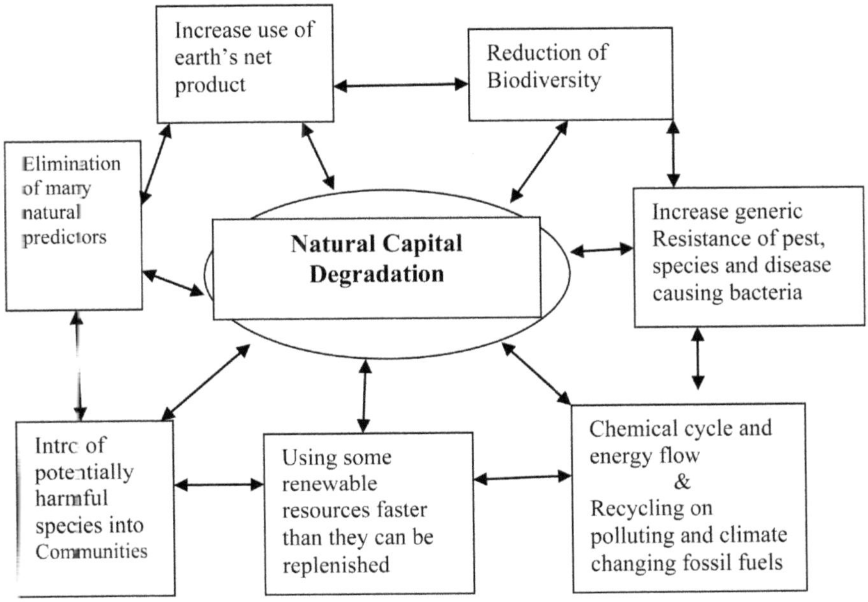

Figure 13.1 Natural Capital Degradation Model. The model also illustrates a very important relationship between population, consumption pattern, and technology in developed and developing nations in the world.

for growth arguments that lead to the conclusion that there is a limited possibility of all people rising to the living standards we take for granted today in developed countries like the United States. Therefore, this is the best time for the citizens of Trinidad and Tobago to start learning how to adopt simpler, less resource-expensive, and sustainable ways of living. Thus, Trinidad and Tobago could easily develop a society in which citizens do much less work but produce and have much more time to enjoy life, without stress and insecurity. It is important to enjoy life without damaging the environment or depriving the nation of its ecosystem (Dibie et al., 2011). Thus, Trinidad and Tobago does not need better technology or more wealth to solve its environmental problems. All it needs is a radical change in system, behavior, lifestyles, and values.

RESEARCH METHODOLOGY

The purpose of this paper is to find out the major economic and social impact of environmental issues and policies in Trinidad and Tobago. The author reviewed several secondary research sources, including government and industrial documents, to find out the level of social impact of environmental policies on the citizens of Trinidad and Tobago. The preliminary primary research design was developed by the author for Trinidad Island. However, the survey was later administered in Tobago Island.

Empirical evidence of the economic and social impact of environmental issues and policies in Trinidad and Tobago was gathered through a number of independent studies drawn from various industrial and commercial sectors. The data generated were complemented by the Delphi method. This approach utilized the opinions of a panel of postgraduate students and fifty industrial workers drawn from a sample of Trinidad and Tobago cities. The pooling of data from a multitude of independent samples was achieved within the multiple comparatives framework. Here analysis was employed to pool common statistics addressing similar research issues in waste management and health issues associated with hazardous waste. More emphasis was put on effect size analysis where summary measures such as proportion were subjected to such analysis.

The questionnaires for this study were administered to 1,000 industrial workers and commercial residents of both Islands of the country. Graduate students from the University of the West Indies at St. Augustine also participated in the survey. Arthur Mol (2001) argued that using a sample of students is considered valid for exploratory studies. Furthermore, students represent the new generation of consumers and have been a growing population of consumers in most developing countries like Trinidad and Tobago. The students completed the survey during class time and were assured anonymity. Participation was voluntary, and no remuneration was offered. The respondents were postgraduate students in the occupational, environmental safety, and health degree (OESH) degree program. The students consisted of both genders and different races. A total of seventy-five completed the questionnaires (yielding a response rate of 15 percent). The majority of the respondents (500) were from the industries (50 percent). The remaining (415) were engaged in commercial business in the country (24 percent). The total respondents' rate was (89 percent).

After data verification, the information was entered onto an SPSS system, which captured all the information from the survey questionnaire. Once computerized, the information was sorted according to the specific environmental impact variables. Finally, this information was used to conduct the data analysis to access the major social and economic impact of environmental policies and issues in Trinidad and Tobago.

ANALYSIS AND DISCUSSION OF FINDINGS

This section presents the study findings on respondents' perception of the social and economic impact of environmental issues and policies in Trinidad and Tobago. Aggregate data are presented first of the major environmental variables. Second, the number of respondents affected by the environmental impact is reported. To demonstrate the dynamic nature of the social and economic impact, the author then examines how the respondents changed during the survey period 2009 to 2011. The analysis that follows provides information on the nature of several of the economic and social impacts of environmental issues and policies in Trinidad and Tobago.

According to the Government of Trinidad and Tobago Report (2010) and the World Bank report (1995), solid waste management encountered by the nation is on an upward trend with a total of 11,452,471 kg of hazardous waste generated in 2003. Other areas with significant amounts of waste include the Pointe-a-Pierre and surrounding areas, Mayaro and surrounding areas, Point of Lisas and surrounding areas, and the St. James, Woodbrook, and St. Clair areas. It is interesting to note that Scarborough represented only 2.4 percent of the national hazardous waste total. On the other hand, Tobago's waste accounts for 1.2 percent of the national total in the same period (Caribbean Environmental Health Institute Report, 2006).

This study also found that the geographical areas that generated the largest amount of waste in Trinidad were those where the oil, gas, and chemical industries had their plants and operations. The most noticeable areas are Fyzabad, Penal, and Santa Flora. The amount of hazardous waste generated in these three areas or locations account for over 99.9 percent of the total in Trinidad and Tobago. It should be noted that the economic activities contributing to these hazardous waste problems in the country include industries in petroleum exploration and companies engaged in oil and gas production. The respondents observed that some of the waste also originated from maintenance activities such as mechanical repairs. It is anticipated that if the current trend of hazardous waste pollution is continued by the oil and gas industries in Trinidad and Tobago, the environment degradation will worsen and that the related health consequences will cost the nation dearly. Therefore, a shift toward more sustainable oil and gas exploration patterns is required to move the nation forward in the twenty-first century.

Eighty-nine percent of the respondents contended that air pollution is an increasingly major problem due to emissions from industrial companies and urbanization in some areas of Trinidad. The respondents complained that apart from the petrochemical industry's pollution, gaseous emissions from vehicles are becoming an incremental environmental problem. Respondents indicated that they are concerned that majority of the elderly people deaths in the cities surrounding Port of Spain were associated with pollution caused by pneumonia, exacerbation of emphysema or chronic bronchitis, or cardiovascular diseases.

Seventy-seven percent of the respondent indicated that they feel that the government of Trinidad and Tobago is not doing enough to promote environmental education in the country. Although the government occasionally organizes environmental contests and displays posters, they have not been able to effectively create public awareness of the danger of an

Figure 13.2 Photo of Garbage Collection in a Street in Trinidad, 2010
Source: Photo taken by author during field research in Trinidad and Tobago, 2012

unclean environment. More needs to be done to promote environmental education from an early age to the university level. Respondents unanimously indicated that environmental science should be one of general education requirements in both the high school and university levels. About 83 percent of the respondents are not aware of the country's national environmental protection policy. This confirmed a major deficiency of the government in the area of appropriate communication strategies and transparency in environmental policy formulation and implementation.

Although the Trinidad and Tobago climate is tropical with six to seven months of rainfall and about four to five months of dry season. The average temperature of the country is generally hot all year. Ninety-six percent of the respondents indicated that the higher overall temperature often increases the range of major disease vectors such as mosquitos. Increases in mosquitos often cause a corresponding increase in vector-borne diseases such as malaria. On the other hand, they stated that changes in the sea level often affect food supplies, claiming agricultural land and disrupting breeding grounds for fish and shellfish. Flooding in the past had enhanced the breeding of mosquitos, a physical disruption that increased contact with rodent vectors, overcrowding that aided the transmission of respiratory illnesses, and damage to the sewage and drinking water system that protected against diarrheal disease in Trinidad and Tobago. Further, the respondents indicated that previous heat waves have produced crop failure, as well as human deaths from heart attack and heat stroke. According to Robert Friis (2012), excessive exposure to the heat of the sun early in life appears to pave the way for skin cancer later.

Before oil and gas were found in Trinidad and Tobago, agricultural production was the breadwinner for the nation. With the discovery of oil and gas, agriculture took a backseat. Currently the nation is known for producing coffee, citrus, coconut, cocoa, sugar, cotton, and tonka beans. Seventy-seven percent of the respondents indicated that the agricultural sources of air pollution in the past three decades has incrementally included smoke and ash from sugarcane and bagasse burning, as well as land-clearing activities by farmers. They contend that the use of pesticides has been a major water pollution problem in the nation's environment. Farmers still use high levels of dicthloro-diphenyl-trichloroethane (DDT) to kill pests because the pests are animals or plants that interfere with human well-being or interest by causing diseases or discomfort in people or in animals (Friis, 2012). Respondents stated that agricultural pollutants such as pesticides, herbicides, and fertilizer run off agricultural land, animals, and other organisms, entering streams, rivers, ocean, and wetland areas. In some cases, pesticide percolates into the groundwater. Excessive fertilizer used in intensive crop cultivation in Aranguez has been found to contain high level of phosphorus and nitrogen, and this has had a great impact on the Caroni River (National Environmental Policy Report, 2009).

The health implication of these agricultural pollutants is that organochlorine pesticides are nerve toxins. They can disrupt the central nervous system of human beings and animals, causing convulsions (Maxwell, 2014). Exposure to pesticides has been linked to neurologic effects, cancer, and reproductive and development outcomes. Other effects of pesticides include headaches, dizziness, nausea and vomiting, muscle weakness, and even convulsion and coma at high exposures (Dye, 2011; Kraft, 2011). The respondents stated that DDT is still being used in Trinidad and Tobago for mosquito control.

Seventy-nine percent of the respondents indicated that part of the Caroni River is used for the filling of solid waste. Industrial pollutants have been found to enter the river from the east and west corridors and from west central Trinidad. The Caribbean Environmental Health Institute Report (2006) pointed out that the sugar refineries emitted phosphate-based industrial waste in the lower reaches of the Caroni basin, The Manusica, Guanapo, Aripo, and Arima rivers flow south into the Caroni River and are high in sewage and industrial waste. In addition, a quarry in the Arima Valley has been noted for generating a large quantity of silt from wash planes, affecting the downstream activities of the Caroni wetland. Respondents also stated that the most common materials in the municipal solid waste are paper and paperboard that together make up about one-half by weight of what the citizens in the country throw away. They observe that municipal waste sometimes contain hazardous waste items, such as drain cleaners or mothballs. Municipal waste also includes food, and this means that it has to be removed on a strict schedule. When municipal wastes are not removed on schedule, trash cans often overflow, odor quickly becomes a problem, and rats and other pests feed on the waste.

Table 13.1 shows the various environmental problems identified by respondents and the nature of the social and economic impact in some cities and towns in Trinidad and Tobago.

It is interesting to note, however, that Trinidad and Tobago's occupational safety and health (OSH) policy is consistent with those of the United States, Canada, Scandinavian countries, and the Republic of South Africa; however, the nation is yet to refine its implementation process to a more efficient level.

Trinidad and Tobago's dependence on nonrenewable fossil fuel has put the nation in a precarious position. The citizens' ability to harness different

Table 13.1 Economic and Social Impact of Environmental Issues

Environmental Problem	Nature of Social and Economic Impact	Affected Cities/ Towns
Agriculture	The introduction of agricultural and domestic pesticides, industrial and domestic cleaners and private and industrial waste into surface water tend to pose an increased risk to human health because many of these chemicals are considered toxic, carcinogenic or genotoxic. There are concern that inorganic nutrients from fertilizers especially nitrogen and prosperous may have stimulated aquatic productivity in some parts of the country. Siting low income housing with septic tanks and absorption pit systems in areas where the water table is high or that are prone to flooding	All over Trinidad and Tobago
Water Pollution	Respondent indicated that there has been presence of harmful radioactive elements in rivers and surface source drinking water. Other pollution problem were due to untreated or partially treated liquid effluents from oil and petrochemical industries, breweries and alcohol distilleries, sugar refineries, canning and packaging plants. Pesticides, fertilizer and herbicides run off from agriculture also pollute the nations land and water.	Mostly Trinidad: There are other pollution problems from sewage coming from the tourism industry in Tobago

Waste Management	There are concerns about the solid waste coming from oil and petrochemical industries, breweries and alcohol distilleries, sugar refineries, canning and packaging plants. Respondents whose communities were served by ground water supply had significantly lower exposure to both pathogens than communities supplied with either a protected or non-protected surface source. The was also concern that waste from the industries were flowing into nearby rivers.	Mostly Trinidad
Deforestation	Some respondents complained about deforestation resulting from logging, land clearing for agriculture, quarrying and property development along hillsides. The deforestation problem has also enhanced erosion in some hill side regions of the country.	All over Trinidad and some part of Tobago
Natural Hazard	In Trinidad and Tobago the average annual rainfall ranges from 2,200 mm in the east to 900 mm in the west. The island of Trinidad lay just south of the hurricane belt and has not experienced several tropical cyclones for decades. The island of Tobago is however located in the lower hurricane belt and sometimes experienced heavy impact. In 1963 hurricane Flora seriously damage the Tobago.	All over Tobago but Trinidad is safe.
Environmental Health Issues	Health consequences of consuming sublethal toxin concentrations are some concerns raised by respondents. Further, liver damage and tumor growth were reported as two of the primary health problem associated with the consumption of cyanotoxin from the industries. Smoke and other bad environmental practices often cause respiratory, intestinal and outbreak of gastroenteritis caused by poor sanitation among children and factory workers in the country	All over the country

Source: Derived from Survey 2009–2012.

forms of energy has played a large part in humans becoming the dominant species in the country. The exploitation of fossil fuel has provided the means to support a rapidly expanding population. These fossil fuels have also caused extensive pollution, which, as we have seen, threaten the nations with catastrophic climate changes. The country's reliance on fossil fuels has allowed the nation to develop an energy-intensive society. Dibie and Offiong (2012) contend that the country's workers may be comfortable today but that, without the effective implementation of the occupational safety and health policy and the development of alternative sources of energy, the transition to a steady state of energy economy will be rough in the near future.

Technological developments in Trinidad and Tobago, such as new synthetic chemicals, have also generated significant environmental externalities in the form of toxic and hazardous waste. The concentration in and removal of wastes from the ecosystems in which they would naturally biodegrade have resulted in massive quantities of refuse, usually referred to as solid waste (Dibie and Offiong, 2012). Most solid waste in the country comes from mining, oil and gas production, agriculture, and industry. City waste presents greater political and managerial problems and delivers serious environmental impacts on land and water resources. Given a nation with limited resources and not a lot of space in which to put all its waste, there is cause for concern in waste management in Trinidad and Tobago.

In addition to the waste management problem, air and water pollution is a major concern. Trinidad and Tobago is also threatened by the petroleum and construction industries, which are composed of multiple and complex facets. On one hand, prevention, through designing out potential risks, is the most effective way to protect occupational safety and health and is crucially important in a relatively new, expanding, and potentially hazardous industry. At the same time, tertiary sector prevention is also very important. The third problem is the ineffective implementation of occupational health and safety policy in the nation's industrial sectors. All three factors require active hazard research assessment and the monitoring of near misses, injuries, and occupational diseases, as well as attention to social and other factors mediating risk such as education, training, regulatory oversight and intervention, compensation, and reporting. Unfortunately, the local and federal government in Trinidad and Tobago have not been very successful in arresting the OSH and environment health problems associated with workplace diseases and injuries as well as appropriately managed environment.

In Trinidad and Tobago, negative human impacts have adversely affected the air, water, and soil with dire consequences on human health and the type of nutrition essential for living. For many years, these actions, whether intended or not, have continued to have a significant impact on the nation's ecosystem. Undeniably, increased economic activity enriches the lives of people in the country resulting in longer life spans, better health, greater material wealth, and more mobility and leisure time. However, these improvements have resulted in the degradation of the natural system sustaining the nation. Evidences of these negative impacts abound in the form of air and water pollution, soil erosion, and the loss of biodiversity, leading to the compromise of human well-being, posing risks to human life, and threatening the nation's ability to build a viable society that will survive and thrive in the long term (Withgott and Brennan, 2008; Wright and Boorse, 2014). Dibie and Offiong (2012) contend that the elements of the Trinidad and Tobago's environment were functioning long before the occupation by humans who have not been insightful enough to keep the elements in a better condition.

Even though the nature of many occupational and environmental health problems is similar in the United States, Canada, Norway, Nigeria, South Africa, and Trinidad and Tobago, workers tend to be more intensively exposed than community residents to various hazards. Thus, whether the environment is workplace, school, home, or a community setting, the pathophysiology of the specific hazards in humans is the same (Levy et al., 2011). For example, the transportation of lead, asbestos, and other hazardous substances on workers' clothes, shoes, skin, and hair may cause the same level of health problem as it would for a worker who was exposed to the agent only in the work environment. There are occupational safety and health policy, environmental laws, and government agencies meant to protect the workers and the environment in Trinidad and Tobago, such as the Occupational Safety and Health Act of 2004 and 2006. However, socioeconomic and environmental issues have continued to undermine the development of the country, even on a sustainable basis.

ALTERNATIVE STRATEGIES FOR IMPROVING ENVIRONMENTAL POLICIES

A review of the National Environmental Policy of Trinidad and Tobago reveals many positive ideas on how to address all the relevant environmental issues and problems facing the country. While some scholars have argued that the policy should be conservative in keeping with the scarce financial resources of the country, others stressed the need to act sustainably immediately (Dibie et al., 2011; Dye, 2011; Gobin, 1999; Kraft, 2011; Miller and Spoolman, 2014). The nation is still considered one of richest countries in the Caribbean Islands.

For sustainability in Trinidad and Tobago to become operational, governments will need appropriate policies to produce incentives for its citizens and society to change unsustainable behaviors. There are two ways to implement such policies. The first is to enact policies that would emphasize the use of subsidies and tax incentives to change behavior (Kraft, 2011; Peters, 2013). For example, in the United States renewable energy development is promoted with such incentives. The second approach is for the government to provide leadership to develop regulations and policy that promote conservation. This approach will encourage a change to renewable energies and to sustainable practices by the citizens (Dye, 2011; Vig and Kraft, 2013).

Environmental policies in Trinidad and Tobago, as in in any other country, are politically motivated gestures. At best, the nation would need to generate specific strategies, define instruments of effective and efficient implementation, identify a local action plan, and ultimately meet the government mission. The nation needs to embark on a fundamental shift in focus to the type of economic growth and health values associated with the protection of its surface water. It is also very necessary for the national government to start to provide incentives and justification for the protection of surface water (Davies and Mazumder, 2003; Maxwell, 2014).

The history of the country shows an interesting pattern. The relationship between individual decision makers that control the government and the financial gurus tends to make it difficult to take certain actions regarding environmental management and protection. The main challenge continues to be moving from a well prepared environmental policy document to a strategic and effective implementation of the policies to promote healthy citizens and biodiversity (Ahmed, 1998). For this to happen, the political players and other individuals must have the will to curb any development that is done at the expense of environmental degradation. If the foregoing limits to social and economic growth analysis are basically valid, then some of the key principles for a sustainable society are clear and indisputable.

Material living standards must be much less burdensome on natural resources. In a sustainable society or nation, per capita rates of resources must be a small fraction of the ones in Trinidad today. According to Robert Dibie (2010), there must be small-scale, highly self-sufficient local economies. In addition, there must be mostly cooperative and participatory local systems whereby small communities control their affairs, independently of the national economies. While these suggestions are crucial, there must also be the use of alternative technologies that maximize the use of resources. A very different economic system must be developed that relies on renewable energy.

Another approach to addressing the change in behavior of citizens toward the environment is the market-based model, which requires government to introduce carbon taxes and cap-and-trade policies (Van den Eynden et al., 2008; Waller-Hunter and Jones, 2002). In spite of allowing the market to operate, these policies require governmental leadership in development and implementation. Other alternative approaches to live a simpler life must be adopted in Trinidad and Tobago (Dibie et al., 2011). The citizens should be encouraged to live well with a much smaller amount of production, consumption, work, resources use, trade, investment, and gross national product (GNP) than there is now (Miller and Spoolman, 2012) . Adopting these principles will allow Trinidad and Tobago to escape the economic treadmill and devote their lives to more important things than producing and consuming. Living more simply does not mean deprivation or hardship. It means focusing on what is significant for comfort, hygiene, health, efficiency, and so on (Ramon, 2009; Schueler, 2000; Selin and Pierskalla, 2005).

The government of Trinidad and Tobago has succeeded in establishing the Environmental Management Authority (EMA); however, there are inadequate experts in this public agency to effectively regulate the policies that have been enacted by the nation's parliament. The lack of experienced inspectors is one of the major problems facing the EMA, and the inability to attract experienced professionals is another. Therefore, the time has come for the government of the country to help the agency provide attractive compensation packages for its employees. Addressing the manpower issue may open the way to the appropriate regulation of significant development projects such as the Aluminum Smelter in La Brea, the Steel Plant in Point Lisas, and the human-made industrial island in the Gulf of Paria. It is understandable that the government should require a certain level of economic development to meet the needs of the population. However, these requirements should not be implemented at the expense of the environment and natural resources of Trinidad and Tobago.

The existing National Environmental Policy appears to be a good document that could be regarded as the road map to address the environmental concerns of the country. However, what is required is the building of solid political and institutional capacities for effective democratic governance of environmental policies (Dibie, 2009). The Government of Trinidad and Tobago should improve on it current democratic system. There should be a genuine participatory democracy system in Trinidad and Tobago. Most of its local policies and program could be worked out by elected paid and nonpaid committees. Citizens should be allowed to vote on the important decisions concerning their small areas at regular town meetings. There should be some functions for the regional and national governments, but these tasks should be relatively few or limited.

Although the nation has occasionally promoted environmental issues, the effort to educate citizens, especially at an early age, must involve more than conducting essay competitions and putting up posters if the nation is going to change behaviors and lifestyles. So many of the societal problems facing the country are related to the environment. However, environmental policy issues are regarded as the last thing on citizens' minds. Of course, citizens may have heard about climate change, greenhouse gases, flooding, depleted fish reserves, and the impending end of the nation's oil and gas reserves. Yet I observe that the behavior of the citizens has not really changed. Students who probably won the essay contest or saw the posters still probably throw coke cans and other garbage from moving vehicles. Thus, the message of environment cleanup and protection has not really hit home. The government of Trinidad and Tobago needs to do more to promote environmental science at a very early stage in its educational system. One of the options available for behavior change is that of financial incentives for recycling (Dye, 2011; Kraft, 2011). Every truckload of plastic bottles finds its way into the drains of Port of Spain, the capital city. A simple properly planned and executed recycling project for plastic bottles could do wonders for the nation's environment.

It is important to note that the issue of monitoring environmental policies in Trinidad and Tobago is crucial as a mechanism for achieving sustainable development in the country. The promotion of economic and appropriate sustainable industrial development may occur through the implementation of a combination of many types of environmental policies. The successful coordination and definition of the main objectives for these environmental policies are crucial for achieving the expected results or goals. Although Trinidad and Tobago seems to be on the right direction for establishing these policies, there is the need for collaboration and partnership between the government and the business sector in providing the necessary scientific, technological, and innovative strategies for achieving the appropriate sustainable development goals. Trinidad and Tobago's environmental strategies are on the right path, but much still needs to be done. Appropriate policies need to be formulated and effectively implemented for the country to achieve heightened levels of economic and industrial development.

It is interesting to note that the environmental problems identified by Trinidad and Tobago's petrochemical companies in generating sustainable innovation, such as pollution control, high costs, and risk management, are mainly structural issues that need consistent and persistent policies to be solved. The promotion of cooperation among all stakeholders also needs to be maintained and expanded through a reduction in bureaucracy and the establishment of clear rules for defending intellectual property. In addition to this interaction, it is very important for the government of Trinidad and Tobago to stimulate the financing of innovative activities. The government should also reinforce the importance of these activities with the industrial sector and all public agencies.

Also, the Government of Trinidad and Tobago needs to consider both the short- and long-term implications of water resource management. The effective management of drinking water sources is especially critical for citizens and institutions that do not have the financial resources to build and maintain treatment facilities other than the use of chlorine as a disinfectant (Maxwell, 2014). Many environmental health scholars have argued that the prevention of water pollution seems a wiser and more prudent management directive than does cleaning the nation's water source in the future. The political leaders of Trinidad and Tobago must realize that water is essential for sustainability; therefore, it is crucial to manage and use water in a sustainable and cost-effective way (Davies and Mazumder, 2003; Peters, 2013; Vig and Kraft, 2013).

Finally, Trinidad and Tobago may have all the high-tech and modern ways to foster sustainable development, but such technologies could take many forms—windmill design, medicine, public transport, household appliances. The nation should also have a national system for some things, such as railways, telecommunication, and taxes, but on nothing like the present scale. It should also have far more resources for science, research, education, and the arts than it does now because it would help to decrease the wasting of vast quantities of resources. This will also discourage the production of unnecessary items. Sufficient and simpler ways should be adopted for most people now in the country. It is important to note that if the limits of growth analysis are basically correct, then the nation has no choice but to work for the sort of alternative society as outlined here. In Trinidad and Tobago, a sustainable nation should be conceived only in terms of simpler lifestyles, mostly in highly self-sufficiency and participatory democratic, steady-state economic conditions. The government alone cannot help with this task. It has to be done at grassroots level, as citizens begin exploring how they can come together to build the state of the environment. The overarching aim is to transform the existing nation gradually into a sustainable and self-sufficient country.

SUMMARY

This chapter examined the impact of environmental issues in Trinidad and Tobago. It argued that each environmental issue is unique and has a huge and special potential to contribute individually and in association with its community members to sustainable development and environmental well-being in various ways. These potentials must be discovered and realized to ensure sustainable development and environmental improvement. The analysis of environmental problems reveals that individuals, not organizations, have to actually perform the actions, and this requires a shift from macro- to microlevels of policy implementation in Trinidad and Tobago.

Biodiversity is a renewable natural resource that can perpetuate itself if human interference is limited to below tolerable level. However, biodiversity degradation and destruction in Trinidad and Tobago have been widespread and above the threshold level due to the oil and petrochemical industry, breweries and alcohol distilleries, sugar refineries, canning and packaging plans, and pesticides and fertilizers from agricultural sources. Biodiversity organisms from the environment are used by human beings for more food, cloth, and fuel. Consequently, the nation of Trinidad and Tobago now characterized by denuded landscapes and huge biodiversity loss. Specie and ecosystem disappearances have had enormous potential consequences for sustainable development and environmental maintenance.

Sustainable development is more about the future than the present. The prevailing attitude of discounting the future and maximizing present gains at the expense of the coming generations has to be reversed. Widespread and concerted biodiversity conservation and enrichment have to be accomplished at an accelerating rate to ensure future sustainable development and environmental well-being (Harris, 1992; Kraft and Furlong, 2013;

Sonnenfeld, 2000). The remaining biodiversity still has enormous potential of propelling sustainable socioeconomic growth and development in the country. That potential is now by and large dormant. It requires activation and value optimization through a concerted effort using ever advancing science and technology.

With respect to climate change, the nation could preserve lives and livelihoods and improve citizens' health if it takes immediately and collaborative action to protect past gains and to ensure that future development is climate resilient. The government, through its partnership with the business sector, could help people adapt to new weather patterns and cope with extreme events. It could also help citizens find new economic opportunities in a changing climate and work toward meeting energy and other environmental needs sustainably. At the same time, Trinidad and Tobago could leapfrog past outdated technologies and move forward with untapped hydropower, wind power, solar power, and other renewable source of energy.

The government of Trinidad and Tobago should passionately follow two approaches: (1) conservation, enrichment, bioprospecting, and sustainable harvest; (2) harvestable material use diversification and value maximization. These approaches require adequate knowledge of the biodiversity under consideration: the type, extent, distribution, and character at all levels, as well as continuous research on the innovative and creative applications of the identified characters into various end uses. For example, weeds and otherwise useless organisms have to be transformed into economical commodities, as has been done with *Vernonia spp.*, which was a weed until very recently and has become a multibillion-dollar plant in the paint industry. The limited traditional uses of species and varieties have also to be diversified through innovative and creative research, as has been done in the United States (Easton, 2012; Ehrlich and Holden, 1970).

Trinidad and Tobago should make more efforts to regulate the oil and gas industries effectively. The Environmental Management Agency should be enhanced to hire more experts to coordinate their regulatory functions all over the country. In addition, biodiversity (plants and animals) coming into and going out of the country has to be regulated through the development and use of effective instruments to ensure the safety of the resource base and to fairly and equitably share benefits accruing from genetic materials taken out of the country.

Biodiversity will surely serve as an engine of sustainable development in Trinidad and Tobago if all relevant stakeholders are effectively engaged in environmental conservation, research, development, and marketing, and if they all adequately capacitate themselves and work in tandem in synergic and reinforcing ways to attain and sustain these recommendations. The direction in which the country is headed, in both socioeconomic development and biodiversity conservation, is correct. The balance and synchrony between conservation and socioeconomic development needs adjustment. It is recommended that the government of Trinidad and Tobago should train its citizens in addition to developing the foresight and call for a sustainable way of living in the country that is linked not only by the transport of pollutants and pathogens, but by the very climate that will frame the future of the country.

REFERENCES

Ahmed, B. (1998). "Role of the Environment in the Sustainable Development of the Caribbean." *Social and Economic Studies, 47*(4), 83–97.

Alba, A. (2008). National Forestry Plan (NFP) Update. Information as of 2004, 1–9.

Basiago, A. (1995). "Methods of Defining Sustainability. Sustainable Development," *3*(3), 109–119.

Caribbean Environmental Health Institute Report. (2006). "A National Inventory for the Generation of Hazardous Waste in Trinidad and Tobago." Environmental Management Authority Case Study Series, 2005–2006.

Clifford, M., and Edwards, T. (2012). *Environmental Crime*. Burlington, MA: Jones & Bartlett.

Coke, L. B., Weir, C. C., and Hill, V. G. (1992). "Environmental Impact of Bauxite Mining and Processing in Jamaica." *Social and Economic Studies, 36*(1), 289–334.

Davies, J. M., and Mazumder, A. (2003). "Health and Environmental Policy in Canada: The Role of Water Management in Sustaining Clean Drinking Water Quality at Surface Source." *Journal of Environmental Management, 68*, 273–286.

Dhillon, B. S., Kataria, P., and Dhillon, P. K. (2010). "National Food Security vis-à-vis Sustainability of Agriculture in High Crop Productivity Regions." *Current Science* (January), *11*(2), 37–58.

Dibie, R. (2010). "Social Impact of Environmental Policies in Ethiopia." *Journal of International Politics and Development, 8*(1), 1–23.

Dibie, R. (2009). "Environmental Governance and the Political Economy of Sustainable Development in Sub-Saharan Africa." *Journal of International Politics and Development, 7*(1), 1–23.

Dibie, R., Mello, W., and Benitez, G. (2011). "Social Impact of Environmental Health Policy in Brazil." *Journal of International Politics and Development, 9*(1), 1–23.

Dibie, R., and Offiong, J. (2012). "The Nature and Impact of Environmental Policy in Nigeria." *Journal of International Politics and Development, 10*(1–2), 113–144.

Dye, T. (2011). *Understanding Public Policy.* (13th ed.). New York: Longman, 210–231.

Easton, T. (Ed.). (2012). *Taking Sides on Environmental Issues*. New York: McGraw-Hill.

Ehrlich, P. R., Ehrlich, A. H., and Holden, J. P. *Human Ecology: Problema and Solutions*. 304 Seiten, 8, 79 Abb., 26. Cited in T. Miller and S. E. Spoolman. (2010). *Environmental Science*. Belmont, CA: Brooks & Cole, 23–34.

Friis, R. (2012). *Essentials of Environmental Health*. Burlington, MA: Jones & Bartlett.

Gaskell, G., Eyck, T. T., Jackson, J., and Veltri, G. (2005). "Imaging Nanotechnology: Cultural Support for Technological Innovation in Europe and the United States." *Public Understanding of Sciences, 14*, 81–90.

Gobin, J. (1999). "Land-Based Marine Pollution in Trinidad and Tobago." Institute of Marine Affairs EMA Supplement, Sunday. *Guardian* (June 6), 9. Retrieved from http://www.malis.gov.tt/Agri/agri land-Based-marinePoluttion.html. Accessed July 21, 2011.

Gordon, L., Payne-Sturges, D., and Gee, G. (2010). "Environmental Health Disparities: Select Case Studies Related to Asian and Pacific Islander Americans." *Environmental Justice, 3*(1), 21–26.

Government of Trinidad and Tobago. (2010). *National Environmental Policy for Trinidad and Tobago*. Government Press.

Government of Trinidad and Tobago. (2002). *National Wetland Policy for Trinidad and Tobago*. Government Press.

Harris, J. M. (1992). "Global Institutions for Sustainable Development." In *Sustainable Development and Environmental Policy*, edited by F. Dietz et al. Orange Park, FL: Sigma Press, 37–51.

Johnson, D. G. (2007). "Ethics and Technology in the Making: An Essay on the Challenge of Nanoethics." *Nanoethics*, *1*, 21–30.

Kenny, J. (2008). *The Biological Diversity of Trinidad and Tobago: A Naturalist's Notes*. Port of Spain: Prospect Press.

Kraft, M. (2011). *Environmental Policy and Politics*. New York: Longman.

Kraft, M., and Furlong, S. (2013). *Public Policy: Politics, Analysis, and Alternatives*. Washington DC: Congressional Press.

Levy, B., Wegman, D., Baron, S., Sokas, R. (2011). (Eds.) *Occupational and Environmental Health* (6th ed.). New York: Oxford University Press.

Maxwell, N. (2014). *Understanding Environmental Health*. Burlington, MA: Jones & Bartlett.

McConnell, R., and Abel. D. C. (2008). *Environmental Issues: An Introduction to Sustainable Development*. Upper Saddle River, NJ: Prentice Hall.

Miller, T., and Spoolman, S. E. (2012). *Living in the Environment*. Belmont, CA: Brooks & Cole.

Miller, T., and Spoolman, S. E. (2014). *Environmental Science*. Belmont, CA: Brooks & Cole.

Myers, N., and Spoolman, S. (2014). *Environmental Issues and Solutions*. Belmont, CA: Brooks & Cole.

Mol, A. (2001). *Globalization and Environmental Reform: The Ecological Modernization of the Global Economy*. Cambridge, MA: MIT Press.

Peters, G. (2013). *American Public Policy: Promise and Performance*. Washington, DC: Congressional Quarterly Press.

Ramon, E. (2009). "Prospects for the Oil Importing Countries of the Caribbean." Inter-American Development Bank CCB Working Paper CSI-114.

Schueler, T. (2000). "Microbes and Urban Watersheds: Concentrations, Sources, & Pathways." *Watershed Protection Techniques*, *3*(1), 554–565. Retrieved from http://www.cwp.org/Resource_Library/Center_Docs/PWP/ELC_PWP17.pdf

Selin S., and Pierskalla, C. (2005). "The Next Step: Strengthening the Social Science Voice in Environmental Governance." *Society and Natural Resources*, *18*, 933–936.

Smith, Z. (2004). *The Environmental Policy Paradox*. (4th ed.). Upper Saddle River, NJ: Prentice Hall, 82–98.

Sonnenfeld, D. (2000). "Contradictions of Ecological Modernization: Pulp and Paper Manufacturing in Southeast Asia." *Environmental Politics*, *9*(1), 235–256.

Spencer, D. (2006). "Trinidad and Tobago's NPA, Progress and Lesson Learned," Paper presented at the 4th World Water Forum, Mexico City (March 18), 9–13.

Tikka, P., Kuitunen, M., and Tynys, S. (2000). "Effects of Educational Background on Students' Attitudes, Activity Levels, and Knowledge Concerning the Environment." *Journal of Environmental Education*, *31*(3), 12–19.

Trinidad and Tobago Biodiversity. (2009). *Republic of Trinidad and Tobago Biodiversity Clearing House Report*. Government of Trinidad and Tobago.

United Nations. (2011). *The Economic of Ecosystems and Biodiversity in National and International Policy Making*. New York: UN Environmental Program Publication.

Van den Eynden, V., Oatham, M. P., and Johnson, W. (2008). "How Free Access Internet Resources Benefit Biodiversity and Conservation Research: Trinidad and Tobago's Endemic Plants and Their Conservation Status." *Oryx*, *42*(3), 400–407.

Vig, N., and Kraft, M. (2013). *Environmental Policy: New Direction for the 21st Century*. Los Angeles, CA: Sage Press.

Waller-Hunter, J., and Jones, T. (2002). "Globalization and Sustainable Development." *International Review for Environmental Strategies, 3*(1), 53–62.

Withgott, J., and Brennan, S. (2008). *Environment; The Science Behind the Stories*. San Francisco: Pearson.

Woods, N., and Potoski, M. (2010). "Environmental Federalism Revisited: Second-Order Devolution in Air Quality Regulation." *Review of Policy Research, 27*(6), 721–727.

World Bank. (1995). "Trinidad and Tobago—Memorandum on Recent Economic Developments and Prospects." Retrieved from http://documents.worldbank.org/curated/en/1971/05/1559509/trinidad-tobago-memorandum. Accessed November 20, 2013.

Wright, R., and Boorse, D. (2014). *Environmental Science: Toward a Sustainable Future*. Boston: Pearson.

14 Environmental Policies and Issues in Belize

Emmanuel Iyiegbuniwe, PhD

INTRODUCTION

Belize, the only Anglophone (English-speaking) country in Central America (previously known as British Honduras), had a population of 312,698 as of 2010 (Statistical Institute of Belize, 2011). Belize is a very small country, only slightly larger than Massachusetts, with a total area measuring about 22,966 square kilometers (22,806 and 160 square kilometers of land and water, respectively). The World Bank also regards Belize as a small, ethnically diverse middle-income country with a stable democracy (World Bank, 2000).

Environmental governance is rooted in the creation of the Belize Environmental Protection Act (BEPA) in 1992. BEPA was modeled after the United States National Environmental Policy Act (NEPA), which emanated from the U.S. environmental movement of the 1960s and 1970s and was signed into law in 1969 following the publication of Rachel Carson's *Silent Spring*. The book discussed the impact and environmental effects of DDT and other pesticides and their responsibility in causing the endangerment of the bald eagle (due to the thinning of the eggshells, preventing the hatching of the chicks). Environmental policy agencies have increasingly attempted to base their rule in evidence, through the use of quantitative risk assessment. Belize's Department of the Environment (DOE) was established under BEPA with a mandate to ensure the protection and sustainable use of natural resources in the country. To this end, the DOE is charged with the power to "issue the necessary licenses, with or without conditions, for use for the exercise of activities that may cause pollution" and to "prevent and control pollution by coordinating all activities relating to the discharge of wastes into the environment (Government of Belize (GoB), 2001; GoB, 2000).

BEPA also created the environmental impact assessment (EIA) for each major infrastructure project in Belize (Hershowitz, 2008). Environmental impact assessment (EIA) reports are evaluated by the National Environmental Appraisal Committee (NEAC), chaired by the head of Belize's Department of Environment (BDOE). NEAC is an eleven-member interdisciplinary technical group that consists of nine designated government officials from various agencies and two civil society representatives. The main responsibilities of the NEAC include reviewing all EIAs, advising BDOE of the "adequacy or otherwise" of EIAs reviewed, and advising BDOE when a public hearing on an EIA is "desirable and necessary" (Hershowitz, 2008). Similar to the EIA is the health impact assessment (HIA), which in some ways may be viewed as a natural extension of the EIA because it examines the effect that a policy, program, or project may have on public health and encourages decisions that protect and enhance health and health equity. BEPA was signed into law with specific requirements. The responsible official (both private and government projects) must prepare a detailed environmental impact statement for large projects, any adverse environmental effects that cannot be avoided, alternatives to the proposed action, the relationship between local short-term uses of the human environment and long-term productivity, and any irreversible commitment of resources. An EIA must contain "measures which a proposed developer intends to take to mitigate any adverse environmental effects and a statement of reasonable alternative sites (if any), and reasons for their rejection."

"Every project, program or activity shall be assessed with a view to the need to protect and improve human health and living conditions and the need to preserve the reproductive capacity of ecosystems as well as the diversity of species." Public participation in the form of consultation both during and after the preparation of the EIA, as well as the provision for a public hearing in some circumstances, is also required.

Protected areas are generally defined to include any area where specific measures of some form are put in place to conserve the natural and cultural heritage of the country (i.e., nature reserves, national parks, wildlife sanctuaries, national monuments, marine reserves, and forest reserves). These include all management approaches that are in the internationally accepted meaning of the term, ranging from strict nature reserves and national parks through areas designed to conserve biodiversity while delivering a steady supply of environmental goods and services as well as areas where conservation is fully integrated with human settlement and economic activity.

As defined by the Government of Belize, the key statement on the role and management of protected areas is the National Protected Area Policy (Government of Belize (GoB), 2000; GoB, 2001). Under international agreements to which Belize is a signatory, the National Protected Area System (NPAS) was created with a view to developing a coherent framework that meets all obligations. The NPAS stipulates a comprehensive representation of all ecosystems in the country, including areas providing important environmental services, possessing exceptional scenic values, and providing critical habitat for species of conservation concern or economic importance. It also requires the integration of regional and national approaches in order to effectively promote biological connectedness. One of the NPAS policy goals is to enhance the economic, social, and ecological sustainability of Belize. The system also seeks to optimize socioeconomic benefits that are compatible with maintaining biodiversity values and sustainable resource management. It is also expected to ensure the equitable distribution of these benefits and public awareness of their importance.

The NPAS policy stipulates that the management of the system should be transparent, geared toward the delivery of measurable benefits, and emphasize public participation at all levels. This applies to the establishment, management, modification or dereservation of all the protected areas included in the national network. According to the National Protected Area Policy, "A comprehensive Protected Area Policy is drafted that ensures increased social and economic benefit while guaranteeing core protected area objectives." Also, the National Protected Area System Plan was designed to implement the National Protected Area Policy and focuses on the following strategic actions, reproduced here from the executive report submitted by the Protected Areas Task Force (Meerman and Wilson, 2005). The goals of the policy include:

1. Establish a National Protected Areas Commission to ensure coordinated action in system development.
2. Revise the National Parks System Act as the National Protected Areas System Act in order to give legislative underpinning to the plan. Present legislation governing marine reserves requires no modification. This places overall responsibility for the system to the Ministry of Natural Resources but leaves responsibility for individual sites unaltered.
3. Provide support services to protected area managers across the system in order to enhance management capacity. This includes a national training program and site-specific assistance in administration, financial management, survey and research, as well as management and business planning. The aim of the policy is to make for more cost-effective and -efficient management plus a business-like management culture and approach to securing support.
4. Secure comprehensive coverage by concentrating attention on gaps in the present network. Action directed toward the following areas will greatly strengthen the system: northern Belize to the south and west of Shipstern, the central northern coastal plain, the east-central Belize Valley, the karst hills of western Toledo, Turneffe Atoll, the Moho River and parts of the Rio Hondo and New, Belize, and Temash rivers. These areas include private lands, and suitable incentives must be explored to bring them into the national system.
5. Simplify the existing system by consolidating adjacent protected areas into single, multizoned, management units. This creates a smaller number of sites that are individually more important. Indeed, some will be of exceptional importance on a regional, even global, scale. Special attention

> will be devoted to the Barrier Reef, the Maya Mountain-Mountain Pine Ridge massif, and the northwestern forests. The Protected Areas Task Force believes that the plan would create a more effective protected area system and produce tangible benefits in a simpler and more cost-effective manner (Meerman and Wilson, 2005).

This chapter examines the impact of environmental health policies in Belize, a country on the Caribbean coast. It uses survey data to analyze the social impact of environmental health and policies in Belize. The challenges of biodiversity loss in Belize and pollution must be addressed by policy makers with equal priority and in close coordination, if the most severe impact of each is to be avoided in the near future. The chapter reviews social and economic evidence that speaks to the various environmental problems and the policy implications. It describes the natural resource policies and the many conflicts between natural resources preservation and economic development in Belize. New directions for policy change in the country are presented.

ENVIRONMENT, HUMAN BEINGS, AND SOCIAL ADVANCEMENT

Environment is central to human advancement. The relationship between the environment and human beings continues to be radically altered by innovation in economic scientific, technological, economic, and social advancement (Dibie and Offiong, 2012). Environmental social theories developed in the past decades view human beings as both "biologically embodied" and "ecologically embedded." A perennial theme in developing these theories lies in the significance and value of this relationship, which is as old as human society. A long established thought in environmental theory is that humans have maintained a very strong relationship with the natural environment. This relationship has been radically altered by our scientific, technological, economic, and social advancement, and thus it has outstripped our ability to cope with environmental changes (Vanderheiden, 2011). However, recent developments in environmental theory and research offer hope on how human behaviors contribute to environmental problems and the need to build a stronger understanding in order to effectively alter such behaviors (Rosa and Dietz, 1998; Stern, 2000; Stern et al., 1992 & 1999). These scholars generally agree that the manner in which humans understand value impacts health behavior and have developed theories to guide human actions and align facts with value.

Human actions can potentially impact the environment, and the environment will, in turn, adversely affect public and environmental health, resulting in the implementation of controls or remedies for such harmful actions (Stern et al., 1999). Political debates on environmental protection are often strongly influenced by the economic-instrumental value of the environmental landscape (Vanderheiden, 2011).

Environmental theories are generally classified into two categories: explanatory and normative (Vanderheiden, 2011). These theories generally prescribe some set of values and recommended actions or remedies for those values. Explanatory environmental theory is based on providing an account of the nature of environmental phenomena and their relationship or causal connection to humans or public health, and vice versa (Vanderheiden, 2011). This theory is clearly illustrated by an explanation of the consequences of global warming and climate change on public and environmental health (Frumkin, 2010). For example, global warming impacts glaciers and the oceans, leading to potential rise in sea level, and affects agriculture and food productions as well as potable water supplies for large cities (Rom, 2012). Normative environmental theories seek to explain or predict what would happen under theoretical constraints.

An application of the normative environmental theory is the argument in favor of establishing biodiversity laws for the protection of endangered species (Vanderheiden, 2011). In a coherent theoretical system, a combination of fact-based explanation and normative value judgments results in distinct discourses that characterize environmental problems (Vanderheiden, 2011). Summaries of a few environmental theories and their applications follow.

Ecological limits theory is a first-generation normative environmental theory based on the premise that natural resources are finite and the environment has a limited capacity for pollution. Public health has its origins in ecology and relates environmental factors to disease causation and possible endpoints. A good illustration of this theory is the 1962 publication of *Silent Spring* by Rachel Carson, which ushered in the modern environmental movement in the United States (Vanderheiden, 2011). Carson discussed the public health and environmental effects of DDT, an organochlorine pesticide that was widely used in agriculture since World War II. She noted that DDT had entered the environment, causing thinning of the eggshells of the bald eagles and other song birds, thus preventing the hatching of the chicks (Rom, 2012).

The widespread use of DDT not only threatened the populations of bald eagles but was associated with increased rates of cancer and birth defects in humans. Carson's legacy included her observation that public health was largely influenced by large environmental processes, and a few decades later, the EPA named its main meeting hall Rachel Carson Great Room (Rom, 2012).

The ecologic limit theory is also illustrated by the 1968 landmark bestseller titled *The Population Bomb* by Paul Ehrlich and Ann Ehrlich, in which they discussed environmental pollution and resource depletion and the critical choices that humans still face today (Vanderheiden, 2011). The ecologists proposed an agenda for avoiding global ecocide in the 1990s by prescribing sustainable population growth. Another classic case of this theory was Garrett Hardin's (1968) analysis in "The Tragedy of the Commons," in which he stated that "[t]he population problem has no technical solution; it requires a fundamental extension in morality." According to Hardin, in medieval land tenure in European, herders shared a common parcel of entitled land to graze their cows. By adding additional cows to graze on the land, each herder received all the benefits whereas damages to the commons were shared by the entire group. This is akin to modern-day ranchers grazing their cattle on government lands in the American West or overfishing of the world's oceans (Hardin, 1968).

The environmental value theory represents the second generation of environmental theories. It focuses on an ecocentric approach and was developed to include nonhumans or whole ecosystems as having value and requiring respect (Eckersley, 1992; Ibtissem, 2010). This theory has been successful in critiquing the anthropocentric value theory that views nonhumans as having no morally relevant interest or good of their own. It assumes that natural resources are limited and valuable and offers accounts of what is genuinely important and why. In her critique, Eckersley (1992) noted that anthropocentric theories were irreconcilable with strong environmental protection.

The environmental value theories have been successfully applied in advocating for the need to base environmental policy decision making on values that recognize the importance of our reliance on the natural world (Vanderheiden, 2011). It argues in favor of reforming unsustainable environmental practices and the need to implement a more reflexive environmental decision-making process since most contemporary environmental problems have resulted from actions of narrowly defined instrumental value to human welfare (Vanderheiden, 2011). The rise of the modern environmental movement was attributed to the growing acceptance of a new ecological paradigm that implicates human actions as adversely affecting our fragile biosphere (Stern et al., 1999).

NATIONAL APPROACHES TO ENVIRONMENTAL HEALTH

Belize has a density of approximately 10 persons per square kilometer; it has one of the lowest population densities in Central America and one of the lowest in the world (GoB, 2007). Geographically, it is located between 15°52′ and 18°30′ north latitude and 87°28′ and 89°14′ west longitude in the heart of the Caribbean Basin (on the east coast of Central America). It is bordered by Mexico to the north and by Guatemala to the south and west, and it is east of the Caribbean Sea. The country consists of two physiographic or topographic regions (Young, 2008). These are the northern lowlands (primarily composed of limestone and sandy soils, semideciduous forests, wetlands, and savannahs) and the southern coastal plains

Table 14.1 Belize: Factsheet at-a-Glance

Location	Situated on the Caribbean coast of Central America: Guatemala to the west and south and Mexico to the north
Area	22,966 km^2 total; 22,806 km^2 land and 160 km^2 water (slightly larger than Massachusetts)
Natural resources/ land use	Arable land, timber, and marine products (2% arable land, 1% crops, 2% pastures, 92% forests and woodlands, 3% other uses)
Climate	Tropical
Population	321,115 (July 2011 est.)
Population growth and fertility rates	2.056% and 3.21 children/woman (2011 est.)
Birth and Death Rates	26.43 births/1,000 and 5.87 deaths/1,000 (2011 est.)
Life expectancy	68.23 years total; male = 66.53 years; female = 70.02 years (2011 est.)
Infant mortality rate	21.95 deaths/1,000 live births (2011 est.)
Maternal mortality rate	94 deaths/100,000 live births (2008)
Net migration rate	0 migrant(s)/1,000 population (2011 est.)
Ethnic groups	48.7% Mestizo, 24.9% Creole, 10.6% Maya, 6.1% Garifuna, and 9.7% other (small populations of East Indians, Chinese, and other)
Religion	49.6% Catholic, 27% Protestant, 9.4% none, 14% other (2000 est.)
Languages	Spanish 46%, Creole 32.9%, Mayan dialects 8.9%, English 3.9% (official), Garifuna 3.4% (Carib), German 3.3%, other 1.4%, unknown 0.2% (2000 census)
Legal system/government	English Law/parliamentary democracy: 2 parties: United Democratic Party (UDP) and People's United Party (PUP)
Capital/districts	Belmopan/6 districts (Belize, Cayo, Corozal, Orange Walk, Stann Creek and Toledo)
GDP per capita	$4900 (22.7% agriculture, 24.5% industry, 52.8% services)
Industries	Garment production, tourism, construction, and agriculture
Currency equivalent	BZ $ 2.00 = US $1.00 (fixed rate)

Sources: United Nations (2010); Belize Demographics Profile (2013).

and the Maya Mountains (composed of broad-leaf forest interspersed with pine and savannahs). Belize has a very diverse ethnicity and is made up of 48.7 percent Mestizo, 24.9 percent Creole, 10.6 percent Mayans (Q'eqchi', Yucatec, and Mopan), 6.1 percent Garifuna, and 9.7 percent other (i.e., small populations of East Indians, Chinese, and a sizable migrant population from nearby countries) (CSO, 2000). Table 14.1 presents basic summary information on Belize.

The population of Belize, with a median age of twenty-two years, has historically been rural (Statistical Institute of Belize, 2011). However, recent external and internal migration from surrounding Central American countries, notably Guatemalans, Salvadorans, and Nicaraguans, has been reported in cities and towns, with the majority of the people living in the coastal regions (World Bank, 2000; Young, 2008). Housing stock, particularly in the poorer districts of Belize, has not kept pace with the influx of immigrants. Unfortunately, this ethnically diverse and sparsely distributed population often presents an enormous challenge for the delivery of social services (World Bank, 2000).

The administrative capital of Belize is Belmopan. The name "Belmopan" comes from the union of two words *Belize*, the name of the longest river in the country, and *Mopan*, one of the rivers that empty into the Belize River. There are six distinct political districts in Belize: Belize, Cayo, Corozal, Orange Walk, Stann Creek, and Toledo. An elected official from each district is a member of the House of Representatives. Each district has a unique characteristic with a major town serving as its center of government. For example, the Belize district is the most populous and considered the center of commerce, whereas the Toledo district is home to the largest concentration of Mayas in the country with small Mayan villages inhabiting the foothills of the mountain range known as the Maya Mountains, the most prominent physiographic feature of Belize's mainland. Belize has a unique combination of topography, soils, and vegetation and has been divided into nine land systems, each with a distinctive landscape (King et al., 1992). However, as part of a recent classification scheme for the Protected Areas Systems Plan (NPASP), forty-nine distinct types of natural vegetation have been identified and recognized.

Belize is endowed with unique and abundant natural resources, including the largest barrier reef in the Americas and pristine tropical forests. It has a very rich and interesting history and civilization since it was part of the great Mayan empire that dates back more than 4,000 years. The more advanced Mayan civilization was reported to have existed in the third century CE, reaching its peak between the sixth and eighth centuries. In recorded history, the Mayan civilization was in such serious decline by the fourteenth century that many of the cities had been deserted when the Spanish arrived in the sixteenth century. It is important to note that the Spanish "owned" Belize but did not rule it and thus considered it a backwater suitable only for cutting the logwood used for dye making. It is equally interesting to recognize that during the seventeenth century, a lack of effective government in Belize, coupled with the safety due to the barrier reef, attracted other European invaders, notably English and Scottish pirates.

Belize has over 200 cayes (pronounced "keys"), consisting of offshore atolls or islands, mangroves, and barrier reef. The cayes are found between the Belize mainland and the barrier reef, on the barrier reef, and on or within the barrier reef perimeters of the offshore atolls (ISLP, 2005). As a main attraction, Belize has one of the most extensive reef ecosystems (Gibson, McField, and Wells, 1998). The 230-kilometer long barrier reef is the largest and longest Atlantic reef system in the Western Hemisphere and the second longest reef in the world, with atolls and a complex network of inshore reefs. The reef is found on Belize's Caribbean coast, extending the entire length of the country from the Mexican border in the north and to the Sapodilla Cayes in the south. In 1996, the barrier reef was declared a world heritage site by the United Nations Educational, Scientific and Cultural Organization (UNESCO), an important environmental event with far-reaching implications for the sustainability of marine resources and the perpetuity of potentially endangered species. Table 14.1 shows the environmental and economic profile of Belize.

The coastal zone contains a rich and diverse array of unique marine ecosystems, including patch reef, faros, coastal lagoons, and estuaries. Suffice it to note that the cayes and atolls serve as natural habitats for birds, turtles, and other marine organisms. Belize retains 93 percent of its tropical forest and has a very rich endowment of both natural and cultural resources with global importance (Earthwatch Institute, 2006). Additionally, the coastal zone provides significant recreational and cultural opportunities (water sporting and leisure activities, including snorkeling, scuba diving, fishing, sea kayaking, sailing, and boating) for Belizeans and visitors alike. It must be noted that, although Belize has been able to preserve its environmental resource capital during the past several years, it still faces substantial and serious environmental public health threats.

Environmental health policy is used in a broader general term to include policies that affect the relationship between public health and the environment. A number of nongovernmental environmental organizations and entities in Belize have actively participated or provided support or opposition on major environmental issues and policies of public health importance. They include the Belize Audubon Society (BAS), Belize Zoo, and Tropical Education Center (TEC), Monkey Bay Wildlife Sanctuary, and Wildlife Conservation International (WCI).

Several existing key environmental health policies and legislation are promulgated by the Government of Belize. Some of these policies (e.g., the National Biodiversity Strategy Action Plan, 2009 Water Resources Management Act, 1992 National Lands Act, and 1999 Coastal Zone

Management Act) may directly or indirectly target public health and general environmental outcomes. In addition, Belize is a signatory to several environmental agreements, international legal obligations, and a number of binding agreements to address both national and international environmental issues (Belize Audubon Society, 2008). Examples are the Caribbean and Central American Integration Systems—the Caribbean Community Single Market and Economy (CSME), based on the revised Treaty of Chaguaramas and the Central American Integration System (SICA). In this chapter, two important environmental policies are discussed: BEPA and the National Parks System, Fisheries and Forest Acts. There are many environmental health issues in Belize, but this chapter will discuss just three important issues: (1) waste management; (2) habitat alteration, deforestation, overfishing and unregulated development; and (3) natural disasters and global warming.

WASTE MANAGEMENT

Waste management is one example of an important environmental health problem that has become part of life in modern-day Belize. Waste is an important by-product of human activities and therefore constitutes every citizen's business. It is a natural consequence of consumption and has been particularly problematic in several urban populations in a number of municipalities in Belize. The waste stream is everything that human beings throw away, and the main categories include solid waste (or garbage), liquid waste (or wastewater from washing and bathing), and sewage. Other classifications are municipal solid waste, special waste, hazardous waste, and liquid effluents resulting from both domestic households and commercial establishments that may ultimately lead to water and soil pollution. Historically, solid waste and sewage disposal in Belize, as with most developing or underdeveloped countries, occurred by directly dumping them into open waters (canals, rivers, lakes, or sea) in unsanitary conditions.

Waste management remains to this day a serious problem of environmental public health importance in Belize. According to a 2008 report by the Belize Audubon Society, approximately 130,000 tons per annum of solid wastes are generated in Belize (Belize Audubon Society, 2008). Unfortunately, the Belize Audubon Society's report noted that most of the solid wastes are still being inadequately disposed of at various illegal and unacceptable dump sites throughout the country. Many parts of the country still depend on septic systems and pit latrines (Figure 14.1).

In 1997, a Japanese government–funded technical assistance project aimed at developing a national solid waste management plan for Belize failed in its implementation due to various reasons including "political will and neglect" (Belize Audubon Society, 2008). With an increasing population and major focus on ecotourism, there was need to address the solid waste

Figure 14.1 Typical Pit Latrine in a Mayan Village of Red Bank, Belize
Source: Photo taken by author during field research in Belize, 2010

management problems, especially in four municipalities: San Pedro Ambergris Caye, Caye Caulker, Belize City, and San Ignacio/Santa Elena (SWMA, 2011).

As part of its support for environmental protection, natural resource conservation, and protection of public health, safety, and welfare, the Government of Belize has proposed, enacted, and amended several pieces of legislation to deal with its ever burgeoning solid waste problems (Belize Audubon Society, 2008).

In the early 1980s, the Government of Belize undertook a major project that involved the development and implementation of a central waste treatment system in Belize City. Additionally, a number of legislations govern the production, storage, collection, transportation, treatment, and disposal of solid wastes. They include Littering Offences Regulations (Amendment), 1996; Pollution Control Regulations, 1996; Solid Waste Management Authority Act, 1991; and Derelict Vehicles Regulations, 1999. These solid waste legislations are managed by the Belize Solid Waste Management Authority (SWMA), which is part of the Ministry of Natural Resources and the Environment. SWMA has representatives drawn from both the Government of Belize and the private sector and has recently proposed a project that would involve a new infrastructure for solid waste management in the Western Corridor of Belize. SWMA anticipates that the new infrastructure project will meet both operational and environmental performance standards and will be implemented under a design-build-operate (DBO) contract as follows (SWMA, 2011):

- Closure of dumpsites at the four municipalities
- Design, build, and operate transfer stations at the four municipalities
- Incorporate recyclable material drop-off areas and designated areas for white goods/bulky waste and HHW at the transfer stations
- Design, build, and operate the Regional Sanitary Landfill at Mile 24 Western Highway
- Transportation between transfer stations and the Mile 24 Regional Sanitary Landfill
- Barging of waste from San Pedro Ambergris Caye and Caye Caulker to the Belize Port for long haul to the Mile 24 Regional Sanitary Landfill

HABITAT ALTERATION, DEFORESTATION, AND OVERFISHING

Deforestation, overfishing and loss of biodiversity, due to the continued alteration of both terrestrial and aquatic habitats, affect the indigenous people of Belize. Worldwide, there are more than 33,000 protected areas, and 30 percent of Belize is under some sort of protected status (Wyman and Stein, 2010; Eagles, McCool, and Haynes, 2002). Despite this protected status, there still exists illegal extraction of the forests in Belize, resulting in the rising levels of pollution from cruise ship tourism and leading to the degradation of mangroves and coral reefs. This has made it essential to emphasize the role that the local communities living within or around protected areas play in the decentralization of resource management and in conservation, Community-based organizations could contribute to resource management by proposing revenue-generating schemes and incentives as alternative methods. In general, the poor people, especially in the districts of Toledo and Cayo, live in dwellings constructed of wooden walls, thatched roofing, and earthen floors.

The protected Maya Mountains of Belize are surrounded by a diverse range of habitat including lowland moist rain forest, moist forest, and cloud forest, all containing several endangered species (Kaiser, 2011; IUCN 2010). All these areas are farmed with the unrestricted and rampant use of pesticides. Because agricultural chemicals tend to bioaccumulate in the montane cloud forests around farm areas, it is essential to evaluate the type and concentration of pesticide drift that may potentially impact these protected areas. In a preliminary study to evaluate the effect of possible pesticide drift from the numerous multinational farms conducted over a period of three years, common organochemicals, including glycophosphates, organophosphates, and carbonates, were tested and found to be present in the protected Maya Mountain area (Kaiser, 2011). This poses a threat to public and environmental health due to downstream drift and potential contamination of municipal water sources and streams used for bathing, swimming, and livestock. It must be noted that pesticides are toxic and tend to bioaccumulate and biomagnify in fish and other aquatic organisms.

The production and use of most organochlorine pesticides (OCs) have been banned or discontinued in the industrialized countries of North America (United States and Canada) and Europe. However, many of these OCs (e.g., DDT, HCH, chlordane, heptachlor, toxaphene, Dieldrin and Aldrin, etc.) are currently used for agriculture and vector control in a number of tropical and subtropical countries, including Belize (Alegria, Bidelman, and Shaw, 2000). In general, OCs and persistent organophosphates (POPs) are known to undergo global distillation into the ambient atmosphere and are often transported to colder climatic regions of the world. Alegria and colleagues (2000) conducted a study to determine the levels of OCs in the ambient air in Belize between December 1995 and August 1996. Samples were collected from Belize City (highly urbanized) and Belmopan (rural with extensive farmlands). The results showed that airborne levels of selected OCs were elevated in Belize, mostly in the inland areas of Belmopan when compared to levels in the Great Lakes region of North America.

The study supported the notion that pesticide use and application in the Central American region constitutes a potential source of ambient air emissions of pesticides to North America. It also suggested that the continued indiscriminate use of these chemicals in Belize would constitute a potential source of OCs and POPs to North America (Alegria et al., 2000).

In a related article, Murray and colleagues (2002) reviewed key potential environmental health issues and progress with regard to the surveillance of pesticide-related illness in Central America. The study focused on the benefits of collaborative programs such as the Occupational and Environmental Aspects of Exposure to Pesticides in the Central America Isthmus (PLAGSALUD) pesticide health project. PLAGSALUD was developed over a ten-year period (between 1994 and 2003) through the collaborative efforts of Central American Governments, the Pan-American Health Organisation (PAHO), and the Danish Agency for International Development (PAHO, 2002; Murray et al., 2002). The participating Central American countries included Belize, Costa Rica, El Salvador, Guatemala, Honduras, Nicaragua, Panama and Regional Level. The PLAGSALUD project was aimed at addressing certain environmental and public health issues, including river basin pollution, deforestation, loss of species and contamination of the food chain caused by the indiscriminate and extensive use of pesticides in Central America (PAHO, 2002). Key subject areas for the project were identified as epidemiologic surveillance, developing alternatives to pesticide use, intersectorial and interinstitutional coordination, pesticide legislation, and education and research (PAHO, 2002; Murray et al., 2002). The authors noted that policy makers have continued to recycle previous imprecise and misinformed surveillance results. Hence, the need for improved and coordinated use of available PLAGSALUD data by policymakers for analysis, interpretation, presentation and increasing awareness of the general public cannot be overemphasized (Murray et al., 2002).

Konradsen and colleagues (2003) summarized the results of a study that focused on the problems of self- and occupational pesticide poisoning in developing countries. A number of studies have shown that pesticide self-poisoning is one of the most common methods of poisoning in rural areas, thus highlighting the inadequate use or complete lack of safety equipment (Eddleston, 2000; Konradsen et al., 2003). The problems were attributed to the cost of purchasing the safety equipment as well as the discomfort of wearing it during pesticide application in humid tropical climates, thus leading to increased incidence of occupational illness from pesticide poisoning (Karalliedde, Eddleston, and Murray, 2001; Konradsen et al., 2003). Some of the suggestions that have been made are restricted access to the most harmful pesticides; develop, promulgate, and enforce national policies, laws, and regulations that would phase out World Health Organization's Classes I and II pesticides; encouraging the use of cheaper and safer pesticide alternatives; and the development of adequate and effective health programs to address some of the factors attributed to self-poisoning, including social, mental, and physical distress, substance abuse, and financial problems (Konradsen et al., 2003).

As a financial incentive for additional income for smallholder growers and to alleviate their dependence on the forest, Rosenberg and Marcotte (2005) emphasized the usefulness of cultivating shaded cacao (*Theobroma cacao*) in Belize. The authors also noted the need to develop policies and research to promote the use of land use systems (LUS). Accordingly, LUS would make available goods and services that enhance and sustain the source of revenue for rural smallholders. The authors stated that an LUS model was previously developed to show the financial gains of shaded cacao cultivation in the Toledo district of Belize. The results were positive and showed that, to enhance cacao financial performance and encourage implementation, the use of labor-saving technology, better shade management practices, grafting, and incorporating nonhardwood shade trees and laurel proved to be good strategies (Rosenberg and Marcotte, 2005).

Tourism is a good example of a community-based revenue-generating conservation program. This stems from the fact that, through tourism, the community would preserve its natural resources while reaping its benefits (Wyman and Stein, 2010). Tourism is the lifeblood and mainstay of the economy of Belize and is the country's largest foreign exchange earner. Tourism is also vital to the interests and prosperity of the people of Belize with the Barrier Reef serving as the main source of attraction. According to the Belize Tourism Board, approximately 60 percent of visitors participate in a marine experience, exerting a heavy burden on the coastal resources, especially the reefs and terrestrial and marine protected areas. As part of the national system, Belize has an extensive network of 94 protected areas that represent approximately 36 percent of terrestrial and 13 percent of marine areas, respectively. Marine and terrestrial protected areas, including statutory sites and private protected areas, are considered key repositories for sustainable development (Meerman and Wilson, 2005). Traditionally, Belize's economy was based on the agricultural sector, which is dominated by the sugar and citrus industries. Agriculture accounts for over 50 percent of total exports and about 10 percent of the country's gross domestic product (GDP). However, during the past several years, especially in the 1990s, the service sector has surpassed agriculture and accounted for 60 percent of GDP (GoB, 2007).

As a shared regional ecosystem, the Central American Commission on Environment and Development (CCAD) executed the Conservation and Sustainable Use of the Mesoamerican Barrier Reef System (MBRS I) Project Phase I, which was aimed at assisting Belize, Guatemala, Honduras, and Mexico in strengthening and coordinating national policies, regulations, and institutional arrangements for the conservation and sustainable use of the MBRS. The project served as a model for regional coordination and joint management of transboundary resources. Additionally, it was successful in catalyzing international cooperation among the four affected countries. It is anticipated that the next phase (MBRS II Project) will consolidate ongoing efforts to coordinate national policies for the conservation and sustainable use of the shared reef system, to broaden its multisectorial focus on tourism and agriculture, and to promote the adoption of biodiversity-friendly activities in both the local communities and the private sector.

In Belize, overfishing has resulted in serious declines in the levels of fish caught and the varieties of fish species in various aquatic habitats. In a 2005 article titled "Lobster and Conch Fisheries of Belize: a History of Sequential Exploitation," Miriam Huitric lamented the problems associated with overfishing in Belize. According to Huitric, the application of new technology and organizational techniques have drastically increased the area of fishing grounds and maintained the yield for small fisheries. In 2001, Hurricane Iris caused damage estimated at $475 million (Huitric, 2005). Hurricane Mitch of 1998 destroyed reefs throughout Belize (Andrefouet et al., 2002; Huitric, 2005), and in 2000, Hurricane Keith destroyed fishing grounds for lobster and conch after hitting the northern lagoons and cayes, with a 60 percent decline in lobster yield in Caye Caulker alone (Huitric, 2005). This has encouraged exploitation and discouraged the need for adopting sustainable methods of fishing, proper regulations, and learning the effects on the ecosystem. Additionally, the widespread dependence on fossil fuels and the impact of hurricanes have contributed to the depletion of stocks. Huitric further noted that during the last hundred years, six hurricanes have had a huge impact on the infrastructure and fishing grounds in Belize. These included Hurricanes in 1931, 1961, 1978, 1998, 2000, and 2001.

The good news is that a number of programs are in place to empower the local people in Belize and other Central American countries to develop and manage their lands in a sustainable manner. Such a program is the

Integrated Ecosystem Management in Indigenous Communities Regional Program (IEM), which is coordinated by the Central American Indigenous and Peasant Coordination Association for Community Agroforestry, the Central American Indigenous Council, and Central American Commission for Environment and Development (CCAD) (Kessler, 2005). The program targets communities in Belize and other Central American countries by providing needed financial support for small projects amounting to $11.5 million from the Inter-American Development Bank (IDB), the World Bank, and CCAD, along with other indigenous groups. According to Kessler (2005), the main objective of the program was to assist indigenous people to develop and manage conservation areas, as well as to finance income-generating projects such as tourism, forestry, and the production of traditional products like coffee, cocoa, and handicrafts. The long-term implication of this program will include enhanced public health; sustainable development due to the reduction in deforestation, erosion, and loss of biodiversity; and the preservation of indigenous lifestyle in Belize (Kessler, 2005). Marine protected areas (MPAs) have increasingly become the best method of protecting fisheries from overfishing.

In December 2010, a ban was instituted on bottom trawling in territorial waters because of its documented destructive impact on the marine ecosystems. It must be noted that policy is different from practice because practice does not always follow the implementation of legislation. Often leaders believe that they have established environmental health policy, only to realize those policies may not work as desired or may have unintended consequences. The enforcement of such environmental policies and management actions may be inadequate due to a lack of operational capability, poor implementation, or weak institutional capacities (Belize Audubon Society, 2008). On the other hand, environmental civil organizations are very strong and have continuously participated actively in the formulation of environmental policies for the management of protected areas.

The massive Chalillo Dam, a large multinational corporation infrastructure project, is a good example of a controversial or inadequate implementation of environmental policies in Belize. The BEPA Chapter 328 covers issues relating to EIA, and it states, among other requirements:

> Every project, program or activity shall be assessed with a view to the need to protect and improve human health and living conditions and the need to preserve the reproductive capacity of ecosystems as well as the diversity of species. When making an environmental impact assessment, a proposed developer shall consult with public and other interested bodies or organizations.

In a 2008 article published in the *Ecology Law Quarterly*, Hershorwitz discussed the implications and potential environmental impact of such foreign infrastructure projects in Belize. The proposed Chalillo Dam over the Macal River in the Cayo district of western Belize was the brainchild of a Canadian-based company. The building of the dam was originally approved in 2005, but only after a long drawn-out debate that was eventually settled in the highest court of appeal for Belize, the Judicial Committee of the Privy Council in London (Hershowitz, 2008). This project is clearly recognized as the first environmental lawsuit in nearly a decade after the BEPA legislation was enacted in 1992. Suffice it to note that the Government of Belize had accepted the environmental impact assessment submitted by the management, citing the need for cheaper power in Belize for their positive support.

Environmentalists and other opponents of the project have pointed out the consequences of dam failure on endangered wildlife and the geology of the area. Environmental groups in Belize have argued that the EIA approved for the dam was fundamentally flawed because it contained errors describing the geological composition of the dam's foundation. They were particularly concerned that there existed a strong possibility of water leakage from the dam's reservoir, impacting one of the largest underground cave formations in the world (Hershowitz, 2008). Furthermore, they noted that the area is one of the most biologically rich and diverse habitats in Central America. Additionally, the opponents noted that the proposed dam would cause flooding and impact the terrestrial flora and fauna habitat of various species, including more than a dozen threatened and endangered species.

Other issues include the elimination of drinking water sources for the downstream communities, a decrease in tourism, and the possibility of a foreign company having a majority control of Belize's power sector. Unfortunately for the group, the Privy Council, in a divided 3–2 ruling, discounted the geological flaws in the EIA and upheld the Belize government's decision to allow the dam to be constructed, thus. The Privy Council's ruling noted that "the decision to build the dam was essentially a sovereign one for Belize to make." However, the environmental groups maintained their position regarding the geological flaws, arguing that the "Belizean authorities should be held to the same standards as those in the developed world. To do any less, the dissent reasoned, would undermine good governance and the rule of law" (Hershowitz, 2008).

The Community Baboon Sanctuary (CBS) was developed in 1985 to protect one the largest remaining populations of black howler monkeys (Wyman and Stein, 2010). These black howler monkeys have lost 56 percent of their natural habitat, and over the next decade, an estimated 70 percent loss is anticipated due to deforestation (IUCN, 2003; Wyman and Stein, 2010). The sanctuary was developed with the help of local Belize nongovernmental organizations (NGOs) and scientists. Private landowners in seven Creole communities located along 33 kilometers of the Belize River have set aside strips of riparian forest and forested corridors as habitats for the black howler monkeys (Wyman and Stein, 2010). They participated through nature-based tourism and a voluntary pledge to provide these habitats. As reported by Wyman and Stein (2010) in a study conducted among 135 resident communities to evaluate local communities' perception of the value of protected areas and to monitor the effectiveness of conservation initiatives, a significant relationship was found between community involvement, perceived benefits, and attitude toward riparian forests and conservation. Though the community sees the importance of the riparian forests for area awareness and water quality, they do not believe they obtain a significant economic benefit from the tourism. Despite this, the study underlined the importance of community involvement in conservation as a worthwhile basis for planning future conservation initiatives and improved economic benefits (Wyman and Stein, 2010).

In a related study on MPAs in Belize, a capacity-building exercise was conducted among local NGO community workers and a government fishery officer (Crabbe et al., 2010). The purpose of the program was to directly interface community engagement with fisheries operation and policy. Most capacity-building programs involve empowerment and building competencies in the local community, but a limited number involve policy makers or government officials (Crabbe et al., 2010; Mequanent and Taylor, 2007). Crabbe and colleagues (2010) demonstrated the importance of engaging the capacity-building team for positive outcomes by developing personal action plans to aid the future of sustainable MPAs in the Mesoamerican barrier reef. According to the authors, implementation of individual action plans reflected the priorities and abilities of the communities they represented. Additionally, the capacity-building program was useful in producing new ideas to facilitate organizational management, educational support, and policy development in southern Belize.

In a study of Maya tree–dominated home gardens, Nanci Ross (2011) stated that in order for the Mesoamerican conservation programs to work, there is need to understand the complex interactions between ecological and social entities within an ecosystem that developed in intimate association with humans. Because the ancient Maya gardens are found all over Mesoamerica, there is the school of thought that this accounts for the dominance of species useful to Maya in the contemporary forest, but this pattern may be confined to a small area depending on ancient land use. There was a significant difference in the composition of species in the dense residential area structures and areas that have no or very small ancient settlement. By examining the long-term impacts of Maya forest gardens, Ross provides useful data for the future development of conservation models. To this end, an adequate knowledge of the ecosystem would assist in the development of biodiversity recovery strategies and conservation goals (Ross, 2011).

The environmental public health implications of drilling for petroleum, natural gas, and oil exploration are often difficult to assess and therefore cannot be overemphasized. Who can forget the largest marine oil spill in history (also known as the Deep Water Horizon Oil Spill or the Gulf of Mexico Oil Spill or the BP Oil Spill)? It started on April 20, 2010, when over 4 million barrels of crude oil spilled into the Gulf of Mexico: it killed 11 workers, destroyed uncountable wildlife, and caused extensive ecological change to the environment (Encyclopedia of Earth, 2012). By extension, there has been significant mitigation of environmental risks from oil exploration in "environmentally sensitive areas" in Belize during the past decades.

NATURAL DISASTERS AND GLOBAL WARMING

In recent times, climate change has become a fiercely debated and divisive topic because of its potential impact on international trade and commerce. Scientific data provide clear evidence that the global climate is changing rapidly, at a rate that currently exceeds 2°C per decade (Coppola, 2007). The earth's climate is a system with many interdependent parts and is determined by a variety of factors, including atmospheric gases, earth's orbit around the sun, and solar variability. The notable effects of climate change include higher temperatures, changes in precipitation patterns, rising sea levels, and more frequent weather-related disasters. Worldwide, the nature of disasters is rapidly changing, due in part to human activities and development patterns. More and more disasters are occurring each year, they are becoming more costly, and the number of people affected is increasing. Additionally, poorer countries of the world are disproportionately impacted by a much greater proportion of natural and technological disasters (Coppola, 2007). The current trend in disasters indicates that greater variations in precipitation are predicted to affect droughts and floods. Additionally, storm hazards are expected to become stronger and develop more rapidly, thus adversely impacting vulnerable populations in developing countries such as Belize. For example, the Earthwatch Institute (2006) reported that in the last thirty years, coral reef cover in the Caribbean region dropped by 80 percent due to climate change, coastal runoff, and coral diseases (World Health Organization, 2004).

Belize is extremely vulnerable to the risks of natural disasters and climate change, and there are many other environmental concerns relating to coastal development, industrial growth, overfishing, and tourism (Earthwatch Institute, 2006). The vulnerability of Belize to natural disasters is compounded by the inherent effects of climate change. In this context, vulnerability refers to the degree to which a system (such as a socioecological system) is likely to be impacted or experience harm from a perturbation or stress in the natural or social environment (Turner et al., 2003). Therefore, vulnerability may result from a combination of processes that shape the degrees of exposure to a hazard, sensitivity to its stress and impacts, and resilience in the face of those effects. Although the level of vulnerability varies widely, it is considered a characteristic of all people, ecosystems, and regions confronting environmental or socioeconomic stresses and is generally higher among poorer people (Kasperson et al., 2001). Aronson and Precht (2006) suggested that the biggest problem to coral reefs is not from overfishing but from global climate change and emergent marine diseases. They pointed out that macroalgae that kill the corals was not a result of decreasing number of herbivore fish from overfishing but that the proliferation of macroalgae was due to the exposed surfaces of already dead corals form climate change and marine disease (Centers for Disease Control and Prevention [CDC], 1997). The authors also noted that the establishment of MPAs will have a limited impact unless additional government policy is directed toward addressing the disastrous impact of global climate change and emerging marine diseases.

The countries of the Caribbean, including Belize are among the most vulnerable to global climate change (IPCC, 2007). Suffice it to note that Belize is particularly vulnerable to the impacts of climate change due to its geographic location (AFB, 2011). Belize has experienced frequent natural disasters of catastrophic proportions due to its long and low-lying coastal areas, the ever increasing storm surges, and rising sea levels. The coastal regions are especially vulnerable to intense and frequent tropical storms and hurricanes, flood damage, and the ever increasing sea levels. Belize lies in what is considered the hurricane belt, and much of the coastal region is below sea level. Inevitably, Belize is very prone to frequent flooding, and in recent years, several natural disasters have affected primarily the coastal regions.

A few examples of such flooding-associated disasters include Hurricane Hattie's destruction of Belize City in 1961, thus prompting the government to relocate and build a new federal capital in Belmopan. In 1998, Hurricane Mitch resulted in devastating floods in the coastal regions. In two consecutive years, 2000 and 2001, Hurricanes Keith and Iris struck Belize with a vengeance, each causing damages in excess of 45 percent and 25 percent of GDP, respectively. In 2008 and 2010, Tropical Storm Arthur and Hurricane Richard, respectively, caused widespread flooding and extensive damage to infrastructure and the agricultural sector, thus highlighting the ongoing vulnerability of Belize to natural hazards and climate change. In addition to human, economic, and social losses, natural disasters in Belize have contributed to large fiscal deficits and debt accumulations that required it to restructure its public debt in 2007 (AFB, 2011).

The World Bank (2011) has previously noted that climate change is expected to hit developing countries the hardest. Suffice it to note that climate change poses significant risks for agriculture, food, tourism, and water supplies. Experts believe that the short-term impacts of natural disasters and the long-term effects of climate change may undermine the resilience of our natural ecosystems and humanity thus increasing the urgent need to tackle these challenges. As part of the Hyogo Framework for Action created by the January 2005 World Conference on Disaster Reduction, it was recognized that unprecedented cross-border global cooperation by the world's nations is needed to address the problem of climate change. The implementation of such international cooperation would require the integration of comprehensive disaster management and risk reduction activities into overall national development. To this end, the World Bank Group is in the forefront of contributing to a global solution by providing support to developing countries and tailoring their approach to the differing needs of its partners (World Bank, 2011). In the past few decades, through the Roads and Municipal Drainage Project, the World Bank has committed funds for critical projects including the construction of road and drainage systems to reduce flooding-associated disaster in six rural areas in Belize. It is anticipated that the completion of the project will enable local farmers to have access to regional markets (World Bank, 2000). In September 2011, the World Bank Board of Directors endorsed the Belize Country Partnership Strategy by supporting the country's efforts to adopt a sustainable natural resource-based economic model. Additionally, the World Bank believes that this strategy will enhance Belize's resilience to climate change and natural hazards. Specifically, the strategy would provide support in the following key areas (World Bank, 2011):

1. Enhancing the institutional capacity of the government in natural resources management and climate change-related activities. This includes incorporating environmental sustainability into development policy and strategy formulation, planning, coordination, implementation and monitoring.
2. Investing in the country's resilience to climate change and natural hazard impacts. The program will finance natural resource-based adaptation measures and physical infrastructure designed to protect the most vulnerable.

In 2011, the World Bank submitted a project proposal titled "Belize Marine Conservation and Climate Adaptation Initiative" to the Adaptation Fund Board (AFB). The purpose of the proposed project of US$100 million was to bring innovative financing to bear on the climate change and fiscal challenges facing Belize with a view to setting up a trust that would finance ecosystem-based adaptation measures that enhance the resilience of the critical Barrier Reef. The three parts of the proposal were (1) financing the initial climate adaptation measures during the first five years, (2) setting up the Marine Conservation and Climate Adaptation Trust (MCCAT), and (3) demonstrating how to address the challenges of climate change and high debt in vulnerable developing countries (AFB, 2011).

Recognizing that Belize is prone to natural disasters, policy makers must integrate, at all levels of government, proactive and comprehensive disaster risk reduction into regular development planning and poverty reduction programs. For such programs to be successful, the Government of Belize policy makers who are charged with the development and reduction of the poverty sector of the population need to shift the focus of disaster risk reduction to benefit the poverty-stricken and most vulnerable areas by encouraging community-based and community-driven intervention programs. In this regard, during the planning stage of all new development projects, processes must be changed or modified to include disaster risk assessment so as to ensure that subsequent natural disasters will encounter well prepared and resilient communities capable of withstanding the impact. Additionally, policy makers must recognize mitigation as the cornerstone all future disaster management interventions. A formidable disaster mitigation process should focus on reducing the impact of disasters in all affected communities.

SUMMARY

This chapter examined environmental health issues in Belize. It argues that although the country has undeniably experienced frequent natural disasters of catastrophic proportions due to its geographic location, Belize still effectively implements its environmental health policies. The nation could be described as a small but endowed with abundant natural resources (e.g., the second longest Barrier Reef System in the world). It, however, suffers from a mixed blessing of ever increasing storm surges, coupled with rising sea levels. In Belize, industries such as tourism, sugar, bananas, citrus, forestry, and aquaculture are the main sources of growth and therefore the basis for the nation's capital. There are a number of environmental health issues of public health importance, including inadequate waste management, habitat alteration, deforestation, overfishing, loss of biodiversity, and natural disasters and climate change. In Belize, environmental governance is rooted in the creation of the BEPA and other regulations. BEPA established the environmental impact assessment, among other programs. The construction of the Chalillo Dam over the Macal River in the Cayo district provided a test case for BEPA's EIA program. The 1996 National Environmental Action Plan (NEAP) provided insights as to how the indigenous poor people are affected by inadequate land and waste management, water pollution, and natural disasters.

In this regard, urgent action is needed to address multiple environmental threats simultaneously because the interdependent and harmful effects of such stressors as coastal pollution, overfishing, poor land use practices, climate change, and global warming. It is anticipated that with assistance from multilateral organizations through the establishment and capitalization of the Marine Conservation and Climate Adaptation Trust (MCCAT), Belize can establish a sustainable financing mechanism for climate adaptation measures for the Barrier Reef System and the marine and coastal ecosystems. Prompt implementation of proven practical solutions would lead to both short- and long-term benefits, including the return of keystone species and the economic benefits they entail. Concerted efforts should be aimed at mitigating risks related to natural disasters and the impacts of climate change through the implementation of effective environmental policy, capacity building, and investments. Also, there is need to focus on environmentally sustainable outcomes for the development of these resources, especially in the protected areas.

The World Bank (2000) has noted that using natural capital in a sustainable manner is critical. Global environmental problems are intimately linked to economic conditions and problems of social justice. Therefore, the protection of historical sites, such as the Mayan ruins and other protected areas, are very critical and important in sustaining and preserving the cultural heritage and livelihood of the people of Belize. When people are poor and the national economy of a developing small country such as Belize is weak, the environment ultimately suffers. Conversely, if the environment is abused and the available natural resources are depleted through unsustainable consumption, poor people suffer, and eventually the economy declines. Therefore, to maintain sustainable outcomes in the long term, all three factors—the social, environmental, and economic needs of the people—must remain in balance with one other.

REFERENCES

Adaptation Fund Board (AAFB). (2011). "Proposal for Belize: AFB/PPRC.6/4." Project and Programme Review Committee Sixth Meeting, Bonn, September 14, 2011. Retrieved from from http://adaptation-fund.org/sites/default/files/AFB.PPRC_.6.4%20Proposal%20for%20Belize.pdf, Accessed on May 7, 2012.

Alegria, H. A., Bidelman, T. F., and Shaw, T. J. (2000). "Organochlorine Pesticides in Ambient Air of Belize, Central America." *Environmental Science & Technology, 34*(10), 1953–1958.

Andrefouet, S., Mumby, P. J., McField, M., Hu, C., and Muller-Karger, F. E. (2002). "Revisiting Coral Reef Connectivity." *Coral Reefs, 21*(1), 43–48.

Aronson, R. B., and Precht, W. F. (2006). "Conservation, Precaution, and Caribbean Reefs." *Coral Reefs, 25*, 441–450.

Belize Audubon Society. (2008). *An Environmental Agenda for Belize 2008–2013*. Retrieved from www.belizeaudubon.org. Accessed on November 18, 2011.

Belize Demographics Profile. (2013). Retrieved from http://www.indexmundi.com/belize/demographics_profile.html. Accessed on October 4, 2013.

Carson, R. (1962). *Silent Spring*. Boston: Houghton Mifflin.

Centers for Disease Control and Prevention (CDC). (1997). "Update: Blood Lead Levels—United States, 1991–1994." *Morbidity and Mortality Weekly Report, 46*(7), 141–145. Retrieved from http://www.cdc.gov/mmwr/preview/mmwrhtml/00048339.htm. Accessed on November 6, 2011.

Central Statistics Office (CSO), Ministry of Budget Management. (2000). "Belize National Population and Housing Census 2000." Retrieved from http://www.cso.gov.bz/publications.html. Accessed December 22, 2011.

Coppola, D. (2007). *Introduction to International Disaster Management*. Burlington, MA: Butterworth-Heinemann–Elsevier.

Crabbe, M. J. C., Martinez, E., Garcia, C., Chub, J., Castro, L., and Guy, J. (2010). "Is Capacity Building Important in Policy Development for Sustainability? A Case Study Using Action Plans for Sustainable Marine Protected Areas in Belize." *Society & Natural Resources, 23*(2), 181–190.

Dibie, R., and Offiong, J. (2012). "Analysis of Environmental Policy in Nigeria." *Journal of International Politics and Development, 10*(1–2), 53–82.

Eagles, P. F. J., McCool, S. F., and Haynes, C. D. (2002). "Sustainable Tourism in Protected Areas. Guidelines for Planning and Management." *Best Practice Protected Area Guidelines Series*. No. 8. Cambridge, UK: IUCN.

Earthwatch Institute. (2006). "Conservation Research Initiative: Sustainable Southern Belize." *Earthwatch Institute Journal, 25*(2), 12–1.

Eckersley, R. (1992). *Environmentalism and Political Theory: Toward an Ecocentric Approach*. Albany, NY: SUNY Press.

Eddleston, M. (2000). "Patterns and Problems of Deliberate Self-Poisoning in the Developed World." *Quarterly Journal of Medicine, 93*, 715–731.

Ehrlich, P., and Ehrlich, A. (1995). *The Population Bomb*. Cutchogue, NY: Buccaneer Books.

Encyclopedia of Earth. (2012). "Deepwater Horizon Oil Spill. December 5, 2010, Updated: February 8, 2012." Retrieved from http://www.eoearth.org/article/Deepwater_Horizon_oil_spill. Accessed on May 7, 2012.

Frumkin, H. (2010). *Environmental Health: From Global to Local*. San Francisco: Jossey-Bass/Wiley.

Gibson, J., McField, M., and Wells, S. (1998). "Coral Reef Management in Belize: An Approach Through Integrated Coastal Zone Management." *Ocean & Coastal Management, 39*(3), 229–244.

Government of Belize (GoB). (2007). "The People." Belize: Government of Belize. Retrieved from http://www.governmentofbelize.gov.bz/ab_people.html. Accessed on December 26, 2011.

Government of Belize (GoB). (2001). "NEMO Continues Relief Efforts." Belize: Government of Belize Press Office. Retrieved from http://www.governmentofbelize.gov.bz. Accessed on December 26, 2011.

Government of Belize (GoB). (2000). "Laws of Belize: Belize Environmental Protection Act, Chapter 328), Revised December 31, 2000." Belize: Government of Belize Press Office.

Hardin, G. (1968). "The Tragedy of the Commons." *Science, 162*, 1243–1248.

Hershowitz, A. (2008). "A Solid Foundation: Belize's Chalillo Dam and Environmental Decision-Making." *Ecology Law Quarterly*, 35(1), 73–105.

Huitric, M. (2005). "Lobster and Conch Fisheries of Belize: A History of Sequential Exploitation." *Ecology & Society, 10*(1), 26.

Ibtissem, M. H. (2010). "Application of Value Beliefs Norms Theory to the Energy Conservation Behavior." *Journal of Sustainable Development, 3*(2), 129–139.

Intergovernmental Panel on Climate Change (IPCC). (2007). "Summary for Policymakers." In S. Solomon, D. Qin, M. Manning, Z. Chen, M. Marquis, K. B. Averyt, M. Tignor, and H. L. Miller (Eds.), *Climate Change 2007: The Physical Science Basis. Contribution of Working Group I to the Fourth Assessment Report of the Intergovernmental Panel on Climate Change*,1–18. Cambridge and New York: Cambridge University Press.

International Union for Conservation of Nature (IUCN). (2010). "IUCN Red List of threatened species. Version 2010.2." Retrieved from http://www.iucnredlist.org/. Accessed on November 4, 2011.

International Union for Conservation of Nature (IUCN). (2003). "IUCN Red List of Threatened Species." Retrieved from http://www.iucn.org/themes/ssc/redlist_archive/redlist2003/English/profilesEn.htm. Accessed on November 3, 2011.

Kaiser, K. (2011). "Preliminary Study of Pesticide Drift into the Maya Mountain Protected Areas of Belize." *Bulletin of Environmental Contamination & Toxicology, 86*(1), 56–59.

Karalliedde, L., Eddleston, M., and Murray, V. (2001). "Global Epidemiology of Organophosphate Insecticide Poisoning." In Karalliedde, L., Marrs, T., Feldman, F., Henry, J. A. (Eds.), *Organophosphate Pesticides and Health*. London: Imperial Press, 93–122.

Kasperson, R. E., Kasperson, J. X., and Dow, K. (2001). "Vulnerability, Equity, and Global Areas Environmental Change." In Kasperson, J. X., and Kasperson, R. E. (Eds.), *Global Environmental Risk*. London: Earthscan, 67–98.

Kessler, R. (2005). "Empowering Indigenous Peoples." *Environmental Health Perspectives, 113*(9), A588.

King, R. B., Baillie, I. C., Abell, T. M. B., Dunsmore, J. R., Gray, D. A., Pratt, J. H., Versey, H. R., Wright, A.C.S., and Zisman, S.A. (1992). "Land Resource Assessment of Northern Belize, vols. 1 & 2." *Natural Resource Institute Bulletin, 43*, 1–513.

Konradsen, F., Van der Hoek, W., Cole, D. C., Hutchinson, G., Daisley, H., Singh, S., and Eddleston, M. (2003). "Reducing Acute Poisoning in Developing Countries—Options for Restricting the Availability of Pesticides." *Toxicology*, 192, 249–261.

Meerman, J., and Wilson, J. R. (2005). "Task Force on Belize's Protected Areas Policy and Systems Plan: The Belize National Protected Areas System Plan, November 30, 2005." Retrieved from http://www.inbio.ac.cr/web-ca/biodiversidad/belice/appendix3.pdf. Accessed on May 7, 2012.

Mequanent, G., and Taylor, D. R. F. (2007). "The Big Push Approach to African Development and Local Capacity Building: Understanding the Issue." *Canadian Journal of Development Studies, 28*, 9–26.

Murray, D., Wesseling, C., Keiffer, M. Corriols, M., and Henao, S. (2002). "Surveillance of Pesticide-Related Illness in the Developing World: Putting the Data to Work." *International Journal of Occupational and Environmental Health, 8*(3), 243–248.

Pan-American Health Organization (PAHO). (2002). "Bilateral Partnerships: Denmark." Retrieved from www.paho.org/english/dec/op11-denmark.pdf. Accessed on November 12, 2011.

Rom, W. (2012). *Environmental Policy and Public Health: Air Pollution, Global Climate Change, and Wilderness*. San Francisco: Jossey-Bass/Wiley.

Rosa, E. A., and Dietz, T. (1998). "Climate Change and Society: Speculation, Construction and Scientific Investigation." *International Sociology, 13*, 421–425.

Rosenberg, D. E., and Marcotte, T. P. (2005). "Land-Use System Modeling and Analysis of Shaded Cacao Production in Belize." *Agroforestry Systems, 64*, 117–129.

Ross, N. J. (2011). "Modern Tree Species Composition Reflects Ancient Maya 'Forest Gardens' in Northwest Belize." *Ecological Applications, 21*, 75–84.

Stern, P. C. (2000). "Toward a Coherent Theory of Environmentally Significant Behavior." *Journal of Social Issues, 56*(3). 407–424.

Stern, P. C., Dietz, T., Abel, T., Guagnano, G. A., and Kalof, L. (1999). "A Value-Belief-Norm Theory of Support for Social Movements: The Case of Environmentalism." *Human Ecology Review, 6*(2), 81–97.

Stern, P. C., Young, O. R., and Druckman, D. (1992). *Global Environmental Change: Understanding the Human Dimensions*. Washington, DC: National Academy Press.

Solid Waste Management Act (SWMA). (2011). "Belize Solid Waste Management Authority." Retrieved from http://www.belizeswama.com/default.aspx. Accessed on December 27, 2011.

Statistical Institute of Belize (2011). National Population of Belize. Retrieved from www.statisticsbelize.org.bz/dms20uc/dynamicdata/docs/20121214153635_2.pdf Accessed June 20, 2012.

International Senior Lawyers Project (ISLP) (2005). "Support for NGOs Addressing Global Development Issues." Retrieved from http://islp.org/content/economic-development. Accessed June 20, 2012.
Turner II, B. L., et al., (2003). "A Framework for Vulnerability Analysis in Sustainability Science." *PNAS*, *100*(14).
United Nations. (2010). *2010 Human Development Index*. New York: Oxford University Press.
Vanderheiden, S. (2011). "Environmental Philosophy and Theories." In Butts, J. B., and Rich, K. L. (Eds). *Philosophies and Theories for Advanced Nursing Practice*. Sudbury, MA: Jones & Bartlett.
World Bank. (2000). "Memorandum of the President of the International Bank for Reconstruction and Development to the Executive Directors on a Country Assistance Strategy of the World Bank Group for Belize, August 2, 2000." Retrieved from http://ctrc.sice.oas.org/trc/Articles/Belize/WorldBank_Country_Strategy.pdfAmerica. Accessed on December 20, 2011.
World Bank (2011). "Belize." Retrieved from http://data.worldbank.org/country/belize. Accessed on December 27, 2011.
World Health Organization. (2004). "Protection of the Human Environment." Retrieved from http//www.who.int/phc/en, 2004. Accessed November 21, 2012.
Wyman, M., and Stein, T. (2010). "Examining the Linkages Between Community Benefits, Place-Based Meanings, and Conservation Program Involvement: A Study Within the Community Baboon Sanctuary, Belize." *Society & Natural Resources, 23*(6), 542–556. Retrieved from http://web.ebscohost.com/ehost/pdfviewer/pdfviewer?sid=c30800fa-e0b8–4db1–80ef-4199b32 bb0e5%40sessionmgr114&vid=2&hid=126. Accessed on November 4th, 2011.
Young, C. (2008). "Belize's Ecosystems: Threats and Challenges to Conservation in Belize." *Tropical Conservation Science, 1*(1), 18–33. Retrieved from tropicalconservationscience.org. Accessed December 26, 2011.

15 Environmental Policies and Issues in Some Arab Countries

Robert Dibie, PhD
and Mauwya Hussein, MA, MBA

INTRODUCTION

The Arab world is comprised of twenty-two countries; twelve from West Asia (Bahrain, Iraq, Jordan, Kuwait, Lebanon, Oman, Occupied Palestinian Territories (OPT), Qatar, Saudi Arabia, Syria, United Arab Emirates (UAE), and Yemen) and ten from Africa (Algeria, Comoros, Djibouti, Egypt, Libya, Mauritania, Morocco, Somalia, Sudan, and Tunisia). It extends from the Atlantic Ocean in the west to the Arabian Gulf in the east and from the Mediterranean and the southern Anatolia in the north to Lake Victoria in the heart of Africa. The total area of the Arab countries is approximately 10.3 million square miles (United Nations, 2012).

The Arab world has undergone huge changes in the last century. Its population has risen from below 50 million a century ago to over 331 million according to 2007 estimates, and it will be home to some 385 million people by 2015 (United Nation Human Development Index, 2008). During this same period, the environment has deteriorated, and natural resources have dwindled due to development patterns that were largely unsustainable. In most cases, policies were overwhelmingly only provisional short-term measures, meant to tackle momentary challenges rather than to engage in long-term planning. Some parts of the region have seen unprecedented growth, bringing both economic and social prosperity to millions of Arabs during the last decades, thanks largely to the rising income from oil. Has this economic development, however, come at a cost? Can the patterns of development that some Arab countries are experiencing continue while sustaining livelihood and quality of life for future generations?

Today, the state of the Arab environment stands at a pivotal juncture, with numerous environmental problems—both current and imminent—threatening the region. At the same time, awareness of the issues, as well as signs of the political and social willingness to act, provide hope for timely intervention.

The growth of Arab cities and towns poses particular challenges. Accelerating urban drift in the region is straining already overstretched infrastructure and creating overcrowded, unhealthy, and insecure living conditions in many cities. In 1970, 38 percent of the Arab population was urban. By 2005, this had grown to 55 percent, and it is likely to surpass 60 percent by 2020 (United Nation Human Development Index, 2008). The two major environmental threats perceived in the Arab countries are water scarcity/desertification and land degradation. There are other environmental hazards in the Arab world, such as degradation of the marine environment, threats to biodiversity, a decline in the quality of the atmosphere, and climate change. Coastal zones in the Arab world are under stress as a result of demographic shifts from rural to urban areas, landfills, and dumping untreated waste (Abdel-Gadir and Abu-Ismail, 2009).The marine environment is increasingly threatened by land-based sources of pollution and by the heavy ship traffic in the region. Out of the world's exported oil, 60 percent is transported through the Straits of Hormuz. The contaminated water could cause irreversible damage to marine ecosystems. Further, the unique biodiversity of the Arab world is at serious risk from increased human activities. The main issues are the degradation and/or destruction of habits and loss of species. This is mainly the result of population growth, agricultural and urban expansion into ecologically important areas, poverty and unsustainable use of biota, and industrial pollution. Further, urban air pollution is emerging as a serious threat facing most of the cities in the region. Cities are experiencing air pollution with gases, particulates, and lead at levels often exceeding global standards. The main sources of air pollution in the region are energy and industrial production and vehicular emissions, which account for almost 90 percent of air pollution in urban centers and which are mainly due to poor maintenance, aged cars, low-quality fuel, and poor traffic management. Stationary sources, such as outdated power generation stations, are also a major source. Finally, the region is also experiencing climate change as a result of burning fossil fuels (UNEP, 2004).

ENVIRONMENT JUSTICE FRAMEWORK

The concept of environmental management in the Arab Region has undergone significant transformation over the last three decades. Much of this parallels the concept of global reorientation of environmental justice. The bone of contention has been the distribution of environmentally friendly among citizens and the use of natural resources. The key to environmental justice is on human health and well-being and on how they are influenced by particular types of production and consumption (Clifford and Edwards, 2012). According to White (2008), the interplay between humans and the environment is very important. He uses the anthropocentrism and ecocentric concepts to explain why the human relationship with the environment is very crucial. White contends that the anthropocentrism concept postulates that humans and animals have the natural right to dominate the world and the environment. On the other hand, the ecocentric approach suggests an enlightened human self-interest that is informed by the notion of the interrelationship between humans and the natural environment (Clifford and Edwards, 2012; White, 2008). Thus, everybody in the Arab countries is influenced by environmental issues and conditions.

The social learning theory postulates that differential association with other people, such as parents, schools, peers, and teachers might expose an individual to other behavior that may include techniques, rationalizations, motivation, and drives (Akers, 1998, Pratt et al., 2010). The process of social learning theory may be used to understand how individuals could emulate environmental protection practices from other people directly, indirectly, and reciprocally. Appropriate environmental practices are expected to result from an individual's high level of exposure to appropriate ecological attitudes. Thus, individuals are exposed to an appropriate environmental model, thereby helping the individual to define good environmental practices as morally or professionally justified (Clifford and Edwards, 2012).

The social learning theory, in both the anthropocentric and ecocentric versions, is used to provide insights into the development of understanding about environmental policies and practices in some Arab countries. Environmental concepts in most Arab countries have evolved into three distinctive phases: (1) support for sanitary engineering, municipalities, and public health (1920s–1960s); (2) a shift from public and environmental health to environmental management (1970s–mid-1980s); and (3) a gradual move from environmental management toward SD (mid-1990s–present). This transformation can be witnessed at the national as well as the regional levels. Despite this historical perspective, most national environmental institutions in the region suffer from the relatively recent establishment of environmental agencies or the restructuring of health and agricultural ministries to accommodate the regulation of their ecological system. Further, power politics tend to sideline public environmental agencies relative to economic and social benefits to citizens. One other challenge is the overlapping institutional jurisdictions that often lead to policy conflicts, program duplication, and inefficiency.

The application of these theories and concepts to Arab countries makes one wonder whether the environmental theories and the sustainable development concepts provide any real advancement in the understanding of environmental sustainability and good practices beyond the academic debate (Dibie, Melo, and Benitez, 2011). The bone of contention, however, is that the management of the ecosystem and biodiversity in Arab countries seems to be a reformulation of the current environmental policies and a pragmatic implementation strategy.

PLANNING FOR ENVIRONMENT IN THE ARAB REGION

Most Arab countries have begun or completed the task of developing a national environmental strategy (NES) and/or national environmental action plan (NEAP), but progress in formulating umbrella national sustainable development strategies (NSDSs) or National Agenda-21 remains obscure. This is because preparations for NES and NEAP were (conceptually) considered adequate substitute for NSDS formulation.

This misconception has in a few cases resulted in national strategies and action plans focusing on environmental management rather than on sustainable development (SD). However, on the positive side, most of the NES and NEAP, particularly in the West Asia side of this region, were prepared using the participatory bottom-up approach with the involvement of most stakeholders and relevant sectors, including the national socioeconomic development sectors (UNEP, 2011).

Great strides have been taken in the Arab Region over the past two decades in the developing and strengthening of environmental institutions and legislation. Initiatives at the national level to protect the environment have depended mainly on command-and-control mechanisms, particularly legislation. The main avenues for the implementation of environmental policy in the region have been national institutions cocoordinating environmental management and enforcing laws (e.g., ministries, general directorates, and the environment protection councils or secretariats) and the setting of standards and norms through legislation (UNEP, 2011). Recent socioeconomic changes have also brought policy changes that had environmental implications. Unprecedented urban and industrial growth in the region, particularly in the Gulf States, has resulted in increased demand for natural resources and rates of waste generation (both domestic and industrial). In addition, structural adjustment programs have led the governments of some countries in the region to suspend many government-supported activities, including environmental planning. In addition, the hostilities in the region over the last two decades have caused large population migrations toward marginal land and water resources. This, along with the lack of adequate waste disposal and/or treatment, has also posed a serious threat to the environmental and human health in the region.

ARAB LEGISLATION ON ENVIRONMENT PROTECTION

A number of laws promulgated in the Arab Region are concerned specifically with the environment. These laws contain provisions stipulating a full-fledged legal system of environment protection. Concurrently, various Arab countries were prompt in setting up cabinet portfolios for environmental affairs, in addition to their normal realm of activities, including issues like water, electricity, agriculture, and health. A case in point was the experience of some Arab countries' assigning the environmental responsibility to ministries of water, electricity, agriculture, and health. Not only portfolios were established but also other bodies, councils, and institutional entities concerned with the environment, either pursuant to independent decisions or by providing for their establishment within the context of environment protection codes.

In addition to passing legislation on environmental protection in general, Arab countries were well disposed to enact other environmental laws that are relevant to different environment issues. As part of this legislative interest in environmental affairs in the Arab world, a broad-based spectrum of laws was enacted, including the list in Table 15.1.

Laws on natural reservations were also created. In Egypt, Law no. 102/1983 on natural reserves was enacted. Qatar passed Law no. 19/2004 on natural life, and its criteria were also enacted. In Bahrain, Law no. 2/1995 on wildlife protection was promulgated as well as decreed Law no. 12/2000 on the Establishment of the National Agency for Natural Life Protection. Oman enacted Law no. 6/2003 on Natural Reservations and Maintenance of Natural Life. In Jordan, Law no. 29/2005 on Natural Reservations and National Recreational Places was enacted (United Nations Food and Agriculture Organization (UNFAO), 2002b). Legislation also exists in a number of Arab countries on the protection of water resources and the protection of coastal regions.

Though each of these laws has its own specificity, Arab nation laws share common principles that tend to promote the goal of environmental protection. The primary assertion is the right of humans to live in a healthy, clean, and balanced environment without destroying the ecosystem. That assertion also includes the responsibility of the official state authorities, public and private institutions, and individuals for keeping the environment safe, safeguarding its natural resources, fighting pollution, and protecting land and marine life. This also includes a definition of objectives that the relevant legislation attempts to achieve and a statement of fundamentals whereby competent authorities are assigned to carry out tasks entrusted to them in this respect. Among these, the following goals can be singled out: protection of the national environment against the negative impact of activities conducted beyond the borders and waters of the region and the commitment to global environmental protection such as biodiversity, ozone protection, climate change, combating desertification, and others (United Nations Food and Agriculture Organization (UNFAO), 2002a). Furthermore, obligations are placed on state authorities, especially those charged with developing plans of economic development, to take into account environmental considerations in all stages of economic development and planning. This is to ensure that development planning embraces environmental concerns as part of the overall strategy, be it industrial, agricultural, urban, or tourist.

Table 15.1 Selected Environmental Policies in Arab Countries

Policy Number	Name of Policy	Country
Law no. 4/1994	Environmental Protection	Egypt
Law no. 26/1995	Environmental Protection	Yemen
Law no. 3/1997	Environmental Protection	Iraq
Law no. 24/1999	Environmental Protection and Development	United Arab Emirates
Law no. 21/1995 modified by Law no.16/1996	Environmental Protection	Kuwait
Law no. 21/1996	Environmental Protection	Bahrain
Law no. 114/2001	Environmental Protection	Oman
Republican Decree no.34 on 28/7/1422 Hij	Environmental System	Saudi Arabia
Law no. 30/2002 on Environment Management in Qatar;	Proposed Environmental Management Policy but not yet enacted	Qatar
Law no 50/2002	Environmental Protection	Syria
Law no. 444/2002 on 29/7/2002	Environmental Protection	Lebanon
Law no. 15/2003	Environmental Protection and Improvement	Libya
Law no. 52/2006	Environmental Protection	Jordan
Law no. 11-3 relating to Environment Protection and Reclamation issued by Royal Decree no. 50-3-1 on 15 Rabei Awal 1424 Hij corresponding to 12 May 2003	Environmental Protection	Morocco
Law no. 3/10	Environmental Protection and Sustainable Development	Algeria
Law no. 7/1999	Environmental Protection	Palestine
Law no. 91/1988 on Establishment of a National Agency for Environment Protection in Tunisia, modified by two Laws no. 115/1992 and 14/2001	Environmental Protection	Tunisia

Source: United Nation Human Development Index (2008). New York: Oxford University Press; United Nations Environment Program (UNEP). (2011).

IMPACTS OF ENVIRONMENTAL ISSUES TO THE SOCIAL WELFARE OF THE CITIZENS WATER SCARCITY

The issue of water scarcity is the most serious threat to Arab nations' security because virtually all the countries are well below the level of "adequate clean water supply." The International Bank has classified twenty-two countries as below the water poverty line (that is, the per capita water availability is below 1,000 cubic meters/year). Fifteen of these countries are Arab (see Table 15.2). Per capita water in Qatar, Kuwait, Libya, and Bahrain is 91, 95, 111, and 112 cubic meters/year, respectively. In the cases of Saudi Arabia, Jordan, Yemen, Tunisia, Algeria, and Oman, the figures are 241, 318, 340, 434, 517, and 874, respectively (United Nations Human Development Index, 2008). If this is the case today, one can imagine the water famines that the Arab nations will have to confront in ten years with the present levels of population increase. An increasing population's demand for water would reduce the per capita share to 460 cubic meters by 2025, lower than the extreme water poverty level according to international classifications. As a result of poor technologies, the agricultural sector remains the major user of water sources in most Arab countries. A low level of efficiency is noted in the utilization of water in all sectors that use water. Water use efficiency is relatively low in the region, typically ranging between 37 and 50 percent, respectively (Abu-Thallam, 2003). This has created a range of problems such as water logging and salinity, low productivity and infertility of the soil, and the deterioration of the quality of groundwater. Further, water governance remains fragmented among various institutions, which creates another set of prob-

Table 15.2 Level of Water Stress in Four Arab Countries

Slight Water Stress (less than 2,500 persons per million cubic meters)	**Significant Water Stress (between 2,500 and 5,000 persons per million cubic meters)**	**Serious Water Stress (between 5,000 and 10,000 persons per million cubic meters)**	**Critical Water Stress (more than 10,000 persons per million cubic meters)**
Egypt	Jordan	Bahrain	Kuwait
Lebanon	Saudi Arabia	Iraq	United Arab Emirates
Oman		Occupied Palestinian Territory	
Syria		Qatar	
		Yemen	

Source: UN-ESCWA (2007).

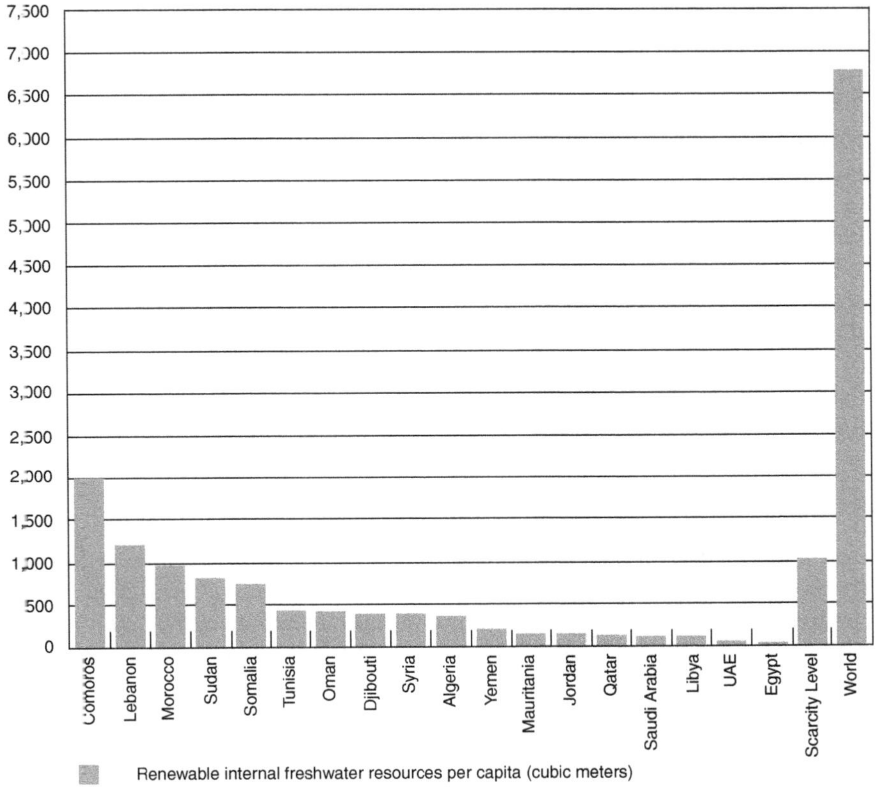

Figure 15.1 Arab Countries' Internal Freshwater Resources
Source: United Nations Environment Program (UNEP). (2011).

lems with the rational use of water. The problem is further aggravated by the high rate of population increase (Abu-Thallam, 2003). The geographical location of Arab countries in the Great Desert Belt and the lack of national programs to rationalize water consumption have created yet another problem. For example, almost 60 percent of water resources in the region originate from outside the borders, giving rise to tensions in using jointly shared water. This is acutely clear in the cases of the Nile, the Euphrates, and the Tigris rivers. Figure 15.1 shows that Arab internal freshwater resources are often below scarcity levels and the world average.

Poor distribution and heavy demand, especially of ground resources, characterize water use in the Arab countries. This leads to a lack of clean water for much of the population and the waste of significant amounts in the agriculture, industry, and tourism sectors (AOAD, 2003; UNFAO, 2002a). A report from the United Nations Economic and Social Commission for Western Asia (UN-ESCWA, 2007b) applies the question of water stress to that national level in the Arab states (UN-ESCWA member countries are Bahrain, Egypt, Iraq, Jordan, Kuwait, Lebanon, Oman, OPT, Qatar,

Saudi Arabia, Syria, United Arab Emirates (UAE), and Yemen. It distinguishes among four levels of water stress as gauged by the ratio of population to renewable freshwater: slight, significant, serious, and critical (UNFAO, 2003). Table 15.2 shows some Arab countries that are facing shortages of freshwater supply. In the table, the study shows that four countries are facing slight water stress, two are facing significant water stress, five are facing serious water stress, and two—Kuwait and the UAE—are facing critical water stress.

A particularly striking example of the conflict that exists between rapid economic development and scarce water resources is the recent boom in the construction of golf courses in certain parts of the region. In fact, most of the current and planned golf courses are in Egypt and the Gulf region, particularly the United Arab Emirates, where water resources are already low, even by regional standards (UN-ESCWA, 1995).

The expansion of water-intensive projects like grass golf courses cannot go on unchecked, especially with meager investments to develop sustainable desalination technologies. There are plans to increase the sixteen golf courses operating in the GCC countries now to forty in the near future. In most cases, golf courses in the region are irrigated with desalinated seawater, treated effluents, or a combination of the two. A 2007 report, released by the international consultants KPMG, estimated the use of water for each golf course in the region at an average of 1.16 million cubic meters per year, reaching 1.3 million cubic meters in Dubai, enough to cover the water consumption of 15,000 inhabitants. Currently, the quality of water resources in the Arab Region is affected by pollution, urbanization, floods, and overuse of water resources.

In the northern part of the Jordan Valley, Abu-Thallam (2003) has estimated the impact of water shortage on the planted area, income, and labor at the regional level, and it has been found that by decreasing the irrigation water quantity, the amount of crop decreased. This will, in turn, decrease the cultivated area. This has resulted in the reduction of the total net income and consequently of the labor used in the area. For instance, decreasing water supply by 20 percent will be followed by a reduction in the total cultivated area by about 14 percent. This will lead to a decrease in the total net income generated by 15 percent (Abu-Thallam, 2003). As a matter of fact, the reduction in employment will be accompanied by a direct and indirect loss in income too (Abu-Thallam, 2003).

DESERTIFICATION

A joint study of the League of Arab States and the United Nations Environment Program (LAS-UNEP Study, 2004) indicates that the highest ratio of desert to total land area is in the Arabian Peninsula (nine-tenths, or 89.6 percent). This is followed by North Africa (over three-fourths of the land, or 77.7 percent), the Nile Valley and the Horn of Africa (less than a half, or 44.5 percent), and the Mashreq (35.6 percent). Ongoing desertification threatens about 2.87 million square kilometers, or a fifth of the total area of the Arab countries.

The ratios of land facing peril proceed in the opposite direction to the preceding figures, with 48.6 percent of the land area in the Mashreq facing peril, 28.6 percent in the Nile Valley and the Horn of Africa, 16.5 percent in North Africa, and 9 percent in the Arabian Peninsula (LAS-UNEP Study, 2004). The amounts of desertified land or land threatened by desertification vary greatly from one country to another within these regions. In North Africa, for example, they are the greatest in Libya and the least in Tunisia; in the Nile Valley–Horn of Africa region, they are the greatest in Egypt and Djibouti and the least in Somalia. In the Mashreq, they are the greatest in Jordan and the least in Syria. Further, the Arabian Peninsula, Bahrain, Kuwait, Qatar, and the UAE are the most affected countries and together form the most desert-covered area in the Arab Region, in contrast with Syria, which is the least desert covered. The most apparent impacts of desertification are aridity due to the depletion of groundwater or subterranean water resources, shrinking forest areas and their further degradation through felling, cattle raising, burning, and the loss of soil fertility and composition in arid lands.

The Arab countries' large proportion of desert land is due in large part to the arid natural environment of the region but is not solely the product of natural causes. Desertification can also result from harmful activities undertaken by humans who, for the most part, are unaware of the detrimental impact of their activities on the natural environment. A study by the Arab Organization for Agricultural Development has attributed the causes of desertification in the Arab countries to the following developments (AOAD, 2003):

- *Population Growth:* The enormous growth in population, its mounting needs, its increasing use of modern technologies and methods in cultivation, and the overexploitation of the earth's various resources have greatly weakened the rejuvenating capacities of the region's ecosystems, distorted the environmental equilibrium, and propelled it toward degradation.
- *Changing Social Systems:* Large portions of the population of the Arab countries, particularly in arid and semiarid areas, were once engaged in nomadic or seminomadic pastoral or crop-gathering lifestyles. However, since the nineteenth century in some countries and the mid-twentieth century in others, these populations gradually became sedentary. The transition from one lifestyle to the other in many cases brought with it overgrazing, tree-felling, and the unregulated use of the soil in both rain-fed and irrigated agriculture, altogether contributing to the despoliation of vegetation, soil degradation, and erosion.

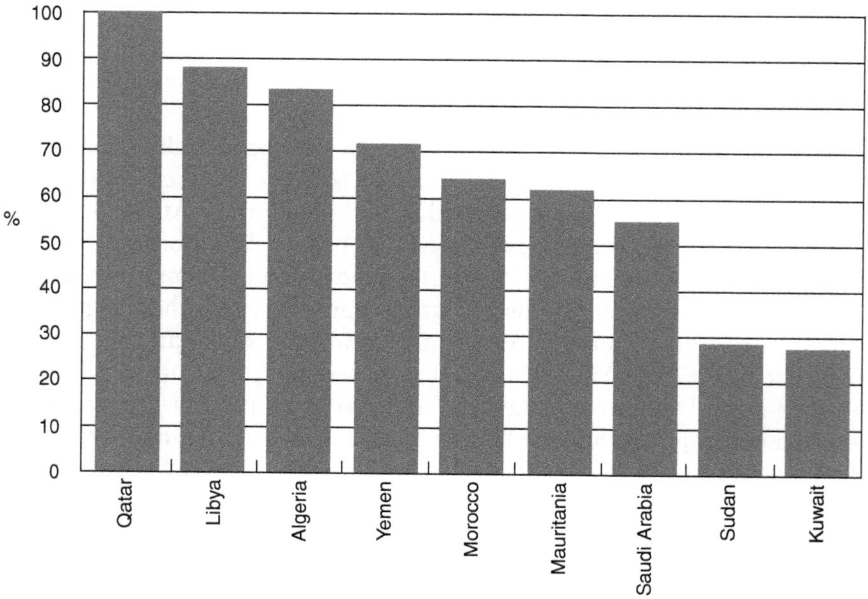

Figure 15.2 Desertification Trend in Selected Arab Countries

Sources: LAS (2008); UNEP (2004) (in Arabic).

- *Changes in the Mode of Agricultural Production:* As traditional modes of production failed to meet the needs of the growing population, modern agricultural tools and technologies were increasingly brought into play. Many of these technologies, particularly in plowing, are inappropriate for arid and semiarid lands because they cause soil to decompose and disintegrate, rendering it vulnerable to erosion. Figure 15.2 shows the extent of desertification in nine affected Arab countries.

POLLUTION: NATURE OF EMISSIONS IN SOME ARAB COUNTRIES

Most countries in the region, particularly in the capitals and other major cities, suffer from different degrees of air pollution. About 90 percent of total emissions of carbon monoxide (CO) in Arab countries is due to transportation activities. It is estimated that Arab countries emit collectively about 16 million tons per year of CO (Hamdan, 2007). The Arab vehicle fleets emit 1.1 million tons per year of nitrogen oxides (NO_x). Between 70 and 80 percent of total hydrocarbon (HC) emissions originate from the transportation sector and play an important role in the formation of photochemical oxidants (UN-ESCWA, 1996). Lead, used as an additive in petrol, accounts for more than half of the total lead atmospheric emission in the Arab countries and almost 100 percent in urban areas. Diesel engines also emit sulphur dioxides (SO_2) and fine particulates. Stationary sources, such as thermal power plants, refineries, smelters, fertilizers plants, cement manufacturers, and water desalination plants, also significantly contribute to air quality deterioration. Industrial compounds and manufacturing facilities emit gases such as CO_2, methane, volatile organic compounds (VOCs), and nitrogen oxides (NO_x). In the Gulf Cooperation Council (GCC) countries, total atmospheric emission loads are about 3.85 million tons per year and made of 28 percent CO, 27 percent SO_2, and 23 percent particulates (UNEP, 2004). Recent studies have indicated that the Gulf countries emit about 50 percent of the total of Arab countries' emissions of CO_2 (254 million metric tons of carbon). Recent reports have included several Arab countries among the top 50 countries with the highest CO_2-emitting power sectors (Center for Global Development, 2007). These Arab countries include Saudi Arabia (75.9 million tons and ranked twenty-second), Egypt (45 million tons and ranked thirtieth), Kuwait (19 million tons and ranked forty-eighth), and Algeria (17.7 million tons and ranked forty-ninth).

Table 15.3 shows the changes in the carbon dioxide per capita emissions over the past three decades. In most of the listed countries, a substantial increase has been recorded between 1980 and 2003. These figures are also compared to the Middle Eastern and world averages. These figures clearly indicate the wide variations among regional nations in energy consumption intensities and patterns. In general, GCC and other oil-exporting nations have emission rates higher than the world average. Indoor air pollution was responsible for 250 deaths a year in the United Arab Emirates.

WATER POLLUTION

With the Arab Region's quickly increasing population and pressured, fragile environment, pollution is increasingly becoming a concern of policy makers and civil society. Pollution continues to threaten the water and air of the Arab Region. Water pollution is primarily attributed to the increased use of chemical fertilizers, pesticides, and horticultural and veterinary medical treatments that leave long lasting traces that eventually find their way into the water. The influx of domestic and industrial wastewater has also considerably raised levels of water pollution (UN-ESCWA, 2005a & 2005b).

In several parts of the Arab Region, water pollution is manifested in low levels of access to clean water, a resource already constrained by general water scarcity, as previously analyzed. The lack of access to sufficient clean water in particular threatens human security in many ways. It can lead to the spread of disease, such as dysentery, among children and affect their regular attendance at school and academic achievement. It deprives women, for example, of long hours of the day that they could devote to

Table 15.3 Arab Region's National Per Capita CO_2 Emissions from the Consumption and Flaring of Fossil Fuels

Region/Country	1980	1990	2000	2003
Algeria	0.94	0.90	0.75	0.70
Bahrain	6.11	7.68	8.66	9.13
Egypt	0.27	0.44	0.46	0.53
Iraq	1.04	1.03	0.88	0.78
Jordan	0.67	0.85	0.84	0.85
Kuwait	6.12	3.48	8.15	8.16
Lebanon	0.58	0.40	1.24	1.17
Libya	2.83	2.74	2.21	2.77
Mauritania	0.10	0.13	0.33	0.29
Morocco	0.21	0.24	0.28	0.29
Oman	0.88	1.90	2.33	2.17
Qatar	16.37	10.54	12.64	10.78
Saudi Arabia	4.79	3.53	3.39	3.74
Somalia	0.05	0.04	0.03	0.07
Sudan	0.05	0.04	0.05	0.07
Syria	0.52	0.81	0.86	0.80
Tunisia	0.36	0.43	0.56	0.57
United Arab Emirates	8.09	10.99	12.61	14.45
Yemen	0.21	025	0.15	0.14
Middle East average	1.43	1.48	1.76	1.89
World average	1.12	1.11	1.07	1.11

Source: US-EIA (2009).

personal and income-generating activities rather than fetching water for their families. In addition, water scarcity and pollution threaten agriculture and food production and precipitate domestic rivalries over scarce water resources, as in the ancestral dispute between farmers and herders in Darfur (Sudan) (Center for Global Development, 2007). Water shortages can also cause tensions between neighboring countries. Access to clean water for domestic or economic purposes reflects power relationships. In general, the poor do not get clean water, and the rich consume enormous amounts and have no problem obtaining the quantities they want. Unsurprisingly, the parts of a country that have the most difficult access to water are the rural areas and the poorer quarters of the cities.

The World Bank's World Development Indicators provide water pollution data for fifteen Arab Countries, showing that Egypt, Algeria, Tunisia, Morocco, and Iraq are among the largest polluters in the Arab Region, in terms of daily emissions of organic water pollutants. However, the data also show that these rates pale in comparison to the daily rates of countries such as the United States, which showed in 2003 an absolute rate of pollution more than ten times that of the highest Arab polluter, and the Russian Federation, which in that same year "out-polluted" the highest Arab polluter by a factor of nearly 7.5.

Nevertheless, these absolute comparisons offer no basis for complacency, as the per worker ratio of pollution is relatively higher in Arab countries than in industrialized ones. Of the Arab countries, Tunisia succeeded in reducing this rate, reaching levels comparable with the United States, from 0.18 kilogram per worker in 1990 to 0.14 kilogram per worker in 2003. Yemen and Syria also succeeded in reducing these rates, respectively, from 0.27 and 0.22 kilogram per worker daily in 1990 to 0.23 and 0.20 kilogram per worker in 2003.

Also connected to water pollution is the region's uneven and cumulatively lagging progress in providing its people with access to sanitation services. Based on the most recent data, in some Arab least developing countries (Comoros, Mauritania, and Sudan), over 60 percent of the population has no access to improved sanitation services. It also shows that, at the regional level, approximately 30 percent of the population goes without such access (United Nations, 2010). Not only does poor access to sanitation services infringe on the health and dignity of human beings, but it is also a factor contributing to water pollution, with widespread consequences for human security. According to the Human Development Report (UNDP, 2006b), Egypt continues to experience high levels of pollution from raw sewage in the Nile Delta region. This plague undermines the potential environmental health problems as well as the benefits of access to Nile water.

CLIMATE CHANGE

The Arab Region will face an increase of 2–5.5°C in the surface temperature by the end of the twenty-first century (Tolba and Saab, 2008). In addition, this temperature increase will be coupled with a projected decrease in precipitation of 0–20 percent. The results for the region include shorter winters, drier and hotter summers, a higher rate of heat waves, increased weather variability, and a more frequent occurrence of extreme weather events (Hamdan, 2007). Clearly, adaptation and mitigation strategies need to be researched, discussed, and implemented.

Sea level rise (SLR) due to rising temperatures has the potential to cause the loss of significant portions of agricultural land in the Arab Region. As an example, even a 1-meter SLR could potentially cause the loss of 12–15 percent of agricultural land in the Nile Delta region and could reduce Qatar's land area by 2.6 percent (Center for Global Development, 2007; Mougou and Mansour, 2005). In addition to the agricultural, industrial, and tourism sectors, urban areas and the gross domestic product (GDP) in a number of Arab countries are threatened by the negative impact of sea level rise. The most serious impacts would be in Egypt, Tunisia, Morocco, Algeria, Kuwait, Qatar, Bahrain, and the UAE. The effects on the region's agricultural sector would mostly be felt in Egypt, where a 1-meter rise would put 12 percent of the country's agricultural land at risk. It would also directly affect 3.2 percent of the population in the Arab countries, compared to a global percentage of about 1.28 percent (Mougou and Mansour, 2005).

Higher temperatures will also increase the incidence and impact of drought in the region, threatening water resources and productive land. Drought frequency increased during the last twenty to forty years in Morocco, Tunisia, Algeria, and Syria and changed in Morocco from an average of one year of drought in every five-year period, before 1990 to one year of drought for every two-year period (Abbas, 2002; Hamdan, 2007; Mougou and Mansour, 2005). In Lebanon, varying conditions of water shortage (regarding the availability of the hydraulic resources) have been experienced in the last decade (Hamdan, 2007).

In Morocco, of the twenty-two drought years in the twentieth century, ten occurred during the last two decades and included the three successive dry years of 1999, 2000, and 2001. Drought is also a recurring event in the Near East. Jordan, for example, is predominantly arid and has experienced chronic water shortages and suffered from severe shortages since the 1960s. The recent droughts in Jordan and Syria were the worst ever recorded in recent decades (FAO, 2002b).

Recent droughts in Jordan and Syria were the worst recorded in many decades. In addition, increased precipitation variability and water resource availability directly related to climate change affect a number of the countries in the region. A warmer climate brings with it increased climate variability and higher risk of both floods and droughts, and it exacerbates the already precarious situation created by chronic water scarcity faced by most Arab countries. This report recognizes an alarming deficiency in scientific and technological capabilities, as well as the political will to address and face problems posed by climate change in the Arab Region. Not enough scientific facilities exist to study this phenomenon, insufficient funds are allocated to such research, and the studies that are undertaken still leave gaps to be filled. Climate change mitigation and adaptation need to be integrated into development strategies, and issues of planning, scientific capacity, stakeholder involvement, and public awareness need to be urgently addressed (UNDP, 2002).

Individual Arab countries will be affected differently under various climate change–related sea level rise projections. Qatar, the UAE, Kuwait, and Tunisia are the most vulnerable in terms of their landmass: 1–3 percent of land in these countries will be affected by a 1-meter SLR. Of these, Qatar is by far the most exposed: under various different SLR projections, the figure rises from approximately 3 percent of land (1 meter) to 8 percent (3 meters) and even up to more than 13 percent (5 meters) (United Nations, 2010). As for SLR's effect on GDP, Egypt's economy is by far the most vulnerable: for an SLR of 1 meter, more than 6 percent of its GDP is at risk, which rises to more than 12 percent for an SLR of 3 meters. Qatar, Tunisia, and the UAE are also exposed; over 2 percent of their respective GDPs are at risk for an SLR of 1 meter, rising to 3–5 percent for an SLR of 3 meters (Mougou and Mansour, 2005). When it comes to the agricultural sector, Egypt will be most impacted by SLR. More than 12 percent of Egypt's best agricultural lands in the Nile Delta are at risk from an SLR of 1 meter, and this figure rises dramatically to 25 percent (SLR of 3 meters) and even almost 35 percent (extreme SLR of 5 meters) (United Nations, 2010).

The limited research conducted in Arab countries has shown that climate change plays an important role in the spread of vector-borne infectious diseases, such as malaria and schistosomiasis (Egypt, Morocco, and Sudan). It also affects the seasonal concentrations of some allergens in the atmosphere, causing allergic reactions and pulmonary diseases (Lebanon, Saudi Arabia, and the UAE), and worsens the public health impact of heat waves, especially in Arab countries with hot summer climates (UNDP, 2006b). Heat waves are projected to become more intense, frequent, and prolonged due to climate change. A number of studies in the region have looked at heat-related mortality rates and have consistently found a significant association between temperature and mortality. The link between infectious diseases—which globally kill 14–17 million people every year—and climatic conditions has been studied extensively (Mougou and Mansour, 2005).

Malaria, for instance, which infects about 3 million people in the Arab Region yearly, may become more prevalent because higher temperatures reduce the disease's incubation period, spread the range of malaria-bearing mosquitoes, and increase mosquito abundance (United Nations, 2010). Indirectly, a number of climate change impacts discussed in various sections of this report may also have health ramifications. For instance, sea level rise and coastal flooding may impact food security and lead to malnutrition and hunger, and reduced precipitation and increased temperatures may aggravate water scarcity, increasing its negative impact on human health (United Nations, 1994; UNDP, 2007).

Projected climate changes may have disastrous effects on agricultural production in the Arab world. As a number of studies have shown, increased temperatures cause much higher water needs in summer crops. Water scarcity in the Arab Region is projected to increase rather than decrease, and therefore agriculture and in turn the Arab Region's food security are highly vulnerable to climate change, with the risk of a 50 percent decrease in food production if current practices continue.

In New Halfa, Eastern Sudan, a time-series analysis study (Himeidan et al., 2007) showed that temperature and rainfall are the driving forces of the spatial distribution of the malaria vector *Anopheles gambiae*. In northern

Sudan, surveys of breeding sites of the *Anopheles arabiensis* revealed a seasonal pattern for the larval population that appeared to be linked to the rise and fall of the Nile River level (Ageep et al., 2009). Similarly, Hamad and colleagues (2002) reported seasonal transmissions of malaria in Eastern Sudan, increasing with the onset of the rainy season and high humidity.

ENVIRONMENTAL DEGRADATION COSTS IN SOME ARAB COUNTRIES

The costs of environmental degradation reached about US$9 billion in six Arab countries at the beginning of the millennium (2000). The indicators of these countries with respect to environmental sustainability are very low (United Nations, 2010; UNDP 2006a). Estimates include environmental damage from such sources as indoor and outdoor air pollution, lack of access to water and sanitation services, land degradation, coastal zone degradation, and waste management (Carew-Reid et al., 1994; El-Laithy and McAuley, 2006). In addition to the damages to the national economy. Table 15.4 shows data on the damage to the global environment through greenhouse gas emissions.

Table 15.4 shows the average annual damages cost of environmental degradation in Arab countries. Overall, the damage cost estimates for seven countries vary from 2.1 percent of GDP in Tunisia to as high as 4.8 percent of GDP in Iran. This high cost of environmental degradation negatively affects

Table 15.4 Average Annual Damage Costs of Environmental Degradation, in MENA Countries (Percentage of GDP, 2002)

Source	Algeria	Egypt	Lebanon	Morocco	Syria	Tunisia
Air pollution	1,0	2.1	1.0	1.0	1.3	0.6
Lack of access to water supply and sanitation	0.8	1.0	1.1	1.2	0.9	0.6
Land degradation	1.2	1.2	0.6	0.4	1.0	0.5
Coastal zone Degradation	0.6	0.3	0.7	0.5	0.1	0.3
Waste management	0.1	0.2	0.1	0.5	0.1	0.1
Subtotal	**3.6**	**4.8**	**3.5**	**3.7**	**3.4**	**2.1**
Global environment (CO_2 emissions)	1.2	0.6	0.5	0.9	1.3	0.6
Total	**4.8**	**5.4**	**4.0**	**4.6**	**4.7**	**2.7**
Total cost of degradation (US $ million/year)	1,693	4,280	565	1,230	624	440

Source: United Nations Environment Program (UNEP), (2011).

public health, household budgets, the competitiveness of the economy, and intergenerational equity, and the rate of mining and degradation means that many of the resources will not be available in the future (LAS, 2007; Olivier and Tell, 1995). The environmental degradation cost estimated in this chapter shows only one side of the coin. Any policy action that causes environmental damages may also produce some benefits to the society (SEDAC, 2005).

These results are underestimates. Because of data limitations, the results do not include damage stemming from untreated industrial, hazardous, and hospital wastes or the losses of forest cover and biodiversity (Stern, 2006; Tabutin and Schoumaker, 2005). Also owing to data constraints, the impact of the inadequate treatment of industrial and municipal wastewater is limited to coastal recreational and tourism losses.

In terms of the deterioration of environment conditions, the Arab countries show low ranks in the Environmental Performance Index studies that were conducted in 2006 and 2008. Most of them deteriorated further if one compares the two periods. Table 15.5 shows the status of Arab countries in 2006 and 2008.

Table 15.5 Environmental Performance Index for Arab Countries, 2006 and 2008

SN	Country	Rank in 2006	Rank in 2008	Remarks
1.	Tunisia	82	59	Up 23
2.	Algeria	63	66	Down 3
3.	Jordan	64	70	Down 6
4.	Egypt	85	71	Up 14
5.	Saudi Arabia	59	78	Down 19
6.	Morocco	68	81	Down 13
7.	Lebanon	36	90	Down 54
8.	Oman	60	91	Down 31
9.	Syria	97	99	Down 2
10.	Kuwait	NA	111	NA
11.	United Arab Emirates	47	112	Down 65
12.	Sudan	124	129	Down 5
13.	Iraq	NA	135	NA
14	Djibouti	NA	139	NA
15.	Yemen	122	141	Down 19
16.	Mauritania	NA	146	NA

Source: Environmental Performance Index (2008).

POLICY RECOMMENDATIONS FOR THE FUTURE

The magnitude of environmental problems in several Arab countries, as previously discussed, is mind-boggling. Several millions tons of hazardous waste are produced in the region every year, and many dump sites continue to pose serious threats to public health. Some of the government environmental agencies in the region have reported that less than 12 percent of the waste in the region was disposed of in an environmentally sound manner (United Nations, 2010). The newly created environmental protection agencies are weak due to a lack of capable manpower or lack of adequate funding. Below are some policy recommendations to ensure continuing protection of citizens' health and the environment.

Environmental Institutions

Environmental institutions must be strengthened financially and empowered with executive authority to plan, coordinate, and enforce legislative programs for environmental protection (Dibie, 2010; Dibie et al., 2011). Environment agencies in the Arab countries should be empowered to provide long-term planning strategies that are necessary to cope with global changes, such as changing economic growth patterns and the use of crops for the production of fuels and its effects on food availability, national oil incomes, and development plans in general.

Integration of Environmental Considerations into Development Planning

The incorporation of market mechanisms, natural resources, ecological accounting principles, and the various types of environmental assessments (strategic, cumulative, project) are urgent needs. A responsible approach to using nuclear power in some Arab countries must also consider its whole life cycle, including final disposal (Clifford and Edwards, 2012). It is very important that Arab countries seeking nuclear energy power should at least have a set of technologies available to avoid expanding geologic disposal sites in the region because of the danger to citizens' health and to biodiversity.

Environmental Legislation

Provisions should be included in Arab environmental legislation that commit to harnessing market forces and mechanisms for compliance with environmental law. Other provisions must be incorporated to professionalize environmental occupations and vocations in a manner that permits only those who are qualified to practice. Finally, environmental legislative mandates and laws must be made effective, and any impediments to enforcement have to be removed.

Education

The revamping of academic curricula and the incorporation of environmental content aimed at including environmental education at all levels (Dibie, 2010). It is also crucial and a major priority to raise the level of environmental consciousness of all students in elementary, secondary, and university levels in all Arab countries.

Scientific Research

Allotting more funding for scientific and environmental research is long overdue in the region (Dibie, 2009). Moreover, researchers and centers of research networks need to be created at the national and regional levels. The governments of Arab countries need also to provide incentives for scholars to conduct research in specific environmental issues on how these issues could benefit the citizens.

Communications and Media

Environmental training programs for reporters and media experts should be developed to promote related issues. In addition, committing space and programs in the print and audiovisual media is needed in order to raise public awareness about environmental degradation (Dibie, 2009). The costs and benefits of environmental protection and the rational use of natural resources and biodiversity need to be effectively communicated to the citizens of the region. The use of media is also very important in the communication of change in the behavior of the citizens toward the environment and toward the use and disposal of consumer products (Dibie, 2010).

SUMMARY

This chapter examined the social impact of environmental issues in several Arab countries. It argues that several Arab countries have not shown a level of environmental stewardship that is consistent with protecting environmental security. Indeed, only three Arab countries score among the top hundred countries on the Environmental Sustainability Index (Carew-Reid et al., 1994; El-Laithy and McAuley, 2006; Taylor, 2012).

The index is a composite measure that ranks countries by their plans for managing environmental and natural resources and population and development. Significantly greater efforts are required to formulate and implement national sustainable development strategies and to monitor the state of the environment within and across countries. Arab governments have several means available to them to encourage the participation of key social forces and the private sector in environmental protection efforts. Such means include the taxation system, incentives to use environmentally friendly technology, campaigns to adopt nonpolluting renewable energy sources, policies that encourage sustainable uses of energy sources, the encouragement of mass transport over private automobiles, and the implementation of strong measures to combat desertification and deforestation.

The ecosystems in the Arab Region are interrelated; any disruption in one ecosystem could have negative effects on others, thereby destabilizing the entire region. It is interesting to note that water scarcity is one of the most acute environmental challenges facing the Arab Region, which is the most water-scarce of all regions in the world. This challenge needs to be met by applying integrated water resources management (IWRM) principles to development policies, institutional frameworks, and the regulation of water supply and demand. These principles call for the integrated management of land and water resources and other related natural resources in a coordinated approach that aims to maximize the socioeconomic welfare equitably without sacrificing the sustainability of ecological systems.

Environmental issues are global in nature and initiatives to address them need to be global as well. Further, environmental policy cannot be discussed in the Arab countries in isolation from energy policy or from policies that are related to economic growth (AFED, 2008; ADAD, 2003; Arab Water Council, 2008). Failure to combine environmental policy and economic growth may lead to a higher cost of fixing the outcome in the near future (Peters, 2013). The Arab countries have taken part in the global concern for environmental security and ratified most international environmental conventions. While the larger share of responsibility for some critical aspects of climate change lies with the highly industrialized countries, the Arab countries can make their contribution by showing commitment to these conventions (Asian Development Bank, 1996). They could also develop strong national and regional authorities, early warning systems, and action plans to address water scarcity, desertification, air pollution, and land degradation on a regional scale. It would be advisable for Arab countries to introduce various climate change adaptation and mitigation measures through the formulation of disaster risk reduction strategies and risk management practices as a critical element for adaptation.

REFERENCES

Abbas, A. (2002). "Drought Suppression Procedures for Dry Lands, The First Technical Workshop of the Mediterranean, and Component of CLIMAGRI Project on Climate Change and Agriculture." Rome (September 25–27).

Abdel-Gadir, A., and Abu-Ismail, K. (2009). *Development Challenges for the Arab Region: A Human Development Approach*. United Nations Development Programme. New York: Palgrave McMillan.

Abu-Thallam. K. Y. (2003). "Assessment of Drought Impact on Agricultural Resources in Northern Jordan Valley, University of Jordan, Amman, Jordan." *Water Resource Management, 14*(5), 327–338.

Ageep, T. B., Cox, J., Hassan, M. M., Knols, B. G., Benedict, M. Q., Malcolm, C. A., Babiker, A., and El Sayed, B. B. (2009). "Spatial and Temporal Distribution of the Malaria Mosquito *Anopheles Arabiensis* in Northern Sudan: Influence of Environmental Factors and Implications for Vector Control." *Malaria Journal, 8*(1), 123.

Akers, R. (1998). *Social Learning and Social Structure: A General Theory of Crime and Deviance*. Boston: Northeastern University Press.

Arab Forum for Environment and Development (AFED). (2008). "Conference Recommendations." Arab Forum for Environment and Development First Annual Conference. (October 26–27). Manama. Retrieved from http: //www.afedonline.org/en/inner.aspx?contentID=348.

Arab Organization for Agricultural Development (AOAD). (2003). *Dirassa Hawl Mo'asheerat Rased al-Tasahor fil Watan al-'Arabi (Study of Desertification Monitoring Indicators in the Arab region)*. Khartoum: League of Arab States. Retrieved from http://www.aoad.org/ftp/desertification.pdf. Accessed in June 2008.

Arab Water Council. (2008). "Messages for the Ministerial Process." MENA/Arab Region Contribution to the Political Process of the 5th World Water Forum. Retrieved from http://www.arabwatercouncil.org/administrator/Modules/spotLight/Messages for the Ministerial Process.pdf. Accessed in May 2008.

Asian Development Bank (ADB). (1996). *Economic Evaluation of Environmental Impacts: A Workbook*. Manila: Asian Development Bank.

Asian Development Bank (ADB). (1988). *Guidelines for Integrated Regional Economic-cum-Environmental Development Planning,* Vols. I and II. Manila: Asian Development Bank.

Carew-Reid, J., Prescott-Allen, R., Bass, S., and Dalal-Clayton, B. (1994). *Strategies for National Sustainable Development—A Handbook for Their Planning and Implementation*. London: Earthscan.

Clifford, M., and Edwards, T. (2012). *Environmental Crime*. Burlington, MA: Jones & Bartlett Learning Press.

Dibie, R., Melo, W., and Benitez, G. (2011). "Social Impact of Environmental Health Policy in Brazil." *Journal of International Politics and Development, 9*(1), 1–24.

Dibie, R. (2010). "Social Impact of Environmental Policies in Ethiopia." *Journal of International Politics and Development, 8*(1), 1–23.

Dibie, R. (2009). "Environmental Governance and the Political Economy of Sustainable Development in Sub-Saharan Africa." *Journal of International Politics and Development, 7*(1), 1–23.

El-Laithy, H., and McAuley, A. (2006). "Integrated Social Policies in Arab Countries." In *Integration and Enlargement of the European Union: Lessons for the Arab Countries*, edited by N. El Ahwany. Cairo: Center for European Studies, Cairo University and Conrad Edenhawar.

Environmental Performance Index. (2008). Retrieved from http://sedac.ciesin.columbia.edu/data/set/epi-environmental-performance-index-2008. Accessed June 19, 2012.

Hamad, A. A., Nugud, A. E., Arnot, D. E., Giha, H. A., Abdel-Muhsin, A. M., Satti, G. M., Theander, T. G., Creasey, A. M., Babiker, H. A., and Elnaiem, D. E. (2002). "A Marked Seasonality of Malaria Transmission in Two Rural Sites in Eastern Sudan." *Acta Tropica. 83*(1), 71–82

Hamdan, F. (2007). "Arab States Ignore Climate Change." *Executive Magazine* (April). Retrieved from http://www.klima der-gerechtigkeit.de/wpcontent/fouad-hamdan-arab-states-ignoreclimate-change-english.pdf. Accessed in May 2008.

Himeidan, Y. E., Hamid, E. E., Thalib, L., Elbashir, M. I., and Adam, I. (2007). "Climatic Variables and Transmission of Falciparum Malaria in New Halfa, Eastern Sudan." *Eastern Mediterranean Health Journal, 13*(1), 17–24.

League of Arab States (LAS). (2008). The Council of Arab Ministers Responsible for the Environment (CAMRE), The Arab Center for the Studies of Arid Zones and Dry Lands (ACSAD). Retrieved from http://www.unep.org. Accessed in May 2010.

League of Arab States (LAS). (2007). "The Arab Ministerial Declaration on Climate Change." 19th Session of the Council of Arab Ministers Responsible for the Environment at the Headquarters of the League of Arab States. Cairo. (December 5–6).

Mansour, M, Mougou, R, and Mougou, A. (2005). "Effect of Several Modes of Irrigation and Fertigation on Artichoke Crop." *Acta Horticulturae, 681,* 127–134.

Olivier R., and Tell, S. (1995). "The Development and Integration of Environmental Information Management, Assessment and Decision Support Capacities in West Asia." Internal UNEP report from fact-finding mission.

Peters, G. (2013). *American Public Policy*. Washington, DC: Congressional Quarterly Press.

Pratt, T. C., et al. (2010). "The Empirical Status of Social Learning Theory: A Meta-Analysis." *Justice Quarterly, 27*, 765–802.

Socioeconomic Data and Applications Centre (SEDAC). (2005). "Environmental Sustainability Index." Retrieved from http://sedac.ciesin.columbia.edu. Accessed in June 2008.

Stern, N. (2006). "The Economics of Climate Change." *The Stern Review.* Cambridge and New York: Cambridge University Press. Retrieved from http://www.hm-treasury.gov.uk/stern_review_report.htm, Accessed in December 2008.

Tabutin, D., and Schoumaker, B. (2005). "The Demography of the Arab World and the Middle East from the 1950s to the 2000s—A Survey of Changes and a Statistical Assessment." *Population, 60*, 5–6.

Taylor, R. (2012). *Taking Sides: Clashing View in Sustainability*. New York: McGraw-Hill.

Tolba, M., and Saab, N. W. (Eds.). (2008). *Arab Environment: Future Challenges.* Beirut: Chemaly & Chemaly, Arab Forum for Environment and Development (AFED). Retrieved from http://www.afedonline.org/afedreport/Full English Report.pdf. Accessed in June 2008.

United Nations. (2012). *Human Development Index*. New York: Oxford University Press.

United Nations. (1994). "Part I, Introduction, Article 1(a)." In *Convention to Combat Desertification in Those Countries Experiencing Serious Drought and/or Desertification, Particularly in Africa.* Retrieved from http://www.unccd.int/convention/text/convention.php?annexNo=-1#art1. Accessed November 21, 2013.

United Nations Development Program (UNDP). (2007). *Human Development Report 2007/2008. Fighting Climate Change: Human Solidarity in a Divided World.* New York: Palgrave Macmillan.

United Nations Development Program (UNDP). (2006a). *Arab Human Development Report 2005: Towards the Rise of Women in the Arab World.* New York: Palgrave Macmillan.

United Nations Development Program (UNDP). (2006b). *Human Development Report 2006. Beyond Scarcity: Power, Poverty and the Global Water Crisis.* New York: Palgrave Macmillan.

United Nations Development Program (UNDP). (2002). *Arab Human Development Report 2002. Creating Opportunities for Future Generations.* New York: Palgrave Macmillan.

United Nations Economic and Social Commission for Western Asia (UN-ESCWA). (2007). *Water Development Report 2, State of Water Resources in the ESCWA Region.* December 4. New York: United Nations. Retrieved from http://www.escwa.un.org/information/publications/edits/upload/sdpd-07–6-e.pdf. Accessed on January 2008.

United Nations Economic and Social Commission for Western Asia (UN-ESCWA). (2005). "Towards Integrated Social Policies in Arab Countries, Framework and Comparative Analysis." Retrieved from http://www.escwa.un.org/sp-readings/Final%20 Towards Integrated Social Policy in Arab Region-En.pdf. Accessed in June 2008.

United Nations Economic and Social Commission for Western Asia (UN-ESCWA). (1996). Personal correspondence with the UNEP/GEO Team. "United Nations Commission on Economic and Social Commission for Western Asia."

United Nations Economic and Social Commission for Western Asia/Food and Agriculture Organization (ESCWA/FAO). (1995). "Wildlife Conservation for Sustainable Development in Arab Countries." Report No. E/ESCWA/AGR/1994/11. New York: ESCWA and FAO.

United Nations Environment Program (UNEP). (2011). *State of Desertification in the Arab World: Updated Study.* Damascus. Retrieved from http://www.unep.org. Accessed on May 24, 2012.

United Nations Environment Program (UNEP). (2004). *Halat al-Tasahhur fil- Watan al-'Arabi: Dirasa Mustahdatha (State of Desertification in the Arab World: Updated Study*). Damascus. Retrieved from http://www.unep. Accessed in May 2011.

United Nations Food and Agriculture Organization (UNFAO). (2003). *Review of World Water Resources by Country, Water Reports (23).* Rome: Food and Agriculture Organization of the United Nations.

United Nations Food and Agriculture Organization (UNFAO). (2002a). AQUASTAT database. Retrieved from http://www.fao.org/ag/agl/aglw/aquastat. Accessed November 21, 2012.

United Nations Food and Agriculture Organization (UNFAO). (2002b). *The State of Food and Agriculture 2002.* Rome: Food and Agriculture Organization.

United Nation Human Development Index Annual Report. (2008). New York: Oxford University Press.

White, R. (2008). *Crime Against Nature: Environmental Criminology and Ecological Justice.* Devon, United Kingdom: Willan.

16 Environmental Sustainability and Solutions

Robert Dibie, PhD

INTRODUCTION

Environmental change is the most complex challenge facing nearly every country in the world today. The United Nations Environmental Program (UNEP) agency argued that no country alone can handle the social and economic impact of the interconnected challenges posed by environmental problems. Global climate change, affected by the release of carbon dioxide (CO_2) and other greenhouse gases into the atmosphere, could change weather patterns, food supplies, tree growth, water levels, and patterns of disease around the world (Perrotta, 1999; Peters, 2013). A serious environmental problem that confronts many countries in the world is the air pollutants released from the burning of fossil fuels such as gasoline, coal, and oil. These pollutants are commonly present in the air at levels that have been linked to increases in respiratory infections, hospitalizations for asthma, lung and heart disease, and nontraumatic deaths (Rosenbaum, 2014).

According to Michael Kraft and Scott Furlong (2013), evidence collected around the world has demonstrated consistently that human health is affected by the low levels of air pollutants that are present in the atmosphere in many countries. For instance, smog is composed of a mixture of air pollutants, including ozone, carbon monoxide, sulphur dioxide, nitrogen oxide, inhalable particulates (PM10), reparable particulates (PM2.5), and sulphates. These air pollutants are sometimes emitted as by-products when fossil fuels such as coal, oil, and gasoline, are burned in cars, furnaces, electrical generating stations, and industrial plants (U.S. EPA, 1999).

In both developed and developing countries, use many chemicals, with organochlorine compounds. These compounds have been detected in the waters of rivers and lakes. Leaking toxic dumps, agricultural runoff, air pollution, and industrial and municipal discharges have all contributed to the pollution of the many rivers and lakes. Further, ground-level ozone is produced when nitrogen oxides (NO_2) that are emitted from cars, furnaces, and industrial process, react in the presence of sunlight, with volatile organic compounds (VOC) that can be released from paints, degreasers, dry cleaners, and pesticides (Davies and Mazumder, 2003; U.S. EPA, 1999).

The world ecosystem is also threatened by the depletion of the ozone layer in the upper atmosphere. The ozone layer protects humans, animals, and plants from the harmful ultraviolet (UV) fraction of sunlight. It has been estimated that for every 1 percent decrease in the ozone layer, there is a 1 percent increase in the UV index, and 1–3 percent increase in the number of nonmelanoma skin cancer cases (Maxwell, 2014; Friis, 2013). In addition, some environmental health studies have indicated that organochlorine compounds such as PCBs, DDT, and dioxins, as well as heavy metals such as mercury, cadmium, and lead, can affect the immune system, resulting in decreased resistance to infections and chronic diseases such as cancer or in an increase in allergic reactions (Davies and Mazumder, 2003; Friis, 2013; Shendell and Ana, 2011). According to Nancy Maxwell (2014), toxicological studies have also demonstrated that a number of environmental pollutants including organchlorine compounds, pesticides, phthalates, and alkylphenol compounds have the ability to mimic the female hormone estrogen. Further, wildlife research suggests that environmental contaminants that are capable of mimicking hormones such as estrogen could be responsible for the elevated levels of birth defects, reproductive disorders, behavioral abnormalities, thyroid disorders, and compromised immune systems observed in fish and wildlife around the world (Davies and Mazumder, 2003; Health Canada, 1999; U.S. EPA, 1999). Many of these health effects have now been reproduced in laboratory test animals using the suspected environmental contaminants (Kraft, 2010; Rosenbaum, 2014). There is a growing body of literature demonstrating that human health can be affected by low levels of exposure to the persistent toxins that are accumulating in our food chain. Although this literature indicates that a causal relationship between pesticides and cancer remains uncertain, a growing body of health literature suggests that there are reasons to reduce our use of and reliance on chemical pesticides (Davies and Mazumder, 2003; Friis, 2013; Maxwell, 2014; Rosness et al., 2012; Shendell and Ana, 2011). In recognition of this health concern, many countries have adopted policies to phase out the use of pesticides on agricultural farms (Maxwell, 2014). Governments in both developing and developed countries have also recognized the need for greater education of the public on both the hazards of and alternatives to pesticides.

This book uses five perspectives of sustainability to examine the social and economic impact of environmental issues and policies in several countries. The perspectives include (1) the natural capital perspective; (2) the soft and hard perspectives of sustainability; (3) the way in which different countries of the world view the concept of sustainability; (4) the controversy over localism versus globalization; and (5) the difference between sustainability as a science and sustainability as a social movement. Chapters 12 and 13 argue that ecosystems and biospheres are complex systems and that changes in biotic composition and the relationship among elements have often had consequences for biodiversity. On one hand, the chapters reported a positive correlation between environmental changes and the loss of biodiversity. On the other hand, they contend that poverty and sustainable development remain core priorities (Chapters 5, 6, and 7).

The book argues that the level of negative ecofriendly practices of multinational corporations all over the world make scholars argue that corporate sustainability is more public relations than real. The major questions addressed in the countries examined in the book are:

1. In what ways has the growth of material consumption caused environmental problems?
2. How has environmental problems affected the social and economic welfare of people in some countries?
3. In what ways have major corporations fostered environmental problems?
4. Is the failure of government policies the cause of environmental problems in most countries?

In addition, Chapters 2 and 3 in the book pointed out that growth in material consumption may cause waste management and other environmental problems. On one hand, Chapters 7–10 assert that a corporate performance measurement system that integrates sustainability into corporate culture would strengthen and galvanize strategic thinking (Barbier, 1987; Davies and Mazumder, 2003; Dernbach, 2012). The environmental problems that most corporations have created in several countries tend to have affected the social and economic welfare of the people living in those nations. On the other hand, corporations can genuinely focus on sustainability issues and creatively increase their sales margins simultaneously. This is because corporate sustainability is an essential initiative for the companies, nations, society, and human beings all over the world to adopt.

Under the right conditions, an ecosystem often will conserve itself until some human-made or natural event causes it to reorganize or adapt. The health of an ecosystem depends on its ability to withstand disturbance or shock. A diverse and redundant ecosystem can resist severe climate change and maintain its health. This conclusion suggests that the government's environmental policies could present a resilience challenge in a number of countries (Rosenbaum, 2014).

This book explores the intriguing challenges when democratic advancements and structural adjustments in several countries around the world have propelled the growth of decentralized governance systems. Decentralized government systems will enhance the implementation of environmental policies at the federal, state, and local levels. These new governance systems have given birth to new actors in urban ecosystems management, particularly solid waste management, and air and water pollution at the local, state, and federal government levels (Dibie, 2003). A significant proportion of the actors in this new governance system, such as the nongovernmental organization (NGOs), operate informally. The practice is embraced in the conceptualization of Earth System Governance as constituting the institutional environment (IE); the informal component agency in the solid waste management (SWM) collection subsystem is ignored and not built upon in the institutional arrangements (IAs) of urban waste management. Empirical evidence of waste management practice operating outside the so-called transaction sector was examined through a number of primary researches in several countries, and secondary analysis is used to explore institutional environment incongruity and the appropriateness of institutional arrangement (Dibie and Gadzekpo, 2003; Dobkowski and Walliman, 2002). Thus, human activities often inflict harsh and irreversible damage on the environment and on critical resources. If the government does not check these harsh behaviors toward the environment, many current human practices may put at serious risk the future that the human societies anticipate. These negative behaviors toward the environment may also affect the plant and the animal kingdoms, as well as alter the living world in such a way that it will be difficult to sustain life as human beings now know it.

It is also interest to note that the failure to incorporate informal waste collection systems hinders the science-driven vision of waste management and masks the volume of generation and growth trends. Additionally, this weakens government regulation of the industrial and agricultural sectors and results in negative progress toward sustainable urban futures. Sustainability in this respect is related to ecological system integrity, or the relationship of people to their natural environment (Elliott, 2007; Endres, 2004). The issues are human natural resources use, the physical impact of humans on the environment, and the capacity of people to adapt to complex environmental conditions through scientific knowledge and technological innovation. A common argument found in all the countries examined is that ecosystems develop better under appropriate climatic conditions.

The impact of political support for sustainable environmental policies is also discussed in several of the countries examined. The environmental policies that have been enacted in some of the countries covered in the book tend to promote smart growth, planning of more efficient land use, and more rigid management of urban development. Over the past three decades, there have been various international, national, and regional government mandates to ensure that the locations of hazardous waste sites and the undesirable use of the environment are equitably monitored. The book also contends that social and economic behavior in some societies may not necessarily be compatible with sustainable development because human consumption patterns have increased drastically over the past six decades (Guzmand et al., 2009; Kraft, 2010). It noted that during this period there has been growth in human consumption patterns; more fossil fuel, metal, and minerals have been mined from the environment; more trees have been cut; and the exploitation of these resources to maintain ever increasing levels of consumption has put avoidable pressure on the ecosystem around the world. Pressure resulting from unrestricted population growth tends to put demands on the natural environment (Rosenbaum, 2014). This growth in population could overwhelm any efforts to achieve a sustainable future. Developing and developed countries must make a commitment to population stability and resources conservation. The world may be better able to meet the challenges of sustainable development if human beings could invest more in protecting the environment. Thus, slower population growth could speed economic growth and conserve natural resources in both developing and developed countries. A better environment is within human reach, but success hinges on changing behavior and public opinion.

The book argues that sustainable development activities in some developing countries might not be useful or reasonable for every rural community. The forces of globalization or sustainable development do not respect national, regional, rural, or community boundaries. This book clearly states that developing nations will succeed if they adopt alternatives to centralized power in the national government. In addition, the transformation of developing nations in the twenty-first century will be a function of many factors. These factors include changing consumption patterns, controlling waste, and pragmatic environmental policies, among others. Developing nations must disengage themselves from the notion that good public and political leaders are born, not made, and emphasize the development of political and managerial skills through formal education, industrial training, and seminars. Developing nations should embrace the assistance of NGOs and other civil societies in seeking this transformation.

This book stresses that sustainable development will be difficult unless developed and developing nations transform their political and managerial talents. If sustainable development is not achievable, it will hold back the trends in human development and health improvement. Nations of the world can improve environmental quality through economic growth by collaborating to adopt new ecofriendly habits and sustainable practices (Rosenbaum, 2014). Thus, practicing sustainable development requires a combination of wise public investment, effective natural resources management, cleaner agricultural and industrial technologies, less pollution, renewable energy sources, and slower population growth. Nations of the world need to protect human life and ecological resources. Citizens of the world must act together in a differentiated and equitable way. Much more is need in reshaping our energy and environment strategies for the future and in helping countries all over the world to strengthen their risk management practices and expand their safety nets to cope with risks that cannot be fully mitigated. There are opportunities for us to shape our environment's future for an inclusive and sustainable development if we act now in the spirit of shared governance. What is crucial, however, is the fact that without practicing sustainable development, humanity would face a deteriorating environment and may even invite ecological disaster.

PARTNERSHIP FOR SUSTAINABILITY SOLUTION

The previous chapters present arguments that sustainable development potentially involves human interaction with the natural environment at all levels and over an indefinite time frame. The question that we like to ask at this point is: Can sustainable development be achieved in both developing and developed countries? A very important issue that needs to be addressed especially by developing countries is the matter of energy and environmental policies that have been neglected for several years. The developed countries will have to set good examples for the developing nations in order for them to learn how to effectively and efficiently implement policies that they have enacted.

According to Norman Vig and Michael Kraft (2013), the problem of traditional regulation in each of the environmental mediums (air, water, soil) and in each type of problem (air emission, water discharges, toxic waste) is dealt with under different laws and by different agency. For sustainable development to take place, a more integrated approach is needed that focuses more holistically on the health of natural systems that are subject to multiple stresses and whose boundaries will have to coincide with single legal jurisdiction (Vig and Kraft, 2013). There is also the need for a broader collaboration between the lead national government agencies and multiple public and private stakeholders (including citizens groups and nongovernmental organizations), working together to try and reach consensus on plans for the future management of the ecosystem or resource base.

This collaboration of multiple public and private stakeholders is one of the key principles of the social marketing of environmental policies and issues. Such a holistic approach will help to develop strategies and interventions that will resonate with each country's values, desires, and needs as well as increase chances of success. They will also be better able to communicate in the ways preferred by the citizens. This will help citizens to be sensitive to their unique culture and to relate to their goals and priorities, motivation, and stage of environmental problem recognition. In addition, improving environmental health as well as sustainable conditions cannot be achieved by any one organization or by addressing one individual or nation at a time. Eliminating poverty and disparate outcomes requires participation from key public and private institutions working in partnership with community-based organizations.

It is paramount for developed and developing countries to provide incentives for institutions such as business, schools, banks, government, health care facilities, and corporations and nongovernmental organizations to have a major influence on environmental improvement issues (Dibie and Offiong, 2012). Thomas Easton (2010) contends that, as employers, investors, and purchasers, the public, private and nonprofit sectors have an impact on the local economy. Further, as providers of services and purchasers, they influence what is and is not available to citizens and the nation. Consequently, as prominent facilities within the nation and communities, the private, public, and nonprofit sectors could help to establish environmental ethic norms for employees, students, and the general public. These activities could help to galvanize a welcoming atmosphere that supports environmental health and change in behavior toward the ecosystem. According to Michael Kraft and Scott Furlong (2013) and Nancy Maxwell (2014), universities and other educational institutions are essential national resources and excellent venues for reaching families and citizens in general. While these educational institutions are meeting the educational needs of the citizens, they could also promote environment health and changes in behavior toward the ecosystem. Current students who may become our future leaders can be taught by these institutions how to preserve and nurture the environment. Therefore, engaging all stakeholders in the nation, communities, and the world in shaping solutions and taking action for change in how human beings treat the environment is crucial to achieving sustainable development.

One of the most important considerations for the national government of any country in implementing its environment policies is the encouragement of citizens and community-based organizations to give input. Strong social networks and community-based input, trust, and respect among citizens and environmental health promotion traditions, such as active lifestyles, sustainable values, and recycling practices, are all requisite factors that need support and enhancement for reducing ecological health disparities. A good example of collaboration in the sustainable development processes of several cities is the smart growth project. Growing smart guidelines specifically emphasizes regional planning for the control of sprawl, environmental protection, and fiscal responsibility rather than growth at all cost (Vig and Kraft, 2013). The practice in the city of Seattle in the United States shows that sustainable community planning often uses highly participatory processes to create livable areas while enhancing local democratic, community, and civic life. Further, Seattle's climate change initiative also involved tree planting and urban reforestation, enhanced bicycling opportunities, green roofs on city buildings, improved walkability, technical assistance to builders, zoning changes that discourage sprawl, shoreline and wetlands protection, reduced use of fossil fuels in city-owned vehicles, and additional increases in the recycling and composting of waste (Kraft and Furlong, 2013; Vig and Kraft, 2013). Partnership for the environment is very essential for human empowerment and prosperity. This book is calling for a new environmental ethic of caring for creation. It reiterates that practicing sustainable development requires a combination of wise public investment, effective natural resource management, cleaner agriculture and industrial technologies, less pollution, and slower population growth.

ENVIRONMENTAL SUSTAINABILITY AND GENDER EQUALITY

Women have a very important role to play in environmental management. In many developed and developing countries, men and women have different roles and responsibilities within the social structure and fabric of society. The impact of economic adjustment manifests differently between men and women (Dibie and Dibie, 2008). This can be seen in the way men and women respond to economic incentives and shift in economic policy. For example, in many African countries, women's access to land is restricted because of prevailing exclusionary land tenure policies (Ako–Nai, 2013; Dibie and Dibie, 2008; Sam Okere, 2013; World Bank, 2010). Further, access to agricultural land is gained through marriage and forfeited through divorce proceedings. Cultural and traditional practices in some countries also dictate that women cannot inherit assets from their deceased husband, including agricultural land for production and growing of food to feed their families (Ako–Nai, 2013; World Bank, 2010).

Economic structural adjustment should therefore recognize the social costs and impacts associated with economic adjustment within the context of women's empowerment (Taylor, 2010). According to Ronke Ako-Nai (2013) structural adjustment programs or policies in developing countries are intended to improve the allocation and effective use of resources, as well as the trade regime or environment, including the implementation of incentive schemes. The sociocultural status of women and their consequent orientation and contribution toward development are made problematic by the policy environments created as a result of the social role differentiation between men and women.

The crucial role of women in sustainable development all over the world was reiterated in the World Bank report (2010, 116). In the past few years, the emphasis has been more gender mainstreaming, that is, incorporating women's concerns as an integral part of the planning, implementation, and monitoring of all development and environmental management programs. Justina Sam Okere (2013) contends that by developing specific targets, the role of women is seen in terms of the gender imbalances that prevail in many developing countries and is incorporated into a network of objectives to ensure sustainable development.

Studies conducted by the World Bank (2000) suggest that where the influence of women in public life is greater, the incidence of corruption is lower. Although more extensive research is needed in this regard, a comparison among countries with the same human development, economic development, and engendered rights confirms that the inclusion of women results in improved governance and the application of the rule of law (World Bank, 2010). Ronke Ako-Nai (2013) contends that the contribution of women to the informal economic sector activity by indicating that about 60 percent of the informal economy is made up of women and that they provide approximately 70 percent of total agricultural labor in most developing and developed countries.

Public policies that focus on raising public awareness of women's rights and the benefits of women's participation in environmental issues and sustainable development are long overdue in several countries. Macro policies should not only contribute toward women's access to land, but they should also create an environment conducive to awareness creation through the implementation of gender awareness projects, with specific reference to the role of women in agriculture, preservation of the ecosystem, and other environmental issues. It is therefore essential that, apart from national policy and initiatives, institutional mechanisms are instituted for the advancement of women and institutional development funds be made available specifically earmarked for women's capacity building in environmental health and sustainable development.

Public policies should therefore be enacted to ensure the systematic access to basic technologies and resources to assist women to fulfill their economic roles effectively. This will in turn markedly affect macroeconomic supply, response policies, and performance. The World Bank (2000) report indicated that putting additional income in the hands of women results in a larger positive impact in intergeneration sustainability. Unfortunately in many countries, intergeneration inequality in household structure prevents this. Public policies that encourage investing in female education not only yields the highest social return but also affects the economic sector through its positive impact on agriculture, industry, and infrastructure (Ako-Nai, 2013; Sam Okere, 2012; Wright and Boorse, 2014). Both nature and managed ecosystems are being pressed to provide increasing goods and services without consideration of the limits of the ecosystem to continue to produce. Human beings may have reached the limit in some resources, and most of the time the ways we use the other resources, including women, are currently unsustainable.

GREEN ECONOMY

The United Nations Environment Program (UNEP) defines a green economy as one that results in "improved human well-being and social equity, while significantly reducing environmental risks and ecological scarcities" (UNEP, 2010). In its simplest expression, a green economy is low-carbon, resource efficient, and socially inclusive. In a green economy, growth in income and employment is driven by public and private investments that reduce carbon emissions and pollution, enhance energy and resource efficiency, and prevent the loss of biodiversity and ecosystem services (Myers and Spoolman, 2014; Pearce et al., 1989).

The historic United Nations Earth Summit on environment and development, held in Rio de Janeiro, Brazil, in 1992, did not fail the world.

Far from it (Kraft, 2010). It provided the vision and important pieces of the multilateral machinery to achieve a sustainable future. A green economy and sustainable development were outcomes of the United Nations General Assembly decision to hold a summit in Rio de Janeiro in 2012 (Rio+20) in order to celebrate the twentieth anniversary of the first Rio Earth Summit in 1992. Two of the agenda items for Rio+20 were "Green Economy in the Context of Sustainable Development and Poverty Eradication" and "International Framework for Sustainable Development." With the green economy now firmly established on the international policy agenda, it is useful to review and clarify the linkages between a green economy and sustainable development (Steiner, 2011).

As a result "Towards a Green Economy" is among UNEP's key contributions to the Rio+20 processes and the overall goal of addressing poverty and delivering a sustainable twenty-first century. The report makes a compelling economic and social case for investing 2 percent of global GDP in greening ten central sectors of the economy in order to shift development and unleash public and private capital flows onto a low-carbon, resource-efficient path (Pearce et al., 1989).

It is essential to note, however, that a green economy does not favor one political perspective over another. It is relevant to all economies, be they state or more market-led. Neither is it a replacement for sustainable development. Rather, it is a way of realizing that development at the national, regional, and global levels and in ways that resonate with and amplify the implementation of Agenda 21.

The major goal for a transition to a green economy is to enable economic growth and investment while increasing environmental quality and social inclusiveness (Taylor, 2012). Critical to attaining such an objective is to create the conditions for public and private investments to incorporate broader environmental and social criteria. In addition, the main indicators of economic performance, such as growth in gross domestic product (GDP), need to be adjusted to account for pollution, resource depletion, decline ecosystem services, and the distributional consequences of natural capital loss to the poor.

Mounting evidence also suggests that transitioning to a green economy has sound economic and social justification. There is a strong case emerging for a redoubling of efforts by both governments and the private sector to engage in such an economic transformation (Pearce et al., 1989; UNEP, 2011). The implication for governments is that green economy would include leveling the playing field for greener products by phasing out antiquated subsidies, reforming policies and providing new incentives, strengthening market infrastructure and market-based mechanisms, redirecting public investment, and greening public procurement (Pearce et al., 1989; UNEP, 2011).

Green economy for the private sector, would involve understanding and sizing the true opportunity represented by green economy transitions across a number of key sectors and by responding to policy reforms and price signals through higher levels of financing and investment. A major challenge is reconciling the competing economic development aspirations of rich and poor countries in a world economy that is facing increasing climate change, energy insecurity, and ecological scarcity (UNEP, 2011). A green economy can meet this challenge by offering a development path that reduces carbon dependency, promotes resource and energy efficiency, and lessens environmental degradation. As economic growth and investments become less dependent on liquidating environmental assets and sacrificing environmental quality, both rich and poor countries can attain more sustainable economic development.

The concept of a green economy does not replace sustainable development, but there is a growing recognition that achieving sustainability rests almost entirely on getting the economy right (Wright and Boorse, 2014). Decades of creating new wealth through a brown economy model based on fossil fuels have not substantially addressed social marginalization, environmental degradation, and resource depletion. In addition, the world is still far from delivering on the Millennium Development Goals by 2015 (Gallai et al., 2009; Global Footprint Network, 2010). The next section looks at the important linkages between the concept of a green economy and sustainable development

According to Barbier (2010a), the green economy concept could motivate policy makers to create the enabling conditions for increased investments in a transition to a green economy in three ways. It has the potential to be a crucial strategy to eliminate perpetual poverty, a new engine of growth, and a generator of decent jobs. Implementing the green economy grand policy program would not be easy. The policy reform path would require both public and private investment to transform key sectors that are critical to greening the global economy. It illustrates through examples how added employment through green jobs offsets job losses in a transition to a green economy. Further, the report shows how a green economy can reduce persistent poverty across a range of important sectors—agriculture, forestry, freshwater, fisheries, and energy. Sustainable forestry and ecologically friendly farming methods help conserve soil fertility and water resources. This is especially critical for subsistence farming, upon which almost 1.3 billion people depend for their livelihoods (UNEP, 2011; Withgott and Brennan, 2009).

In addition, the UNEP report also provided guidance on policies to achieve this shift by reducing or eliminating environmentally harmful or perverse subsidies, addressing market failures created by externalities or imperfect information, creating market-based incentives, implementing appropriate regulatory frameworks, initiating green public procurement, and stimulating investment (Eliasch, 2008; UNEP, 2011).

Many scholars' interpretations of sustainability take as their starting point the consensus reached by the World Commission on Environment and Development (WCED) in 1987, which defined sustainable development as "development that meets the needs of the present without compromising the ability of future generations to meet their own needs" (Gallai et al., 2009; Global Footprint Network, 2010; Roorda, 2012; WCED, 1987).

Public policy analysts and economists are generally comfortable with this broad interpretation of sustainability because it is easily translatable into economic terms: an increase in well-being today should not result in reducing well-being tomorrow. This means that future generations should be entitled to at least the same level of economic opportunities and the same level of economic welfare as are available to current generations (Roorda, 2012).

According to Guy Peters (2013) and Niko Roorda (2012), economic development today must ensure that future generations are left no worse off than current generations. Or, as some economists have succinctly expressed it, per capita welfare should not be declining over time (Pezzey, 1989). According to this view, it is the total stock of capital employed by the economic system, including natural capital, that determines the full range of economic opportunities and thus well-being. Table 16.1 shows the relative impact of a green economy.

Eliasch (2008) and Gallai et al. (2009) contend that a green economy would provide essential benefits to human beings in areas such as freshwater; capture fisheries; water purification and waste treatment; wild foods; genetic resources; biochemicals; wood fuel; pollination; spiritual, religious, and aesthetic values; the regulation of regional and local climate; erosion; pests; and natural hazards. The economic values associated with these ecosystem services, while generally not marketed, are substantial. The gains that have been achieved, however, come with growing costs in the form of degradation of many ecosystems' services, the exacerbation

Table 16.1 Impact of Green Economy

Biodiversity	**Ecosystem Goods and Services (examples)**	**Economic Values (examples)**
Ecosystems (variety and extent/area)	Recreation • Water regulation • Carbon storage	Avoiding greenhouse gas emissions by conserving forests: US$3.7 trillion (NPV)
Species (diversity and abundance	Food, fiber, fuel • Design inspiration • Pollination	Contribution of insect pollinators to agricultural output: ~US$190 billion/year
Genes (variability and population	Medicinal discoveries • Disease resistance • Adaptive capacity	25–50% of the US$640 billion pharmaceutical market is derived from genetic resources.

Sources: Eliasch (2008); Gallai et al. (2009); TEEB (2009).

of poverty for some people, and the growing inequality and disparities across groups of people (Kraft and Furlong, 2013; Taylor, 2012). As a result of the crucial relationship between the environment and the economy, there is an essential need for a different kind of economy. Therefore, the path to a sustainable future will require a green economy.

RENEWABLE ENERGY

A general consensus among environmentalist is that all sources of energy used for production and exploration often have a negative impact on the environment. In the past three decades, many nations across the world have experienced significant industrial expansion. This expansion has also caused severe environmental damages that can be rationalized as the cost of industrial progress and economic development. In most developed and developing countries, fossil fuels and nonfossil fuel by-products, such as petrochemicals, have been creating a lot of damage to soil, waterways, and air.

It is important to know that renewable energy is not synonymous with sustainability (Raskin et al., 2012). Renewable, as used in this book, refers to the naturally regenerative nature of the energy source. Generally, renewable energy technologies are perceived as being more sustainable compared to nonrenewable sources, due to their smaller environmental impacts, but there is still a need to develop agreed-on standards to reduce and manage these impacts. On the other hand, sustainability has a broader scope, including economic, social, and environmental considerations. The environmental and social impacts of large hydropower reservoirs are one prominent example of the potential to release carbon dioxide and methane from decaying biomass in tropical locations. Concern about minimizing impacts has led to the development of policy principles and guidelines under the coordination of the World Commission on Dams (UNEP, 2011).

NONRENEWABLE ENERGY SOURCES

There are two types of energy sources. On the one hand are nonrenewable energy sources, which include coal, natural gas, oil, and nuclear power. On the other hand, renewable energy sources include hydropower, solar power, wind power, nuclear energy, and biomass.

Among all the nonrenewable energy sources, *natural gas* is regarded as the cleanest-burning of all fossil fuel and relatively inexpensive. Carbon dioxide is given off as it burns in lower amounts per unit of energy produced compared to other fossil fuels (Kraft and Furlong, 2013; Peters, 2013). According to Withgott and Brennan (2009), natural gas is very easy to transport through pipelines. Despite the easy transportation profile, natural gas pipes cannot be efficiently built across oceans. As a result, alternative methods for transporting natural gas have to be explored. One of the alternative methods created by scientist is liquid natural gas. Natural gas may be converted to a concentrated form called liquid natural gas (Wright and Boorse, 2014). The good news is that natural gas can now be transported by refrigerated tanker ship with special care. If care is not well taken or the gas is mishandled, an explosion on board a tanker ship could result in a massive fireball, causing fires and third-degree burns as far as miles away (Dye, 2011; Smith, 2004).

In the past three decades, the exploration and transportation of *oil* have caused many environmental problems, in addition to damaging surrounding ecosystems. Oil is particularly well-known for polluting groundwater through the injection of brine solution into aquifers (Smith, 2004). According to Kraft and Furlong (2013), carbon emissions contribute to the greenhouse effect. Oil spills from tanker ships, such as the Exxon Valdez in 1989, were reported to have released 10.1–12.6 million gallons of crude oil into Prince William Sound, Alaska. This area was known for its teeming marine and bird life. In 2010, there was another oil accident in an offshore drilling rig in the Gulf of Mexico. In both incidents, toxic petroleum hydrocarbon killed marine life and destroyed the local fishing industry. It is interesting to note that cost-effective oil and gas resources are becoming scarce as requirements continue to rise. Any future gap between supply and demand for oil and gas will need to be filled by unconventional sources, such as shale oil, tar sands, biomass oils, and gas from coal (Peters, 2013; Raskin et al., 2013).

Coal is regarded as the most harmful nonrenewable energy source. Coal mining itself has many attendant hazards. Coal dust can produce coal workers pneumoconiosis (CWP), though it has declined dramatically over recent years due to improved ventilation and dust control. According to William Rom (2012) and Robert Friis (2013) progressive massive fibrosis (PMF) can occur in heavily dust-exposed miners. It is characterized by fibrosis in the mid-upper lung zone and reduced pulmonary function. Coal also produces annoying and debilitating smoke, and any accident could have both short-term and long-term effects on human health and environment. Anne Nadakavukaren (2011) contends that coal mine explosions come from methane seepage that can mobilize large amounts of coal dust, which has even greater explosive characteristics. It has also been reported that methane is often released to the environment during mining, and the EPA estimates that 27 percent of all energy-related methane emission in the United States comes from coal mining (Kraft and Furlong, 2013). Twelve thousand miners died from black lung disease (CWP and PMF) between 1992 and 2002 in the United States (Rom, 2012). According to Nancy Maxwell (2014) and Robert Friis (2013), scientists have come up with a new instrument that could be used to measure the level of damage caused by coal. For example, coal mining has to use underground techniques for deep coal seams, but surface mining is much cheaper, safer, and productive.

Further, coal is low in sulfur; as a result, it helps to reduce the problem of acid rain and sulfur dioxide (Maxwell, 2014; Rosenbaum, 2014; Vig and Kraft, 2013).

A new report by the U.S. Environmental Protection Agency indicated that air pollution, land devastation, thermal pollution, acidification of streams and rivers, erosion, polluted groundwater, subsidence of land caused by underground mines, hazards to miners, and, of course, global warming are all recognized as part of the destructive impact from the human usage of coal (Smith, 2007; Wright and Boorse, 2014).

ALTERNATIVE SOURCE: RENEWABLE ENERGY

Most environmentalists would recommend alternative sources of energy because they are considered to be green as a result of their having lower carbon emission and creating less pollution. Among the alternative sources of energy that could be substituted for and eventually replace fossil fuels are biomass or biofuel, hydropower, nuclear energy, solar energy, geothermal energy, and wind power.

Biomass is the product of photosynthesis. Biomass is used as a fuel (firewood, biodiesel, biokerosene and ethanol). In addition, biomass is a raw material for the paper and pulp lumber and construction industries. While it may lower the requirements for traditional firewood, the rapid economic growth in market forces increases biomass demands in modern societies; thus prosperity for some can cause pain for others (Raskin et al., 2012). Biomass is plant matter that acts as a versatile fuel capable of producing high-quality gaseous, liquid, and solid fuels, as well as electricity. According to Myers and Spoolman (2014), using biomass feedstock to create electricity does not contribute to global warming and helps avoid the release of other harmful emissions. Also, viable fuels are produced in many industries with no additional processing costs involved, further strengthening the financial feasibility of new plants on favorable sites (Smith, 2004; Peters 2013). In most cases, biomass is made from forestry and wood products, crop residues, animal waste, and energy crops. William Rom (2012) contends that biomass fuels include a wide variety of material. Most of the byproduct of biomass are forestry by-products, agricultural wastes, municipal wastes, landfill gas, and syngas. There are several new next generation biomass feedstocks in different stages of commercialization and this innovation trend is beginning to attract many energy companies. These new fuels have the potential to dramatically scale plant sizes and the feedstock supply chain, creating significant opportunities for additional generation fleet expansion, especially in the reduction of green gas effect. In countries such as Brazil and the Philippines, wood is the most widely used for biomass energy. In both developed and developing countries, wood is burned to produce energy. Some industries also use firewood as their source of energy. According to Ferrell and colleagues (2013), Brazil is the largest exporter of bioethanol. Biofuel production in other countries such as the Philippines has been criticized because it has contributed to the rapid deforestation of ecologically sensitive areas. The ecosystem and the biomass that have been most significantly altered globally by human activity are marine and freshwater ecosystems, temperate broadleaf forests, temperate grasslands, Mediterranean forests, and tropical dry forests. The degradation of fisheries is also reflected in the fact that the fish being harvested are coming from the less valuable lower trophic levels (Smith-Sebasto, 2012).

To solve the problem of deforestation, scientists are researching alternative sources of biomass fuel by using algae and nonedible plants such as grasses. Since grass and algae are not food sources and do not require the destruction of trees, ethanol proponents are looking forward to the adoption of these alternative biofuel sources. It should be noted, however, that land-intensive biofuel production competes with food production. This is because it tends to put additional pressure on agricultural innovations to increase yields and stabilize the prices of food staples. It would take the moderation of energy demands for biomass demand eventually to decline from current levels, thereby sparing land for agricultural and natural farm production (Raskin et al., 2012).

Hydropower involves the use of turbines that are turned by the natural forces of water. Dams are built across rivers to enhance the natural forces of water. Hydropower is an important source of electricity power generation in many developed and developing countries. This is because hydroelectricity power plants produce electricity far more cheaply than any other source. Unlike coal, the water resources use for hydroelectricity is free, reliable, and relatively nonpolluting. Further, the lifetime of the plant or dam is much longer (Kraft, 2010). The World Energy Council estimates that the world has 45,000 large dams that have generating capacity of 800 gigawatts, and they currently supply one-fifth of the electricity consumed worldwide. In addition, one hundred and sixty countries use hydropower to some extent. Brazil, Canada, China, Russia, and the United States currently produce more than half of the world's hydropower (Rom, 2012). Myers and Spoolman (2014) contend that Europe has already exploited about 76 percent of its feasible hydropower, but Africa and Asia have large unharnessed capacity. New dams are constructed every year all over the world.

Another very important advantage of hydropower is that it decreases greenhouse emission and air pollution. To be ecofriendly and effective alternatives to fossil fuel, hydroelectricity plants should be built to minimize the destruction of wildlife, habitats, and human inhabitants close to the facilities or dam. The construction of dams often causes adverse environmental effects. For example, in the past four or more decades millions of people have been relocated in Nigeria, India, and China because of flooding upstream of dams. It could there be argued that dams have ecological effects on the ecosystem upstream and downstream as well as present a barrier to migrating fish and other biodiversity creatures in the rivers.

Nuclear energy has been considered a controversial form of alternative energy. This is because it is sometimes classified as green energy due to its pollution-free nature. One other characteristic of nuclear energy is that requires uranium, which is an abundant energy resource. According to Myers and Spoolman (2014), William Rom (2012), and Wright and Boorse (2014), there are several technologies involved in the production of nuclear energy. The most used method is pressurized water reactors that generate steam to drive turbines. These are light water reactors using ordinary water to slow neutrons and cool the reactor. Uranium is the fuel, however, 99 percent of this is unburned, and long-lived nuclear waste such as plutonium, americium, and curium are produced that require geologic isolation.

Uranium could be used for building weapons of mass destruction as well for generating electricity. The consequences of the 1986 meltdown of the Chernobyl plant in the Ukraine and the 2011 nuclear crisis in Japan has increased concerns about the safety of nuclear energy. The Three Miles Island nuclear plant mishap in the United States also caused a major disaster. As indicated previously, nuclear energy does not emit greenhouse gases, but it does produce tons of radioactive waste. Standing a few feet from uncontained nuclear waste would kill a human in seconds. It is estimated that nuclear energy provides 20 percent of U.S. electricity. There are about 104 operable nuclear energy plants and 27 new ones are under consideration by the U.S. federal government. Further, there are 439 nuclear reactors worldwide with a capacity of 370 gigawatts, and 20–30 percent of today's reactors will be decommissioned between 2013 and 2030 (Rom, 2013).

Solar energy is another alternative power source. Solar energy uses light and heat from the sun to generate electricity. It originates with thermonuclear fusion in the sun. Solar thermal uses solar stored in special heat-retaining fluid to drive a turbine and create power (Myers and Spoolman, 2014). Solar energy ranges from ultraviolet light to visible light to infrared light, or heat energy. As a result, full sunlight can deliver about 700 watts per square meter to earth's surface when the sun is directly overhead (Wright and Boorse, 2014) According to Ferrell and colleagues (2013) and Wright and Boorse (2014), some Walmart stores in California are already using solar energy to generate their electricity. Solar thermal plants are being envisioned for the Sahara Desert that can ship power to Europe. According to Michael Kraft and Scot Furlong (2013), the U.S. government now offers federal tax credits of $2,000 to homebuyers who wish to install solar panels in their houses. Some state governments in the country may also give additional credits.

Wind energy is becoming very popular in Europe and North America. Wind has a very promising future to provide green, renewable energy. It is estimated that about 50 percent of Denmark's electricity needs are generated by wind farms. India also has one of the most ambitious wind energy programs in the world. The United States is on the cutting edge of wind energy as well. The U.S. wind power capacity represents more than 20 percent of the world's installed wind power (Rom, 2012). There are, however, some concerns about wind energy because of the lack of an economic means of storing the electricity generated from the wind. Wind energy also requires backup from another source of energy. Wind turbines are sprouting up everywhere in Argentina, Mexico, New Zealand, and many other countries. According to William Rom (2012) and Myers and Spoolman (2014), China is the world's leader in the development of wind power, with 43, 900 megawatts by the end of 2013. Spain and Germany are leading in the European Union. Denmark sells its wind power to Norway and Sweden. Further, Sweden and Norway use this extra wind power to pump underground water into upstream reservoirs to produce more hydropower. Regiments of wind turbines standing in rows are found in California. These wind farms are capable of generating electricity power equivalent to that produced by large coal-fired power plant. It is reported that, at the end of 2015, wind energy capacity in the United States will reach 47 gigawatts, enough to power 15 million homes (Wright and Boorse, 2014). Some of the shortcomings of wind power are that it needs backup power sources. This is because wind turbines tend to generate only about 20 percent of their capacity. Wind turbines are also responsible for approximately 41,000 American bird deaths per year, far less than the million that succumb to domestic cats. It has also been reported that several thousand migratory bats are killed every year by windmills (Rom, 2012).

Geothermal energy comes from natural heat inside the earth and is extracted by drilling into steam beds. One advantage of geothermal energy is that it can provide a constant supply of energy every day (Ferrell et al., 2013; Kraft, 2010). Geothermal energy is reported to come from the heat contained in underground rocks and fluid. The renewable energy source can be tapped by drilling three or more miles deep underground in areas where 300-degree Fahrenheit rock exists, followed by applying high water pressures to open fractures in the rock. Injection wells are then drilled to circulate water in the human-made reservoir and extract steam to the surface to rum electric turbines (Rom, 2012). According to Wright and Boorse (2014) and Myers and Spoolman (2014), there are new and more innovative ways to generate geothermal power. For example, geothermal energy can now be harnessed at lower temperature than is required for boiling water by using closed circuit and liquid compounds with lower boiling temperature.

According to Smith (2004), about 10 percent of the world's landmass is suitable for developing this heat into usable energy. Another type of geothermal energy involves the extraction of subterranean hot water or steam that is then used for direct heating or conversion into electricity through the use steam turbines. The other type of geothermal energy involves the pumping of water into the ground through a piping system that is near a source of heat such as hot dry rock or magma. It is reported that naturally occurring wet steam and hot water geothermal deposits are the most common but are more expensive and difficult to convert to electricity (Withgott and Brennan, 2009; Smith 2004). About twenty countries around the world are exploiting geothermal deposit for supplying their heating systems (Smith, 2004). The U.S. produces 2,800 MW with forty six power plants at seven sites in California, including the Geyser at 1,000 MW. It is reported that this provides 0.3 percent of the electricity in the United States, however, 17 percent of its renewable energy (Myers and Spoolman, 2014). According to William Rom (2014), hydro-fracking for tapping geothermal energy of magnitudes envisioned would need to be tested so that it does not touch off earth tremors nearby. The Chena Hot Spring in Alaska is a good example of how geothermal energy works. The Chena Hot Spring is located about 66 miles from the electricity grid and uses a geothermal closed circuit to generate about 600 kilowatts of geothermal electricity per day to heat the resort's forty-five buildings, run a greenhouse, and a café made of carved ice (Rom, 2012).

Many developed and developing countries are beginning to enact policies to promote the use of renewable energy. This is because they want to become less dependent on energy imports and the volatile prices of oil and gas. The current U.S. President Barack Obama is particularly committed to spending more public revenue on renewable energy research.

Individual policies to overcome various barriers to renewable energy development and deployment are most effective when they are part of a broad enabling policy framework, which builds on complementarity among a range of measures operating at multiple stages of the chain from research and development through to deployment and diffusion.

An enabling policy framework for renewable energy includes clear commitments to long-term development of the sector. Such commitment can be manifested by targets for investment in additional capacity and penetration rates within the energy mix. When supported by other enabling policies, setting targets to achieve these goals can send a strong signal to potential investors.

Renewable energy producers, for example, may be granted exemptions from general energy taxes. Such measures are potentially most effective where overall energy taxes are high, such as in Nordic countries (Friis, 2013; Myer and Spoolman, 2014; Rom, 2012) The United States and Sweden, for example, provide a 30 percent tax credit for solar photovoltaic systems, France offers a 50 percent income tax credit, and Australia provides rebates up to AUS$8 per watt (Barbier, 2010a).

Policy targets for renewable energy have also been established in many developing countries. Developing countries with targets for 2020 or beyond are, among others, Brazil, China, Egypt, India, Kenya, the Philippines, and Thailand. It is interesting to note that Tunisia, a country in North Africa, has been encouraging the use of renewable energy since 2004 (UNEP, 2011).

The Brazilian Alcohol Program (Proalcool) was established in 1975 for the purpose of reducing oil imports by producing ethanol from sugarcane. Incentives aimed at both the production and the consumption of ethanol, including vehicle technology advancement through flexible fuel engine development, made petroleum substitutes competitive on the Brazilian energy market (UNEP, 2011). Other green economy policies for supporting renewable energy technologies include:

1. Development and infrastructure planning: R&D financing to support large-scale experimentation.
2. Stable, technology-specific incentives: Feed-in tariffs, credits, loan guarantees
3. Technology-neutral but declining support: Green certificates, greenhouse gases (GHG) trading
4. Accelerate adoption by addressing market barriers: Measures to stimulate demand, such as building codes

SUSTAINABLE CITIES

To achieve sustainable development, human beings will have to embrace John Rawls principles of social justice (1971). Citizens of the world should also embrace the values and understanding of how the natural world works, as well as a realization that what is happening to ecosystems is a result of human activities. According to Wright and Boorse (2014), value-based stewardship could result in the establishment of structures within the community that show great promise in helping the entire world move toward a sustainable future.

According to Michael Kraft (2011), cities in the United Sates just as in other countries are responsible for around two thirds of the energy used, 60 percent of all water consumed and 70 percent of all greenhouse gases produced worldwide. As a result of these urban problems, sustainable cities are looking at ways to improve their infrastructures to become more environmentally friendly, increase the quality of life for their residents, and cut costs at the same time. Sustainable cities are also exploring intelligent traffic solutions, green buildings, water management, and smart grid infrastructure (Myers and Spoolman, 2014). Modern technologies are helping interested cities to steer today's urbanization toward sustainability (Peters, 2013). The United States's longstanding expertise has given rise to innovative technology for sustainable solutions in energy efficient buildings, water treatment facilities, transportation infrastructure, public safety systems and healthcare imaging and diagnostics (Baumer and Horn, 2014; Roorda 2012; Wright and Boorse, 2014). What is at stake, however, is that, if human being can succeed in introducing and establishing many real sustainable cities or communities the world could be transformed into an increasingly stable eco-friendly place for both current and future generations.

The partnership between local governments, citizens, nonprofit organizations, private sectors, and community-based organizations has led to more sustainable modern cities. Sustainable, or ecofriendly, cities are designed with a consideration for environmental impact and are inhabited by people dedicated to minimizing required inputs of energy, water, and food, as well as the waste output of heat, air pollution (CO_2, methane), and water pollution (Vig and Kraft, 2013). These modern ecofriendly cities are able to feed themselves with minimal reliance on the surrounding countryside and power themselves with renewable sources of energy (Wright and Boorse, 2014). The reasons for creating sustainable cities are to establish the smallest possible ecologically friendly residential town, to produce the lowest quantity of pollution possible, to efficiently use land, and to compost used materials, recycle them, or convert waste to energy. The environmental aim is help minimize these cities' overall contribution to climate change.

According to Michael Dobkowski and Isidor Wallimann (2002) and Basten and colleagues (2011), the challenges cities face can be overcome in ways that allow them to continue to thrive and grow, while improving resource use and reducing pollution and poverty. In the twenty-first century, people tend to want to live in cities with all their opportunities and access to basic services, energy, housing, transportation, and more.

In the past two decades, city governments have been able to bring together the business community, civil society organizations, and various levels of government to tackle urban issues. They have also been able provide multidisciplinary forums for collaborative problem solving and idea generation around all aspects of the design, implementation, and regulation of sustainable urban energy systems. It is however very refreshing to note the growth of sustainable communities and cities in the United States and some European countries. The desire to seek a sustainable cities has given citizens in hundreds of communities in the United States and some European countries the opportunity to affect significant local decision on transportation, energy use, land use, urban sprawl, and quality of life. The successful efforts in working towards sustainable cities across the United States indicates a great potential for these innovative efforts (Davies and Mazumder, 2003; Kraft, 2010).

According to the United Nations Human Development Index report (2012) and the United Nations Environmental Program report (2011), the portion of the world population living in urban areas reflects the following:

- The world's cities occupy just 2 percent of the earth's land but account for 60–80 percent of energy consumption and 75 percent of carbon emissions.

- 828 million people live in slums today, and the number keeps rising.
- Rapid urbanization is exerting pressure on freshwater supplies, sewage, the living environment, and public health.
- 95 percent of urban expansion in the coming decades will take place in the developing world.
- 3.5 billion people—half of humanity—live in cities today.
- By 2030, almost 60 percent of the world's population will live in urban areas.
- The high density of cities can bring efficiency gains and technological innovation while reducing resource and energy consumption.

The types of advance technologies that have been developed over the years have facilitated the creation of smarter and more sustainable neighborhoods throughout the world. Sustainable cities are classified as proven solutions to the predicaments that rapid urbanization has created. Sustainable cities are already in use in the United States, Canada, Brazil, Australia, Switzerland, France, and many other countries around the world.

Sustainable cities could help to maximize the social, economic, and environmental impact that these sustainability changes can provide by using a phased implementation. The aim is to deploy technologies with the shortest economic payback first and then move toward solutions with longer-term returns (UNEP, 2011). Innovative initiatives in both developed and developing countries all over the world are beginning to reap benefits for the current generation and, in years to come, for the millions of people still living urban and rural areas.

Sustainable cities have a mix of building types, ranging from residences to commercial spaces to public facilities—in some cases, all at the same address (Peters, 2013; Vig and Kraft, 2013). One of the goals of sustainable cities is to balance energy consumption with personal comfort across a variety of building types, requiring the use of smart building controls connected to an intelligent building automation system that utilizes the building's real-time environmental information. Intelligent traffic solutions, green buildings, water management, and smart grid infrastructure are just a few of the technologies helping to make urbanization more sustainable.

According to the United Nations Environment Program report (UNEP, 2011), the challenges presented by sustainable urban development are immense. In 2010, 82 percent of Americans lived in cities; by 2050, the number will be 90 percent. Cities are responsible for around two-thirds of the energy used, 60 percent of all water consumed, and 70 percent of all greenhouse gases produced worldwide. Sustainable cities are looking at ways to improve their infrastructures to become more environmentally friendly, increase the quality of life for their residents, and cut costs at the same time (UNEP, 2011).

One of the major benefits of sustainable cities is that they are designed in a collaborative process with building owners, school districts, and city agencies to set—and achieve—sustainability goals. They include obtaining nationally recognized classifications of the Environmental Protect Agency by optimizing energy and water efficiency; by reducing greenhouse gas emissions; by improving indoor air quality and occupant comfort; by installing meters, submeters, and energy management systems; and by modernizing heating systems and boilers as well as providing renewable energy (Kraft and Furlong, 2013).

Another very important cost-saving benefit of creating sustainable development is remodeling aging infrastructure. Aging buildings typically use more energy while posing potential risks associated with older materials. By using advanced technology to remodel old infrastructure, such facilities are transformed into healthier building environments, while newer technologies can reduce energy consumption and lower greenhouse gases. Buildings can also be optimized using intelligent control systems to provide light only when needed as well as to reduce heating and cooling (Barbier, 2010b). Some banks in the United States, such as the Shore Bank, permit buyers of homes in sustainable cities or transit-friendly locations to carry larger mortgages because their transportation cost is likely to be lower. Further, an increasing number of cities have adopted Leadership in Energy and Environmental Design [the LEED standard for new municipal buildings, including some in Chicago, Los Angeles, and Toronto in Canada (Vig and Kraft, 2013)].

Solar energy (via the photovoltaic, or PV, technology) is one of the cleanest forms of power generation. Several sustainable cities have significant numbers of rooftops that could support solar installations. However, solar installations depend on incentives, including federal tax credits, which typically account for some 50 percent of any solar installation's capital costs (Wright and Boorse, 2014). To break even in fewer than five years, state and local incentives are required (TEEB, 2010; Roorda, 2012). Without these incentives, the payback period can exceed fifteen years, making solar power economically unattractive for investors. To optimize the tax incentives, there needs to be an investment entity with the means to monetize the federal tax credits over the duration of the credit.

In accordance with the concept of sustainable development, the public, private, and nonprofit sectors and community-based organizations should be encouraged to work together to make the changes that human beings need so that the lives of those in the generations after us are even better off than our we are. Citizens of this world must work together to share ideas, to promote innovation, to make pragmatic change, and to balance short-term costs and long-term benefits. Both developed and developing nations should continue to make progress to make sustainable living and homes affordable. Other notable benefits of sustainable cities include transportation that is powered by renewable energy. For example, electric vehicles and their charging infrastructures help to reduce greenhouse gas emissions while minimizing fossil fuel consumption. The amount of emissions created by a nonrenewable energy source–generating power for an electric vehicle is 30 percent less than the emissions emitted by a similar gas-powered vehicle. The United Nations Environmental Program report indicated that for every 300 kilowatts of solar power generated, 20 tons of greenhouse gases are avoided (UNEP, 2011).

City governments have also initiated innovative incentives to enhance sustainability. Such initiatives include land trusts, transferable development rights, and location efficiency mortgages. The city of Boston uses gas taxes to fund transit and has kept fares exceptionally low (Peters, 2013; Vig and Kraft, 2013). These innovative initiatives have led to the decline of automobile use by more than half while transit use has increased.

It is very important for the government and media to effectively communicate how citizens are consuming power, heat, or water. Environmental transparency and efficient communities could produce less waste while preserving the environment (Vaughn, 2011). However, by providing information on how a neighborhood is impacting the environment in real time, residents can take appropriate actions to become more efficient as a community. This change in consumption patterns and behaviors could further lead to a higher quality of life and a deeper civic engagement (Steiner, 2011). According to Robert Taylor (2012), it is becoming very certain that citizens and organizations from all walks of life will have to pay far more attention to sustainable imperatives in the future.

STEWARDSHIP

The chapters in the book reveal that environmental concerns are creating ethical issues for citizens all over the world. The impact of environmental issues also affects individuals, organizations, and public policy makers. In addition to the impacts of environmental issues is the desire for sustainable development. This sustainable quest means that ethics, social responsibility, and special initiatives must be used to implement effective changes. Stewardship of the actions and programs that manage natural resources and human well-being for the common good are required by all nations that wish to achieve sustainable development. Stewardship is crucial for the common good because it could lead people to stop the destruction of the environment or to stop the polluting that is degrading human neighborhoods and health (Wright and Boorse, 2014). According to John Rawls (1971), a just society is one in which social institutions conform to the principles of justice, which are based on a hypothetical contract that rational individuals form to obtain the benefits of social cooperation. Rawls argued that fundamental rights and liberties should be distributed equally but that social and economic goods may be distributed unequally, provided that the unequal distribution benefits all members of the society, and there is fair equality of opportunity. Stewards are likely to be the ones to distribute the common good as specified by John Rawls. Apart from leading people to stop the destruction of the environment and to prevent pollution, stewardship is required because society may redistribute resources to compensate its least advantaged members for undeserved disadvantages, such as social or economic liabilities resulting from genetic heritage.

It is very important for developed and developing countries, despite their differences, to seek stewardship. Human beings should see stewardship as a fundamental way people think about caring for the ecosystem and for fellow human beings. If we have true stewardship, there will not be the widespread degradation of essential ecosystems and the devastating impact of poverty and injustice. In the twenty-first century, we should seek to work toward caring for the environment and its human inhabitants. Rachel Carson, for example, is known for her tireless efforts in promoting stewardship and advocating for policies that would better reflect sustainability (Weaver et al., 1997; Wright and Boorse, 2014). Rachel Carson was a steward who alerted the public about the danger of pesticides and their role in decimating bird populations. If all of us could be a little bit like Rachel Carson, the world environment would be better than it is today.

ECONOMIC GLOBALIZATION AND INTERNATIONAL REGULATION

According to Norman Vig and Michael Kraft (2013), governments all over the world, especially those of developing countries, are ill equipped to resolve many long-term and severe problems in the global environment. As a result of this predicament, institutional reforms and new methods of decision making will be critical to success in effective regulation all over the world.

One thing is certain: citizens and organizations from all walks of life will have to pay far more attention to sustainability imperatives in the future. This is because the linked challenges of biodiversity loss and climate change must be addressed by policy makers with equal priority and in close coordination, if the most severe impact of each is to be avoided. On one hand, reducing the loss of carbon-storing ecosystems such as tropical forest, salt marshes, and peat lands will be a crucial step in limiting the buildup of greenhouse gasses in the atmosphere. On the other hand, reducing other pressures on the environment can increase their resilience, make them less vulnerable to those impacts of climate change that are already unavoidable, and allow them to continue to provide services to support people's livelihood and help adapt to climate change.

The globalization of environmental issues continues to receive more awareness since 1992. This awareness has galvanized environmental legislation (international, national, state, and local), and international agreements, such as the Kyoto, have not only raised the profile of environmental change but have also begun to drive global policy change (Adams, 2012).

In addition to global trends, public awareness of international and social issues is in many cases now well developed. According to Adams (2012) and Raskin (2012), citizens in almost all countries not only know the issues but also tend to feel that the quality of the environment is very important both to their own well-being and to the common good. In the midst of practicing the concept of globalization, the greening of business has grown to be a central issue in corporate social responsibility for many global companies. Although for many it is a boutique concern, for others a sustainable development relationship is a mechanism for achieving structural changes to human and biodiversity wellbeing. The good news, however, is that the dilemma presented by globalization offers huge risks and outstanding opportunities at the same time. Thus, the need to create a sustainable post–fossil fuel society and economy has never been more widely recognized, although the challenges on the road to achieving it remain breathtaking.

ERADICATION OF POVERTY AND GREEN ECONOMY

United Nations Human Development (2012) index data show that the majority of people living in developing countries depend directly on natural resources as farmers. The livelihoods of many of the world's rural poor are also intricately linked with exploiting fragile environments and ecosystems (Barbier, 2010c). It is also estimated that more than 1.7 billion of the rural poor currently live on lands prone to degradation and water stress and in upland areas, forest systems, and dry lands that are vulnerable to climatic and ecological disruptions (Myers and Spoolman 2014; World Bank, 2010; Wright and Boorse, 2014). As a result of local traditions and cultures, most people living in developing countries' rural communities are likely clustered on marginal agricultural lands and in fragile environments. People having to live under such conditions is likely to be a continuing problem for the foreseeable future, given current global rural population and poverty trends (Gallai et al., 2009). It is interesting to note, however, that in spite of constant migration to urban areas and rapid urbanization, the rural populations of several developing countries have grown at a slower rate in recent decades (United Nations, 2012). The United Nations data confirm the argument made by Chen and Ravallion (2007) that around three-quarters of the developing countries' poor still live in rural areas. One thing is clear about poverty in developing nations: their poor people live in rural rather than urban areas because they cannot afford the rent and standard of living in the big cities.

It is crucial for developing and developed countries to find ways to protect global ecosystems, reduce the risks of global climate change, improve energy security, and simultaneously improve the livelihoods of the poor people around the world. These efforts could pose some difficult challenges in the transition to a green economy, especially for developing countries. However, the author is optimistic that poverty around the world could be eradicated if there is an honest political will.

According to the United Nations Human Development report, one out five people in the developing world lacks access to sufficient clean water, and about half the developing world's population, 2.6 billion people, do not have access to basic sanitation (United Nations, 2012). More than 660 million of the people without sanitation live on less than US$2 a day and more than 385 million on less than US$1 a day (UNDP, 2006).

While people living in developing countries take electricity and water supply for granted, billions of people in developing countries have no access to modern energy services, and those consumers who do have access often pay high prices for erratic and unreliable services. Among the energy poor are 2.4 billion people who rely on traditional biomass fuels for cooking and heating, including 89 percent of the population of sub-Saharan Africa and the 1.6 billion people who do not have access to electricity (UNEP, 2011; Wright and Boorse, 2014).

As result of these poor living standards in several developing countries, a pro-poor orientation must be superimposed on any green economy initiative. Investments in renewable energy, for example, will have to pay special attention to the issue of access to clean and affordable energy. Payments for ecosystem services, such as carbon sequestration in forests, will need to focus more on poor forest communities as the primary beneficiaries (Vig and Kraft, 2013). The promotion of organic agriculture can open up opportunities, particularly for poor small-scale farmers who typically make up the majority of the agricultural labor force in most low-income countries, but it will need to be complemented by policies to ensure that extension and other support services are in place.

It is therefore essential for the developed and developing countries to form a humanitarian partnership in working toward eradicating extreme poverty and hunger, including halving the proportion of people living on less than US$1 a day by 2020. The green economy concept could be adapted as one of the instruments to accomplish this task (United Nations, 2012). Public policies related to a green economy should be enacted. These policies must not only be consistent with that objective but must also ensure that resources and investments are geared toward reducing environmental risks and that scarcities are compatible with ameliorating global poverty and social inequity.

ENVIRONMENTAL POLICY IMPACT

The overall aim of environmental policy is sustainable development. This entails the handoff to the next generation of a society in which the major environmental problems have been solved. It is recognized that economic growth, a healthy environment, and high public health standards are interdependent. Efforts to support sustainable development have involved changes in polluting systems and human behavioral change with regard to the use of the ecosystem. Areas affected include road traffic, waste management and disposal, carbon dioxide emissions reduction, reduction of the use of fossil fuel, deforestation, and degradation.

The distribution and redistribution of income are among the social justice aims of the industrialized world. In the developing countries, government does not have any direct wage ceilings but indirectly forces companies to keep wages low with extensive tax pressure. Extremes of wealth and poverty have been reduced through the efforts of social democratic governments and trade unions. Most developed countries have adopted the concept of the advanced welfare state and communal institutions to ensure the well-being and economic security of their citizens. No other countries have as low a rate of poverty and social exclusion as in developed countries. Health, education, and social welfare programs are comprehensive and universal. Coverage for all citizens prevents, to some extent, the development of an underclass social economic system.

For this reason, it is relatively easy for developed and some developing countries' governments to implement environmental policies. The government and its people in some countries tend to be extremely sensitive about their environment, and many government programs are designed to help promote and encourage sustainable development. The government therefore promotes environmental policies through areas that positively benefit the social and economic lives of its citizens. They do this through high environmental protection standards, environmental awareness and education particularly in schools, introduction of more environmentally sound production processes, and encouraging new environmentally sound technology and innovations. Consumers are also targeted through campaigns because changing demand patterns can help stimulate more sustainable production of goods and services. In developing countries, the reverse is the case.

The political and administrative culture strongly influences the making of environmental policies in many developing countries. This is because decision making is centralized with very limited public participation. This also means that government and its regulatory agencies bear the full responsibility of protecting the environment with no involvement from its citizens. The political culture in several African and Asian countries is one in which the government decides the best interest of its people, be it socially or economically. Adherence to traditional values and the maintenance of social stability in the context of rapid economic change have been the consistent goals of environmental policies. A limited amount of information is allowed out through media sources for public knowledge and discussion in some developing countries. As a result, strong environmental policies can only partially affect citizens of many developing countries in Africa, Asia, and Central and South America.

SELECTED ENVIRONMENTAL POLICY RECOMMENDATIONS

Most developed countries are viewed as global leaders in positive environmental practices, both in terms of public policy and business sector activities. After several years of study and deliberation, new environmental reform structures were created. In 1998, existing environmental protection laws were consolidated into a single environmental code, the implementation of which was to be guided by environmental quality objectives, for ecological sustainability to be fulfilled by 2020. Indeed, the concept of sustainable development had been presented in the 1987 Brundtland Report by the then Swedish prime minister. This shows that the European Union has taken the initiative with respect to environmental policy and sustainable development and has been ranked as the top environmentally friendly economic union in the world.

The overarching point, however, is that, if humanity musters the will for a great transition, a new and vital phase of human history could open up in countries all over the world. A planetary civilization will be created that pursues peace and justice, that delivers material sufficiency and rich lives, and that understands humanity to be respectful members of a wider community of life and sustainability of natural resources. The dilemma, however, is that the each scenario raises crucial questions of feasibility. The laissez-faire optimism of market forces would invite a host of environmental and social crises that could feedback on and amplify the undercutting of this rosy assumption of perpetual economic and environmental growth. According to Raskin and colleagues (2012) and Roorda (2012), the necessary political will for policy reformation is nowhere in sight. The organized cooperation of the global political and business leaders in the fortress world are still facing unfolding crises, and the resistant of the excluded masses would be no doubt extremely difficult. Despite these assumptions, optimists might believe that sustainable development and green economic would be accomplished in many areas of the world. What environmental policy recommendations can be made after more than two decades since the Rio de Janeiro United Nations Earth Summit in Brazil? Areas where change is needed to effectively improve on nations' environmental policies include:

- The public, private, nonprofit sectors, community-based organizations, media, and citizens of all walks of life must be involved in the environmental policy decision-making process. Improving education and communication is essential to achieving the objectives of environmental conservation. This means education and transparency of government are needed.

- There is a need for the enactment of public policies that achieve a balance between sustaining ecosystem services and pursuing the worthy short-term goals of economic development.
- A national strategy for sustainable development with specific goals and priorities would harness all sectors of society to achieve nations' economic, social, environmental, and security goals.
- Tax incentives must be used in such a way that the policy would encourage a shift of taxes from labor and income on one hand to material and energy consumption on the other. This would encourage both greater efficiency and reduced negative environmental impacts.
- Groups must be empowered, particularly those dependent on ecosystem services or those affected by their degradation, including women, indigenous peoples, and young people. Women's participation in environmental decision making should not be restricted by economic, social, religious, or cultural structures.
- Appropriate public policies, social infrastructure, and institutions should be designed and operated to further economic, environmental, and social goals at the same time. In addition, public and environmental health services and, at a minimum, basic medical services should be available to all citizens. Further, transportation infrastructures should be more efficient and diverse, and they should provide people with more choices.
- The conservation and protection of natural resources and environment must focus more holistically on resources to be protected and on understanding those resources. National and regional governments need to ensure that these resources are protected from all significant threats and are protected from those threats to the same degree. Further, the type of substantive goals of the air and water pollution control programs, as well as supportive implementing mechanisms, should be applied to biodiversity, climate change, forest, rivers and oceans under a nation's control and to other natural resources.
- Stronger and more environmentally friendly legislation must be developed with respect to all areas of production, emission standards, EIS, and pollution regulations.
- Each agency's role and functions must be clearly defined with regard to monitoring, policing, and enforcement
- Individuals found guilty of breaking environmental laws must be charged and punished. Nobody should be above the law.
- Technologies need to be promoted, in all nations, that enable increased crop yield without the harmful impacts of pesticides use on water and nutrients and that increase energy efficiency and reduce greenhouse emission.
- Governments should promote measures that reduce aggregate consumption of an unsustainably managed ecosystem. It is also paramount for governments to promote behavioral changes that could reduce the demand for degraded ecosystem services and products.

Although these recommendations are crucial for the accomplishment of sustainable development and green economy, it is highly unlikely that they will be followed up or implemented with immediate effect in countries all over the world. To reiterate this point, there is the need to assume that the core of mainstream sustainability thinking has become the idea of three dimensions: social, economic, and environment. The political leadership in some countries and the environmental degradation that so commonly accompanies economic growth are where human beings should try to provide solutions. Certainly, no one knows what our environmental future will be. The future may bring changes to our behavior towards the environment or it could come with different set of problems and possibilities. The choice we may have could vary as well. According to Myers and Spoolman (2014) and Guy Peters (2013), one choice may be to continue degrading our life-support eco-systems. On the other hand, we may choose to learn how to live more sustainably (Rosenbaum, 2014). Finally, as citizens of the world, we all have the obligation to take part in meeting the environmental challenges of the future as well as working together to deal with the problems and find solutions. We are strongly encouraged to start by thinking about and making environmentally beneficial changes in our life and in our communities and cities.

SUMMARY

This chapter examined the historically unprecedented global economic growth of recent decades and how it has resulted in equally unprecedented improvement in living standards, life expectancy, education, and the quality of life for billions of people. However, it argued that global economic growth has been associated with serious environmental problems. The problems include global climate change, acid rain, depletion of the ozone layer, rapid rates of deforestation, and a significant increase in the rate of species loss (OECD, 2001b). Therefore, the author argued that the costs of global development are rapidly rising every day. At the national level, deteriorating air and water quality, scarce water and fuel, soil erosion, and deforestation are also exacting a high price (Dibie and Offiong, 2012). New ideas are by their very nature disruptive but far less disruptive than a world running low on drinking water and productive land, set against the backdrop of climate change, extreme weather events, and rising natural resource scarcities.

The challenges posed to the global community and national governments in terms of energy security, climate change, health impacts, and energy poverty are pressing. The nature of the environmental predicament justifies why the greening of the energy sector is imperative. The existing challenges are exacerbated by the expected growth in the global demand for energy, as population and incomes rise (UNEP, 2012). Shifting from fossil fuels to renewable energy plays a critical role in greening the energy sector, along with other changes, particularly raising energy efficiency.

The cost-effectiveness of renewable energy technologies has evolved considerably in recent decades. Many renewable energy technologies are maturing rapidly, and their costs are becoming competitive with fossil fuel alternatives. Consequently, the investments in deploying renewable energy have increased dramatically in the last decade.

These environmental conditions make it clear that the basic model needs to be amended to account for the impact of large-scale economic activities on the environment. The solution to these predicaments can be accomplished by enabling the ecosystem to interact with the scale of economic activities in two ways. First, all countries and their economic activity generate waste, such as air and water pollution. Since waste cannot escape the ecosystem, the ecosystem acts as a sink to absorb and reprocess the pollution. Second, the ecosystem is a source of raw materials such as air, water, fossil fuels, metals, and forest, which are used to produce marketable goods and services. It is therefore very important to realize that there is no real equivalent for the sustainable yield of an ecosystem. In its place, we refer to an ecosystem's carrying capacity, which is the ability of an ecosystem to absorb and process the waste dumped into it.

The overall aim of environmental policy is sustainable development—to hand on to the next generation a society in which the major environmental problems have been solved. The most threatening environment hazard is acid rain from all over the world. Based on the environmental goals adopted by some countries. The U.S. Environmental Protection Agency and other similar regulatory agencies have focused their efforts on combating environmental threats. Among the goals are to prevent the greenhouse effect, the depletion of the ozone layer, the acidification of water and soils, the effects of photochemical oxidants and ground-level ozone, and urban air pollution and noise.

To fulfill all the goals of sustainable development, governments have imposed heavy taxes on everything possibly associated with the environment. For instance, 80 percent of the $5-per-gallon price that gas costs in some European countries is tax. Trucks and buses are heavily taxed, and oil-heated households also pay high taxes. These sound environmental policies have led to a very clean environment. Successful efforts toward socially, economically, and ecologically sustainable development are predicated on measurable indicators. These must be coordinated not only at the international, national, and local level; if they are to generate public interest and involvement, they must be followed up and evaluated.

This centralization of power limits public participation in decision making and places the whole responsibility of protecting the environment on the government and its agencies. Green economy and sustainable development are facets of human development that are ecologically sustainable. Its aims are human freedom, a higher quality of life, and opportunity for all. It is not another name for economic development, although it includes economic development. Thus, a few people should not decide what is in the best interest of the people—socially, economically, or environmentally.

REFERENCES

Adams, W. M. (2012). "The Future of Sustainability: Re-thinking Environment and Development in the 21st Century." In *Sustainability*, edited by N. Smith-Sebasto. New York: McGraw-Hill.

Ako-Nai, R. (2013). *Gender and Power Relations in Nigeria*. New York: Lexington.

Ayittey, G. (2005). *Africa Unchained: The Blueprint for Africa's Future*. New York: Palgrave Macmillan.

Baumer, D., and Van Horn, C. (2014). *Politics and Public Policy*. Los Angeles, CA: Sage.

Barbier, E. B. (2010a). *A Global Green New Deal: Rethinking the Economic Recovery*. Cambridge, UK: Cambridge University Press and UNEP.

Barbier, E. B. (2010b). "A Global Green Recovery, the G20 and International STI Cooperation in Clean Energy." *STI Policy Review, 1*(3), 1–15.

Barbier, E. B. (2010c). "Poverty, Development and Environment." *Environment and Development Economics, 15*, 635–660.

Barbier, E. B. (1987). "The Concept of Sustainable Economic Development." *Environmental Conservation, 14* (2), 101–110.

Basten, S., Herrmann, M., and Lochinger, E. (2011). *Population Dynamics, Poverty and Employment Challenges in the LDCs*. Laxenburg: IIASA and UNFPA.

Center for Global Development. 2007. Annual Report. Retrieved from http://issuu.com/cgdpubs/docs/cgd_2007_annual_report_lores. Accessed November 21, 2012.

Chen, S., and Ravallion, M. (2007). "Absolute Poverty Measures for the Developing World, 1981–2004." *Proceedings of the National Academy of Sciences, 104*(43), 16757–16762.

Davies, J., and Mazumder, A. (2003). "Health and Environmental Policy in Canada: The Role of Watershed Management in Sustaining Drinking Water Quality." *Journal of Environmental Management, 68*, 273–286.

Dernbach, J. (2012). "Synthesis." In *Sustainability*, edited by N. Smith-Sebasto. New York: McGraw-Hill.

Dibie, J., and Dibie, R. (2008). "Current and Future Roles of NGOs in the Empowerment of Women in Africa." In *Non-Governmental Organizations (NGOs) and Sustainable Development in Sub-Saharan Africa*, edited by R. Dibie. Lanham, MD: Lexington.

Dibie, R. (2003). *Public Management and Sustainable Development in Nigeria*. Aldershot, UK: Ashgate.

Dibie, R., and Gadzekpo, L. (2003). "Managing Public Sector Ethics in Ghana and Nigeria." *Politics, Administration and Change, 39*, 1–24.

Dibie, R., and Offiong, J. (2012). "Analysis of Environmental Policy in Nigeria." *Journal of International Politics and Development, 10*(1–2), 53–82.

Dobkowski, M., and Wallimann, I. (2002). *On the Edge of Scarcity: Environment, Resources, Population and Conflict*. Syracuse, NY: Syracuse University Press.

Dye, T. (2011). *Understanding Public Policy*. Upper Saddle River, NJ: Prentice Hall.

Easton, T. (2010). *Taking Sides: Clashing View on Environment Issues*. Boston: McGraw-Hill.

Eliasch, J. (2008). "Climate Change: Financing Global Forests." *The Eliasch Review*. Retrieved from http://www.official-documents.gov.uk/document/other/9780108507632/9780108507632.pdf.

Elliott, L. (2007). "Improving the Global Environmental Policies, Principles and Institutions." *Australian Journal of International Affairs, 61*(1), 7–14.

Endres, A. (2004). "Games Theory and Global Environmental Policy." *Poiesis Prax, 3*, 123–139.

Ferrel, O., Fraeedrich, J., and Ferrel, L. (2013). *Business Ethics*. Mason, OH: Cengage Learning.

Friis, R. (2013) *Essentials of Environmental Health*. Boston: Jones & Bartlett.

Gallai, N., Salles, J.-M., Settele, J., and Vaissière, B. E. (2009). "Economic Valuation of the Vulnerability of World Agriculture Confronted with Pollinator Decline." *Ecological Economics, 68*(3), 810–821.

Global Footprint Network. (2010). *The Ecological Wealth of Nations: Earth's Biocapacity as a New Framework for International Cooperation*. Retrieved from http://www.footprintnetwork.org/images/uploads/Ecological_Wealth_of_Nations.pdf. Accessed on October 6, 2013.

Guzmand, J. M., et al. (2009). "The Use of Population Census Data for Environmental and Climate Change Analysis." In *Population Dynamics and Climate Change*, edited by J. M. Guzmand. New York and London: UNFPA and IIED.

Health Canada. (1999). "Canadian Environmental Protection Act. Priority Substances List. Supporting Documentation. Health-Related Section (Exposure) for 1,3-Butadiene." Draft. Ottawa, Ontario.

Hudson, J., Lowe, S., Oscroft, N., and Snell, C. (2007). "A Case Study of Local Environmental Policy-Making in the United Kingdom." *Policy Studies, 28*(1), 56–70.

Kraft, M. (2010). *Environmental Policy and Politics*. New York: Harper Collins, 198–210.

Kraft, M., and Furlong, S. (2013). *Public Policy: Politics, Analysis, and Alternatives*. Washington, DC: Congressional Quarterly Press.

Maxwell, N. (2014). *Environmental Health*. Boston: Jones & Bartlett.

Migdal, J. (1988). *Weak States: States–Society Relations and State Capabilities in the Third World*. Princeton, NJ: Princeton University Press.

Myers, N., and Spoolman, S. (2014). *Environmental Issues and Solutions*. Belmont, CA: Brooks/Cole.

Nadakavukaren, A. (2011). *Our Global Environment: A Health Perspective*. Long Grove, IL: Waveland Press.

Organization for Economic Cooperation and Development (OECD) Sustainable Development: Critical Issues. (2001a). Paris: OCED.

Organization for Economic Cooperation and Development (OECD) (2001b). *Sustainable Development: Critical Issues*. (2001b). Paris: OCED.

Pearce, D., Markandya, A., and Barbier, E. (1989). *Blueprint for a Green Economy*. London: Earthscan.

Perrotta, K. (1999). *Environmental Health Issues in the City of Toronto, Canada*. Toronto, Ontario: Health Promotion & Environmental Protection, Toronto Public Health.

Peters, G. (2013). *Public Policy*. Los Angeles, CA: Sage.

Peters, G. B. (1988). *Public Bureaucracy in Comparative Perspective*. Tuscaloosa, AL: University of Alabama Press.

Pezzey, J. (1989). "Economic analysis of sustainable growth and sustainable development." Washington DC: World Bank.

Raskin P., Electris, C., and Rosen, R. (2012). "The Century Ahead: Searching for Sustainability." In *Sustainability*, edited by N. Smith-Sebasto. New York: McGraw-Hill.

Rom, W. (2012). *Environmental Policy and Public Health*. San Francisco, CA: Jossey-Bass.

Rawls, J. (1971). *A Theory of Justice*. Cambridge: Harvard University Press.

Roorda, N. (2012). *Fundamentals of Sustainable Development*. New York: Routledge.

Rosenbaum, W. (2014). *Environmental Politics and Policy*. Los Angeles, CA: Sage.

Rosness, R., Blakstad, H. C., Forseth, U., Dahle, I., and Wiig, S. (2012). "Environmental Conditions for Safety Work: Theoretical Foundations." *Safety Science, 50*, 1967–1976.

Sam Okere, J. (2013). "Faith Based NGOs and Women Empowerment in Nigeria." *Journal of International Politics and Development, 11*(1–2), 73–102.

Shendell, D., and Ana, G. (2011). "Promoting Environmental Public Health in Rapidly Urbanizing Areas of Less Developing Countries in Africa." *Journal of Environmental Health* (July/August), 37–62.

Smith, Z. (2004). *The Environmental Policy Paradox*. Upper Saddle River, NJ: Prentice Hall.

Smith-Sebasto, N. (2012). *Sustainability*. New York: McGraw-Hill.

Steiner, A. (2011). "Towards a Green Economy: Pathways to Sustainable Development and Poverty Eradication." Nairobi, Kenya: United Nations Environment Programme.

Taylor, R. (2012). *Taking Sides: Clashing Views in Sustainability*. New York: McGraw-Hill.

The Economics of Ecosystems and Biodiversity (TEEB). (2010). *The Economics of Ecosystems and Biodiversity: Mainstreaming the Economics of Nature: A Synthesis of the Conclusions and Recommendations of TEEB*. Bonn, Germany: TEEB.

The Economics of Ecosystems and Biodiversity (TEEB). (2009). *TEEB for National and International Policy Makers. Summary: Responding to the Value of Nature*. TEEB. Retrieved from http://www.teebweb.

Thin, N. (2002). *Social Progress and Sustainable Development*. Bloomfield, CT: Kumarian.

United Nations. (2012). *Human Development Index*. New York: Oxford University Press.

United Nations Development Program (UNDP). (2008a). *Human Development Index 2007–2008*. New York: Oxford University Press.

United Nations Development Program (UNDP). (2008b). *Human Development Report 2007/2008. Fighting Climate Change: Human Solidarity in a Divided World*. New York: United Nations Development Program.

United Nations Development Program (UNDP). (2006). *Human Development Report. Beyond Scarcity: Power, Poverty and the Global Water Crisis*. New York: United Nations Development Program.

United Nations Environment Programme (UNEP). (2011). "Towards a Green Economy: Pathways to Sustainable Development and Poverty Eradication." Nairobi, Kenya: UNEP Press.

United Nations Environment Programme (UNEP). (2010). *Green Economy in Developing Countries Success Stories*. Geneva: UNEP.

United States Environmental Protection Agency (U.S. EPA). (1999). "Extrapolation of the Benzene Inhalation Unit Risk Estimate to the Oral Route of Exposure." Washington, DC: National Center for Environmental Assessment.

Vaughn, J. (2011). *Environmental Politics: Domestic and Global Dimension*. Boston: Wadsworth.

Vig, N., and Kraft, M. (2013). *Environmental Policy: New Direction for the Twenty-First Century*. Los Angeles, CA: Sage.

Weaver, J., Rock, M., and Kusterer, K. (1996). *Achieving Broad-Based Sustainable Development*. Bloomfield, CT: Kumarian.

World Bank. (2010). *World Development Indicators*. Washington, DC: World Bank.

World Bank. (2000). *Can Africa Claim the 21st Century?* New York: Oxford University Press.

World Commission on Environment and Development (WCED). (1987). Brundtland Report. *Our Common Future*. Oxford: Oxford University Press.

Wright, R., and Boorse, D. (2014). *Environmental Science: Toward a Sustainable Future*. New York: Pearson.

Index

accident 9
accountability 174, 179, 180
acid 276, 277
act 8, 13
adverse mental health 272
African continent 21, 271, 400
agricultural land 272
agriculture 5, 42
agriculture forest 2
agricultural productivity 2
agrarian: heritage 275; population 275
AIDS/HIV 11
airborne CO_2 320
air pollution 71, 304, 341
air quality 317
aluminum 331, 340
Amazon forest 300, 304, 311, 314
animals 11, 40, 268, 353
anthropocentrism 401, 402
appropriate: communication 360; correction 60; environmental policies 1; technology 150
asbestos 275
asthma 309
Atlantic ocean 234
atmosphere 2, 424
atmospheric change 71
automobiles 40, 332

bauxite 332, 341
behavioral change theory 70
behavior lifestyle theory 120, 122
Belize 374, 380, 381, 384
Biafra conflict 284
biodiversity 63, 72, 273, 377
biofuel 77, 305, 438
biological factors 271
biomass 438
biomass energy 438
biomedical science 303
biosphere 201, 202
Bisphenol A (BPA), 140
Brazil 274, 303, 304, 310
Brazilian Amazonian agenda 21, 323
brewing 332, 351
British colonial regime 275
British Petroleum 269, 283
bronchitis 309
Brundtland Commission 26
burning of refuse 339

Canada 36, 117, 130, 134, 147
cancer 74
carbon dioxide 2, 429
carbon emission 432
carbon monoxide 5, 353
cardiovascular morbidity 6
Caribbean 3, 21, 351, 377
Carson, Rachel 75, 198
cement factory 277, 332
centralization 160, 455
chemical: factory 5; waste 270
Chevron 284
children 1, 37, 134, 135; environmental health of 134
China 274
chromium 6, 136, 250, 276
citizens 3, 4, 11, 60, 83, 161, 177, 182, 272, 303, 321, 332
city 277, 278, 304, 315
climate 20
climate change 2, 280, 317, 401, 413, 424
climatological station 240
coal 437
coast 271, 352, 377
cocoa 202, 361
collaboration 51, 161, 162, 345, 429

combustion 5, 40, 353
common good 7, 9; *see also* public good
common pool 12
communication 185, 419
communities 13, 275, 277, 291, 308, 331, 332
community-based organizations 48, 430
compartment relationship model 30
compassionate environmental theory 63
conflict 184, 269, 296, 304, 377
consumption 2, 43, 60, 170, 273, 274, 277, 332, 333, 334, 428, 446, 447
contamination 202, 305
corporations 4, 51
corruption 161, 163, 181, 311
corrupt practices 273
crude oil 269
culture 33, 282, 308

DDT 75, 361, 378, 386
decentralization 162, 177, 313
deforestation 2, 42, 62, 71, 303, 314, 329, 331, 338, 439, 451
degradation 4, 20, 47, 171, 200, 273, 343, 416
democracy 78, 63, 83, 159, 160
democratization 268
demography 171, 315
depletion 2, 40, 160
desert encroachment 275
desertification 275, 320, 401
developed countries 11, 40, 57, 429
developing countries 5, 11, 156, 429
development 2, 3, 36
development administration 179
Dibie, Robert 5, 268, 419
disaster 159, 240, 280, 303, 392
discrimination 63
disease 1, 43, 74, 277, 353
distilleries 351
distribution 43, 64
diversification 161, 408
diversity 46, 303
domestic solid waste 5
drought 20, 159, 240, 279, 285, 414
dumping 270, 401
dump sites 331

earth 2, 432
ecocentric concept 401, 402
Ecofeminism theory 33, 36
eco-friendly behavior 4
ecological 2, 11, 37; degradation 60; deterioration 170; equilibrium 3; feminism theory 33, 45; imbalance 42; limits theory 378; modernization theory 32; scarcity 432; theory 32
economic growth 354
ecosystem(s) 3, 13, 20, 268, 450; vitality 3; fragility 269
education 50, 51, 81, 177, 359, 419
electoral legitimacy 184
electricity 60, 404
electrocution 3
emission 78, 320, 352
empowerment 33, 47, 166
endangered species 20
energy 2, 5, 9, 11, 52, 60, 76, 81
energy conservation 330
enforcement 182, 185
enforcement power 289
environmental: awareness 174; degradation 286; externalities 60; health 1, 48; impact 1; justice theory 34, 78, 401; law 418; policy 1, 2, 4, 10, 272, 274
Environmental Protection Agency (EPA) 67, 69, 74, 133
epidemiology 309
equity 180
erosion 20, 277, 280
ethanol 438
ethical environmental values 273
ethical justification 7
exacerbation 270, 278
Explanatory environmental theory 378
exploration 273; of petroleum 275
exportation 75, 202
extinction 20
Exxon Valdez (or Exxon Mobile) 269

factories 5, 42
famine 240
farming 273, 275, 361
female circumcision 46
fertility rates 37
fertilizer 75, 276
firewood 42, 275
fisheries 273
flood 285, 307, 329
focus theory 120, 124
food 5, 60, 307
food production model 272
food safety 320
forest 2, 20
forestry 81
fossil fuel 2, 5, 73, 76, 78, 277, 353
freedom of expression 183

freshwater 133, 278, 304, 408
future generation 2, 26

garbage: collection 360; waste 20
gas 6, 43
gasoline 43
gender 176, 431, 358, 432
gender equity 431
gender inequalities 46
geophysical environment 20
geothermal energy 438, 441
global 13, 64, 424
global corporative approach 126
globalization 13, 40, 182, 448
global warming 280, 392
governance 43, 159, 161, 268, 313, 431
government 1, 3, 37, 161
government institution 7, 174
grazing land 42
green economy 432, 433, 449
greenhouse effect 40, 331, 416
greenhouse gasses 305, 331, 424
groundwater 76, 171, 277
growth 37, 38
Gulf of Biafra 271
Gulf of Mexico 269

habitat 20, 42, 329
hazard 9
hazardous: environment 3; exposure 5; generators 42; waste 2, 5, 59, 72, 321
health 1, 3, 6, 9, 10, 47
health belief theory 88
HIV/AIDS 11
housing 2, 52, 270
human health relation theory 120, 121
humanitarian catastrophe 280
hurricanes 72
hydro-power 438, 439

illegal burning 352
impaired driving 3
impediment to environmental health 306
implementation 6, 290, 336, 360
inclusiveness 7
India 274
industrial countries 6, 273
industrialization 38, 44, 62, 171, 268, 270, 332
industrial waste dump 296
infectious disease 3
infrastructure 161, 270
inhabitation 277
instability 2, 159, 170, 279
institutional change 268
interconnection, human and environmental 29
international judicial system 159
international organizations 311
inter-state compact 290
irrigation 40, 331
Islam 33

Jamaica 15, 329
Japan 21, 36, 45
Jordan 16
judicial system 159
judicial theory 1
judiciary 180
justice theory 1

Kenya 13
Kealeboga Maphunye 14
Kendal 20, 23
Kentucky 2
Kraft, Michael 2, 3, 22
Kuwait 16

land degradation 44, 47, 332, 364, 409, 416
landfill 2, 59, 171
land use 306, 322
land use planning 81, 322
Latin America 309
latrine 384
leadership 184
legislation 303, 418, 448
legislature 59, 180
legitimacy 163, 184
liberalization 174, 185
life expectancy 170
lifestyle 4, 7, 37, 161
limestone 329
livestock 42
lung disease 3, 6

macroeconomics 2, 10, 314
management 175, 271, 378
Mangrove 271, 352
man-made disorder 12
manufacturing 38, 59
market 268, 418
marketplace mechanism 418
Marxist 159
mechanism 162, 418
media 20, 419
medical care 5
Mediterranean 439

men 20
mental health 3
methane 321
military regime 271
mining 331, 417
mismanagement 240
modernization 63, 240
moral education 185
multinational companies 264; corporations 20, 22, 160, 273
municipalities 270, 384
municipal solid waste 331
musculoskeletal disorder 3

national cooperation 59
national revenue 271
natural capital model 357
natural disaster 42, 392
natural endowment 279
natural resources 20, 40, 42, 43, 177, 271
neo-colonialism 268
neo-liberalization 268
new public management 181
Niger Delta 269, 273, 280
Nigeria 2, 269, 270
noise 41, 289
nongovernmental environmental organizations 7
nongovernmental Organizations (NGOs) 7, 161, 182, 240, 282, 251, 334
nonrenewable energy 77, 269, 436
normative environmental theories 378
normative value judgment 378
nuclear energy 418, 440, 438

Obama, Barack 442
occupational and environmental health 317; health 314, 323; injuries 3
oil spill 21, 269
opinion poll 304
organic production methods 272
organic waste 351
overcrowding 360, 401
overcultivation 275
overexploitation 271
overgrazing 409
overpopulation 62, 230
ozone 40, 62, 74, 352, 455

participation 159, 177, 455
particulate matter 40, 70
partnership 50, 429
pest control 321
pesticides 304, 361
petroleum industry 270
phosphate-based industrial 5
planning 278, 418
pluralism 59, 166
pluralist theory 165
poisonous waste 276
policy formulation 7; policy making debacle 7; impact 450
policy recommendations 272, 418
political economy 166; ideology 67; leaders 159, 185; power 67
pollutant 5, 74
polluter-pay theory 120, 127, 128, 129
pollution 12, 36, 38, 51, 59, 74, 159, 416
polygamy 33
poor drainage 320
population 2, 11, 37, 38, 170, 191, 409; bomb 378; growth 276
poverty 12, 20, 21, 43, 48, 59, 92, 176, 401, 449; alleviation of 44; eradication of 433, 449
precautionary principle 127, 128
precipitation 320, 343
pregnancy abnormalities 3
preservation 307, 432
preventive mechanism 317
principal-agent theory 120, 125
profit 36, 271
property damage 280
public: good 162 (*see also* common good); health 3, 5; institution 8; leaders 160; participation 330, 354; policy 8, 73, 124, 183; transportation 284

Quarries Act of 1990 286
questionnaire 12, 335, 354, 358

race 303
rain 40, 72, 280, 303, 317
rainwater 330
recycling 38, 321
redistribution 451
refineries 332, 351
regulation 268, 286
relationship 7, 11, 20, 61, 332
religion 46, 303
renewable energy 294, 436, 438, 442
resettlement 240

respiratory ailment 277
responsiveness 179
risk 10
risk society theory 64
river 59, 60, 75, 277, 330
River Niger 277
rubbish dump 2
rule of law 182, 185

Sachs, Jeffery 210, 295, 300
safety 10
safety legislation 321
sand pipe 320
sanitation 171, 304
savanna 304
scarce commodity 40
self-control theory 273
self-determination theory 120, 125
self-organizing 163
sewage 331
shared governance 177
sickness 1, 4
smog 60
smoke 40, 136, 277
smoking 4
social: capital 177; cognitive theory 88; culture 431; distribution 34; economic 10, 92; economic variable 270; equity 13; justice 52, 313; learning theory 88, 402; norm 176; system 409; transportation 178
social learning theory 402
society 2, 33, 46, 50; infrastructure 7
soil degradation 71
solar energy 440, 438
solid waste 2, 81, 321, 331, 359, 361
stakeholders 51
stewardship 7, 9, 197, 447
sub-Saharan Africa 13
substandard housing 322
supranational agencies 159
sustainability 62
sustainability future 7
sustainable: approach 5; cities 443; communities 45; development 4, 7, 8, 25, 36, 51, 148, 165, 170, 177, 242, 317, 429, 451
syngas 438

tar sand 143
technology 36, 164, 182, 333
Texaco 269, 283
textile 276, 315, 331
tornadoes 72
tourism 329
toxic 6, 59, 75, 331
toxicology 73
toxic waste 270
trade-off 6, 29
traditional communities 20
transformation 50, 185, 277
transnational corporations 274, 296
transparency 174, 179, 376
transportation 5, 38, 52, 81, 277, 306
trans-theoretical models 88
traumatic injuries 3
Trinidad and Tobago 16, 351
Trokosi system 46
tsunami 159
typologies 70

unemployment 270, 271, 277
United Nations 41
United Nations Development Programmme (UNDP) 25
United Nations Environmental Programmme (UNEP) 25, 425, 432, 444
United States 2, 269, 216
United States Environmental Protection Agency (EPA) 67, 69, 74, 133
urban design 322
urban development 306
urbanization 38, 78, 171, 270, 277
urban life 162
urban planning 343
urban poor 43

value-belief-norm theory 120, 124
value of ecology 332
vector: control 323; disease 353
vegetation 275, 320, 409
volatile organic compound (VOCs) 5, 353, 424
volcanic eruption 21, 40
voluntarism 50, 51
volunteer 7, 50

warming 2
waste: disposal 38; management 416, 388; water 50, 331
water pollution 71, 411; quality 308; scarcity 406
weather 274
welfare 83

wildlife 42
wind 2, 307, 438
wind energy 440
wind power 438
women 20, 33, 177, 176
wood 43, 331
world 2, 20, 433
World Bank 41, 345, 359, 374, 432
World Commission of Environment and Development 26
World Health Organization (WHO) 1, 303, 306
workers 3, 4, 5, 6, 10, 35, 42, 69, 93, 121, 125, 250, 262, 284, 316, 317, 331, 358, 363, 364, 365, 391, 392, 437
world population 2